ROSENHAIN CENTENARY CONFERENCE

THE CONTRIBUTION OF PHYSICAL METALLURGY TO ENGINEERING PRACTICE

WALTER ROSENHAIN, 1875–1934

ROSENHAIN CENTENARY CONFERENCE

THE CONTRIBUTION OF PHYSICAL METALLURGY TO ENGINEERING PRACTICE

A JOINT SYMPOSIUM HELD 22–24 SEPTEMBER 1975
OF THE METALS SOCIETY, THE NATIONAL
PHYSICAL LABORATORY AND THE ROYAL SOCIETY

LONDON
THE ROYAL SOCIETY
1976

ISBN 0 85403 081 6

This report was first published in *Philosophical Transactions of The Royal Society*, series A, volume 282, no. 1307, pages 1–483

The conference was held on 22–24 September 1975; the preprints of the invited papers were issued in August 1975 and the MSS for keynote speeches and discussion at the meeting were received on 4 December 1975

Printed in Great Britain
at the
University Printing House, Cambridge
(Euan Phillips, University Printer)

Preface

As our technology becomes more complex the relation between it and the science germane to it also becomes more complicated. Good communication between teams of technologists, scientists and engineers working in related fields is then a necessity. This meeting was organized with the aim of improving communication between physical metallurgists in the laboratory and in the class room, and those technologists engaged in design, construction and maintenance of structures and machines.

We took the occasion of the one hundredth anniversary of the birth of Walter Rosenhain, a noted metallurgist of his day, to organize this Conference with the assistance of the Metals Society and the Royal Society. Rosenhain was a Fellow of both and in the simpler world of more than fifty years ago both made himself, and stimulated in others, notable advances in the design of alloys for engineering practice. We chose as our theme to consider in general the contribution of physical metallurgy to engineering practice in the civilian field and specifically examined bridges, pipelines, offshore structures, airframes, aircraft undercarriages and gas turbines. For each of these we asked an engineer to define his material requirements and then discussed how far these are at present met with modern materials at competitive cost and how far the essential requirements can be well characterized in terms that the engineer trusts and the metallurgist recognizes.

We opened the Conference with a review of the life and work of Walter Rosenhain and we closed it with an attempt to look forward to how the (physical) metallurgist could assist the engineer in the future and how the education of metallurgists, particularly at the University, may need to change in order to bring this about most effectively.

The organization of the sessions of the Conference was novel, and because authors, keynote speakers and chairmen worked so well and so generously with us in preparing for the Conference, was found to be very successful in producing clear and stimulating discussion. The method was for each of the invited authors to write a paper and to send it to us six months before the Conference. Copies of these papers were sent to keynote speakers immediately and printing of them put in hand. Three weeks before the date of the Conference, copies of the papers were sent to all attending. During the formal sessions the keynote speaker summarized his view of the papers presented by the authors and the Chairman of that session then introduced discussion of the papers. The authors of the papers contributed to this of course but did not give a formal presentation of the papers. Thanks to the authors' generosity, the keynote speakers' judgement and brevity, and the Chairmen's sagacity and peparedness, the whole worked extremely well.

The volume contains all of the papers written by the invited authors before the Conference, the written statements of the keynote speakers and the edited discussion. Those collected in this volume under the headings Sessions 1, 2 and 3 contain the message of the Conference. The account of those sessions in this volume is set between an account of Walter Rosenhain and of his work, and two sessions, 4 and 5, devoted to the future rôle of physical metallurgy and the task set the future educator respectively. In those sessions a single author presented his paper and in the latter two this was then discussed.

A final session, 6, gives a summary of the Conference made at the time. This has not been rewritten but merely edited slightly to make an extemporaneous verbal presentation at least partly syntactical in print. It is included because it represents one man's view of the highlights of what had been said, given at the time, and so gives something of the flavour of the Conference.

We are both indebted to Dr David Miles for the enormous amount of work he has put into the preparation of this volume.

R. G. Baker
A. Kelly

April 1976 *National Physical Laboratory*

CONTENTS

[Frontispiece and Plates 1–29]

PART 2

Walter Rosenhain and materials research at Teddington

By A. Kelly, F.R.S.
Deputy Director, Materials Group,
National Physical Laboratory†

1. Introduction

At this Conference we aim to honour the memory of Walter Rosenhain who initiated formally the study of materials at the N.P.L. and we wish also to look forward by examining the contribution of scientific studies of the structure and constitution of materials to engineering practice. To produce an impact was one of Rosenhain's aims and I have been told that the concept of such a Conference as this, and its method of organization, would have appealed to Rosenhain were he here.

I have designed this written presentation, of which a shorter version will be presented at the Conference, to cover three themes. The first is that of Rosenhain the man and metallurgist and what he and his colleagues brought about. The second sketches the historical development of the scientific study of materials on the Teddington site and hence must touch upon the organization of this and with the growth of the Engineering Department at the N.P.L. – which gave rise to the National Engineering Laboratory; and with the story of the National Chemical Laboratory. The third part is very much a personal view and deals with the definition of a technologist.

Several memoirs of Walter Rosenhain appeared shortly after his death on 17 March 1934; obituary notices in the *Journal of the Institute of Metals* (D.H. 1934); *Obituary Notices of Fellows of the Royal Society* 1932–35 (Desch 1934) and the Institute of Metals Autumn Lecture in 1934 delivered by Dr J. L. Haughton, one of his colleagues under the title 'The work of Walter Rosenhain' (Haughton 1934). Haughton's account contains a list of all of Rosenhain's published papers. A recent reappraisal of his work appeared in the *Dictionary of scientific biography* (Cahn 1975). I give the main facts concerning his life and the honours he received in Appendix 1.

2. First researches

Walter Rosenhain was born in Berlin‡ on 24 August 1875. The family emigrated to Australia when he was five and he learned English on the voyage out there. He revisited Europe in 1890 and again in 1894, interrupting his University course to do so. He graduated from Queen's College, Melbourne in civil engineering and went to Cambridge as an 1851 Exhibition scholar at St John's in 1897 at the age of twenty-two. He had published one paper in Victoria on determining the specific heat of a liquid, and worked initially on steam jets. However, advice from (the later) Sir Alfred Ewing, F.R.S. led him to examine plastically deformed metals under the optical microscope.

Sorby (1826–1908) had first looked at metals but was primarily a geologist and though a descendant of the first master cutler was not interested in metals themselves but principally

† Present address: Vice Chancellor's Office, University of Surrey, Guildford.

‡ This fact is ascertained from his daughter Mrs N. Kirsner – the published memoirs, with the exception of Cahn's are in error on this point.

as a geologist. He examined iron and steel because of his interest in meteorites. His own view of his work on metals in 1897 was 'What I really proved was that various kinds of iron and steel are varying mixtures of well defined substances and that this structure is in many respects analogous to that of igneous rock'. Though Sorby first investigated metals in 1863–64 it took him a long time to publish all the results as befitted a man of private means and his work at high power ($\times 650$) was published in 1886 more than twenty years after it was done. In 1877 Adolf Martens began a series of microscopical studies of iron and steel, unaware of Sorby's work. By the nineties of the last century a number of eminent metallurgists were using the microscope to study phase changes in iron – Roberts Austen, Stead and Arnold in the U.K. and Osmond in France. Rosenhain looked at a polished surface of a plastically deformed polycrystalline iron and discovered slip within the grains. Slip in crystals was first discovered by Reusch (1867) in rock salt but Rosenhain does not refer to this. Rosenhain and Ewing gave an account of their work in several papers including the Bakerian Lecture in 1899, Ewing & Rosenhain (1899, 1900*a*, *b*, 1901). In those days the *Proceedings of the Royal Society* were not divided between physical and biological sciences and the first paper appears between one on a sugar bacterium and another on the physiological action of choline. The paper reported slip in iron, steel, copper, silver, lead, bismuth, tin, gun metal and brass and in nickel in an added note.

The authors recognized the need for a grain to accommodate itself to its envelope of neighbouring grains as strain proceeds, but say that the region between slip bands appeared to be rigid, and from the observations on nickel deduced a lack of any foreign matter in the intergranular junctions. In later papers they extended the observations to zinc, cadmium, gold and to lead eutectics. They claimed to find deformation twins in lead, gold, silver and copper but these, I believe, were either due to recrystallization or were deformation bands (see, for example, Barrett 1949). They certainly found deformation twins in cadmium, zinc and tin. They concluded that slip and twinning were the modes of plastic deformation of metals. Ewing and Rosenhain also studied recrystallization and deduced some of the laws of recrystallization much as we know them today (e.g. Burke & Turnbull 1952), Ewing & Rosenhain (1901). The explanation of how recrystallization occurs involved the notion of a eutectic being formed between the grains which allowed the metal to dissolve on one side and to precipitate upon the other; since recrystallization did not occur in a non-deformed metal the notion was that deformation led to break up of the eutectic film which was regarded as hard and brittle – the notion of hard grain boundaries persisted in Rosenhain's thinking for many years. The deduction was made that two welded pieces of metal will not allow recrystallization across the weld, which is true unless extraordinary care is taken. If clean sand or other material was introduced between the welded pieces again there was no recrystallization. The experiments were very ingenious but came to quite the wrong conclusion, namely that pure metals would not recrystallize. In fact very pure metals do so much more easily than others (Albert & Le Hevicy 1956).

In all of these papers Rosenhain is shown as a very ingenious, imaginative and thorough experimentalist. The experiments to test his ideas on recrystallization indicate this and his proof that slip bands were steps on the surfaces of crystals is classic. This he did by electro-depositing copper upon a surface after slip and then cutting a section through the two metals to photograph the steps in profile. This is quite brilliant; the steps are only at most 10 μm high.

Ewing and Rosenhain were concerned with polycrystalline metals of commercial importance. They make no reference to Mügge's work which appeared in 1899 which was carried out on native crystals of gold, silver, copper, iridosmium and iron, which established the glide plane as {111} and *slip direction* ⟨110⟩. Mügge also *measured* Neumann bands and deduced they were twins, a fact which the metallurgical fraternity did not regard as proven until 1953 (Paxton 1953; Kelly 1953). Mügge was the great classical crystallographer studying crystals for their own sake. Ewing and Rosenhain, of course, were engineers. Their work proved that polycrystalline metals used in the foundry and in the shop were crystalline. This was a great advance in thinking at the time; they only thought of 'metals'. In the same issue of the *Philosophical Transactions* there is a paper by Adams & Nicholson (1900) which reports an experimental investigation into the flow of marble and slip and twinning are found to occur. Rosenhain and Ewing are referred to. The essential modes of plastic deformation in both polycrystalline metals and in rocks were therefore established at much the same time but the two fields have served different technologies and made only peripheral contact since.

3. Glass industry

Rosenhain's initial work then gave him prominence among scientists in the metallurgical and engineering world. A number of opportunities were offered to him on leaving Cambridge in late 1900 – a chair in Poona, a demonstratorship in Engineering at Cambridge, a position as a civil engineer in Australia, and the one he accepted as 'tame scientist kept on the premises' by Chance Bros Ltd, Glass Manufacturers in Birmingham. In 1901 his probation with the firm ended and he then married Louise Monash, with whom he had corresponded weekly since leaving Australia three and a half years before.

At Chance Brothers he worked on optical glass and light-house apparatus and started his study of refractory crucible materials of high purity. He wrote a book on glass manufacture (Rosenhain 1908) which was published after he had left Chance Bros in 1908 and was revised in 1919. During the 1914–18 war there was a shortage of optical glass in this country – most had been imported from Germany – and to offset this disadvantage glass of optical quality had to be produced in Britain. Rosenhain was instrumental in helping this and to support it was commissioned to carry out research work on glass production at the N.P.L. His views are described in his three Cantor Lectures to the Royal Society of Arts (Rosenhain 1916*a*) in which he describes in detail the properties required of optical glass, the defects in it and the methods of production. The lectures also make a number of comments on the organization of the optical glass industry in Britain.

During the time with Chance Brothers his heart remained in physical metallurgical research and in order to sustain this he set up a small private laboratory at his home in Edgbaston, financed by a Carnegie Research Grant from the Iron and Steel Institute. His wife Louise shared his work and he trained her in the preparation of metallurgical specimens. He published papers on the deformation of iron and steel and discussed the mechanism of fracture (Rosenhain 1904, 1905). At this time he began to believe that work hardening was due to destruction of the crystal structure on the slip bands due to the production of a vitreous layer and, in addition, he began to adopt the view that boundaries between grains also were of an amorphous structure.

He continued to develop instruments and his improved form of coal calorimeter (Rosenhain

1902) was exhibited at a Royal Society Soirée in that year. Because of his interest in pure metallurgical research, he was naturally attracted, when in 1906, the post of Superintendent of the Department of Metallurgy and Metallurgical Chemistry at the National Physical Laboratory was offered to him.

4. The National Physical Laboratory

Work on metallurgy and chemistry was started simultaneously in 1902 at the N.P.L. as two sub-departments of the Physics Department. The work on metallurgy was started at the instigation of the Alloys Research Committee of the Institution of Mechanical Engineers. It was in charge of H.C.H. (later Sir Harold) Carpenter F.R.S., who started his professional career as a chemist. By 1905 Carpenter's interests were almost entirely centred round metallurgical research and he resigned from the N.P.L. to take up the newly created chair of metallurgy at the Victoria University of Manchester in the autumn of 1906†.

A separate department of the Laboratory was then formed entitled Metallurgy and Metallurgical Chemistry and Rosenhain was appointed the first Superintendent.

He stayed there twenty-five years and during this time his Department grew from a staff of four to one of seventy. Among the various biographical notes of Rosenhain, the fullest description of the scientific work he directed at the N.P.L. is given by Haughton (1934).

The immediate need was to understand the constitution of alloys. This started with an investigation of the alloys of lead and tin (Rosenhain & Tucker 1908). In the Ninth Report to the Alloys Research Committee (Rosenhain & Lantsberry 1910) phase diagrams and properties of copper–aluminium–manganese alloys are dealt with and their corrosion resistance studied. Remarkably high tensile strength and elongations were obtained, namely 700 MN/m^2 with an elongation of 16 % in an alloy containing 10 % Al and 1 % Mn.

The report mainly deals with copper-rich material but discovered an alloy of aluminium with 3 % Cu and 1 % Mn to have excellent properties. This is close to the composition of Duralumin (3.5–5 % Cu, 0.5–1 % Mn plus about 0.5 % magnesium) which became a most successful light alloy.

There followed investigations of alloys of aluminium and zinc summarized in the Tenth Report to the Alloys Research Committee (Rosenhain & Archbutt 1912) in which the idea of specific strength (called specific tenacity) is first introduced. Further work on the light alloys was continued during the war but much was unpublished until afterwards. A lot of it is published in the Eleventh Report to the Alloys Research Committee (Rosenhain, Archbutt & Hanson 1921). The constitution and properties of Y alloy (4 % Cu, 2 % Ni, 1.5 % Mg) and others, which were much used in airships, aircraft and car engines in the 1920s and 1930s were described.‡ These alloys were used directly and discussions of the papers show an intimate interaction between the N.P.L. work and those fabricating the alloys in manufacturing industry. Even as early as 1921 the suggestion that age hardening of Duralumin is due to precipitation on a scale too fine for resolution in the optical microscope was put forward. At the same time Jeffries & Archer (1921) advanced a similar explanation in the U.S.A. Preston (1938) later proved this at the N.P.L. a few years after Rosenhain's death.

† In 1913 he left Manchester to take up the Chair of Metallurgy at the Royal School of Mines. He was for a number of years a member of the Executive Committee of the N.P.L.

‡ This alloy has been said (Hunsicker – this symposium) to be a precursor of the alloy RR58 selected for use in the airframe of Concorde (2 % Cu, 1.5 % Mg, 1.2 % Ni, 0.1 % Ti).

By the end of the war the number of ternary phase diagrams which had been investigated was becoming large and three-dimensional phase diagrams being widely used. Rosenhain (1920) introduced a neat wire model for displaying the phase fields.

Investigations were also made of bearing metals (Fry & Rosenhain 1919) and on zinc-copper-aluminium alloys (Rosenhain, Haughton & Bingham 1920). This alloy was investigated during the war as a possible substitute for brass for making shell fuses. Rosenhain had suggested zinc alloys as a possible substitute and Zn with 4% Cu and 7% Al was found suitable. According to Haughton the alloy went into production for a short time.

In all of the metallographic work high standards of accuracy were emphasized. Pure materials, as subject of the experiment, pure crucibles, well calibrated instruments, patience in readings, careful microscopy are all insisted upon (Rosenhain 1909). The phase diagrams then produced persisted for many years.

Other work was carried out at the N.P.L. under Rosenhain's direction and he often wrote the introduction to a series of related papers. A notable example is a whole series of papers on the alloys of iron. In his first introduction Rosenhain (1924) describes how he obtained support for the research and gives an outline of what is to be attempted by his team.

Insistence on high standards of accuracy in scientific observation must provide a drive towards improved and novel instrumentation. Rosenhain's design of method and of apparatus is one of his outstanding achievements.

He championed the inverse method of plotting cooling curves. This rests upon measuring the time taken by a specimen to change its temperature by a small number of degrees – say 3 – and plotting this against the temperature. It was simple and reliable though not as accurate as the difference method in which the difference in temperature from a standard specimen was noted. The good measurer or standards man recognizes the appropriate degree of accuracy for the problem in hand.

Advances in metallography were numerous among him and his colleagues. He supported the edge of a specimen during mechanical polishing by depositing a layer of copper before sectioning so that the edge could later be examined under the microscope. This was applied in many ways and became standard for examination of fractures. In order to quench rapidly he used the (then new) silica tube so as to wash a specimen from a hot surface with a stream of water. Mercury is difficult to examine metallographically, and with Murphy a method was developed (Rosenhain & Murphy 1926). A great variety of etchants were produced. Rosenhain designed a metallurgical microscope. This was patented in 1905 and later manufactured by Messrs Beck.

Rosenhain made two large contributions to apparatus for thermal analysis. These were a gradient furnace and plotting chronograph. The second appears very out-of-date today but was important at the time in enabling readings to be taken quickly according to the inverse cooling rate method of thermal analysis. The first arose from the need to subject a specimen to a constant rate of change of temperature over a wide temperature range and was accomplished by arranging the furnace vertically, by heating the top and cooling the bottom with running water and moving the specimen through it at a constant rate. Heating the top prevented irregularities due to convection. Very simple in concept, after it has been conceived! It worked beautifully.

Rosenhain's scientific work covered three aspects. The investigation of the constitution of alloys, the design of apparatus and experiments. These are lasting achievements of a high order though the experimental data will be repeated and the instruments superseded. What motivated him greatly as a man were certain theoretical ideas concerning the crystalline structure of metals. These acted as great stimulants to himself and others and produced much discussion and experiment and some controversy. I hope that it is not too harsh to point out that few of his theoretical ideas have stood the test of time which brings with it more thought and further experiment.

I give below (β iron and amorphous metal) an account of the science of two of these controversies, in which he was engaged, but before doing so, will deal with his other activities. He published a great book *An introduction to the study of physical metallurgy* (Rosenhain 1914) which in its three editions was the standard work on this subject and led to its recognition as a particular field of scientific endeavour. He was in much demand as a consultant, with wide contacts in the metallurgical and mechanical engineering industries. He wrote widely over many years and conducted the *Metallurgist*, a metallurgical supplement to the *Engineer*. This journal reads oddly nowadays since many of the articles were unsigned. A large number are due to Rosenhain and by writing anonymously he exerted considerable influence. He published in the *Engineer* (Rosenhain 1923*a*) a series of reviews of metallurgy in the United States; his views culled from his visit in that year.

Man of energy as he was, he did very much for metallurgy as a profession; a founder member of the Institute of Metals, President in 1928–30, a member of the Iron and Steel Institute and of the Institute of Physics. He made the work of his Department at the N.P.L. widely known and brought it to its 'customers' through technical committees and the Council of the British Cast Iron Research Association, British Non-Ferrous Metals Research Association, as well as technical committees of the Department of Scientific and Industrial Research, the British Standards Institution and the Air Ministry.

He was a profound internationalist. Speaking French and German fluently he was often asked to lecture on the mainland of Europe and he took a special interest in the International Association for Testing Materials, particularly as it was put together again after the First World War.

His account of metallurgy (and science generally) in the U.S.A. in 1923 makes entertaining reading and then (as we would now perhaps) he remarked 'Thus, while England may claim to lead in the development of metallurgical science, its practical applications have been carried much further in America'.

Rosenhain left the N.P.L. in June 1931 and set up as a private consultant with offices in Victoria Street. He was a consultant to three well-known firms; Stones' at Deptford, which was later joined by A. J. Murphy one of his clever young collaborators, British Aluminium and the Broughton Copper Co. At the first named he helped with the manufacture of lead bronze steel backed bearings just introduced in the U.S.A. Besides work for these firms he was frequently called to Court as an expert witness. He continued to write for the *Metallurgist* and in an editorial in February 1933 rebuked a German for national chauvinism, as the left would call it nowadays, for referring to the 'german metal'.

He died after a long illness on 17 March 1934 and as *The Times* said 'was a brilliant metallurgist...covering the whole field of metallurgy both ferrous and non ferrous'.

The University of Melbourne opened the Rosenhain Memorial Laboratory on 3 July 1935 and the Institute of Metals introduced a Silver Medal named after him in 1951.

5. β-Iron

In the first edition of *An introduction to the study of physical metallurgy* 1914 the iron-carbon phase diagram up to 5 % carbon by mass appears on p. 161 as figure 64. There is a line *FG* at a temperature of about 770 °C linking the pure iron axis and the γ phase field. In the third edition, 1935, published after Rosenhain's death – revised and partly rewritten by J. L. Haughton – the line appears dotted as it does in all modern phase diagrams; figure 1. The change represented by this line is the Curie point of body-centred cubic iron. Above this temperature of 768 °C, the modern value, iron is paramagnetic, below it, it is ferromagnetic. In the early years of this century a mighty controversy raged about this change point in iron. A lively discussion of this has been given by Cohen & Harris (1965) and the nature of the controversy is probably most clearly learned by reading a definitive work by Burgess & Crowe (1913) as well as the paper and the discussion of it which finally settled the controversy by Arné Westgren in 1921 (Westgren 1921).

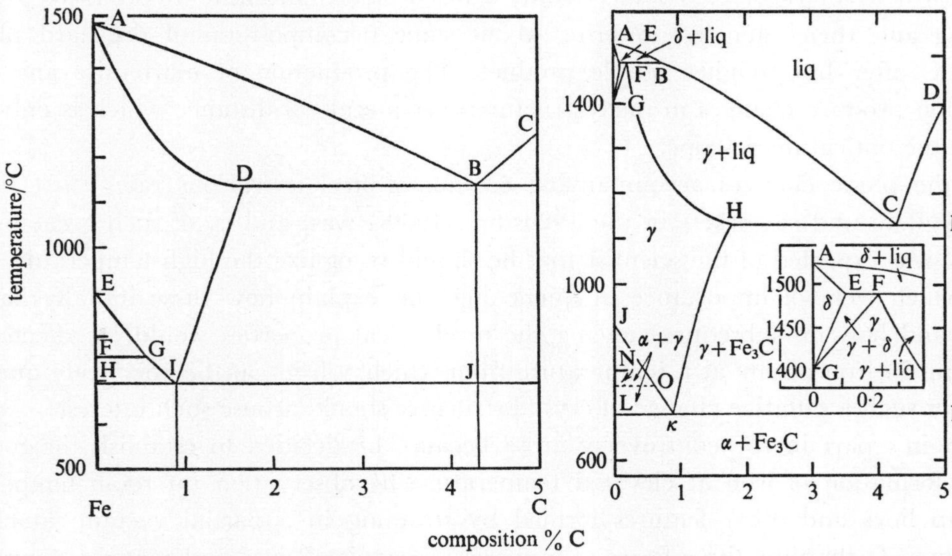

FIGURE 1. Phase diagram of the system iron plus carbon to 5 % carbon as it appeared, left, in the first edition of *Introduction to physical metallurgy* (1914) and right as it appeared in the third edition (1935).

There are two elements in the background to the discussion. Firstly, one of the main thrusts of scientific metallurgy then, as now, is the need to describe the phase changes accompanying change of temperature and composition in an alloy system. At that time many binary alloy fields had to be explored. The main tools to discover phase changes were measurement of the thermal effects – discovery of thermal arrests upon heating and cooling – and observation with the optical microscope of the surface of the specimen. A thermal arrest indicated a phase change. Measurement of physical properties also indicated phase changes since there were often discontinuous changes in these as a particular phase boundary was crossed. To investigate optically, alloys showing phase changes at high temperatures is never easy and specimens are often examined at room temperature following rapid cooling from a temperature of equilibration.

The classification of transitions now followed is to distinguish those in which the first derivative of the Gibbs free energy with respect to temperature (entropy) changes and the first derivative with respect to pressure (volume) also changes; these are called first order transitions. These are separated from those in which entropy and volume are continuous but their derivatives (specific heat and compressibility) are not; these are called second order transitions. If the entropy is continuous there is no latent heat but usually an anomalous specific heat. Such is the case with iron at its Curie point. This sophisticated idea of different types of phase change was formally established by Ehrenfest (1933).

By 1909 it was known that iron showed an anomaly in the specific heat over a range of temperatures near to 770 °C and this was thought to indicate a thermal arrest, i.e. a latent heat – even today such confusion can arise. Careful experimentation showed no volume change, or a very small one, compared with that known to accompany the change from α to γ iron at 910 °C. These differences could have led to a recognition of the anomalous nature of the phase change in iron at its Curie point. However, the second element in the discussion made this difficult at the time. Steel may be hardened by rapid cooling from the γ range when a diffusionless transition occurs and rapidly cooled iron shows a metastable phase known as martensite. To successfully harden steel martensite is produced by rapid quenching and then annealed to bring about some decomposition of the hard phase to produce a softer but tougher usable product. The production of martensite and its decomposition produce changes in microstructure over a scale of distance which is only partly visible in the optical microscope.

Since the phase changes accompanying cooling of steel at various rates (first sketched clearly by Bain and coworkers in the 1920s and 1930s) was, and is, of such great practical import, it was expected of the scientist that he should recognize the high temperature phases of iron which were of importance in quenching and explain how these initially hardened the steel and how on subsequent ageing the mechanical properties would be affected. The Curie point in iron occurs at a temperature from which γ iron can be effectively quenched. No wonder such a putative change of crystal structure should arouse such interest!

Rosenhain's part in the controversy arose because he decided to establish the modes of plastic deformation of iron at elevated temperature by observation (at room temperature) of the slip lines and other features formed by straining in a partial vacuum at elevated temperature. At the time three forms of iron were recognized, α, β and γ at the temperature ranges shown in figure 2. A_1 does not occur in pure iron. Osmond & Cartaud (1900) had very ingeniously established by crystallizing iron through the reduction of ferrous chloride at high temperature that iron was cubic and since γ crystallized differently, that α and β were isomorphous with one another: β was distinguished from α by a 'molecular change' – such a change does in fact occur since the atomic magnets are no longer aligned above A_2†. It was thought by some that β iron was intrinsically hard and that quenching retained this form at room temperature. Carbon was believed to assist the retention of this hard form at low temperature; thus explaining the difference between pure iron and steel. This was essentially the 'allotropist' theory of the hardening of iron. A contrasting theory was that of the 'carbonists' led by Howe (1890) who maintained that a compound called 'hardenite' was formed between iron and carbon at high temperature and that sufficiently rapid cooling enabled the retention of this hard form at low temperature. There were other explanations

† According to Howe, the great American metallurgist, Osmond called β iron decipuum because you knew of its existence as you knew of certain planets (presumably Neptune was in mind) by the perturbation which it caused.

too, namely that quenching γ iron produces amorphous layers particularly at the boundaries between crystals and that these were very hard. It was believed at the time (by Rosenhain and others) that work hardening was due to the production of amorphous material.

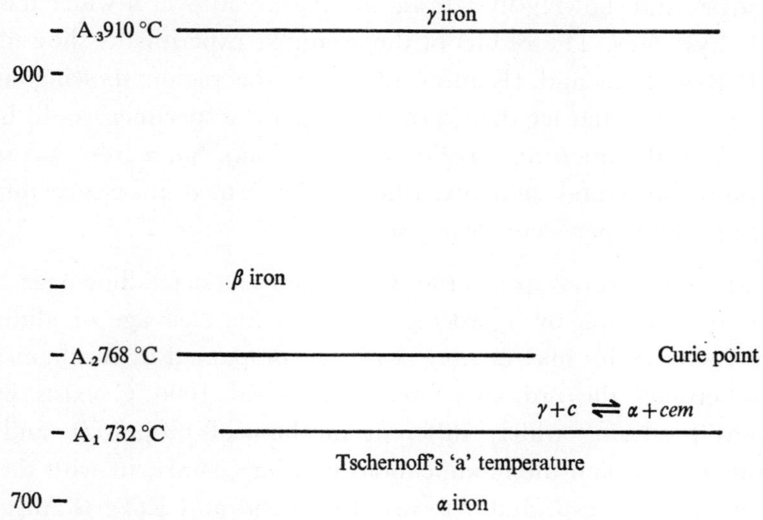

FIGURE 2. The significance of the thermal arrest points in iron and in iron-carbon. A stood for thermal arrest. In very pure iron there is no arrest at 732 °C.

Rosenhain & Humfrey (1909) undertook a series of characteristically ingenious and well conceived experiments on a mild steel. They decided to strain at elevated temperature a suitably shaped specimen with a surface carefully polished for microscopic examination, without allowing the polished surface to be damaged by oxidation. This was achieved by straining in a vacuum (1.33 Pa; 0.01 mmHg). The strain was applied at any desired temperature by releasing a spring which then applied force to the specimen when high temperature was attained. The specimen was subsequently examined at low temperature. Since the temperature varied along the specimen a clever method was devised to measure the temperature at which straining had occurred. Small particles of sodium chloride (melting point 810 °C) and of potassium sulphate (m.p. 1070 °C) were placed upon the back surface of the specimen. Upon recovering the specimen examination showed which particles of salt had melted and which had not. The salts were chosen since the melting points were close to A_2 and to A_3 respectively. Unfortunately, due probably to radiation loss from the specimen the temperature measurement was not exact!

When the specimen was heated without straining a set of interlaced dark lines showing the characteristics of two sets of crystal boundaries were identified at a temperature of 700–800 °C. This was identified with A_2 and so ascribed to the α–β transition. When the specimen was heated to about 1000 °C large crystals were observed which were revealed by a slight amount of oxidation which led to tinting and these were identified as γ iron.

When strained specimens were examined, slip lines characteristic of α iron were found in the cooler parts. Slip appeared to occur more easily at points at which the temperature was higher and more numerous better marked slip lines seen. However, a sudden change apparently occurred at high temperature and along a well defined line running across the specimen no slip appeared to occur. Further along a specimen, i.e. at the hottest parts, slip lines were again seen and twinned crystals, characteristic of γ iron.

Rosenhain & Humfrey also carried out a thermal analysis on their specimens and duly found heat absorption at about 770 °C (770 °C on heating and 763 °C on cooling and at between 900 and 950 °C (941 °C on heating and at 901 °C on cooling) A_2 and A_3 no less, with A_3 much sharper and showing hysteresis, and A_2 spread over a wider temperature range and showing little hysteresis. The results of the straining experiments showed what the allotropists proposed. Rosenhain and Humfrey identified the region showing little slip with β iron. No slip because it was harder than α or γ! In fact the specimen could be fractured and still show, remote from the fracture, a region devoid of slip. So, α iron decreased in strength with increasing temperature and suddenly, when β was formed, it became much harder.

The conclusions to the paper were clearly stated:

'(1) That iron at temperatures up to 1100 °C behaves as a crystalline aggregate and undergoes plastic deformation by a process of slip on the cleavage or gliding planes of its constituent crystals; this may or may not be accompanied by mechanical twinning.

(2) That iron between the ordinary temperature and 1000 °C exists in three distinct modifications possessing widely different mechanical properties, and that the temperature ranges in which the modifications exist are consistent with the view that they are identical with the α-β- and γ-forms of Osmond and Roberts Austen, as indicated by cooling curves.

(3) That β iron although existing at a higher temperature is harder and stronger than α-iron and that the α/β transformation involves a volume change.

(4) That γ iron, as found in approximately pure iron at higher temperatures, possesses the characteristic structure and properties found in the "γ iron" alloy steels.'

The work was elegant and appeared conclusive. It was referred to in Rosenhain's certificate of candidacy to the Royal Society prepared in 1911 which was signed by Ewing, Stead, Heycock, Neville and Sir George Beilby among the metallurgical fraternity – Lord Rayleigh proposed him. The certificate stated that the paper with Humfrey 'gives the final conclusive demonstration of the distinctive properties of the different forms of iron known as α iron, β iron and γ iron.' Rosenhain was elected in 1913 and in a number of short contributions between 1909 and that date expounded the views based on his experiments with Humfrey (Rosenhain 1910*a*, *b*, 1911). He was thus identified as a protagonist for β iron.

In 1913 another paper was published with Humfrey on 'The tenacity, deformation and fracture of soft steel at high temperatures' (Rosenhain & Humfrey 1913). This is a very careful piece of work in which actual stress–strain diagrams were produced for iron up to 1100 °C in a vacuum by the use of a specially constructed apparatus in which the specimen was placed on top of a barometric column of mercury so that this acted as a frictionless stuffing box and the specimen could be strained as in a normal tensile test. The material used was a slightly less mild steel (0.1 % C), heating and cooling curves were again taken and microscopic examination carried out. The curve showing the maximum tensile stress sustained by the specimen, plotted against temperature is shown in figure 3. A small and not very definite break in the curve appears at A_2. A_3 is clear and now γ iron is seen to be harder than α iron. Alas, for hard β. *Hard* β did not exist and Rosenhain partially admitted this 'it seems probable that the microscopic discontinuity observed by the authors in 1909 and ascribed by them to the transition from the α to the β condition, is in reality due to the transition from the β to the γ condition;...as the temperature measurements of the earlier work were admittedly of an approximate character there seems no difficulty in reconciling the observations then made with the facts now established. The theoretical conclusions on the other hand may require some modification.'

The interpretation of the results are then conducted in terms of the amorphous cement theory with little reference to the β iron controversy. Rosenhain still thought of β as a separate crystallographic phase and since he knew that the crystal structure of α and β were clearly the same, presumably thought of Osmond's 'molecular change' as distinguishing the two. He recognized that there was no longer evidence for a volume change in the putative β to α transition.

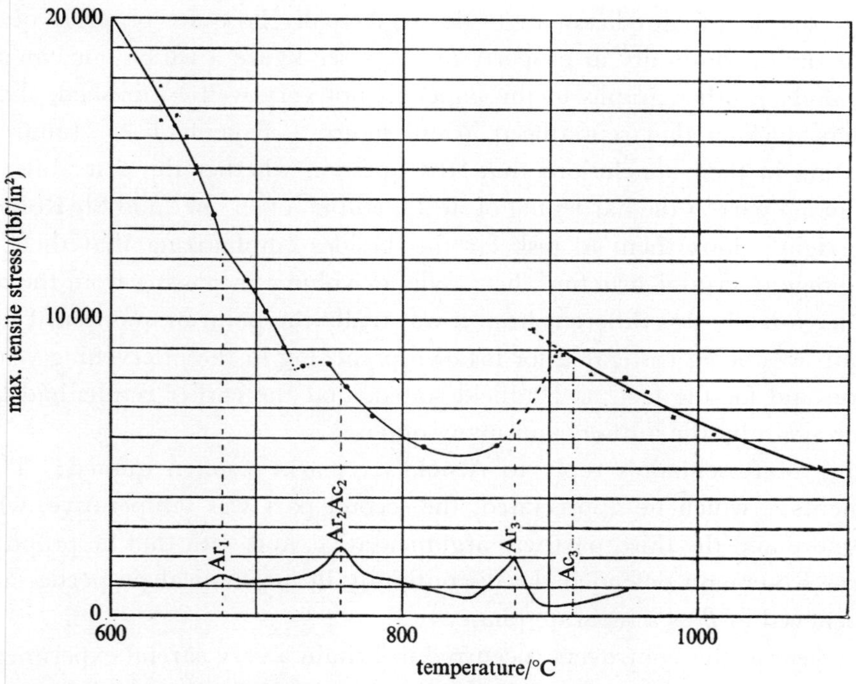

FIGURE 3. Temperature-tenacity curve obtained by Rosenhain & Humfrey (1913) together with their thermal cooling curve.

This paper is in fact a classic in my opinion because, apart from the theoretical views of the authors, it showed experimentally very conclusively that both grain size and strain rate affect the stress–strain curve and that the effects vary with temperature and the practical import of these effects on metal working was clearly recognized. In fact, stating the results in modern form, flow stress σ depends on tensile strain rate $\dot{\epsilon}$ as

$$\dot{\epsilon} \propto \sigma^n$$

where n is an integer, being larger at higher temperatures. Rosenhain & Humfrey found $n = 4.6$ at 836 °C and 6.0 at 948 °C.

The possibility of recrystallization occurring during deformation is also implicit in the results.

The second paper was published in the *Journal of the Iron and Steel Institute* and hence was open to discussion which may not have occurred with the first being published in the *Proceedings of the Royal Society*.†

A furious and vituperative discussion is printed in the *Journal of the Iron and Steel Institute*

† It was, however, read on 25 November 1909 and if there was discussion we have no record of it.

following Rosenhain & Humfrey's second paper. Their retraction concerning the hardness of β iron is not allowed to pass without comment: J. O. Arnold, F.R.S. reproducing side by side conflicting portions of the two papers with the most scathing comments. Arnold was delighted that β iron was no longer thought hard, though he believed in β iron. What angered him – a convinced carbonist – was that the authors did not explain the fall in hardness on cooling from the γ range below A_3 as due to 'hardenite' precipitating out of γ iron and hence leaving the soft β as able to deform. Carpenter, Rosenhain's predecessor at the N.P.L. and now at Manchester University was equally scathing having first said that 'It was a most difficult thing to admit in public that one was wrong'. Certainly his subsequent remarks would not encourage one in this predicament to do so. Actually he quite properly questioned the authors about the discontinuity in properties at A_2 – see figure 3 which one can see from the points there, and on other graphs in the paper, is not very well established, though clearer on some others than on that reproduced in our figure 3. Rosenhain & Humfrey had been unwise in stating in their conclusions that 'the question whether the β modification of iron plays the principal part in the hardening of steel is still an open one' and Sir Robert Hadfield, F.R.S. quite rightly took them to task for this besides emphasizing that their own results showed no change in crystal structure, heat relief or volume in passing from the α to β phase so that whether β really was different from α was really the point in question. I do not doubt that Rosenhain was being castigated for his own preaching in the intervening years in favour of hard β iron and for the fact, as Hadfield stated, that the earlier results had been quoted over and over again by the authors and many others.

The opening of Rosenhain's reply to Arnold's remarks is often quoted: 'The first part was complimentary which he appreciated, the second part was vituperative, which he was inclined to ignore and the third part was argumentative, and with that he proposed to deal'. In his remarks Rosenhain defended the discontinuity in mechanical properties at A_2 and he clearly still believed in β as a separate phase.

In the same year as this controversy occurred in Britain a very careful experimental analysis of whether or not β iron was a separate phase of iron was carried out at the Bureau of Standards in Washington (Burgess & Crowe 1913). They reviewed the literature in depth, citing evidence for critical ranges of the metal deduced from expansion, thermoelectricity, crystalline structure, mechanical methods, electrical resistance, magnetism, calorimetry and then carried out a very careful thermal analysis on specially prepared iron. Their conclusion was certainly not that β did not exist. They were as near what we consider to be the truth as they could be without investigations by using X-ray diffraction. They stated that 'all the physical properties of iron which have been studied with the single notable exception of crystallographic structure, has shown in the hands of one or more skilled experimenters a distinct discontinuity for the A_2 range as well as for the A_3 range. For several of the phenomena, such as electrical resistance, thermoelectricity, specific heat and magnetism it would appear that the discontinuity is at least as great for A_2 as for A_3, while the thermal effect has of course long been recognised as being much the more pronounced at A_3.' Having carried out their very careful experiments and measured with some precision the effects, Burgess in the final discussion of their paper makes the hypothesis that A_2 will probably be found to be an allotropic point and that when sufficiently exact expansion measurements are made on pure iron a minute but abrupt volume change will be found at A_2. He goes on to say that if experiment proves this not to be the case, then the A_2 transformation must be linked entirely with the passage from the ferromagnetic to the paramagnetic state and hence would not be classed as an allotropic transformation 'and β iron would be really dead'.

In 1921 Westgren's X-ray experiments in Gothenburg killed it (Westgren 1921). Hull (1917) showed, using the Debye–Scherrer method, which he independently invented, that α iron is body-centred cubic and following his method Westgren took X-ray photographs up to 830 °C. There was no change in crystal structure. By examining alloys containing 25 % nickel in which γ iron was stabilized at room temperature it was established that this form has a face-centred cubic Bravais lattice. Hence any difference between α and β was not of an allotropic nature. In the discussion Rosenhain tacitly conceded this and used the term β iron 'for a certain condition of iron which existed'. Only F. C. Thompson continued to resist and said specifically that 'he refused to be converted even now to the view that β iron had no existence' and he said that perhaps there was a discontinuous change in the atomic distance in iron as it passed from the α to the β state. For the purist this point was finally dispatched in 1955 when careful experiments showed no abrupt change in lattice parameter of iron as it passed through the ferromagnetic Curie temperature (Basinski, Hume Rothery & Sutton 1955).

The discussion of Westgren's paper throws to me an interesting sidelight on Rosenhain's intellect. A central theme was where was the carbon in α iron and particularly in martensite which Westgren had just revealed to have (approximately) the crystal structure of α. Westgren had also shown that in 12 % manganese steel containing 1.34 % by mass of carbon, of which he had measured both the lattice parameter and the relative density, that the carbon atoms were most likely arranged interstitially rather than replacing iron at the lattice points of the face-centred cubic space lattice. Sir William Bragg took up this point in discussion and suggested that if this were the case the γ iron would be very highly distorted. When Rosenhain entered the discussion he attempted to explain a large number of facts vividly known to him. Because of his own ideas on amorphous layers between crystals and his recognition of the importance of the polycrystal as the true metallic specimen, he suggested that martensite contained α iron as crystals in a mass of non-crystalline material and since Bragg indicated distortion if carbon was in the lattice then the carbon was most likely to be in the amorphous boundary. The amorphous boundary hardened the material; the α-iron was magnetic; at once an explanation of the high coercive force and high remanence of quenched carbon steels struck him. The width of his vision then led him to suggest that it was unlikely to be the case that carbon was interstitially dissolved in iron. In this, of course, he was wrong. His very quickness of intellect and ability to grasp a number of ideas at once as well as his desire to explain a little too much led him to defend a number of false theoretical conceptions.

From the β iron controversy Rosenhain and Humfrey, I believe, emerge rather well. In the second paper they had to admit that the first was in error and this must indeed have been a blow to Rosenhain knowing how important the paper appeared to others and had been to him. Arnold's public attack really does sound vituperative mainly because Rosenhain did not use Arnold's ideas to explain the hardening of steel. One must recall that this whole controversy occurred before X-ray diffraction could be widely used and the α/β transition as Burgess & Crowe summarize had all the recognized features of a phase change. A phase change in the solid state was automatically assumed to indicate a form of allotrophy. Only a volume change was missing and Rosenhain & Humfrey believed they had established this, and Burgess & Crowe accepted their evidence.

Rosenhain & Humfrey's second paper is a classic for its investigation of the high temperature mechanical properties of iron. Rosenhain's defence of his position in the discussion is dignified. Had all protagonists been sufficiently humble to let nature lead them, they would, of course, have discovered the idea of phase changes of two different types, but this would have required an attitude of patient determination which seldom goes with the ability to inspire others and to get experiments done.

There is a marked difference in the attitude to the controversy revealed by the work at the N.P.L. and at the (National) Bureau of Standards. That at the latter is extremely careful, well planned and seriously executed. The experiments are straightforward. Nothing new or fancy here except the serious and patient treatment. Quintessentially the more common view of the work of a standards laboratory. Careful conclusions admirably hedged and, judged on its own, it did not resolve the controversy. A_2 might still be an allotropic transformation or it might not. The N.P.L. work was quite original in conception, the experiments were quite new. They were ingeniously executed; in the first case not quite carefully enough, but in the second case admirably so. They were promulgated with great vigour and aggression. Men could not be neutral on the issue! After Rosenhain & Humfrey's second paper the issue was practically decided. Nobody in the U.K. believed in β iron any more as an important hardener of steel†. The N.P.L. had settled the question, not in the way Rosenhain expected, but nonetheless the function of a standards laboratory to enable controversy to be resolved had indeed been discharged!

6. Amorphous metal

According to one of his close collaborators, Hanson, Rosenhain had two principal objectives in his life, '(*a*) the development of his Department at the N.P.L. and (*b*) to prove and get generally accepted his amorphous metal hypothesis'. The second was the idea, introduced in two papers with Ewen (Rosenhain & Ewen 1912, 1913) that the boundary layer between the unit crystals of a polycrystalline metallic mass consisted of metal atoms in a non-crystalline amorphous arrangement, 'chemically identical with the substance of the metal or alloy in question, but in a widely different physical state'.

The theory was ingenious and comprehensive. It rested on the facts that in contrast to most inorganic crystals such as minerals, fracture of metals does not normally occur at grain boundaries. That the boundaries are in fact stronger than the crystals is suggested by the fact that slip bands in one grain do not penetrate into another. Grain boundary matter was, therefore, assumed harder than the crystals at low temperatures. At high temperatures fracture in metals more often occurs at grain boundaries. These facts were explained by assuming that grain boundary matter is stronger than the crystals at low temperature but weakens more rapidly as the temperature is raised, the flow stress depending on strain rate as in a glass, with the behaviour of which Rosenhain had great familiarity. Rosenhain & Ewen also called on Sir George Beilby's experiments on the surface structure of abraded and drawn metals (e.g. Beilby 1903, 1907) observed under the optical microscope. Beilby found flow to occur so easily and the metal to be smeared in such a fashion that he assumed an amorphous

† There were still some foreign recidivists, e.g.
Howe, H. M. 1916 *Metallography of steel and cast iron*, McGraw Hill.
Sauveur, A. 1935 *The metallography and heat treatment of iron and steel*, Cambridge, Massachusetts.

state. Such heavily worked material also shows differences in electric potential and various thermal properties from annealed material. Without the concepts of sliding friction (for example, Bowden & Tabor 1950, 1964) involving some melting and the idea of large densities of dislocations providing an increased internal energy to the deformed metal, no modern explanation is possible.

Rosenhain & Ewen also drew on the conception of 'crystal molecules' a sort of nucleus as being necessary for the proper crystallization of metals and reasoned ingeniously that between contiguous growing crystals in a solidifying mass, there was not room for a proper crystallization of metal in the interstices and so the material solidified in an amorphous state. The concept of a 'lineage structure' or of a large crystal being composed of a set of blocks with some irregularity between them, a type of analogue of the 'crystal molecule', persisted for many years. It was still being reviewed twenty years after the (theoretical) discovery of dislocations (Hirsch 1956). Rosenhain correctly reasoned that the metastable amorphous cement would have a higher vapour pressure than the normal crystal and hence that the grain boundary would evaporate more easily. He found larger mass losses on heating fine grained specimens than on heating coarsely crystalline ones. He also found that metals heated slowly while sustaining a small mass, broke at the grain boundaries at a temperature a few degrees centigrade below the melting point and he discovered grain boundary grooving when metals are heated at high temperatures. From the discussion of these papers by Desch & Carpenter and others the concept of more rapid diffusion occurring at grain boundaries than that which occurred within the grains, emerged.

When one reads the papers with attention one finds, despite the (to us) prolixity of style, and the discursiveness allowed by editors in those days, that the authors almost cast a spell on the reader. They placed together well remembered and relevant experimental results of themselves and others and almost command assent by presenting an idea capable of rationalizing many observations by a consistent hypothesis.

It is now possible to show that the hypothesis is wrong in detail and in generality but to do so requires knowledge of the short range of the forces between metallic atoms, the concept of the dislocation, the use of X-rays, electron microscopic and Auger electron techniques and, most importantly, experiments on very pure specimens in high vacua. Amorphous metal in any quantity may only be produced metastably by very rapid quenching (Duwez 1967). Many of the effects reported by Rosenhain, his co-workers and others, both protagonists and antagonists of this theory, were due to segregation of impurity atoms to the inter-crystalline regions, a subject much clarified and illuminated by later work at the N.P.L. (McLean 1957; Seah & Hondros 1973).

Rosenhain & Ewen's particular experiment of showing that metals melted at the grain boundaries before doing so in the bulk, occupied metallurgists for a number of years. Rosenhain's nephew repeated their type of experiment in the later 1930s (Chalmers 1940) and claimed to find grain boundary melting. That any real effect must be due to non-obtainment of equilibrium was demonstrated by Shewmon (1957).

Rosenhain thrived on controversy and promulgated his theory with some objectivity as well as much debating skill, in the literature of the learned societies; and it runs throughout the various editions of his book *Physical metallurgy* from the first edition in 1914 to the last in 1934. The theory was wrong in scientific detail but it was of great utility. It enabled the metallurgist to reason and to recognize that at high temperatures grain boundaries are fragile, that heat treatment involving cold or hot work coupled with annealing can lead to benefit in some instances and to catastrophies such as 'hot shortness' in others. The existence of 'burning' and 'overheating' would be avoided. Advances in technology and practice do not always require exact theory. This must always be striven for it is true, but a 'hand waving' argument which calls salient facts to attention, if readily grasped in apparently simple terms, can be of great practical utility.

7. Physical metallurgy and the role of the N.P.L.

From three of his publications in particular we can obtain an idea of Rosenhain's views of the nature and relevance of his subject and of the institution he served for most of his working life. These are his account in the middle of the Great War 'The National Physical Laboratory – its work and aims' (Rosenhain 1916*b*); his presidential address to the Institute of Metals (Rosenhain 1928) and his guest lecture to the German Society for Metallurgy given in Dusseldorf on 7 September 1929 (Rosenhain 1930).

He saw physical metallurgy as the science relating the properties and particularly the mechanical properties of metals, hardness, toughness, strength, elongation and anelastic properties to the chemical constitution and to the microstructure. In this physics and metallurgy were partners. In 1930, as now, there was a feeling among physical scientists that physics is somehow superior to metallurgy because of its more exact and (hence necessarily) abstruse nature. But the physicist must, to obtain his accuracy of prediction, work with pure materials so well characterized and defined that their utility to the engineer and their charm for the metallurgist has been 'researched' out of them. Rosenhain saw the two sciences as a partnership 'Der breite Strom des Wissens und der Forschung wird aber von wielen Quellen gespeist, und unter diesen darf auch die Metallforschung eine anerkannte Stellung beanspruchen'. He saw the metallurgist as leading the physicist so that the latter worked on problems of importance to the engineer – not measuring very accurately the ephemeral 'constants' of a piece of metal of ill-defined microstructure but towards understanding in sufficient detail all the complexity attendant upon casting, forging and joining metal, the necessarily impure metal used in practice.

Thus when addressing physical scientists he had to defend the severely practical so that too much manpower and brainpower was not wasted on the too detailed explanation of trivia, yet when addressing the practical men of industry as President of the Institute of Metals he had to defend the scientific, detailed and analytic approach. Members of the Institute of Metals in 1928 objected to the large number of scientific papers appearing and wanted just to know how to make their alloys better. Rosenhain did this for them. First he explained the 'abstruse' simple stress–strain curve and the even more recondite binary phase diagram. Not things the practical man at the time wished to know about. He goes on to explain age hardening and the basis of the heat treatment of aluminium alloys, so necessary for the burgeoning aircraft industry of the time. He showed how the science gave guide lines to new alloys of commercial advantage and pointed out that had the results of the Ninth Report to the Alloys Research Committee been picked up quickly and prosecuted at the time it was published (1910), Duralumin would have been the commercial property right of an English firm and not of a German. The subject was and is a complex one. The principle of age hardening as deduced by the scientist is not patentable because deducible from prior principles, whereas of course specific age-hardening alloy systems are. Science freely published can aid all, countryman and competitor alike. It is just so today.

Further the 'practical' men were told that defensive research is necessary in a practical commercial situation, when Rosenhain asked 'Is it quite certain that our age-old products 'ordinary' brass and bronze, will always be able to maintain their position?'

Rosenhain's view of the work of the N.P.L. and of its aims can be most easily told by quoting *in extenso* a number of passages from his lecture to the West of Scotland Iron and Steel Institute in 1916 (Rosenhain 1916*b*). He saw the N.P.L. as serving industry by doing applied science, by which he meant enough of the pure variety to enable practical problems to be attacked. The pure science must be of the best not static but 'it must be living growing science' and mindful, not impatient of the difficulties of application.

> 'When, however, we come to apply the results obtained by the worker in pure science to practical ends an entirely different set of conditions is met with. For practical purposes we have to deal with natural phenomena as we find them and have to produce the desired result in an approximate but efficient manner. We have, in fact, to face the extreme complexity of ordinary conditions, and we soon find that every effort to lessen this complexity in order to reduce a process to conditions more nearly resembling those which have been studied in the research laboratory introduces practical complications which tend to increase cost.'

> 'We thus see that on one side the worker in applied science finds the field of his activities not only in the most intimate contact with, but actually overlapping that of the worker in pure science. This contact and overlapping, however, bears good fruit on both sides, since the experience of practical problems frequently serves to open up whole regions of new phenomena to the purely scientific investigator, while on the other hand the stimulus of the most advanced scientific thought prevents the worker in applied science from losing touch with the true spirit of his science and its latest development.'

The difficulty, of course, is to maintain the balance of the two!

He was not unmindful of the real complexity of technical manufacturing industry:

> 'but one further factor requires consideration. I refer to the question – admittedly difficult and delicate – of labour conditions. The growth of the demands of labour is a phenomenon which no observer can miss, and in any rational consideration of the future prospects and policy of industry that factor cannot be left out of consideration. This is not the place to enter into any discussion of the thorny problems with which this subject bristles, and I wish to refer to one aspect only. If increasing demands of labour are to be met, and at the same time reasonable profits are to be maintained, there can be only one sound solution – and that must lie in increasing the efficiency of production. To the labour advocate one might say that if the earning power of the workman is increased by the better, more efficient, more scientific, utilisation of his labour, then can his return for his labour be correspondingly increased, while to the employer the same consideration may be urged in regard to his capital as represented by his plant, stocks, etc. It is not perhaps too much to say that the real hope of peace and prosperity for both lies in the possibilities of the application of science and scientific methods to their common problem.'

Prominent Trades Union leaders would today agree with part of the diagnosis but the touching faith of the last sentence is denied to us. The current view is of the paramount importance of sufficient investment capital.

What a central government research institution can do and which a university cannot do was a question to be answered then as now. Rosenhain thought as follows:

> 'This 'something more', which is additional and supplementary to the work of Universities and Colleges, and in no sense intended to rival or supersede their research activities, is an institution for research in which the educational bias is entirely absent, and whose activities can therefore be directed in an entirely untrammelled manner to the solution of problems suggested by public utility, scientific importance, or industrial demand. Such an institution would possess a staff not burdened by educational duties and selected without reference to teaching ability. Such a staff should constitute a more or less permanent body of investigators and their highly-trained assistants, thus affording opportunities for long-continued researches which can scarcely be undertaken by a Professor working with successive generations of young students. The equipment of such an institution could be designed on lines free from educational considerations; apparatus and experimental plant could be set up, involving, if necessary, the permanent employment of skilled labour as well as scientific supervision. Further, if only one or a very few such institutions were set up in the country, equipment could be carried out on a scale of cost which could not be reproduced at any considerable number of Colleges. Further, appliances could be used which it would be unsafe to place in the hands or even within the reach of students.'

He went on and described the 'remit' as we might call it of the N.P.L., serving as a national base of the measurement system and carrying out the needs of government and assisting industry.

> 'Further, such an establishment would occupy a unique position as a central national institution which would naturally undertake certain duties of a scientific or technical nature required by the Government. Such matters as the national standards and the standardization of measures and measuring instruments of all kinds would be undertaken there under national authority. Then, in another direction, where certain industries require national aid on account of their vital nature as "key" industries, the national central institution would provide the Government with the means of affording the scientific assistance demanded. The same provision of facilities of the best kind under national auspices would prove invaluable in connection with military, naval, and aeronautical matters. Here centralization would be essential in the interests of that degree of secrecy which is essential. Then again there are such matters as arbitration on points of scientific dispute which in modern times frequently arise out of industrial questions; the standardization of testing and in some cases even routine testing for the Government are all activities which would rightly occupy such an institution.'
>
> 'At the same time the institution would also be in hourly touch with industry – helping, advising, assisting, testing – taking up special investigations proposed by industry and dealing with them in the laboratory and with experimental plant. The institution would also be in closest touch with the Government, giving – together with the eminent specialists to be found in the Colleges and Universities – that expert advice which is now so frequently needed and undertaking such tests and investigations as the Government required.'

Although serving governmental needs, Rosenhain did not see the N.P.L. as only drawing its funds from the government.

> 'With regard to the general organisation of the Laboratory, it must be pointed out that the National Physical Laboratory is not actually a "Government" institution in the sense of constituting or forming part of a Government Department. The management of the Laboratory rests with a Committee appointed in accordance with certain Treasury regulations by the Royal Society, and the staff of the Laboratory are employees of the Royal Society and not civil servants. The Treasury makes to the Royal Society an annual grant of £7000 towards the cost of the Laboratory, and in return exacts certain conditions in regard to the work to be undertaken, and, among other things, in regard to the maximum salaries to be paid to members of the staff. I need hardly point out that this annual Government grant is entirely inadequate for the maintenance of the Laboratory; the last balance-sheet issued before the war reached a sum just over £40,000, of which something over £30,000 had been earned by the Laboratory in the form of fees. It is only right to point out, however, that a large part of these fees were derived from Government sources, in the shape of charges for work carried out for or on behalf of various Government Departments, such as the Admiralty, the War Office, the India Office, etc. Private firms and individuals, however, also contributed to a large extent to the fees received for work done, and this aspect of the work of the Laboratory is capable of very great and useful extension once the nature of its facilities becomes better known among those who could avail themselves of its services. Beyond this, technical institutions, like the Institution of Mechanical Engineers, Institution of Civil Engineers, Institute of Naval Architects, and others have contributed to the Laboratory budget, in some cases by simple donations, but more frequently by grants in aid of some definite scheme of research.'

Within two years of this being written the National Physical Laboratory became a civil service institution when it was absorbed into the Department of Scientific and Industrial Research in 1918. The method of funding of the work then changed radically, but gradually; some of the difficulties are described in articles by Hutchinson (1969, 1970).

Rosenhain (1916*b*) goes on to describe the work in his Department current at the time. There were four sections. He regarded physical metallurgy as the most important of these, dealing with methods of metallography, the elucidation of phase diagrams and the effect of thermal and mechanical treatments on the properties of metals. Another dealt with chemical analysis, carrying out a lot of work for the India Office, and as it was wartime assisting manufacturers to deal with requirements formerly filled by German goods, e.g. chemical glass ware, porcelain and filter paper. There was also a section dealing with aircraft fabrics for balloons and airships. There was a section on optical glass, the origin of this again to help deal with the shortage of optical quality glass which had before the war all come from Germany.

8. Origins of present materials work at the N.P.L.

Rosenhain left the Laboratory in 1931. He was succeeded as Superintendent by C. H. Desch who stayed until the Second World War. During the whole of this period at the N.P.L. there was also a flourishing Engineering Department. This, with the Observatory and Physics Departments was one of the three original departments at the Laboratory. Metallurgy under

Rosenhain flourished in concert with the Engineering Department, which for three-fifths of its total period of life, namely from 1901–30, was led by (Sir) Thomas Stanton. The present work on materials at Teddington has origins not just in Rosenhain's work but also in that of the Engineering Department and the work of the Chemical Research Laboratory, which became the National Chemical Laboratory in 1958. Most of the work in the Metallurgy Department under Rosenhain was concentrated upon the elucidation of alloy structure, the accurate determination of phase diagrams and attempts to relate the metallurgical variables, such as composition and heat treatment, to the properties of materials; there was, in addition, a good deal of work on development of instrumentation. This was supported throughout by the work on the properties of materials and on the long term testing of materials carried out in the Engineering Department. There, under Gough and others, work on fatigue and creep of materials under simulated service conditions yielded results which were easily known to the metallurgists and were commented upon by them, as well as stimulating the metallurgists own ideas.

Table 1. Report on the Metallurgy department – 1927

General research

Pure metals; Refractories; Spectrographic analysis; Physical structure of metals and X-ray study; Magnetic phenomena; Metals and alloys under strain and plastic deformation; Surface tension of liquid metals

Government research

Light alloys; Spring steels; Minor metals; Chains and hooks; Alloys for high temperature; Steel castings; Gas Cylinders; Fabrics research

Special investigations

Alloys of iron; Dental alloys; Impurities in copper; Special brasses; die casting

Maintenance of standards, etc.

Metallurgical analysis; Temperature standards; Thermostats; Viscometry; Experimental methods; Furnaces; Test-work, etc.

Report on the engineering department – 1927

General research

Friction of fluids; Hardness tests research; Gear research; Properties of concrete

Government research

For the Lubrication Research Committee
For the Engineering Coordinating Research Board and the Aeronautical Research Committee
Fluid friction and heat transmission; Big-end bearings; Springs research; Materials at high temperatures; Explosions in closed vessels; Rust prevention; Steel strip for aircraft; Fatigue of single metal crystals; Light alloys; Failure of chains; Sizes of nuts
For the Building Research Board
Wind pressure; Vibration of structures
For the Gas Cylinders Research Committee
Roads Research

Special investigations

Tower of London; Various investigations; Test-work; St Paul's Cathedral

A list of the topics which were being studied in the Metallurgy Department in 1927 when Rosenhain was 52 and his Department in the main stream of activity is given in table 1. It is interesting to note that the researches are classified under general research, which meant research suggested by members of the Metallurgy Department themselves, Government research in which work was carried out for various of the Advisory Committees working for the Department of Scientific and Industrial Research, special investigations which were usually paid for by various outside bodies and maintenance of standards which was work in the main stream of the Laboratory and usually meant testing to ascertain whether or not specific standards of performance were obtained. Steels, light alloys, brasses, coppers, were of course there and also some work on fabrics research, which involved the attack of sunlight on cotton. Dental alloys were examined for the Dental Investigation Committee of the D.S.I.R. by means of funds provided by the Dental Board of the United Kingdom. This work, like so much of the others, resulted in alloys which were tested in industry, in this particular case at Guy's Hospital. I have also reproduced in table 1 the headings of research in the Engineering Department. This is broken down into similar though not exactly equivalent headings and the Annual Report for that year states that more of the work of the Engineering Department was carried out specifically for various Governmental bodies. Road research, of course, figures which later left the N.P.L. and was carried out at the Road Research Laboratory; founded in 1930 under the Ministry of Transport and transferred to the D.S.I.R in 1933. There was a particularly close connection between the work of the Engineering Department and that of the Metallurgy Department. Looking at some of the headings in Engineering and comparing them with Metallurgy one sees how similar they are, e.g. springs research and spring steels, materials at high temperature and alloys for high temperatures, rust prevention and steel strip for aircraft, fatigue of metal *single crystals*, light alloys, failure of chain. This was work which clearly could have been carried out either in the Metallurgy Department or the Engineering Department; the precise location depending on how one drew the line or wrote the label on the door. That both Departments benefited becomes very clear from reading the Annual Reports.

An interesting feature which later disappeared from the N.P.L. Annual Report is the statement for whom the work was carried out. This, indeed, would have delighted Lord Rothschild in 1971 and perhaps current attempts to state customer and contractor principles are just a means of going back to what was a rather clearly thought out means of stating within a Department of N.P.L. who the customers for the work might be.

It was in the Engineering Division† that work on plastics at the Teddington site began in 1939 when in collaboration with the Chemical Research Laboratory, (N.P.L. Annual Reports 1940–48) attempts were made to utilize the high strength to density ratio and stiffness to density ratio of engineering plastics and also to investigate the possibilities of large mouldings for aircraft parts. Of course the stiffness of plastics without fillers is not high and this led naturally to a collaborative programme between Engineering Division and the Chemical Research Laboratory on plastics reinforced with fillers – paper was the most useful and important at the time – this was the first work in Britain on the principles of fibre reinforcement. Before the Second World War, the work of the Engineering Division was on two main subjects – the determination and their use in design, of the properties of engineering materials and the development of the principles of fluid dynamics and their application to gas/liquid flow and to lubrication. During the Second World War a good deal of work went on the high temperature properties of metals and it became necessary to build a large creep laboratory to house 50

† Departments at the N.P.L. became named Divisions during the second war, in order to avoid confusion with the *Department* of Scientific and Industrial Research.

creep testing machines. Early in 1940 work began on an extensive study of creep and fatigue properties of high temperature alloys for use in gas turbine blades, in conjunction with the Mond Nickel Company, from whom Dr Allen, the Superintendent of Metallurgy Division, who succeeded Sir Charles Sykes, was also to come. Tests were made on the creep properties of Nicrome alloys 18 % nickel, 20 % chromium, called at that time Nimonic 75 and well known at the N.P.L. to have good creep properties since it had previously been investigated in 1924–25 from German aircraft captured in World War I. When American alloys became available during 1940–43 for investigation, notably vitallium and hastoloid, these were also investigated. During the second war a good deal of work was carried out on the stress distribution and stability of aircraft structures and attention was given to the development of sandwich structures (that is composites), in which two very thin sheets of material were stabilized by filling with a core of very light material. As no material for the core was available at the time, and experimental confirmation of theoretical results was wanted, the collaboration of the Chemical Research Laboratory was sought. A core material was developed for experimental purposes from *foamed* calcium alginate. The successful development of the Mosquito aircraft called in question the shortage of balsa wood and the calcium alginate foam was selected as a possible substitute.

Stress corrosion was one of the subjects in which there was much cooperation between engineering and metallurgy. This arose because the hydrogen needed for barrage balloons was carried in cylinders made from nickel-chromium-molybdenum steel, which had been originally developed for cold cast traction purposes. Trouble was encountered through failures when they were first used for hydrogen; the cylinders burst. The other material-type problems investigated in Engineering Division were the design and failure of helical springs and the development of resistance wire strain gauges. Work in the Metallurgy Division at the N.P.L. during the second war complemented and extended that of Engineering. For instance, the perforation of steel plate by bomb fragments carried out in the Metallurgy Division could have been carried out either in Metallurgy or in Engineering. Another metallurgical problem investigated was the use of Duralumin scrap. Some problems in connection with the loss of ductility in cast alloys made from Duralumin scrap were investigated and solved. Porosity of castings, problems relating to armour piercing shot, electrical tests of hardness of armour piercing caps as well as naval shells are problems of such an engineering flavour, that then, as now, it would be much more likely to find such problems investigated outside the Metallurgy Division. Nonetheless there were metallurgical problems in worrying about how it is that 14 and 15 inch (35–38 cm) shells were retarded by 12 inch (30 cm) carbon steel plate and it was this work carried out under the direction of the then Dr Sykes, Superintendent of the Division, which led Sykes to become not only Superintendent of the Metallurgy Division but also of the Terminal Ballistic Section of the Armament Research Department of the Ministry of Supply. Welding problems were investigated in the Metallurgy Division, cutting tools, gun barrel breeches, high temperature alloys, moisture proof packagings and magnesium alloys. An experimental transmission-type electron microscope was set up for the first time in England in the Laboratory in 1940 – this was destined for the Imperial College of Science and Technology and made by Metropolitan Vickers. Besides this, work on the microstructural changes accompanying deformation of material at high temperature and the cracking of boiler plates and of the wear of refractories was carried out in Metallurgy Division. This recitation of the topics covered illustrates again the close relationship between Metallurgy and Engineering and, indeed, the close relationship between Engineering and the Chemical Research Laboratory in some of the aspects of the C.R.L.'s work.

After the second war the Engineering Division carried out work for the Mechanical Engineering Research Board of the Department of Scientific and Industrial Research and it was planned that this work should be controlled by the Mechanical Engineering Research Board rather than directly by the N.P.L. Advisory Board. In the Annual Report for 1947 it was stated 'A large part of the research programme of the Engineering Division is now carried out for the newly formed Mechanical Engineering Research Board of the Department. The smaller amount of routine work undertaken has enabled fundamental research to take an increasingly prominent place in the programme' (N.P.L. Annual Report 1947). It was decided to transfer the bulk of Engineering Division to a new laboratory. By 1950 progress on the Mechanical Engineering Research Laboratory at East Kilbride had been made and during 1951 the main transfer of equipment and staff was taking place.

The work on materials at high temperatures so intimately interwoven with that of Metallurgy Division was to remain at N.P.L. until 1953. By the end of 1951 the major change had been the transfer of most of the work of the Engineering Division to East Kilbride. That remaining was distributed amongst the Divisions of the N.P.L., e.g. the deadweight standards of load and strain gauges went to the Metrology Division and work on physics of the solid state to Physics Division. This included measurements of the single crystal elastic constants by sonic methods, which today would be regarded as an integral part of materials science. Some work on high temperature materials stayed at the N.P.L. The creep laboratory still exists and there is a programme in the Division of Materials Applications on high temperature materials.

The separation of the work on engineering properties of materials by the move of the Engineering Division to East Kilbride was regretted in the Metallurgy Division at the time and it is clear from Annual Reports that a sense of frustration filled Dr Allen's mind. There was some attempt at continuing cooperation over the years since 1950 but I believe that both organizations have suffered; the materials work from the absence of engineering practitioners on the site and the Engineering Research Laboratory has been unable to draw on first rate scientific work on microstructure.

The other strand in the organization of materials science and engineering at Teddington developed from the Chemical Research Laboratory, set up in 1925, which became the National Chemical Laboratory in 1958 and was amalgamated with the National Physical Laboratory on 1 April 1965. The Chemical Research Laboratory, C.R.L., was set up in 1924 but the first report did not appear until 1934 and contains a summary of the work before that date. The main Divisions were Corrosion of Metals, High Pressure Research, Tar Research, Chemotherapy, Water Pollution, Microbiology, some work on dental materials and on organic and inorganic chemistry. At the period of the amalgamation, the National Chemical Laboratory, which it had been called since 1958, contained work on chemical thermodynamics including thermodynamic measurements and some attendant molecular spectrometry, which enabled one to calculate some thermodynamic properties from observed vibrations of molecules; inorganic chemistry, which contained work on mineral processing besides chemistry of beryllium and electrochemical methods of metal production; a divison of organic chemistry and a division of surface chemistry. The work on mineral processing was transferred to the Warren Spring Laboratory and the remainder of the work on inorganic

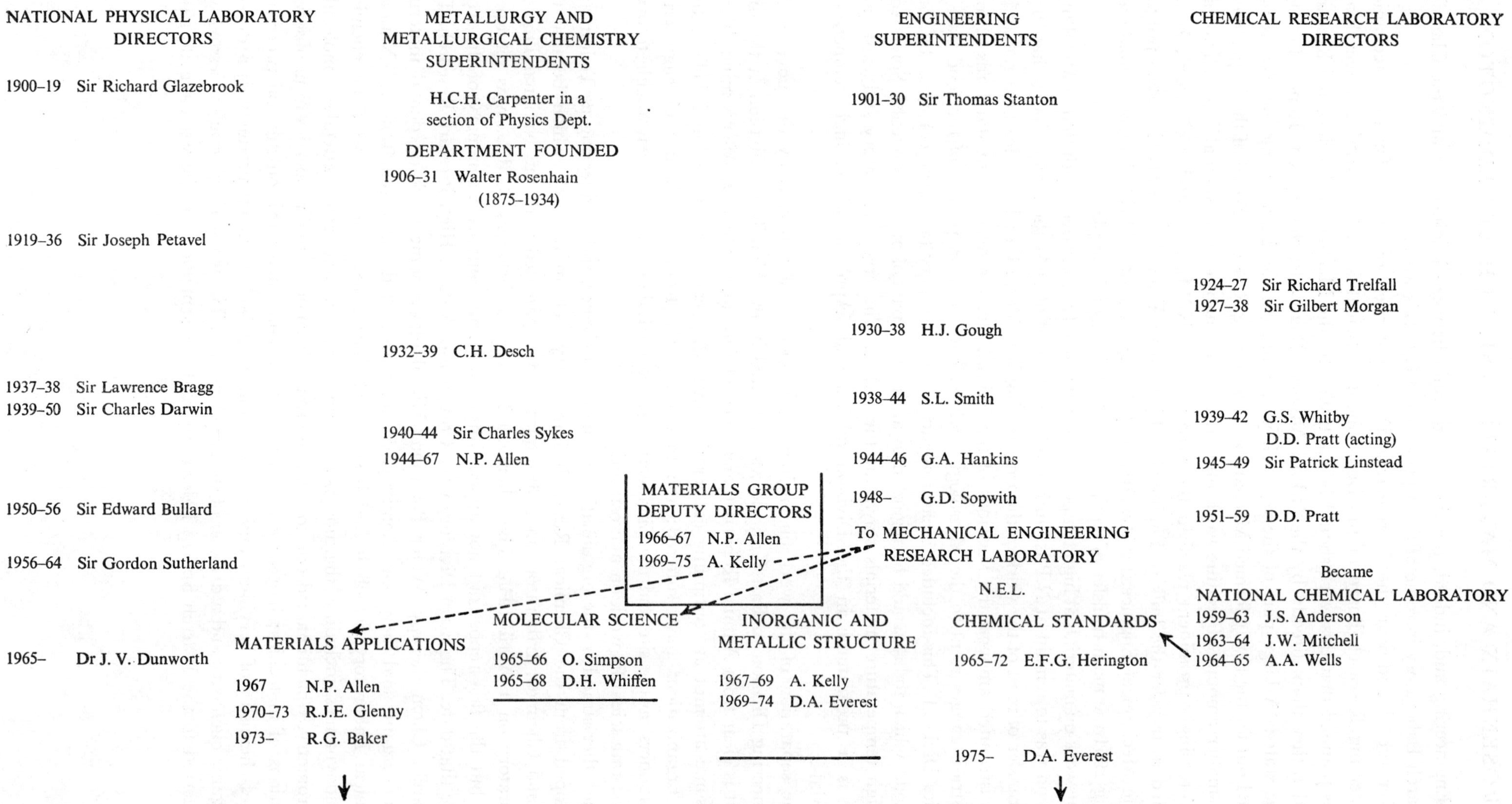

FIGURE 4. Directors of the N.P.L. and of the C.R.L. (N.C.L.) with Superintendents of Divisions concerned with properties of materials.

chemistry came into the Division of Inorganic and Metallic Structure, while that on organic chemistry went principally to a Division of Molecular Science and later was brought together in the Division of Materials Applications with the work on metals. The division of Surface Chemistry had been partly a particular interest of the Director, Dr J. S. Anderson, F.R.S., namely work on field emission and field ion microscopy and also a good deal of work on corrosion mechanisms. This naturally, of course, went into the present Division of Materials Applications. The chemical physics section of the National Chemical Laboratory, which was carrying out accurate physico-chemical measurements on gases and fluids became the Chemical Standards Division of the National Physical Laboratory. The work on organic and inorganic materials was put together with that of the Basic Physics Division of the N.P.L. in a Division called Molecular Science. This included a good deal of work on polymers carried out in the Basic Physics Division of the N.P.L. which had been started under J. A. Pople, F.R.S. when this Division was set up in 1958. All the work on materials was carried out for a number of years under a Deputy Director in charge of a Materials Group. This period has now come to an end and the major strands of the work are as follows. These will be briefly sketched because I think they have indications for us of the importance of the present Conference, but before doing so, it is worth just glancing at figure 4 which attempts to summarize the organizational structures and the personalities involved as head of these over the period from the inception of the N.P.L. and the N.C.L. up until the present time.

The present work concerned with Materials at the N.P.L. is contained in two Divisions. The first, the Division of Chemical Standards, is concerned with the maintenance of secondary standards and the transfer of these to British industry. The topics covered include general chemical reference materials, e.g. surface area standards, metal in oil standards, and standards for characterizing molecular mass distribution in polymers. Reference materials for measuring thermal properties, e.g. benzoic acid for calibrating bomb calorimeters and standards for differential thermal analysis. A data bank of thermal properties including those of metallurgical interest is maintained. A capability to calculate phase diagrams for multicomponent systems, four or more is being developed. All data banks need to be developed in terms of the relevant soft ware for computer manipulation. Materials are sometimes processed at high pressure and so high pressure standards are maintained. Surfaces are characterized in terms of surface segregation and propensity of elements to segregate to grain boundaries. Chemical analysis capability is maintained in this Division. The development of many of Rosenhain's interests is clear.

The second Division presently named Division of Materials Applications contains the majority of the work aimed to support the writing of specifications of materials for engineering use. This involves both the characterization of materials and of engineering needs appropriate to specific applications as well as the provision and development of some standard methods of testing; some standard data is also acquired. The Division carries out work on creep of chromium–molybdenum–vanadium steels for power generation aimed at recognizing micro-structural features to predict creep lives, and work on structural steels and hard metals, also designed to support the best specification of chemical composition and heat treatment. Corrosion prevention and oxidation prevention figure largely and a Corrosion Advisory Service, a part of a National system for advice on corrosion, is maintained. A large section of the Division deals with elucidation of properties which will support the writing of engineering specifications for polymers by measurement of elastic moduli and of anelastic effects. Finally, a section deals with composites and concentrates on understanding the modes of fracture of these and especially of those based upon cements.

The relation between the two Divisions is a close one and the boundaries of the work are not exact. In Chemical Standards some work is carried out on silicate materials, with a view to improving the engineering specification of these.

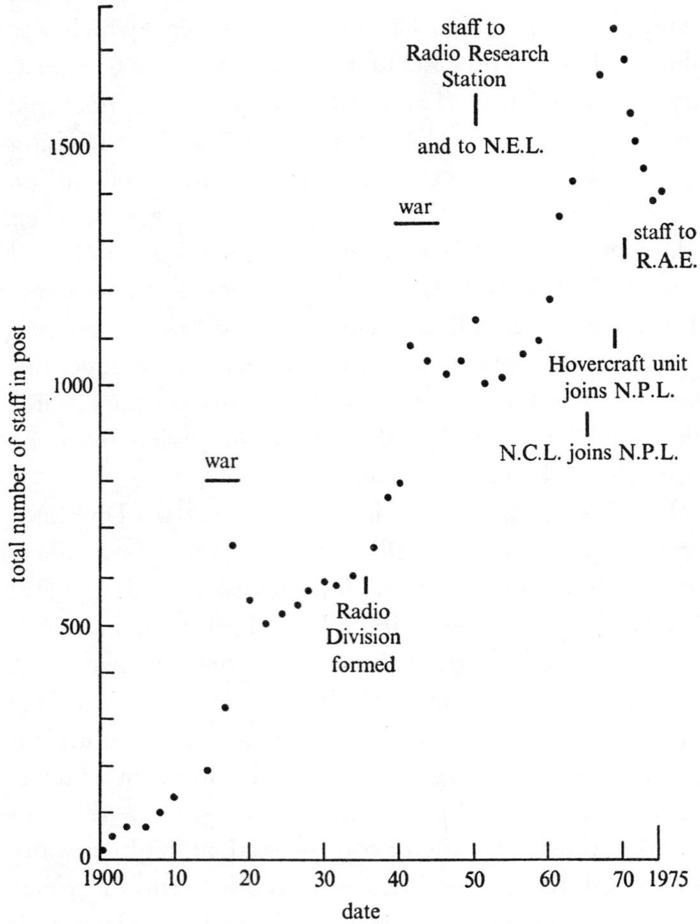

FIGURE 5

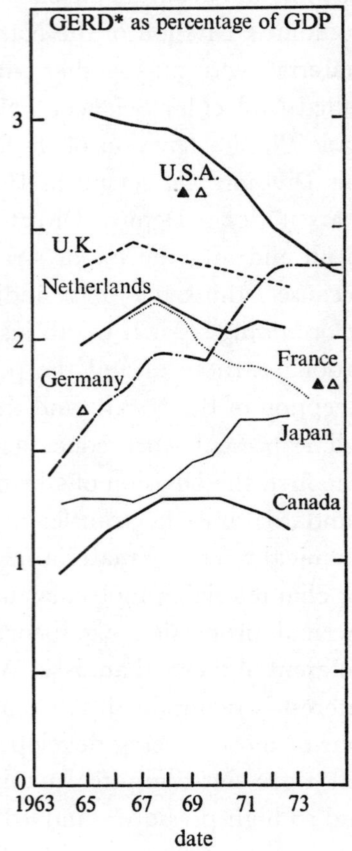

FIGURE 6

FIGURE 5. Total number of staff in post at the National Physical Laboratory from 1900 to the present.

FIGURE 6. Gross expenditure on research and development in a number of industrialized countries as a percentage of gross domestic product from 1963–75. (From *Financial Times* 24 June 1975). *, gross expenditure on research and development (GERD); △, these data include world sciences and/or humanities; ▲, national total approximately GERD.

The Division of Materials Applications will have increasingly close links with the recently formed Division of Mechanical and Optical Metrology. This will recreate, in a somewhat different form, and with specific reference to standards, the valuable interaction which was a feature of the relationship between the Metallurgy and Engineering Divisions.

All of this work is carried out at the behest of the Chief Scientist of the Department of Industry, who finances the work through the Requirements Boards, which contain advisers drawn from industry and the universities.

9. The present scene

When Rosenhain joined the N.P.L. he had a staff of four and when he left that staff was 70. This rise in numbers was paralleled in the growth of the Laboratory. Figure 5 shows the numbers from 1900 until the present day. At the time of the transfer of the N.C.L., the N.P.L. absorbed approximately 196 non-industrial Civil Servants and 43 industrials. This produced a large increase in the size of the N.P.L. which can be seen on the graph of total numbers in figure 5. Figure 5 shows that there have been two large increases in the number of staff at the N.P.L.: one following the first World War in 1920, the second following the outbreak of war in 1940 and there was then a steady rise in numbers during the early 1960s which culminated with the transfer of the staff from N.C.L., so that the total number at the N.P.L. reached about 1750 souls. The largest size of the Laboratory was obtained in 1970 and thereafter there has been a decline in numbers, as there has in many other scientific research establishments.

Figure 6 shows that the growth of spend on research and development in a number of countries – the most advanced technologically – is over and the spend now decreases as a proportion of gross domestic product.

The total amount of science and technology has grown so enormously since Rosenhain's time that the problem of relating the position of the physical metallurgist or of the materials scientist to that of the engineer is almost unrecognizably complex. This may be a truism but I believe it to be worth stating.

Firstly there is a change in the range of materials available. I list for some materials of interest to engineers the date at which they were introduced in tables 2 and 3. I have taken the date as that at which pilot plant scale production started.

Many of these were in their infancy when Rosenhain worked of course, and many have been developed since.

Concurrently with the introduction of new materials, there have been large changes in engineering technology. Some of those listed in *Encyclopaedia Britannica* as occurring between about 1930 and 1970 are:

gas turbine engine	1930–45
catalytic cracking of petroleum	1935–41
atomic power	1942
plastics	1909–54
synthetic fibres	1938
very high rise building	1900–32
automobile mass production	1913–23
television	1920–45
liquid fuelled rockets	1926–67
Man on the Moon	1969
Wankel engine	1968
computers	1944–
computer controlled chemical manufacture	1950s
heavy earthmoving equipment	1945
plastic design and steel framed buildings	1947–55
air cushion vehicles	1958
Xerography	1937–46

TABLE 2. DATE OF DEVELOPMENT OF SOME MATERIALS

iron blast furnace	1735	uranium	1943
crucible steel	1740	tungsten	1909
Bessemer steel	1855	tungsten carbide	1926
oxygen steel	1952	synthetic diamond	1958
copper	1556	fibres	
nickel	1840	nylon	1938
Nimonic alloys	1940–	polyester	1947
tin	B.C.	glass fibre	1938
lead	B.C.	(glass wool)	1930
magnesium	1886	carbon fibre	1968
Aluminium	1886	glass reinforced plastic	1940
titanium	1944		

TABLE 3. DATE OF DEVELOPMENT OF SOME PLASTICS

thermosetting		thermoplastics	
phenolic	1909	low density polyethylene	1933
urea base	1926	high density polyethylene	1954
alkyd	1910–20	polypropylene	1959
polyester-styrenated types	1942	vinyl chloride	1930
epoxy	1939	polyvinyl acetate	1917
		styrene polymers	1935
		cellulose plastics	1900–24

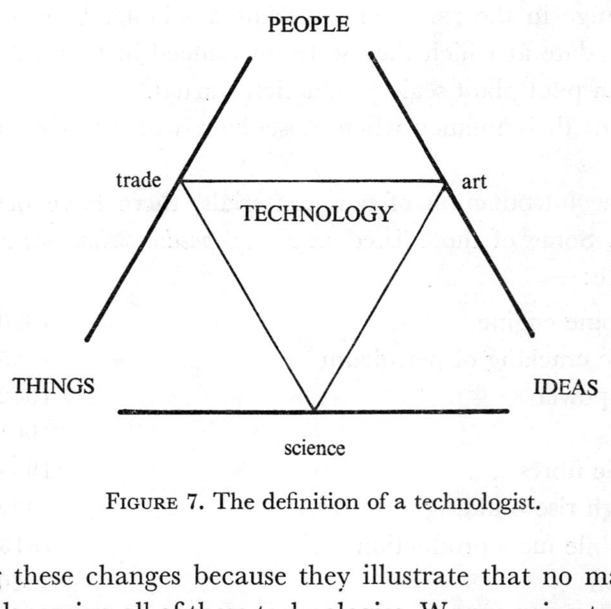

FIGURE 7. The definition of a technologist.

It is worth recalling these changes because they illustrate that no man can encompass now the source of materials serving all of these technologies. We are going to sectorize the discussion and concentrate upon physical metallurgy and what its relation has been, and hopefully may be, to the progress of engineering. Because of the complexity one must inevitably deal with teams engaged in this endeavour; teams of scientists, teams of engineers. Once a team is formed this human grouping develops an inner logic of its own and a good deal of its time is spent in ensuring the proper communication between the component parts of the individual team; i.e. between the individual members of the team. If this is necessary even more time must be spent in ensuring a proper interface for communication between the various teams.

We tend to idealize the team as a single man, say the physical metallurgist or the engineer. But it is important to remember that these are just idealizations.

The *metallurgist* has as his concern to supply the engineer with the alloys he needs, with information about their properties, about how to avoid problems when using them. In order to do this he has to know the engineer's situation, to think with the engineer. To predict the properties he must understand them and this is usually done by scientifically studying the relation between properties and microstructure so as to provide a basis for explanation of the effects of processing, heat treatment, etc.

The *engineer* must design machines, structures and components to achieve a required function and to select and specify the material standards of workmanship and overall quality that must be attained. So he must be aware of properties of different materials relevant to him and of the potential problems which may arise. Overall a value judgement will be exercised by society on the enterprise and this is simply expressed in terms of cost. The alloys must be supplied and the structure made at the price that the customer will pay. Whether or not the structure is supplied at an acceptable cost is the job of all *technologists*.

The job of the *scientist* is easy to define – it may not be easy to do. I have found it helpful to think of the job of the technologist in terms of a diagram which illustrates his relationship to science and to people. This idea has been introduced in a forthcoming book (Davies, Sheahan & Banfield 1976) which I have enjoyed discussing with the authors, and is illustrated in figure 7. The scientist explains things because they are there. He develops a language of explanation by the systematic confrontation of nature with observation and with logic. In terms of the diagram he relates things and ideas but has not very much to do with people. The artist relates man with his idea of himself without the use of language. He relates people and ideas and has little to do with things. The trader brings things to people in an acceptable form and has little to do with ideas. The successful technologist is part scientist, part trader and part artist. The physical metallurgist relates science and technology on these definitions and this is what Rosenhain attempted to do. This Conference will examine further how this business of relating science and technology has been fruitful and will be more so.

I am very grateful to a number of people who have provided me with information; in particular Mrs N. Kirsner, Professor N. Greenwood, Professor R. W. Cahn, Dr E. D. Hondros and to my secretary Mrs P. Lee for so carefully editing the manuscript. I am very grateful to the staff of the library at the N.P.L., in particular Miss H. K. Carter and Miss J. H. Simmons, who have helped me diligently in my research for some rather obscure sources. I am grateful to Drs R. G. Baker, J. V. Dunworth and D. E. Miles for helpful comments on the typescript.

References (Kelly)

Adams, F. D. & Nicholson, J. T. 1900 *Phil. Trans. R. Soc. Lond.* A **195**, 363.
Albert, Ph. & Le Hevicy, J. 1956 *C.R. Acad. Sci., Paris* **242**, 1612.
Barrett, C. S. 1949 *Cold working of metals*, p. 65. Cleveland: American Society of Metals.
Basinski, Z. S., Hume Rothery, W. & Sutton, A. L. 1955 *Proc. R. Soc. Lond.* A **299**, 459.
Beilby, C. T. 1903 *Proc. R. Soc. Lond.* **72**, 218.
Beilby, C. T. 1907 *Proc. R. Soc. Lond.* **79**, 463.
Bowden, F. P. & Tabor, D. 1950 *The friction and lubrication of solids*, vol. I; 1964, vol. II. Oxford University Press.
Burgess, G. K. & Crowe, J. J. 1913 *Trans. Am. Inst. Mining metall. Engrs* **47**, 665.
Burke, J. E. & Turnbull, D. 1952 *Prog. Metal Phys.* **3**, 220.
Cahn, R. W. 1975 *Dictionary of scientific biography* **11**, 548. New York: C. Scribners, Sons.

Chalmers, B. 1940 *Proc. R. Soc. Lond.* A **175**, 100.
Cohen, M. & Harris, J. M. 1965 *History of metallurgy*, p. 209. Am. Inst. Mining. metall. Engrs.
D.H. 1934 *J. Inst. Met.* **54**, 313.
Davies, D. S., Sheahan, R. J. & Banfield, T. L. 1965 *The humane technologist*. Oxford University Press.
Desch, C. H. 1932–35 *Obituary Notices of Fellows of the Royal Society*, p. 353.
Duwez, P. 1967 *Trans. Am. Soc. Metals* **60**, 607.
Ehrenfest, P. 1933 *Leiden Commem. Supp.* 75 f.
Ewing, J. A. & Rosenhain, W. 1899 *Proc. Roy. Soc.* **65**, 85.
Ewing, J. A. & Rosenhain, W. 1900*a* *Phil. Trans. R. Soc. Lond.* A **193**, 353.
Ewing, J. A. & Rosenhain, W. 1900*b* *Rep. Brit. Ass.* p. 698.
Ewing, J. A. & Rosenhain, W. 1901 *Phil. Trans. R. Soc. Lond.* A **195**, 279.
Fry, Hilda E. & Rosenhain, W. 1919 *J. Inst. Met.* **22**, 217.
Haughton, J. L. 1934 *J. Inst. Met.* **55**, 17.
Hirsch, P. B. 1956 *Prog. Metal phys.* **6**, 236.
Howe, H. M. 1890 *Metallurgy of steel.* New York and London: McGraw Hill.
Hull, A. W. 1917 *Phys. Rev.* **9**, 84, **10**, 661.
Hutchinson, E. 1969 *Minerva* **7**, 373.
Hutchinson, E. 1970 *Minerva* **8**, 396.
Jeffries, Zay & Archer, R. S. 1921 *Chem. Metall. Engng* **24**, 1057.
Kelly, A. 1953 *Proc. Phys. Soc.* A **66**, 403.
McLean, D. 1957 *Grain boundaries in metals.* Oxford University Press.
Mügge, O. 1899 Nach. der Königl Gessell der Wissenschaften zu Göttingen *Nat. Phys. Klasse* 56.
N.P.L. Annual Reports 1940–45, 1946, 1947, 1948. London: H.M.S.O.
Osmond, F. & Cartaud, G. 1900 *Annales des Mines* **18**, 113.
Paxton, H. W. 1953 *Acta metall.* **1**, 141.
Preston, G. D. 1938 *Proc. R. Soc. Lond.* A **167**, 526.
Reusch, E. 1867 *Pogg. Ann. Phys.* **132**, 441.
Rosenhain, W. 1902 *Phil. Mag.* series VI, **4**, 451.
Rosenhain, W. 1904 *J. Iron Steel Inst.* **65**, 335.
Rosenhain, W. 1905 *Proc. Roy. Soc.* **74**, 557.
Rosenhain, W. 1908 *Glass manufacture.* London: Constable.
Rosenhain, W. 1909 *J. Inst. Met.* **1**, 200.
Rosenhain, W. 1910*a* *Rep. Br. Ass.* p. 567.
Rosenhain, W. 1910*b* *Physikal. Z.* **11**, 1156.
Rosenhain, W. 1911 *Proc. Instn Mech. Engrs* p. 241.
Rosenhain, W. 1913 *Engineering* **96**, 509, 357; *Int. Z. Metallographie* **5**, 65.
Rosenhain, W. 1914 *An introduction to the study of physical metallurgy.* London: Constable.
Rosenhain, W. 1916*a* Optical Glass. Three Cantor Lectures. *J. R. Soc. Arts* **64**, 649.
Rosenhain, W. 1916*b* *J. West Scotland Iron & Steel Institute* **23**, 213.
Rosenhain, W. 1920 *J. Inst. Met.* **23**, 247.
Rosenhain, W. 1923*a* *The Engineer* **136**, a large number of articles. The last on page 522.
Rosenhain, W. 1923*b* *Third Sorby Lecture*,
Rosenhain, W. 1923*c* *J. Inst. Met.* **30**, 3.
Rosenhain, W. 1924 *J. Iron Steel Inst.* **110**, 85.
Rosenhain, W. 1926 *J. Inst. Met.* **36**, 7.
Rosenhain, W. 1928 *J. Inst. Met.*, **39**, 27.
Rosenhain, W. 1930 *Zeit für Metallkunde* **22**, 73.
Rosenhain, W. & Archbutt, S. L. 1912 *Proc. Instn Mech. Engrs* p. 319.
Rosenhain, W., Archbutt, S. L. & Hanson, D. 1921 *Proc. Instn Mech. Engrs* **2**, 699.
Rosenhain, W. & Ewen, D. 1912 *J. Inst. Met.* **8**, 149.
Rosenhain, W. & Ewen, D. 1913 *J. Inst. Met.* **10**, 119.
Rosenhain, W., Haughton, J. L. & Bingham, Kathleen, E. 1920 *J. Inst. Met.* **23**, 261.
Rosenhain, W. & Humfrey, J. C. W. 1909 *Proc. R. Soc. Lond.* A **83**, 200.
Rosenhain, W. & Humfrey, J. C. W. 1913 *J. Iron Steel Inst.* **87**, 219.
Rosenhain, W. & Lantsberry, F. C. A. H. 1910 *Proc. Instn Mech. Engrs* p. 119.
Rosenhain, W. & Murphy, A. J. 1926 *Proc. R. Soc. Lond.* A **113**, 1.
Rosenhain, W. & Tucker, P. A. 1908 *Proc. R. Soc. Lond.* A **81**, 331.
Rosenhain, W. & Tucker, P. A. 1908 *Phil. Trans. R. Soc. Lond.* A **209**, 89.
Seah, M. P. & Hondros, E. D. 1973 *Proc. R. Soc. Lond.* A **335**, 191.
Shewmon, P. 1957 *Acta metall.* **5**, 335.
Westgren, A. 1921 *J. Iron Steel Inst.* **103**, 303.

APPENDIX 1. (*Chronology*)

date	event	age
1875	Born 24 August Berlin, only son and youngest child of Moritz and Frederike Rosenhain	0
1880	Family went to Australia, lived in Geelong	5
1888	Queen's College Melbourne founded	
1890	Visited Europe with parents	15
1891	Entered Queen's College, University of Melbourne; read Physics and Engineering	16
1894	Visited Europe with parents seeking cure for his lame sister Clara	
1896	Final Honours in Engineering	21
1897	Won 1852 Exhibition to go to Cambridge – St Johns. Worked with Sir Alfred Ewing	22
1900	Ewing delivered Bakerian Lecture based on work with Rosenhain	25
	June – Took up post of scientific adviser to Chance Bros., Smethwick	
	Associated with Stephen Chalmers (died 1918) Clara's husband and Bruce Chalmers' father	
1900–6	Studied refractory materials and glass	25–31
1901	December – Married Louise Monash, daughter of Louis Monash and sister of General Sir John Monash	
	Joined Iron and Steel Institute	
1906–31	Appointed Superintendent, Department of Metallurgy and Metallurgical Chemistry, National Physical Laboratory	
1906	Andrew Carnegie Medal, Iron and Steel Institute	
1908	Institute of Metals founded – Rosenhain original member	
	Published *Glass manufacture* 1st edition, Constable	
1909	Awarded D.Sc. Melbourne	
	Commemorative Diploma, Brussels University	
1911	Commemorative Diploma, Turin University	
	Sir Julius Wernher donates £10 000 for new Metallurgy Laboratory at the N.P.L.	
1911–18	Member of Council, Institute of Metals	
1912	Visited New York for 6th Congress International Association for Testing Materials	
1913	Commemorative Diploma, Ghent University	38
	Elected Fellow of Royal Society	
1914–18	Great War during which half of staff of Metallurgy Division called up	
1914	Published *An introduction to the study of physical metallurgy* 1st edition, Constable	39

	Visited Australia with British Association and returned in December	
1917	Ph.D. established at University of Oxford	
	Department of Scientific and Industrial Research established	
1919	*Glass manufacture* – 2nd edition published	44
	Introduction to Physical Metallurgy – 2nd edition published	
1921	R38 crashed – two staff of the N.P.L. killed	46
1923	Gave third Sorby Lecture. 'Present and future problems of metallurgy' (Rosenhain 1923*b*)	
	Lecture tour of United States	
	May Lecture Institute of Metals, 'The inner structure of alloys' (Rosenhain 1923*c*)	
1926	Autumn Lecture to Institute of Metals 'Modern metallurgy and ancient industries' Liege, Belgium (Rosenhain 1926)	
	Vice President International Association for Testing Materials	
1927	Joined Institution of Professional Civil Servants and helped establish a branch at the N.P.L.	
1928–30	President, Institute of Metals	53–55
1928	Presidential Address Institute of Metals	
1929	Fellow of the Institute of Metals	
1930	Bessemer Medal, Iron and Steel Institute	
1931	Scientific Civil Service established	
	June – Resigned from the N.P.L.; became private consultant	
1932	Honorary member Deutsche Gesellschaft für Metallkunde	
1934	Thomas Turner Gold Medal, University of Birmingham	58
	17 March died at Kingston Hill, Surrey	
	Introduction to physical metallurgy, 3rd edition (with J. L. Haughton) published	

1. ENGINEERING REQUIREMENTS

1.1 The use of steel in bridge construction

By W. J. Harper

British Bridge Builders Ltd, Woodleigh, West Hill, Hessle, North Humberside HU13 0EZ

[Plates 1 and 2]

A very brief history of some of the major changes in the use of steel for bridgework is given, with some discussion on what the author considers are the main areas in which problems remain – distortion and corrosion – with an opinion of the more likely erection methods of the future.

We are so conditioned to the sight of steel being used throughout our every day life that it comes as somewhat of a surprise to remember that it was only in 1867 that the first commercial steel was rolled – for rails – manufactured by the Siemens Martin open hearth process. The new material was first used extensively in bridgework in the Eads Bridge in St Louis, completed in 1874, and the first long span bridge – the Brooklyn suspension bridge – was completed in 1883. Thus it is less than 100 years since this material allowed designers to work with high tensile stresses in addition to the compressive stresses used till then.

As the production of steel increased, and the techniques of rolling it into long lengths became refined, the variety of cross sections, and the size of these cross sections, increased, leading finally to those voluminous handbooks of structural shapes with which all designers were familiar until very recently. These handbooks, with their tables of properties of standard sections became the 'memory banks' of the computers of that time – the human designers – and, it could be argued, were indirectly responsible for some of the troubles which have arisen recently, when, with the advent of welding, the use of non standard shapes and cross sections became economically viable.

The use of standard rolled steel sections, and the standard method of jointing such sections – rivetting or bolting – led, almost automatically to the truss bridge and all its ramifications, whatever the type of bridge considered, whether as a simple beam or as the basis for a cantilever bridge (figure 1), an arch bridge (figure 2), or a suspension bridge (figure 3). It was not until the use of welding became common in the nineteen forties that the truss became somewhat old fashioned, but its demise was due mainly to the economic facts of fabrication costs. Indeed, in those countries with cheaper labour costs, such as Japan, the truss bridge still forms the majority of the major bridges built. Fabrication costs of trusses may also be decreased by the use of highly automated equipment such as the computer controlled punches and drills now in increasing use, and it is possible that the truss is due to be revived, particularly if a cheaper form of jointing members than rivetting can be developed. Cheaper, that is, in overall cost – not just the cost of manufacture and placement of the fastener. Most site erection engineers would welcome such a revival and it would certainly help to take the recent 'anti' out of cantilever (which incidentally leaves c-lever!) methods of erection of large span bridge steelwork.

Practically all the long span bridges built before World War II used the truss form of construction – from 1920 to 1940 being the period of the magnificent American suspension bridges. In the fifty years up to 1940 the longest bridge span increased from 450 to 1250 m, which gives a fairly good idea of the progress of technology in that period. Technology which included better quality steels and higher tensile steels, but which also included the advances made in the construction of foundations under compressed air. Without this last named technique it is doubtful whether many of the large span bridges could have been built in this era, irrespective of the other advances in steel technology.

An interesting exception to this technological advance in the use of steel, and which still applies to this day, is that of the material used to produce the large cables for suspension bridges – generally done by the 'spinning' process invented by Roebling in 1865 and first used with *steel* wire in 1876. Amman designed and built the George Washington bridge in 1927–1931 with galvanized steel wire 0.196 in (4.98 mm) diameter with a minimum ultimate tensile strength of 98 tonf/in^2 (1515 N/mm^2). This could be the same specification for the wire for the cables to be spun for the Humber Bridge almost 50 years later – except that nowadays, the figures are required to be expressed in metric units.

It was in 1940 that a 845 m span suspension bridge at Tacoma Narrows was completed, quickly achieved the soubriquet of 'Galloping Gertie' and collapsed in the same year. This event gave rise to the new science of aerodynamics as applied to bridgework, and it is now standard procedure to carry out an aerodynamic analysis of all major bridge structures at the design stage. This analysis checks the aerodynamic characteristics of the completed bridge, carrying its design loading, but further investigations are needed to check these characteristics during all stages of erection. Indeed, the structure may well have completely different stability characteristics during certain stages of the erection; to such an extent that very expensive additional temporary works may be required to bring the aerodynamic stability within practical limits, thereby allowing the bridge to be erected safely.

Reference has already been made to the tremendous advances in knowledge and techniques of welding which occurred during World War II which, with its direct and indirect effects, have overshadowed all else in the advances in the use of steel for bridges since 1945. Standard rolled sections have almost disappeared, and the use of plate increased enormously. The designer is able to place steel in the structure where, it is calculated, the highest stresses occur, with the net result that a greater proportion of the steel is 'usefully' used, and less is wasted on physically joining up these areas of stress. This leads to a large saving in weight of steel in the structure, and, as the greatest load on any long span bridge is its own dead weight, has a cumulative effect on the overall costs of the structure. It is one of the main reasons for the use of the box girder design (figure 4).

Certainly the dead weight of any structure is one of the major parameters to be borne in mind in design, but it is not the only parameter and total cost must always be borne in mind. It is possible to fabricate steel to tolerances measured in fractions of a millimetre and (less generally) to erect it to such tolerances, but the costs of so doing may far outweigh the savings in cost of material, particularly with the present day escalation of labour costs. This, of course, is a problem applicable to all areas of steel construction but the field of long span bridges is peculiarly sensitive to dead load problems and a study of 'value engineering' could be a fruitful one to any bridge designer.

One effect of this freedom of design to place most steel at the areas of highest stress is that it becomes more important to be able to specify where such stresses actually occur and their magnitude, more precisely. A veritible Pandora's Box emerges with such matters as secondary stresses looming large particularly, as has been shown recently, in the field of stiffened plates in compression. It is only when such problems arise that our basic lack of knowledge of the precise way in which materials carry the loads imposed on them becomes evident, and that the fact that the money spent on basic research into such matters offers a major return on capital becomes more obvious.

One major cause of secondary stresses arises from distortions appearing in the steel during fabrication, and one of the major causes of such distortion is the application of heat to the material. The fact that input of heat into a steel member changes the size and sometimes the shape of the member has, of course, been known since the material was first used, but the distortions produced by the increased use of welding and increased use of thinner plates in the fabrication of bridge steelwork have become such that the problem of distortion during fabrication and erection can only be solved by the application of completely new techniques in both these fields. If the use of relatively thin plate is to continue (as it would appear probable) it could be that some form of joining steel without an input of heat would be economically feasible, more particularly on site than during fabrication.

The answer to this problem of cold jointing of steel may appear from another, rather unexpected, source. With the increased interest being taken in such areas as pollution and world over-population, the new dimension into which man's progress may move might be that of the sea – and not space, which is the more exotic, but probably the less likely. If this is so, it is also probable that more and more structures in steel will be required under and in the water, and the use of adhesives, or some other form of chemical bonding of steel, will emerge, to the advantage also of the bridge engineer.

Thus, it is possible that advances in technology necessarily required for structures built in the water may have a marked effect on methods used for structures over the water. This not only applies to the area of cold-jointing but may also apply to another large problem in the field of steel bridgework – that of corrosion protection.

Corrosion is probably the most difficult of the problems associated with steel used in bridges, but conversely, is the one which, when overcome successfully, will show the greatest effect economically. One approach has been to permute various treatments of the surface of the plain steel with the application of various films with inert or sacrificial corrosion properties. Practically all of these permutations have decreased efficacy in areas of difficult access and in areas where joining of pieces is necessary. In other industries ships, aircraft and machinery have been successfully cocooned for shorter or longer times and it is possible that this may offer the best hope for a reasonable solution from this approach.

The second approach from which the problem of corrosion has been tackled is that of adding trace elements to the steel mix at the production source, but the use of the so-called corrosion resistant steels has not found universal favour, at least with the steels so far produced. This is because such steels are merely corrosion resistant and not corrosion proof, and their use merely decreases the corrosion products formed, rather than prevents their formation. Very careful design detailing is required to deal with these corrosion products and, even then, the results are not to everyones aesthetic taste, to put it mildly. It might be argued that 'stainless' steel is a material which could be used, but the (writers) imagination boggles at the problems and costs of fabricating and erecting a bridge composed of thin plates in present day stainless steel with all joints welded using electrodes depositing an equally stainless steel.

Such problems, however, form the fascination inherent in the planning and carrying out of the engineering involved in the erection of long span steel bridges and the satisfaction derived by site engineers in overcoming them. Mention has previously been made of the fact that cantilever erection imposes a completely different stress pattern on the structure than that for which it was designed – stresses in local areas may be increased and changed in sign, leading to considerable amounts of additional steel being required; steel whose sole purpose is to carry loads imposed by erection.

One method which has found increasing use recently is to erect the structure in such a way that the cantilever is replaced by the much more simply erected beam. It is only the scale of the beam in size and weight that has prevented the economic use of this solution previously, but single lifts of 600 tonnes (Queens Way in America), 900 tonnes (Martigues in France) and 3000 tonnes (Ponte Presidente Costa e Silva in Brazil) (figure 5) have shown what is possible, and even larger single lifts up to 6500 tonnes have been planned. The schemes involving the heavier lifts have applied the technique of climbing jacks, by using experience derived from the developments in the handling of missiles and the use of jack up platforms for offshore oil exploration. It is this latter field which, by its increase in the use and availability of floating cranes with 2000–3000 tonnes lifting capacity, will provide a further alternative for this method, on bridge sites where sufficient water is available. There will still be sites where the cantilever method is required – indeed all the classic methods will continue to be used – the choice depending on the characteristics of the particular site, but all methods will benefit from the availability of larger lifting devices.

The history of the use of steel in bridges shows that progress in materials and methods draws on a very wide field of other sciences, and that development in what is, on the surface, an unrelated field, can have a marked effect on development in steel bridgework. There is no reason to think that this will alter, and it is therefore likely that the next Rosenhain Centenary Conference may contain a paper similar to this, but, reflecting the technical advances made in all engineering fields in the interim. *Plus ça change....*

FIGURE 1. Sheeting the cantilever bridge as a truss; Klang, Malaya.

FIGURE 2. Rohri Channel bridge over the Indus river, West Pakistan, with a main span of nearly 250 m, built by Dorman Long (Bridge and Engineering) Ltd. Sheeting the truss used for the arch rib.

(*Facing p.* 40)

FIGURE 3. Forth road bridge – sheeting suspension bridge as a truss.

FIGURE 4. Tay road bridge – box girder construction.

FIGURE 5. Ponte Presidente Costa e Silva – sheeting erection of centre span.

1.2 Materials requirements for pipeline construction

By W. L. Mercer
British Gas Corporation, Engineering Research Station, Harvey Coombe, Killingworth, Newcastle upon Tyne, NE99 1LH

[Plates 3 and 4]

A broad picture is presented of the current worldwide scene and future plans for oil and gas pipelines. Notwithstanding recent events, oil and gas consumption will continue to expand. As exploration moves into more remote areas of the world, longer and larger diameter pipelines will be required. Such areas are climatically increasingly hostile, throwing great demands on the performance of pipeline materials and equipment. The challenge to the metallurgist to produce low cost pipeline materials with the requisite levels of strength, toughness and weldability is being met. Developments in pipe steels technology which have been substantial during the past ten years will provide the basis for pipeline construction in the years ahead.

The fracture behaviour of pipelines has received considerable attention in recent years. As a fracture can be inadvertently initiated by earth moving machinery, pipeline design is usually based upon demanding fracture arrest criteria. Full scale tests have led the way to the development of effective laboratory scale assessment methods which now find wide use in quality control procedures.

Offshore pipelines present additional problems to the engineer. Relatively thick wall pipes are required to resist buckling stresses during laying and collapse loads during operation in deep water. Apart from problems of attaining the necessary high levels of toughness in such thick wall pipes, the requirements of offshore pipelines are unlikely to present major new challenges to the metallurgist. Overall, the principal requirements are for quality, reproducibility and reliability at reasonable cost.

1. Introduction

Oil and gas provide the main sources of energy used by man at the present time. In most countries they contribute over two-thirds of total energy consumption. Virtually all the gas and much of the oil (or products) are transported over land through pipelines. The capital value of these pipelines has reached enormous proportions. It is believed that the world total of oil and gas pipelines is about 1.8×10^6 km with a current replacement cost exceeding £100 000 million!

Transportation distances for oil and gas have increased remarkably during the past 20 years. Individual pipelines up to 5000 km long have been built. As world demand for oil and gas increases, so does pipeline capacity. Pipelines of 910 mm (36 in) and 1060 mm (42 in) diameter are becoming the norm for both fuels and for gas, diameters of 1500 mm (60 in) and above are receiving active consideration at the present time.

The search for oil and gas is moving into increasingly hostile environments, e.g. offshore continental shelves and polar regions. Pipeline engineering in these areas presents many unknowns. Enormous demands are made on materials of construction, fabrication techniques and test procedures. These must be met by the requisite combination of skills in design and construction engineering and through sound basic knowledge of the associated underlying pipe steel metallurgy.

This paper is concerned with pipelines for the conveyance of gas and oil both on- and offshore. Following a brief review of the pipeline scene today, attention is directed to the principal engineering requirements for these pipelines to ensure a proper level of safety and economy in design, construction and operation. The final section briefly considers probable future trends. The intent is to provide a basis by which the underlying fundamentals and the value of close interaction between basic engineering and metallurgical skills in this vital area of technology can be brought into perspective.

2. The pipeline scene

The first industrial use of a pipeline for conveyance of oil was in Pennsylvania in 1796. From this small beginning, expansion of oil and gas movement has reached tremendous proportions. The world now moves oil at the rate of about 8×10^6 tonnes per day. Already 20% of this comes from offshore sources, a proportion that will increase in the years ahead. However, it is large scale gas movement that provides the main impetus to pipeline construction. Over 90% of the world's total length of fuel pipelines is involved in handling natural gas. This position will be retained in the future.

Worldwide pipeline construction plans are prodigious. The current rate of construction involves expenditure exceeding £5000 million per year. The most spectacular, expensive and demanding projects are those in Arctic and offshore areas. The 1200 mm (48 in) diameter 1250 km Trans-Alaskan oil pipeline, which is intended to convey 0.3×10^6 m^3 per day (0.3M tonnes/d = U.K. current rate of consumption) is estimated to cost $3500M† (i.e. about £1250000 per km). A 1200 mm (48 in) 4200 km long gas pipeline to convey Alaskan gas to major Canadian and U.S.A. markets will cost over $6000M to convey 112×10^6 m^3 (4000×10^6 ft^3) per day at a maximum pressure of 11.5 MPa (115 bar; 1680 lbf/in^2). The picture in the Soviet Union is even more startling. Total reserves are enormous and are increasingly located in distant and inhospitable Siberian regions. Transmission of the gas to European Russia involves individual pipelines exceeding 4000 km in length and up to 1650 mm (65 in) diameter. In the five year period 1971–75, 50% of the 32000 km of gas transmission pipelines laid in the U.S.S.R. were 1200 mm (48 in) or greater diameter (Leyman *et al.* 1973).

In some ways even more demanding are the major undersea pipelines being built in the stormy waters of the northern North Sea. These are breaking records for depth, diameter and length and are very much at the frontiers of knowledge. It is estimated that for gas alone, there will be about 2000 km of pipelines in the North Sea by 1980. The Frigg field in the northern area is the largest offshore natural gas field in the world. Plans are in hand to extract gas from this field at a rate of over 71×10^6 m^3 (2500×10^6 ft^3) per day by 1978. The first of two 380 km long 800 mm (36 in) diameter pipelines is now being constructed at an estimated cost of £150M. This will be the longest offshore pipeline of this diameter in the world. It will be laid in water greater than 125 m in depth (maximum up to 180 m). This will be a world record for a large diameter pipeline.

Finally, feasibility studies are in hand for the conveyance of Canadian Arctic Islands gas to the south at a rate of 130×10^6 m^3 (4500×10^6 ft^3) per day through a 5100 km 1200 mm (48 in) pipeline. This involves construction not only in extreme Arctic conditions but the laying of

† All cost data are given in terms of mid-1974 prices. Technical difficulties and worldwide inflation are producing massive cost escalation. A more recent statement estimated a cost of $6000M!

several hundred miles of 900 mm (36 in) diameter underwater sections in water depths as great as 300 m (Kerr 1974). In many instances, it is necessary to cut through a very thick surface ice layer in order to reach water and then to lower the pipe to the bottom of the sea. Special techniques are being developed to cut through the ice in order to lay this pipeline.

Thus, although well established, pipelining is really in its infancy. World energy demands are pushing men and materials to the limits of endurance. The pipeline scene during the next few years is clearly one of excitement and extreme challenge.

3. On-shore pipelines

Worldwide pipelining is very much based on American technology. Oil and gas pipelines constructed during the twenties and thirties led to the development of the American Petroleum Institute (API) extensive range of standards (see, for example API 1975). Most countries use their own standards to meet their own special conditions for pipelines, but in all cases they can be traced back and seen to be firmly based upon relevant API standards (see, for example, British Gas Standard 1975).

A primary aim of the pipeline construction engineer is to build a high integrity pipeline that will be free from operational faults. To do this he requires a supply of high quality pipes and a sound knowledge of construction and inspection procedures. In considering the role of the metallurgist and his link with the pipeline engineer, it is necessary to concentrate on pipe specification and manufacture. The vital metallurgical parameters can conveniently be examined in terms of strength, toughness and weldability. Of course, other factors are important in aiding the pipeline engineer to construct a high integrity pipeline, e.g. good dimensional control, the provision of sound mill test and inspection procedures, etc. However, in this paper, attention will be directed solely to these three basic parameters.

Correct specification of yield strength is a fundamental aspect of pipeline design. Most large on-shore pipelines are designed to operate at pressures not exceeding 7 MPa (70 bar, 1000 lbf/in^2) with a maximum hoop stress level of 72 % of the specified minimum yield (s.m.y.) of the pipe material (B.S.C.P. 1970; I.G.E. 1970). At this high stress level, it is necessary to be sure that the pipe supplied has uniform and reproducible properties. As recently as 10 years ago a large pipeline would be merely 600 mm (24 in) in diameter with a wall thickness between 5 and 10 mm (0.2 and 0.4 in). It would have been made from a hot rolled carbon manganese steel with a carbon level of 0.2–0.25 % and a yield strength of 240 MN/m^2 (35000 lbf/in^2)†. The growing need for larger diameter systems has called for steels of higher strength to maintain 'reasonable' pipe wall dimensions. Thus, current on-shore pipelines have pipes of yield strength in the range 400–450 MN/m^2 (60–65000 lbf/in^2). Experience has shown that pipe wall thicknesses in the range 10–12.5 mm (0.4–0.5 in) offer important advantages. Firstly, the higher strength steel minimizes pipe weight, with useful savings in initial cost, transportation and field handling. I offers economic and technical benefits during welding and field non-destructive inspection. Finally, it aids the attainment of high levels of toughness during pipe manufacture. All these factors assume particularly high levels of significance in remote Arctic regions.

Most pipes for pipelines exceeding 450 mm (18 in) in diameter are made from hot rolled plate which is cold formed and longitudinally seam welded by a submerged arc process. The demand for high strength, tough and readily weldable pipes has led to the universal exploitation

† API yield strength is defined as the stress to cause 0.5 % *total* strain.

of controlled plate rolling at temperatures as low as 750 °C (Weiner, Mercer & Gibbon 1972). By this means, fine grained steels with carbon contents in the range 0.10–0.15% and yield strengths up to 480 MN/m^2 (70000 lbf/in^2) have become available. New materials with ultra high levels of toughness are now in production. One example is the Mn-Mo-Nb acicular ferrite steel which is being supplied for Arctic applications (Civallero & Parrini 1974). With a carbon level as low as 0.06% this can attain a yield strength of 480 MN/m^2 (70000 lbf/in^2).

Pipes are generally supplied in 12 m (40 ft) lengths. These are girth welded together in the field to form the continuous pipeline which, after appropriate inspection, is lowered into the prepared trench. A lot of weld metal has to be laid down, often under very adverse environmental conditions. A 900 mm (36 in) pipeline may require the deposition of a total of 2 km of individual weld runs for every kilometre of pipeline constructed. A well organized pipeline spread (construction project) will proceed at an average rate exceeding $1\frac{1}{2}$ km per day (Mercer & Needham 1969). In order to achieve the necessary welding speed, most construction teams utilize the high deposition rate, so-called 'stovepipe' welding method. This involves vertical down welding with cellulosic electrodes. Such methods call for good basic weldability of the pipe material.

Weldability is difficult to define and even more difficult to measure in absolute terms. A considerable amount of research has been conducted in this area but, as yet, there is no universally accepted procedure (Hart, Watkinson & Baker 1969; Stout 1974). The pipeline engineer often requests full scale pipe weldability tests carried out under actual site conditions as a qualifying requirement in a pipe supply contract to minimize unexpected problems and expensive delays once the construction job starts (Weiner *et al.* 1972; Phelps 1974). Most specifications tackle the weldability assessment problem by stating a maximum permitted carbon equivalent (c.e.) level. Experience in the U.K. has shown the desirability of maintaining the c.e. ($=C(\%)+\frac{1}{6}\,Mn(\%)$) below about 0.4 (Mercer & Needham 1969). Steels of higher c.e. values will require a degree of preheat, especially in the thicker wall sections. Preheat is not conveniently applied under typical field conditions and is best avoided if possible.

Steel with good weldability offers further benefits during pipeline service. Connexions which are made to a pipeline after it has been commissioned are generally made by 'hot tapping' (De Hertogh 1972; Cassie 1974). This involves fillet welding directly onto a highly stressed operating pipeline (figure 1, plate 3). The internal flowing oil or gas conducts heat away from the point of welding and increases the risk of hydrogen induced cracking in borderline materials. Shutting down the pipeline to make such connexions is inconvenient and can represent a huge loss in revenue – £0.5M per day for a 900 mm (36 in) gas pipeline; £8M per day for a 1200 mm (48 in) oil pipeline such as that being built across Alaska!

One of the more dynamic areas of research into pipeline materials technology during the past two decades has concerned fracture toughness. In common with other major engineering structures a realistic toughness requirement must be built-in at the pipeline design and materials specification stages. However, the unusual geometry of pipelines (they are long thin vessels) and the fact that they can be damaged during service by earth moving machinery, presents special problems. This is especially prominent for pipelines conveying high pressure gas or oil with a high gas content. If a fracture is initiated in such pipelines, it can propagate for long distances unless arrested by the inherent toughness of the pipe steel. A number of such failures occurred in the early 1950s in the U.S.A. One brittle fracture propagated for a total distance of 12 km and several other instances of fractures up to 600 m in length have been recorded.

FIGURE 1. Welding a T on to a live pipeline operating at 6.9 MPa (69 bar) (nominal) pressure.
FIGURE 2. Tough pipe (rear) which arrested a multiple brittle fracture.

(*Facing p.* 44)

FIGURE 4. Ductile fracture propagation and arrest (top left).

These observations led to a 20 year research programme sponsored by the major American gas transmission companies at the Battelle Memorial Institute. This work led to the development of the drop weight tear test (d.w.t.t.) to define a transition between brittle propagation and ductile arrest (McClure, Duffy & Eiber 1965). The test is now widely used throughout the world to ensure freedom from propagating brittle fractures. The studies were further developed by British Gas during the past 10 years (Weiner *et al.* 1972; Fearnehough, Jude & Weiner 1971). The outcome of a full scale fracture test involving a correctly specified pipe welded to one of low toughness is shown in figure 2, plate 3. The multiple fractures propagating through the latter pipe at 500–600 m/s were arrested within 0.3 m on entering the tough pipe.

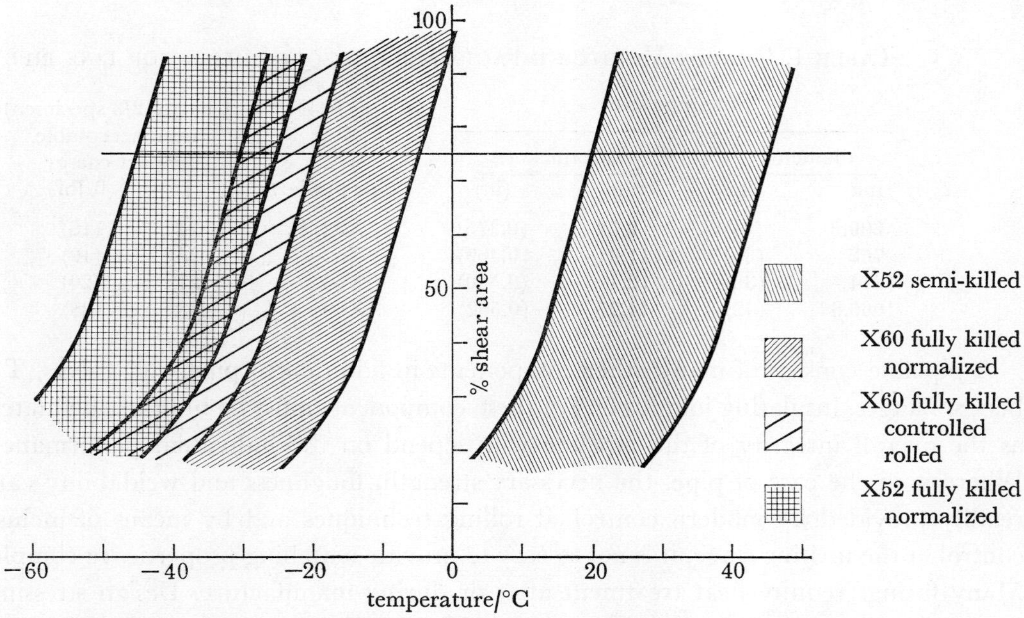

FIGURE 3. Typical ranges of d.w.t.t. curves for a number of pipeline steels of 12.5 mm (0.5 in) wall.

The requirements of the d.w.t.t. have led to the large scale adoption of fully killed steels for high strength large diameter pipelines. Transition curves for pipes produced by various routes are shown in figure 3. The specification requirement for gas pipelines in the U.K. is for not less than 75% shear area at 0 °C (British Gas Standard 1975). The fully killed steels meet this requirement.

Thus it was concluded that propagating fractures could be prevented by a simple transition temperature approach. However, two long completely ductile fractures occurred in the U.S.A. in 1969. These led to extensive further experimental and theoretical studies (Fearnehough 1974; Poynton, Shannon & Fearnehough 1974; Maxey *et al.* 1974; Poynton 1974; Shannon & Wells 1974; A.G.A. 1974). Unlike brittle fractures, which travel typically at 500–600 m/s, ductile fractures have been observed to travel at speeds as low as 60 m/s. Propagation occurs if the toughness of the pipe material on the transition curve upper shelf is insufficiently high, figure 4, plate 4. There are a number of theoretical approaches which attempt to analyse this form of fracture. These have not been wholly successful and the approach remains on an empirical basis. Following a series of full-scale and model tests, British Gas has established specification data in terms of Charpy specimen upper shelf toughness. These values are related to operating stress levels and individual pipeline dimensions, table 1.

The steel maker has made great strides in providing improved materials to meet demanding new toughness specifications derived from these studies. In the case of brittle fracture, it was necessary to reduce and to control grain size. In the case of shear fracture, it was necessary to influence inclusion content, shape and size. This has led manufacturers to follow a variety of routes although the underlying principle in all cases is to reduce the sulphur content of the steel and, in particular, to control the total volume and the size and shape of sulphide inclusions (Weiner 1972; Lowes & Rodgerson 1973; B.S.C. Conference 1974). The British Steel Corporation has had considerable success with rare earth metal additions to the steel. Steels treated in this way can offer Charpy upper shelf toughness levels of 70–80J (50–60 ft lbf) (measured on two-thirds size specimens).

TABLE 1. CHARPY V-NOTCH IMPACT ENERGY REQUIREMENTS FOR LINE PIPE

line pipe size					Charpy (2/3 specimen) min. acceptable impact energy	
diameter		wall thickness				
mm	(in)	mm	(in)	pipe grade	J	(ft lbf)
609.6	(24)	9.52	(0.375)	X52	24	(18)
762	(30)	11.91	(0.469)	X52	25	(19)
914	(36)	12.7	(0.500)	X60	27	(20)
1066.8	(42)	14.27	(0.562)	X60	38	(28)

A pipeline consists of many other components in addition to pipe, e.g. valves, T junctions, bends, flanges, insulating joints, etc. All these components must be built to adequate standards as the overall integrity of the pipeline will depend on the individual performance of each. Whereas, in the case of pipe, the necessary strength, toughness and weldability can be fairly readily provided by modern controlled rolling techniques and by means of inclusion shape control at the melting stage, it is not so easy to provide matching properties in complex fittings. Many fittings require heat treatment after or during manufacture. Design stressing requirements often lead to relatively thick sections which may not respond readily to heat treatment. Indeed, the supply of fittings for larger diameter pipelines is often the major factor hindering a more rapid spread of such pipelines. One approach is to use relatively expensive alloyed steels. This is perhaps justified in view of the small total tonnage of fittings required within a typical pipeline. Age hardening ferritic steels containing copper are showing promise in this area.

4. OFFSHORE PIPELINES

There is no doubt that there will be an enormous effort expended during the next 20 years in the exploitation of offshore oil and gas resources. Pipelines will be built in ever deeper water to convey gas and oil from new fields to the main centres of consumption on shore. Offshore pipelines utilize pipe supplied to similar specifications as those destined for on-shore use. However, wall thickness/diameter ratios are generally two or three times those of on-shore pipelines. Increasing pipe wall thickness can hinder the attainment of high levels of fracture toughness and weldability. At the same time, pipeline failure must be counted in terms of massive pollution as well as the huge cost of repair.

The first oil or gas well out of sight of land was built only 26 years ago. The extensive operations in the Gulf of Mexico were, until a few years ago, limited to water depths of about 50 m. The technological means for exploiting oil and gas in such conditions and for laying the necessary pipelines have been well proved. However, in the North Sea, much of the oil lies in the Northern regions in water depths exceeding 120 m. It is likely that within the next 20 years water depths in excess of 600 m will be tackled. Studies are being initiated to examine the feasibility of laying pipelines up to 760 mm (30 in) diameter in water depths up to 2000 m (Milz & Broussard 1974).

The requirements for offshore pipelines couple those for on-shore pipelines with a need:

(i) to withstand extreme bending and tension stresses during the laying operation.

(ii) to resist collapse and buckling under hydrostatic loads, and

(iii) to provide sufficient negative buoyancy to ensure the pipeline shows no tendency to lift from the sea bed.

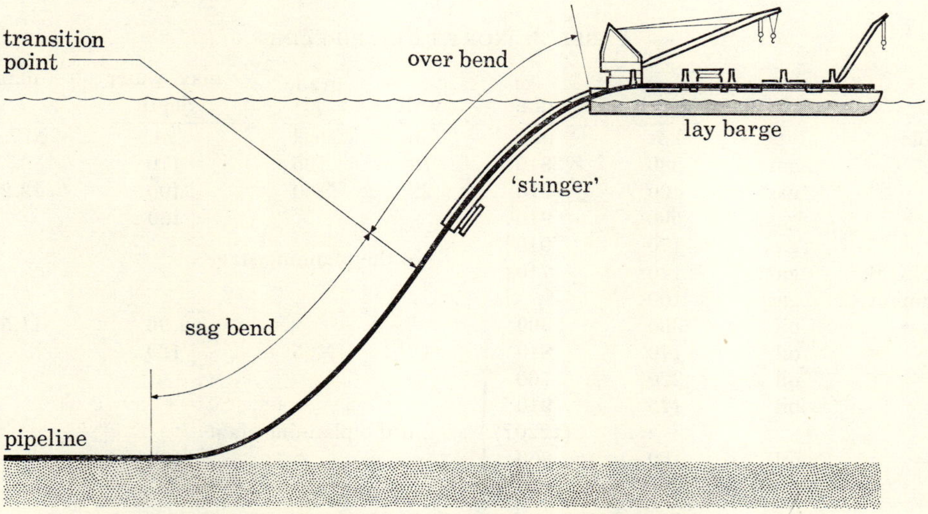

FIGURE 5. Arrangement for laying offshore pipelines.

Laying and buckling stresses are primary factors determining pipe dimensions. This is particularly true for pipelines intended to be laid in water depths exceeding 100 m. Offshore pipeline construction is usually carried out by means of a lay barge on which individual lengths of pipe are welded together to form a continuous string which is then fed out over the stern of the vessel. The pipeline slopes down to the sea bed, as shown in figure 5. The point of contact with the sea bed can be 300 m or more from the lay barge in water of 100 m in depth. The critical areas are the points of maximum bending stress at the over bend immediately behind the barge and also at the sag bend near the sea bottom. In deep water these stresses can cause pipeline buckling, even for pipe of wall thickness/diameter ratio twice that found in equivalent on-shore pipelines, see table 2. The problem is tackled by tensioning the pipeline as it is laid and by the use of lay barges equipped with 'stingers' (figure 5). The stinger is a hinged structure with a variable and controlled buoyancy which supports the pipe leaving the barge and reduces the bending stresses to acceptable levels. Tensioning of the pipeline similarly controls the bending stress in the sag bend. The latter area still presents major difficulties in pipeline construction. It is not easy to observe and to monitor the pipeline and in typical North Sea conditions, severe dynamic loadings due to wave motion aggravate the problem. New semi-submersible barge designs should offset many of these difficulties. The larger of these have impressive dimensions approaching those of a major ocean-going liner (Rosa 1974).

As for on-shore pipelines, it is convenient to consider pipe steel basic requirements in terms of strength, weldability and toughness. The scope for the controlled rolling route to plate manufacture becomes more restricted as pipe wall thicknesses exceed 25 mm. However, remarkable strides have been made and thick wall, low alloy steel pipes are now available in grades up to X65† (table 2) (LeMay, Krishnadev & Galibois 1974). Good weldability is vital to the offshore pipelines engineer. This is both to offset the increased constraint in welding thicker sections and to aid any future repair and maintenance operations, which may require the execution of high quality welds in a far from ideal underwater environment. Extra expenditure at the outset on high quality clean steels with the lowest possible carbon equivalent will pay enormous dividends in an area where the initial pipeline construction cost may approach £0.7M/km and the cost of a single repair can exceed £1M.

Table 2. North Sea pipelines

pipeline	type	length km	dia. mm	wall mm	grade of steel	max. water depth m	max. operating pressure MPa	(bar)
Frigg	gas	380	810	19	X65	170		
Ekofisk	gas	400	910	22	X60	100	13.2	(132)
Brent	gas	560	910	in the planning stage		150		
Placid	gas	170	910	in the planning stage				
Conoco/N.C.B.	gas	140	710	in the planning stage				
Noordwinning	gas	100		in the planning stage				
Ekofisk	oil	350	860			96	11.5	(115)
Forties	oil	170	810	19	X65	120		
Piper	oil	220	760	in the planning stage				
Ninian	oil	175	910 (1220?)	in the planning stage				
Brent	oil	160	860 (910)	in the planning stage		150		

Automatic welding processes come into their own for offshore operations. Following use in the Gulf of Mexico, they are finding extensive application in the North Sea. The equipment which is used was developed for on-shore pipeline construction and is based on the CO_2 process (Phelps & Letchford 1973). Automatic CO_2 welding provides important benefits both in terms of convenience and low cost with thick walled pipes. Furthermore, as a low hydrogen process, it can aid the welding of relatively high carbon equivalent steels. A further possibility is explosive welding, which may find its first large scale pipeline application offshore (Welding Institute 1974).

Underwater pipelines are not normally left exposed on the sea bed. The usual practice is to trench and cover the line by use of high pressure water jets which traverse the completed pipeline. Even so, water currents can cause substantial sea bed displacement and transfer, and pipelines may become exposed. In extreme conditions the pipeline itself can be moved and severely stressed by these currents. In addition, most of the world's continental shelves are important fishing grounds. Bottom fishing makes use of heavy equipment and pipeline damage or fracture is a real and continuing possibility. A fracture, once initiated, may propagate along the pipeline. The hydrostatic head and inertia effects of the surrounding water probably make little difference to the process as compared to the fracture behaviour of an on-shore pipeline. Thus, basic toughness requirements are similar. In addition, a propagating buckle failure could occur under the influence of hydrostatic pressure. This is more likely to occur at lower operating

† API 5LX65 = 450 MN/m^2 (65000 lbf/in^2) yield stress.

pressures, i.e. at times of complete or partial shut down due to damage, etc. Control of this failure mode does not lie in the hands of the metallurgist. The growing practice is to provide collars at suitable intervals along the pipeline and so restrict the length to be replaced if a buckle should form. It is believed that such arresters will also prevent fracture propagation – a useful bonus which, in theory at least, should lessen the demand for 'excessive' and expensive levels of toughness.

5. The future

There is no doubt that pipelines will continue to play an ever increasing role in energy transportation during the remaining part of the 20th century (Lambert 1974). Notwithstanding the recent oil crisis, it is confidently predicted that pipelines will continue to be built at rates exceeding those of the past decade. They will be larger than hitherto; they will be longer; they will traverse many hostile regions of the world. The U.S.A. will remain the world's most important pipeline user, although the U.S.S.R., with its vast resources of Siberian oil and gas, will probably be building pipelines at a greater rate and in larger diameters than the U.S.A. in the years up to 2000. Larger diameter pipelines, especially if built in Arctic regions, will continue to offer a major challenge to the metallurgist. It will be necessary to produce steels of higher strength while maintaining and perhaps increasing attainable levels of fracture toughness. It will remain vital to offer the highest possible levels of weldability to aid fabrication under difficult weather and environmental conditions. Quenched and tempered steels will probably play a growing role during the latter part of this century. In addition to the development of pipe materials, it will be necessary to produce fittings of larger sizes than hitherto in thicker sections, again maintaining good toughness and welding properties.

The developing picture offshore will bring water depths exceeding 600 m into full focus within the next 15–20 years. The demands on the metallurgist are likely to take a slightly different course from on-shore projects. In order to resist the enormous stresses of laying and those arising from hydrostatic pressures in operation, the demand will be for pipe with even greater wall thicknesses than hitherto. It is unlikely that pipe diameters will progress much beyond 900 mm (36 in) in deep water because of the technical difficulties of handling, laying and operation. Thus, it is unlikely that the technological problems in steel production and pipe manufacture will change a great deal. It is not unreasonable to suppose that steels of strength levels not exceeding 410 MN/m^2 (60000 lbf/in^2) will be more than adequate for offshore pipelines for the rest of this century.

6. Summing up

This paper has taken a brief and incomplete glimpse of the current picture and plans for the construction of pipelines on- and offshore. On-shore pipelines of 600 mm (24 in) diameter and above make use of seam welded pipes which are produced from controlled rolled steels with wall thicknesses generally lying in the range 10–15 mm (0.4–0.6 in). There is considerable experience with such pipelines and economic processes have been developed and proved for the manufacture, welding and inspection of these pipelines. Relatively low cost steels capable of producing pipes with yield strengths of 450–480 MN/m^2 (65000–70000 lbf/in^2) are the norm. They offer excellent weldability and very high levels of fracture toughness.

Offshore pipelines are not very different although, in addition to requiring materials with high levels of strength, weldability and toughness, they must cater for enormous laying stresses and the possibility of collapse and propagating buckle failure in service. Hence, offshore pipelines normally require wall thicknesses at least 50% greater than would be found in equivalent on-shore pipelines.

The engineer has met a number of challenges over the past few years in developing the means by which high integrity pipelines can be built, both on- and offshore. He could not have done this without the aid of the metallurgist. What will be required of the metallurgist in the future? Perhaps the most important thing is not the development of something completely new but to continue to offer the engineer reproducibility and quality at the right price. It is a process of learning and gradual improvement. The big breakthrough came with economic means of grain size control through micro-alloying and the large scale introduction of controlled rolling. The message then, is quality and reliability through knowledge and understanding.

The author is grateful to colleagues and friends in both the gas and oil industries for help and assistance in the preparation of this paper. He is also indebted to the British Gas Corporation for permission to publish.

References (Mercer)

A.G.A. American Gas Association 1974 Pipeline Research Committee Fifth Symposium on Line Pipe Research, Houston, Texas, November 1974.

API 1975 API Specification 5LX High test line pipe, 20th Edition, March 1975.

British Gas Standard PS/LX1 1975 Submerged-arc welded line pipe (Supplementary and amending specification to API Specification 5LX).

B.S.C.P. 1970 British Standard Code of Practice CP2010 (Part II). Design and construction of steel pipelines in hand.

British Steel Corporation Conference 1974 Inclusions and their effects on steel properties. University of Leeds, September 1974. (To be published.)

Cassie, B. A. 1974 Hot tap connections on gas pipelines operating at high pressure. *Gas Engng. and Management* **14** (10), 309–313.

Civallero, M. & Parrini, C. 1974 Heavy wall X70 pipe for Arctic applications. *Pipeline and Gas J.* **201** (8), 35–38.

De Hertogh, M. J. 1972 Welding natural gas filled pipelines. *Metal Constr. and Br. Welding J.* **4**, 83–88.

Fearnehough, G. D., Jude, D. W. & Weiner, R. T. 1971 The arrest of brittle fracture in pipelines. I. Mech.E. Conference on Practical Application of Fracture Mechanics to Pressure Vessel Technology, London, 1971.

Fearnehough, G. D. 1974 Fracture propagation control in gas pipelines: a survey of relevant studies. *Int. J. Pressure Vessels and Piping* (2) Vol. 2, pp. 257–282.

Hart, P., Watkinson, F. & Baker, R. G. 1969 Hydrogen-induced delayed cracking in the site welding of line pipe. Paper 32, The Welding Institute Pipe Welding Conference, Autumn 1969.

I.G.E. 1970 Recommendations on transmission and distribution practice – IGE/TD/1 – Steel pipelines for high pressure gas transmission. Institution of Gas Engineers, London, 1970.

Kerr, J. W. 1974 Natural gas challenges of the future. Institution of Gas Engineers 40th Annual Research Meeting, London, November 1974.

Lambert, D. E. 1974 Pipelines to play major roles in future energy development. (Highlights from the 1974 World Energy Conference.) *Pipeline Ind.* October 1974 **41**, 21–26.

LeMay, I., Krisnadev, M. R. & Galibois, A. 1974 Current and potential steels for far north pipelines. *Pipeline Ind.* **41**, 27–30.

Leyman, P. *et al.* 1973 Experience of designing large diameter gas pipelines in the U.S.S.R. IGU/C 27–73, 12th World Gas Conference, Nice.

Lowes, J. M. & Rodgerson, P. 1973 The practical implications of fracture mechanisms in line pipe steels. Institution of Metallurgists – Young Metallurgists Spring Meeting, March 1973.

McClure, G. M., Duffy, A. R. & Eiber, R. J. 1965 Fracture resistance of line pipe. *J. Engng. for Ind.* **87**, 265–278.

Maxey, W. A., Podlasek, R. J. & Duffy, A. R. 1974 Observations on shear fracture behaviour. British Gas/IGE Symposium: Crack Propagation in Pipelines, Newcastle, March 1974.

Mercer, W. L. & Needham, D. 1969 Welding and inspection of high pressure gas transmission pipelines, GC 158, IGE 35th Autumn Research Meeting, London, November 1969.

Milz, E. A. & Broussard, D. E. 1974 Equipment requirements for deep water pipelines, *Pipeline Ind.* **41**, 39–45.

Phelps, B. 1974 Weldability testing of high strength pipeline steels. International Symposium – Weldability of Heightened and High Strength Metals, Zagreb, Yugoslavia, May 1974. Association of Yugoslav Welding Societies, 115–122.

Phelps, B. & Letchford, P. W. 1973 Field experience with automatic welding of large diameter pipelines. IGU 12th World Gas Conference, Nice, 1973, Paper IGU/C 40–73.

Poynton, W. A. 1974 A theoretical analysis of shear fracture propagation in backfilled gas pipelines. British Gas/IGE Symposium: Crack Propagation in Pipelines, Newcastle, March 1974.

Poynton, W. A., Shannon, R. W. E. & Fearnehough, G. D. 1974 The design and application of shear fracture propagation studies. ASME Pressure Vessels and Piping Conference, Miami, Florida, U.S.A., 24–28 June 1974.

Rosa, G. 1974 Saipem building $100M semi-submersible pipelay vessel. *Pipeline Ind.* October 1974, **41**, 31–32.

Shannon, R. W. E. & Wells, A. A. 1974 A study of ductile crack propagation in gas pressurised pipelines. British Gas/IGE Symposium: Crack Propagation in Pipelines, Newcastle, March 1974.

Stout, R. D. 1974 Field weldability of 5LX pipeline steels. API Symposium on Line Pipe and Tubular Goods, June 1974. Paper No. SS-3:1.

Weiner, R. T., Mercer, W. L. & Gibbon, R. B. 1972 Steel pipe for the British gas industry, GC 194, IGE 38th Autumn Research Meeting, London, November 1972.

Welding Institute 1974 Welding in offshore constructions, The Welding Institute International Conference, Newcastle upon Tyne, 26–28 February 1974, Section V.

1.3 Materials requirements for offshore structures

By H. C. Cotton

B.P. Trading Ltd, Britannic House, Moor Lane, London EC2Y 9BU

[Plate 5]

Introduction

The shape and form of marine structures are many and varied: the materials of construction range from wood to high strength steel. Wood and stone have been historically the most popular building materials but the use of these, because of poor availability and difficulty in working has in recent years given way to concrete and steel. The advantages of steel versus concrete are not easy to define because these depend upon so many factors including environmental, social and economic.

Concrete

Concrete has been a preferred building material for marine static constructions for about a century. Harbours, jetties, lighthouses built almost a hundred years ago testify to the durability of that material: steel and wrought iron constructions of similar age are often found to be in an advanced state of corrosion. Yet concrete structures are not without problems both in construction and service although attention to a few simple precautions can often ensure satisfactory performance. It should not be assumed that the great weight of a concrete structure relative to that of steel need confine its use to static installations, indeed many successful ships and racing yachts have been constructed in reinforced concrete and these have often performed well in competition with their steel counterparts. However, one major disadvantage of thin reinforced concrete structures is their vulnerability to impact damage which is not always easy to repair and this is a major drawback where lightness is valued.

Concrete offshore platforms have many advantages (1), but their use is limited to locations where there is a firm foundation within a reasonable depth below the seabed (figure 1, plate 5). Very careful soil investigations are necessary to ensure that there will be no undue long term settlement or tilting as a result of continuous wave action. Some settlement can be tolerated provided that well conductors are allowed to slide through guides attached to the platform. It would obviously be unacceptable for the weight of the platform to be transmitted to the conductors and as the latter are deviating from the vertical at seabed level the allowable long term settlement should be restricted to about 0.5 m.

Given good foundation conditions a concrete structure can be installed rapidly and the period 'at risk' is considerably shorter than is the case for a steel jacket structure. Furthermore the decks and much of the deck equipment can be installed before float-out thus reducing the time consuming and very costly operations of crane barges at locations.

With regard to vulnerability a recent inspection of the Mulberry harbour constructed some 30 years ago to facilitate the invasion of Europe, showed it to be in remarkably good condition; however other coastal defence works erected at the same time often show signs of spalling and corrosion of the reinforcement and these and other effects have been subjected to closer scrutiny in recent years (2).

Long experience, together with observations of structures which have deteriorated, demonstrate however that very little corrosion of reinforcement will take place if adequate concrete cover is provided for the steel and a good dense concrete is achieved.

In addition the designs which have been developed for offshore work are such that all structural elements are constantly in compression. Where necessary prestressing techniques are used to achieve this. The possibility of any hair cracks in the concrete, through which the sea water would penetrate to the reinforcement, is therefore reduced.

Should such penetration take place corrosion will inevitably occur (3) followed in time by concrete spalling. Close attention to concrete mix and workmanship is essential if such effects are to be avoided.

Steel

Irrespective of the type of structure favoured for future offshore installations it appears that these will consume vast quantities of steel whether they are of reinforced concrete, steel/concrete hybrids, or are simply constructed from steel plates and tubes.

There are many advantages for the use of steel plates and tubes in the construction of mobile rigs and platforms and these desirable features extend to static structures especially when the water is shallow. In shallow waters currents are often more severe than in deeper locations and these can intensify the problem of scouring around the foundation of gravity structures. However, by far the greatest advantage of steel is the relative speed with which small tubular towers and mobile rigs can be constructed and the ease of launching and towing them.

Offshore structures in general use were until recently merely extrapolations of pier and jetty traditional designs which involved the use of familiar materials and scantlings and which gave rise only to problems which were well understood and for which, on the whole, answers were readily forthcoming. It is only in the last few years that interest in the recovery of oil and gas from deep sea locations and from remote and inclement seas has necessitated the construction of large ocean-going semi-submersibles for exploratory drilling and static platforms (figure 2, plate 5) of huge size for development drilling and production. Even now this transition is by no means complete and semi-submersibles may come to be regarded as merely a stepping stone to the use of drill ships in deeper water and bottom founded platforms may in similar conditions be replaced by semi-buoyant tension leg platforms linked to seabed well completions (4, 5).

Apart from corrosion effects the problems encountered in using steel in most offshore applications appear to be little different from those encountered in the design and construction of installations on land. Such questions as weldability, brittle fracture and fatigue strength, other than corrosion fatigue effects, are practically unchanged.

It is in the fields of underwater engineering that new problems have arisen and these include *inter alia* underwater maintenance, repair and modifications which mostly invite the use of underwater welding.

Nevertheless the need for enhanced reliability in the performance of costly marine applications has provided a spur towards the search for better solutions to those ever present questions just mentioned and the present stage of progress in tackling these problems will now be reviewed.

FIGURE 1. Concrete gravity platform (designed and constructed by Howard Doris).

FIGURE 2. Jacket section of production platform FA being lowered to the seabed in the Forties Field.

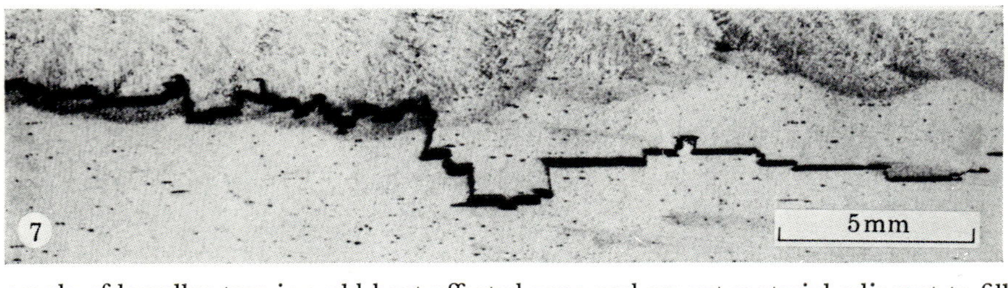

FIGURE 7. Example of lamellar tear in weld heat affected zone and parent material adjacent to fillet butt weld.

(*Facing p.* 54)

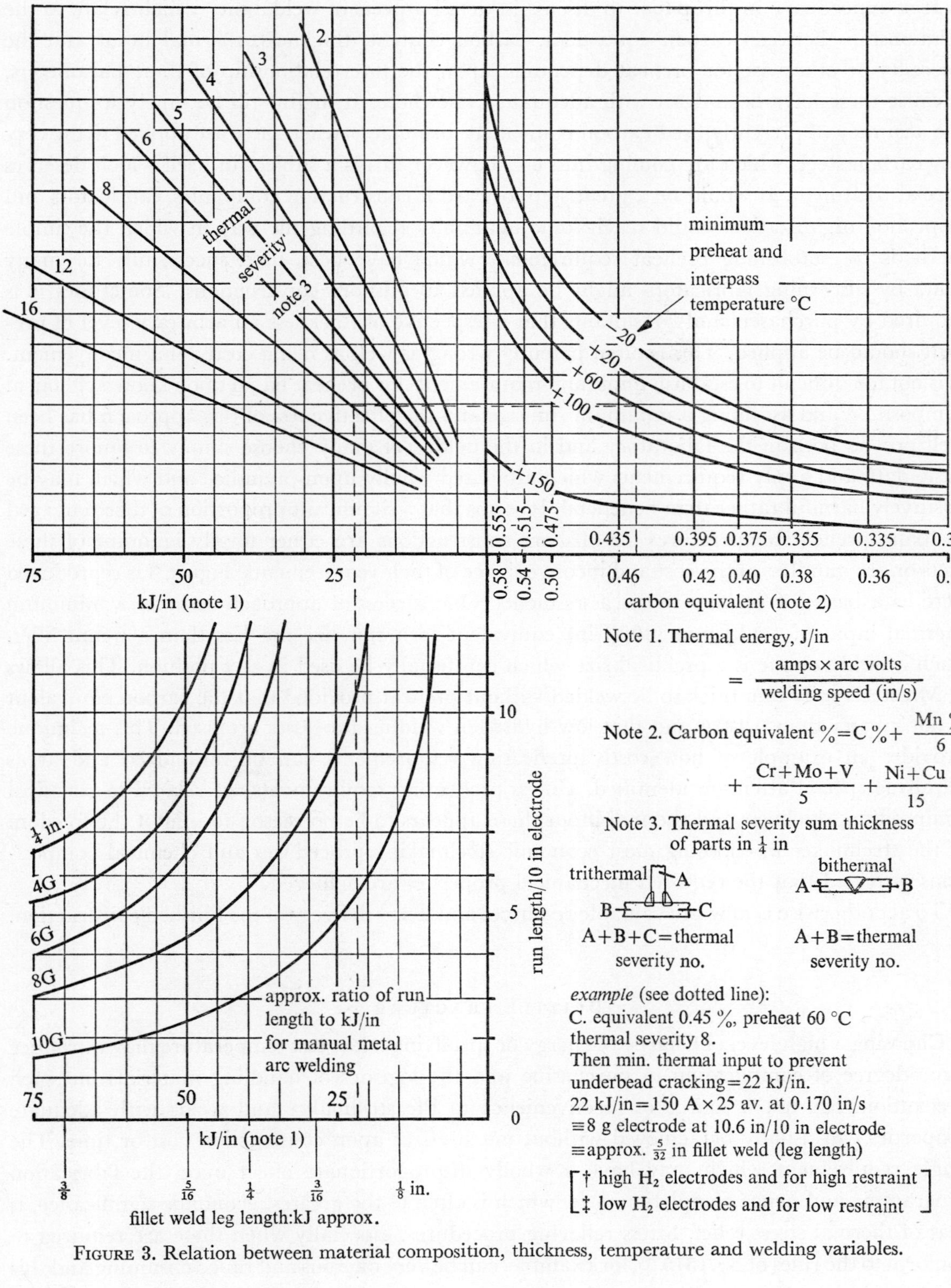

FIGURE 3. Relation between material composition, thickness, temperature and welding variables.

Weldability

It is unnecessary in this paper to discuss in detail aspects of weldability which relate to the relationships between carbon equivalent, carbon content, thickness, thermal input and the necessity or otherwise for preheat depending upon the interrelationship of these parameters, because these have been dealt with adequately elsewhere. It might still be timely to question the accuracy of presently used carbon equivalent formulae (6) when these are applied to the very low carbon steels which are coming into use, however this is a subject in itself which deserves special treatment. It would be a great step forward if construction companies, fabricators and inspection organizations could reach some unanimity regarding the way in which the simple methods of establishing preheat requirements which have been used successfully for many years by discerning fabricators might be applied to offshore constructions. Sometimes it is required by purchasers and consultants that irrespective of thickness an arbitrary level of preheat should be applied. This is often patently wrong as well as being costly and inconvenient. It is not too difficult to establish minimum requirements for preheat based upon known chemical composition and assuming a minimum run size and the effectiveness of this approach has been well proven both in the laboratory and in the field. Yet many choose simply to ignore these basic data and apply requirements which are based wholly upon prejudice and which may be positively harmful rather than beneficial. It seems that a significant proportion of those engaged in fixing preheat temperatures for offshore constructions are either totally ignorant of these facts or are mindless of the cost and inconvenience of their requirements. Figure 3 is reproduced here as a basis for making rough assessments. One successful approach assumes a minimum thermal input of 1.5 kJ/mm (40 kJ/in) equivalent to a run size not less than a 6 mm fillet. Such a weld volume is a practical size which can usually be used in any position. This allows C Mn steels < 40 mm thick to be welded without preheat provided that the carbon equivalent is not greater than 0.42 % and that low hydrogen welding processes are used. This technique provides an example of how costly preheating requirements can be minimized and areas requiring special attention identified. Unless preheating requirements are related to chemical composition, thickness and thermal input there appears to be no reason to inhibit the freedom of the steelmaker in choosing most economic steelmaking procedures and chemical compositions provided that the required mechanical properties are achieved.

To act otherwise is to waste valuable resources and delay delivery of the finished construction.

Brittle fracture

Choosing a high level of Charpy V energy or specifying a low test temperature may introduce some degree of complication in purchasing and the logistics of handling materials but such precautions are not a source of inconvenience to the steelmaker and welds with adequate properties can usually be achieved without much effect upon construction cost or time. The single requirement which may have a wholly disproportionate effect upon the fabrication programme and hence completion date, which is often of the greatest economic significance, is that of thermal stress relief. Stress relieving procedures, especially when these are required to conform to the rules of BS. 1515 (7), for example, can be very onerous and timec onsuming and this is most evident when the diameter of the construction local to the weld is great. The requirement in that standard for a heated band with a minimum width equal to $5\sqrt{(rt)}$ was devised so as to ensure that residual stress levels remaining after the heat treatment of a pressure vessel shell would not exceed about $\frac{1}{2}\sigma_y$.

In the absence of specific guidance in the matter of good practice in the structural field these same rules are now being applied to many offshore constructions. It would be of great value if the improvements in resistance to low stress fracture deriving from the tempering and over ageing processes which are the inescapable concomitants of the post heat treatment process could be distinguished from those accruing from the mere reduction in residual stress. It would then be possible for the necessity of ensuring a wide heated band to be investigated. Considerable economic advantage could be achieved if the use of narrow heat treated band widths could be justified. Lower post heat treatment temperatures and shorter times at temperatures than those presently specified might also be shown to be effective.

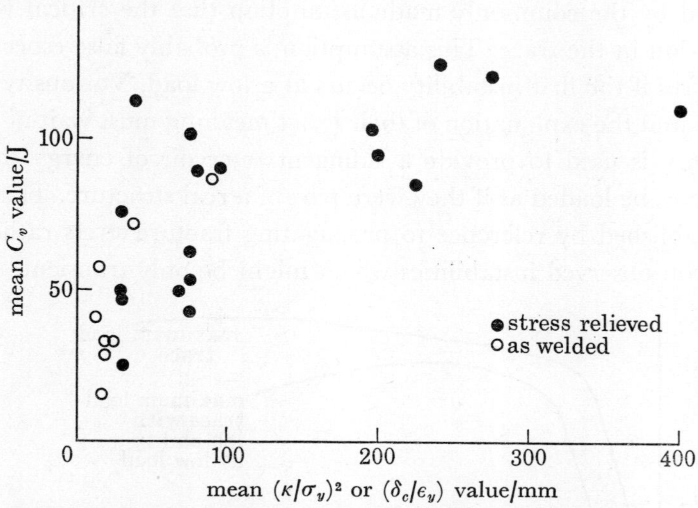

FIGURE 4. Attempted correlation between Charpy V and c.o.d. values for low hydrogen weld metals.

Although it seems that perhaps both the Charpy V and the c.o.d. test can be used to evaluate the tempering effects these relatively small test specimens by virtue of their size cannot quantify the contribution attributable to residual stress reduction, and the interpretation of this effect is largely dependent upon unproven mathematical assumptions.

Deriving flaw tolerance limits from Charpy V data is an extremely if not totally empirical and subjective process because as far as is known there is no generally accepted theoretical basis with which such an assessment could be made. There have been attempts to discover a reliable Charpy V/c.o.d. correlation and the results of one such effort relating to weld metals deposited by low hydrogen electrodes is illustrated in figure 4; it will be seen that the correlation is poor.

Evidently Charpy V absorbed energy alone cannot be relied upon to provide an adequate guide to the likely initiation properties of welds or in many cases of their heat affected zones. Presently there seems to be no way of avoiding costly and time consuming c.o.d. tests on full thickness specimens representing every type of welding procedure which it is anticipated might be used in practice. Furthermore the c.o.d. properties of such joints appear to be greatly influenced by small changes in run size, sequence and preheat etc. as well as of course by thickness; even backgouging technique and extent is suspected as being a significant variable.

Fracture mechanics methods K_1C, J integral, R curves and more commonly c.o.d. are gradually coming into use for flaw tolerance estimation (8). However, precisely how these techniques may be applied on a practical weld basis is still not certain. This might not be obvious to non-specialists in the field who could be excused for thinking that consensus between experts exists regarding this important subject and that the whole process is cut and dried. This is not the case despite the seemingly authoritative reports and draft standards (9, 10, 11) currently in circulation.

The methods presently promoted for deriving limits for allowable flaw size ($\bar{a}_{max}$) from c.o.d. should be given further study.

Apart from the mathematical uncertainties the interpretation of the meaning of the discontinuities in the simple 'stress strain' diagram recording the clip gauge displacement *vs* load is in itself open to question. Once again imponderables relating to machine flexibility, which proved to be the Achilles heel of the Robertson and similar wide plate testing machines and which may present a significant uncertainty in evaluating the performance of Wells type cross welded wide plate tests arise in interpreting c.o.d. traces. Conservatism necessary because of ignorance is illustrated by the commonly made assumption that the critical value of c.o.d. is that of the first hesitation in the trace. This assumption is probably false especially for heterogeneous micro structures if the first instability occurs at a low load. Various types of trace can be observed (figure 5) and the explanation of their exact meaning must wait at least until more sophisticated machinery is used to provide a sufficient reservoir of energy that propagating fractures can continue to be loaded as if they were part of a real structure. Then critical values of c.o.d. could be established by reference to propagating fracture stress rather than making assumptions based upon observed instabilities which might be only transient.

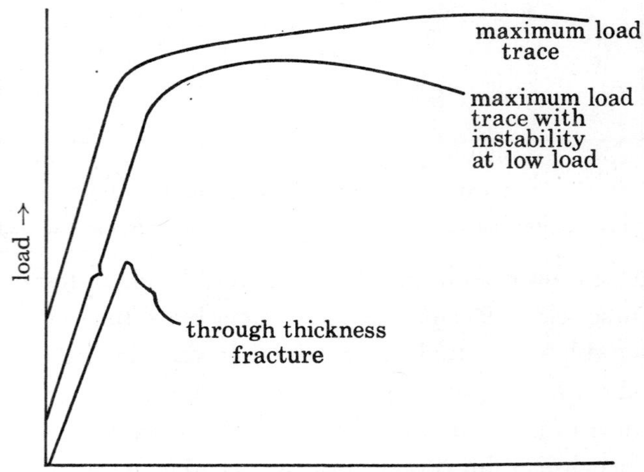

FIGURE 5. Typical load against clip gauge displacements traces for c.o.d. test.

Perhaps the most tantalizing aspect of residual stress relief is that its significance in practical cases appears to be inversely proportional to the probability of brittle fracture. If one assumes that the value of the relief of residual stress is only proportional to the ratio that the residual stress bore to the total stress then the proportional benefit deriving from its relief diminishes as the magnitude of the total stress increases (12). It is rare indeed for a brittle fracture to initiate at a low applied stress without the presence of some local stress intensifier. This is the whole basis for the successful use of cast iron. When applied stresses are low, say $\frac{1}{2}\sigma_y$, it can be seen from figure 6 that the c.o.d. necessary to provide substantial flaw tolerance is relatively small but should this low stress be supplemented by a residual stress of magnitude σ_y, the c.o.d. necessary for an equivalent flaw tolerance will increase by a substantial factor – say about 3. On the other hand when applied stresses are high and therefore local strains are high a proportionately large c.o.d. will be required to achieve an equivalent flaw tolerance. In such a case the additional strain resulting from the residual stress would appear to represent an almost insignificant proportion. Stress relief *per se* seems to be of little value when because of strain intensification, brittle fracture is a serious risk, but provides considerable safety and improved flaw tolerance for low stressed areas where the risk is small unless considerable brittleness exists. Thus stress relief as distinct from post heat treatment has only been shown to be markedly beneficial when rather brittle materials are used at fairly low stresses: this should be investigated.

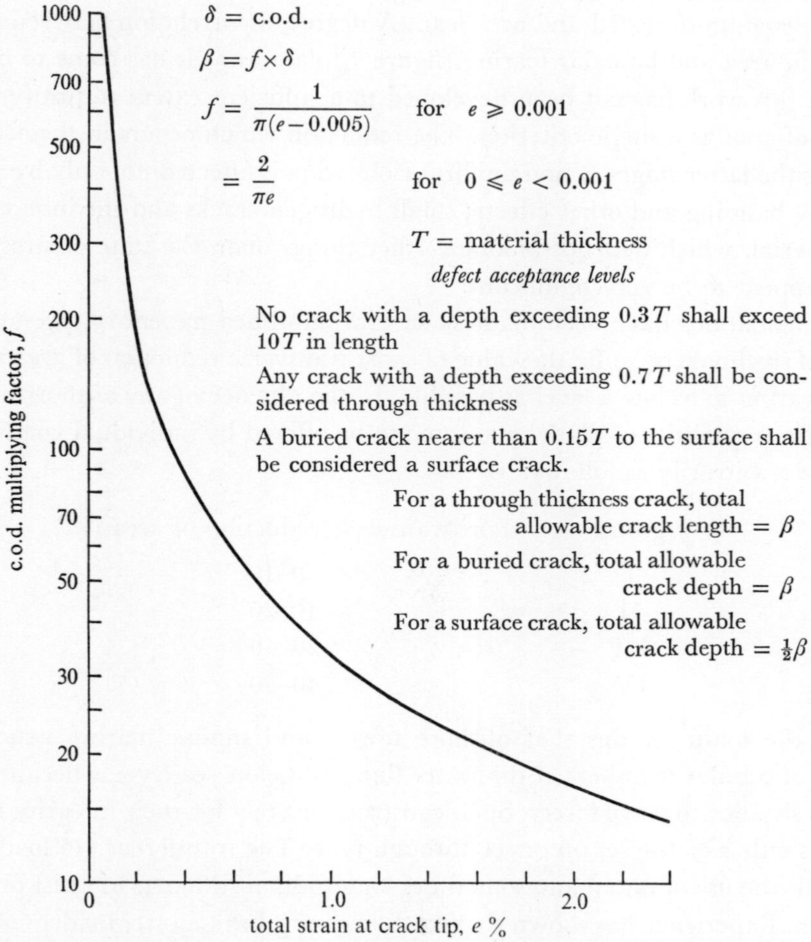

FIGURE 6. Relation between c.o.d. multiplying factor and total strain at crack tip.

SHORT TRANSVERSE PROPERTIES

Perhaps the most costly deficiency that modern steels may display is the tendency for local cracking along planes parallel to the plate surface when attempts are made to weld attachments to that surface.

Such cracking may be attributable to improper preheating or thermal input effects or to excessive restraint but the provision of modern steels with reasonably low carbon equivalent together with some skill in selecting weld sequences is usually sufficient precaution to ensure that preheating requirements will not be onerous and practical welding conditions sufficient to avoid cracking in and around heat affected zones (h.a.z.). On the other hand crystal and inclusion morphology may be so banded and polarized and the effect of the anisotropy so intense that even low through thickness stresses resulting from simple attachment welds supplemented by hydrogen stress may give rise to serious delamination (13). From the point of view of structural integrity small local areas of delamination may be tolerable but the effect of their presence upon ultrasonic inspection assessments may be calamitous because the signals received from such flaws are often indistinguishable from h.a.z. cracks. Frequently expensive repairs are demanded with unwelcome effects upon delivery. Experience has shown that steels with low S content (*ca.* 0.007 % max.) are usually highly resistant to delamination especially if such steels have been vacuum degassed and are clean. A degree of correlation between through thickness reduction in area and lamellar tearing (figure 7, plate 5) as it has come to be called has been shown but this work has not been developed to a sufficient extent to justify the acceptance of reduction of area as a single criterion. The reduction which occurs in the area of a tensile test bar during the latter stages of non uniform elongation is affected not only by inclusion morphology but by banding and other effects; small hydrogen cracks and the inherent notch ductility of the material, which depends amongst other things upon the state of stress at the particular location, appear to be very significant.

Recommendations have been made in several published papers (14), presumably for the convenience of steelmakers, to fix the value of short transverse reduction of area necessary to avoid lamellar tearing to as low a level as possible. In the absence of any authoritive directive in this matter various qualities of plate have come to be offered by individual vendors and these may be grouped arbitrarily as follows:

group	short transverse reduction of area (%)
I	0–10
II	10–20
III	20–40
IV	40–80

One of the main reasons that offshore towers and similar marine structures are usually composed of tubular members at the water line and below sea level is because this shape offers the least resistance to wave forces. Such constructions rely for their integrity upon tube to tube connexions either of the set on or set through type. The transference of load in such locations often entails the use of radial and sometimes longitudinal stiffeners to resist punching shear and other forces. Experience has shown that steels in group I are mostly inadequate for this purpose and it is questionable whether steels in group II are suitable for the fabrication of node joints in, for example, deep sea structures where thick and highly restrained fabrications are usually employed. Normally, short transverse tensile test samples are taken from the end or ends of plates as rolled and the significance of values of reduction of area thus determined depend upon the confidence one may have in the reproduceability of the results at every location throughout the plate area. In establishing the level of confidence one may assume it is necessary to give consideration to the steelmaking, slabbing and plate rolling techniques as well as to chemical analysis; such prudence is especially appropriate when using steel in groups II and III.

The cost differential between steels of group III and IV is not usually evident from commercial bids so that purchasers commonly give preference to the group IV type when availability permits. The whole question of the supply of lamellar tearing resistant steel is inseparable from that of delivery time. When large quantities are required the very best quality is usually readily available but as the production of this grade usually entails close attention to the quality of materials used in production of blast furnace iron it is practically impossible to request the supply of small quantities of these grades which have an ultra low sulphur content unless the use of calcium or cerium treatment is permitted (15, 16). Because rare earth metal (r.e.m.) treated steels have not yet been widely used in commercial production it has not yet been proven that steels with r.e.m. controlled inclusion morphology behave as well in practice as the ultra low sulphur vacuum degassed steels which have been used more extensively. Nevertheless cerium and calcium treated steels have shown a marked improvement over non treated steels with similar composition and rolling histories although the segregation of heavy sulphides is sometimes a problem.

Steels which are sensitive to lamellar tearing are often found to be equally if not more sensitive to delamination by corrosion processes which involve the absorption of hydrogen. Such conditions may occur more frequently in marine applications than on land. The handling of sour gas rich in H_2S is a frequent requirement in the Petroleum Industry and several failures have been reported of pipelines constructed from controlled rolled steel pipes under conditions which would normally have been regarded as safe.

Similar conditions can occur in pipelines and in other constructions exposed to the corrosive action of stale sea water. Under such environments the action of sulphate reducing bacteria can produce atmospheres rich in H_2S. The presence of this poisoner allows the absorption of H, produced by corrosion processes, by interfering with its combination to form H_2. The pressure of this hydrogen upon the surfaces of the steel surrounding the lamellar inclusions may soon result in separations which lead to step type cracking arising out of the rotation required at the tips of the separations to accommodate the strain.

This type of failure is akin in some respects to stress corrosion cracking and the extension of this hazard to what was regarded hitherto as the safe steels of low hardness is unfortunate and can lead to irremedial damage. In many applications at least in the Petroleum Industry the use of controlled rolled and some other steels with sulphur content over 0.007% or thereabouts which have not been r.e.m. treated is open to question and may give rise to unexpected failures. Hot finished and hot finished quenched and tempered steels appear to be relatively immune to these effects.

Fatigue

It appears to be a sad fact that although one of the most important failure mechanisms which might afflict offshore platforms and rigs is that of corrosion fatigue, yet it seems that there is presently very little reliable data upon which estimations of corrosion fatigue strength might be based. As a result Oil Companies and others engaged upon the design and construction of offshore structures are forced to adopt extremely conservative design criteria. Reports on the effect of cathodic protection upon fatigue life are contradictory; some claiming considerable improvements in life, others none. This lack of information is extremely inconvenient because locations at tube to tube node connections, which present the greatest design and fabrication problems and are difficult to non destructively examine, are at greatest risk from fatigue damage.

The fatigue strength of welded high tensile steels in air is known to be little better than that of weak welded steels (17) and presumably this effect will apply to the corrosion fatigue condition but this is not yet certain. As a result steels stronger than about grade 50D cannot be employed usefully in those locations where stronger steel would normally allow useful reduction in thickness. The use of steels stronger than grade 50D e.g. Welten 60 etc. are therefore mostly confined to piling, etc., where strain intensification is low and fatigue is not controlling. Tests are presently in progress with the aim of establishing the fatigue strength of welded tubular T joints with greater accuracy than can be estimated simply by using BS 153 recommendations and applying these to photoelastic model studies and it is hoped that the data so derived will find practical applications when clear relations hips between the air fatigue and corrosion fatigue strengths of typical welded joints are established. The corrosion fatigue strength of welded joints in sea water not yet being established with much certainty and the effects of cathodic and other protection methods upon the fatigue life being open to question, opportunities for the development of optimum materials, welding methods and designs appear to be limited.

Underwater engineering

For various reasons offshore platforms and other marine installations may be found to be in need of repair or modification and this may entail working in the splash zone or at greater depths. For many years manual metal arc electrodes have been used to effect temporary simple repairs underwater but the mechanical properties of such welds have been suspect.

An investigation into this question revealed in a recently published report (18) that manual metal arc electrodes as presently available could not be used to make high quality welds in the wet environment and this was extended to cover also austenitic electrodes of a wide range of composition including some types that are presently in use for the submarine repair of offshore platforms in shallow waters.

The use of a small hand held hyperbaric chamber in conjunction with a self adjusting arc equipment has been shown to be capable of producing acceptable welds under water (19) and on the face of it there seems to be no reason to believe otherwise because water is excluded from the entire welding operation. It is reported however that this process will not operate satisfactorily at depths greater than about 60 m or so but the precise reasons for this deficiency have not been disclosed.

Similar difficulties attend the use of the t.i.g. process in deep water hyperbaric chambers. Changes in ionization and other physical arc effects result in considerable instability and arc extinction accompanied by unwanted transfer of tungsten into the weld. Perhaps this difficulty can be remedied by modification to the design of the characteristics of welding power sources possibly assisted by changes in the composition of the tungsten electrodes. A solution to the problem of how to make satisfactory welds in deep water is required because although damage from normal causes to those parts of platforms which are at great depth is not thought to be probable this view remains to be confirmed, experience of operating the large platforms necessitated by water depth greater than say 100 m being very limited. However the chance of such an occurrence cannot be discounted and in any case it seems that a possibility exists that pipelines at this depth may require local repair. For such simple geometric shapes the large hyperbaric chambers presently proposed appear to be providing an adequate but costly solution. It seems however that the change from air to the helium/oxygen or other atmosphere within the hyperbaric chamber may require changes to the flux composition of m.m.a. electrodes if optimum weld properties are to be achieved (20). The influence of external pressure on the manual metal arc appear to be insignificant in the context of the welding operation and the reasons why an increase in ambient pressure has such a deleterious effect upon both the t.i.g. and m.i.g. systems is presently being investigated.

Explosion welding (21) and friction welding have both been shown to operate well under water. The development of welding with explosives has been retarded to some extent because of lack of funds, and the use of friction welding, for all but very small surfaces, awaits an available rotating power source that can operate efficiently underwater. The possibilities of the application of welding with explosives to the repair and maintenance of offshore structures would seem to be limited but friction welding could possibly, even in the present state of development, be used to advantage to simplify the connection of bolts for patch repairs etc. and for attaching additional supports.

The harmful effects of the rapid quench which underwater welds in the wet are bound to suffer can to some extent be offset by tempering underwater by the use of electric resistance elements together with plastic foam lagging; this can provide considerable assistance in softening welds and h.a.z.'s when wet welding is unavoidable.

In developing suitable processes for underwater welding the necessity of protecting divers from electrical effects which can be lethal, favours the use of automatic equipments which require diver attention only during the machine set up and dismantling periods. Equipments which permit remote operation are preferable also because the close attention of divers to minor details of the welding process in the distracting environment of the deep sea cannot be expected. Similar comments apply to non destructive testing techniques.

Conclusion

In specifying requirements for allowable stress, welded joint configuration, welding consumables and other materials, the necessity for post heat treatment and inspection, etc., engineers are required to take into account the various failure mechanisms which are appropriate to the particular construction under consideration and the environment under which that structure may be constructed, installed and operated.

To do this the engineer needs to be aware of current opinions regarding topics which hitherto were usually the sole province of specialists in the particular field. The success and often the economic viability of large offshore projects may depend upon the ability of the design team to communicate to one another so that problems can be revealed in time for solutions to be provided and acted upon.

The rapid development of offshore engineering over the last few years has overtaken many aspects of engineering design and metalurgical knowledge so that the development of novel techniques is often only marginally ahead of large scale constructions. This is particularly so in the field of physical metallurgy and this report is intended to draw attention to some of the more vital areas requiring urgent attention. Not only are specific answers required to specific questions but general principles need to be revealed and communicated to engineers in language which they can understand.

Permission to publish this paper has been given by The British Petroleum Company Limited.

References (Cotton)

1. Antonakis, C. J. 1973 Concrete offshore structures. Proceedings of the Offshore Scotland Conference, Aberdeen 1973.
2. Gjorv, O. E. 1971 Long-term durability of concrete in sea water. *Proc. A.C.I.* **68.**
3. Gilchrist, J. D. 1970 A report on stress corrosion in prestressing steel. Construction Industry Research and Information Association, London.
4. Panafieu, P. De & Darnborough, E. 1973 The SEAL intermediate system (SIS). Proceedings of the Offshore Scotland Conference, Aberdeen 1973.
5. Garrat, A. 1975 Full fathom five hundred. BP Shield International, February 1975.
6. Winterton, K. 1961 Weldability prediction from steel composition to avoid heat-affected zone cracking. *Welding Res. Suppl.*, June 1961.
7. BS. 1515, Part I, §4; Specification for fusion welded pressure vessels. British Standards Institution.
8. Burdekin, F. M. & Dawes, M. G. 1971 Practical use of linear elastic and yielding fracture mechanics to pressure vessel technology. Institution of Mechanical Engineers, May 1971.
9. DD3 Methods for plane strain fracture toughness (KIC) testing. British Standards Institution.
10. DD19 Methods for crack opening displacement (COD) testing. British Standards Institution.
11. British Standards Institution 1974 Committee WEE/37; Draft British Standard rules for the derivation of acceptance levels for defects in fusion welded joints, December 74/43755.
12. Cotton, H. C., Denham, J. B. & Doble, K. R. 1971 How to conduct COD tests and utilise the results in practice. First International Symposium of the Japan Welding Society, Tokyo, November 1971.
13. Farrar, J. C. M., Dolby, R. E. & Baker, R. G. 1969 Lamellar tearing in welded structural steels. *Welding J. Res. Suppl.*, July 1969.
14. Garland, J. G. & Kirkwood, P. R. 1974 The selection and fabrication of steels for offshore structures. International Offshore Technology Conference, London, October 1974.
15. United States Department of the Interior, Bureau of Mines 1968 Rare earth silicide additions to an alloy steel to increase toughness and ductility. RI 7153 June 1968.
16. Luyckx, L. *et al.* 1970 Sulfide shape control in high strength low alloy steels. Aime Meeting, Denver, Colorado.
17. Gurney, T. R. & Maddox, S. J. 1972 Determination of fatigue design stresses for welded structures from an analysis of data. *Metal Constr. and Br. Welding J.*, November 1972.
18. Stalker, A. W., Hart, P. H. & Salter, G. R. 1974 A preliminary study of underwater manual metal arc welding. Welding Institute Report 3412/5B/74.
19. Levin, M. L. & Kirkley, D. W. 1972 Welding Underwater. *Metal Constr. and Br. Welding J.* **40.**
20. BP unpublished report.
21. Chadwick, M. D., Brown, D. J. & Corbishley, T. J. 1974 Explosive welding in pipeline applications. Proceedings of Conference on Welding in Offshore Construction, Newcastle, February 1974.

KEYNOTE SPEECH

By A. A. Wells (*Queens' University of Belfast*)

Thanks to the careful organization of this conference, the unambiguous definition of its scope, prior circulation of papers, and the example to be followed as outlined by Kelly in his account of the work of Dr Rosenhain, it is possible at this stage to save precious discussion time by addressing the topic of engineering requirements for structural steels almost without preamble.

The three papers in this session, on bridges by Harper, line pipe by Mercer, and offshore structures by Cotton, are completely complementary. Thus Harper, although obviously aware of the necessary inhibitions of the bridge designer, arising simply from the need to ensure safe erection, and long operating lives despite the rigours of wear and weather, and the maintenance of public confidence, shows such an open mind that he would consider the use of adhesive joints in future constructions. In so doing he implies the invitation to consider other far reaching ideas that could be introduced. Mercer's paper is packed with information, in terms of numbers that stretch the imagination. Yet the sources are authoritative, and a fair impression is given of the challenge of pipeline construction to be expected through the next decade. Cotton gives a balanced comparison of the relative merits of concrete and steel offshore structures, not without confirmation that the former embody more steel than the latter. In addition, he mounts a vigorous challenge with regard to existing methods of quantifying notch toughness requirements, ranging from material selection, through fabrication procedures to quality control. He is quite specific in describing particular problems of other types of service failure, in need of the attention of the physical metallurgist.

A thread common to all three papers is the concern with steels at yield points of 450 MN/m^2 (65 $klbf/in^2$) or less, and the projection that these strengths will not much increase with regard to such civil structures as are here discussed at least in the next decade. It is worth remembering that this strength is no greater than that of the steel in the Forth rail bridge and that, also at the turn of this century, penstock pipes longitudinally forge welded by the water gas process were in common use in Germany, and were also exported (Koester, F., *Hydroelectric developments and engineering*, Van Nostrand, 1909). The safe pressures for 107 cm dia. pipes, 19 mm thick, quoted by Koester as 3 MPa (435 lbf/in^2), are not too different from those considered by Mercer. We may very well ask what has been achieved during the intervening 75 years?

It is reasonably certain that in taking stock this question will be extensively answered during this conference. A short answer is that constructional steels have been developed during the period to be compatible with welded fabrication, and that the required new processes of manufacture have been adapted to large scale production and improved consistency. The scientific contribution of the physical metallurgist to this has been the identification of the manifold influences imposed by applied heat and deformation, arising from the constituents of the steel, whether as alloying or residual elements, both singly and in combination. (I was going to demonstrate at least a superficial familiarity with terms in current use by employing the word 'synergistically'. However, in checking, the dictionary definition proved to be 'the doctrine that the human will and the Divine Spirit are two efficient agents that cooperate in regeneration'.)! Some indication of the success of this effort arises in the case of lamellar tearing near welded joints, and the associated short transverse properties (reduction of area at fracture), to which considerable importance is accorded by Cotton. The contributary factors associated with microstructures and inclusions are described by Shenton & Arrowsmith (*Development of steel plate for offshore structures*, Underwater Construction Technology Conference, University College, Cardiff, April 1975), and demonstrated in terms of remarkable increases of short transverse ductility. Here is a good modern example of an engineering requirement (in this case for welded truss joints between large tubular sections) which has been illuminated by the physical metallurgist so that the steel industry can respond to adjust its practices to produce a material by intention rather than selection.

Another important property which has been slowly mastered in structural steels at given strength levels over this period of many years is the notch ductility which proved to be so important when welding was first introduced. It is understandable that this topic is seen to be critical both for line pipe and offshore structures, since these must retain their integrity after being subjected to mechanical damage, possibly even at low temperatures. It is also of interest and somewhat in contrast, that Harper appears to accord a smaller significance to notch ductility in the case of bridge structures. It is tempting to speculate that the trend of the recent past, towards bridges favouring a box girder form, grew in part out of a desire to utilize welded steel in thin plate sections, which avoids the constrained and load concentrating joints in truss constructions, and that this choice was brilliantly successful in this respect in designing away from problems of notch ductility and brittle fracture that had been prevalent in earlier years. That there were instead the subsequent experiences of weld distortion and buckling instability may illustrate that there are limitations to the engineer's valued freedom to design away from problems, even where the scope of conceptual design is large, as it is in bridge building by comparison, for instance, with pipeline construction.

The treatments of notch ductility described both by Mercer for line pipe and by Cotton for offshore structures have a common theme. They embrace in the first instance a destructive test programme conducted at full scale and with realism, having regard to steels and fabrication processes, or at as near to full scale as can be economically conducted (e.g. by using cross welded wide plate tests). Next, the experimentally determined behaviour is synthesized as intelligently as possible by using the mathematical behavioural models of engineering mechanics. These models in turn predict behaviour in terms of idealized notch ductility tests of the materials, but these tests at the present state of development require such a degree of instrumentation and such a scale of specimen size that they are unsuitable as repetitive quality control tests. Hence, there must be a further correlation between the results of the idealized and of the quality control tests. The ranges of materials and pipe sizes, represented on the one hand by the pipelines discussed by Mercer are sufficiently narrow that an overall correlation can be used with some degree of trust. On the other hand, the offshore structural experience reported by Cotton embraces a wider range of structural forms, and attention is drawn in that paper to the mismatches which exist through the various stages of correlation. He accurately reports the current situation with respect to what is otherwise a continuous development of methods of fracture analysis and control, and it is hoped that his criticisms will receive further attention in the discussion period and in Session 2, since the limitation of time forbids that excursion in full at this juncture.

Mercer, and particularly Cotton, draw attention to the overemphasis with regard to fracture control on preheat and postheat treatments of welded joints which are sometimes mandatory and at other times specified by clients or their agents in a defensive posture. This is a good point, well taken, and this conference in the presence of many of those able to give expert and recorded testimony could perform a useful service in a reaffirmation of those circumstances where these heat treatments are genuinely needed, and where alternatively they have become redundant. As one of those originally associated with the quoted heated band width rule for localized treatments I readily confirm that such a treatment designed to relieve residual stresses is not necessarily that most suited to metallurgical conditioning. Even so, where treatment is necessary it would be preferable to limit dwell times rather than heated band widths, where the former are still specified in terms of thickness (e.g. 1 h/in; 24 min/cm). Transient heat penetration is known to be much faster than such empirical rules indicate, and stress relaxation is also rapid at peak temperatures. Metallurgical conditioning is also a rapidly occurring event in many cases. Merely to limit heated band widths is to condone not only ineffective residual stress relief, but the redevelopment of thermal plastic straining during cooling, leading in some cases to a final state which is inferior to that in the untreated case. It is therefore of most relevance to define the circumstances in which these treatments are vital under present conditions and with contemporary materials, and conversely where it would be better altogether to do without.

The relaxation of requirements for heat treatments in materials which have become more tolerant to welding is paralleled by the need, mentioned by Harper, Mercer, and Cotton, for better processes of field jointing and particularly those which can be effected under water. Whereas Harper inclines towards the use of adhesives, Cotton confirms that the welding processes involving thermal cycles are not precluded, although there may be problems of quenching and hydrogen absorption leading to cracking. Familiarity with the clean dry conditions which have hitherto been necessary for the performance of high quality welding would suggest that there would be serious difficulties with underwater welding. Yet it is under just such circumstances that one should keep an open mind. The performance of underwater joining is an engineering requirement certainly to be commended to the attention of the physical metallurgist at this time.

Hydrogen cracking, stress corrosion cracking, fatigue and corrosion fatigue cracking are additional hazards in welded structural steels. Cotton draws attention to their suspected significance in offshore structures and, in particular, in those cases where there is contact with sour gas, rich in H_2S, or stale seawater associated with sulphate reducing bacteria. He draws attention to the serious lack of data on the stress and other conditions for the occurrence of these events, and the restricted lead times in which to obtain it. Test programmes have been started, but an intense effort is necessary.

Although it is reiterated that there is no advantage in using steels of higher yield strengths where high cycle fatigue is dominant, there is insufficient evidence at the present time to confirm that this limitation applies in all cases to the corrosive environment in the relative absence of fatigue loading. Nevertheless, the very existence of such an inhibition may be quite sufficient to explain the caution implied in all three papers with respect to the future prospects of higher strength quenched and tempered steels. This view might be summed up by quoting Mercer, as follows,

'Perhaps the most important thing is not the development of something completely new but to continue to offer the engineer reproducibility and quality at the right price. It is a process of learning and gradual improvement.' And again, 'Quenched and tempered steels will probably play a growing role during the latter part of this century.'

That these are the words of those on whose judgements millions of pounds sterling have to be committed reflects a commendable responsibility. We may nevertheless ask whether such a view should constrain the discussion at this conference, and it is thought not. Commercially available quenched and tempered steels have now been used in civil structures in the U.S.A. for 20 years. They are weldable and economic in their demands on weld heat treatment. Under normal conditions their reserve of notch ductility may seem to be slender, but is offset by low fracture appearance transition temperatures and a remarkably small standard deviation on fracture toughness in the low temperature brittle mode. The principal risks with the use of quenched and tempered steel plates are probably organizational, in the avoidance of undetected substitutions of other grades, and undetected malfunctions of the quenching press. A suitable perspective on their susceptibility to fatigue and environmental cracking is seen in the sustained use of quenched and tempered steels in submarine hulls and artillery without report of serious problems.

The most logical use of higher strength steels of quenched and tempered type is in structures with large components of dead loads, or where the haulage costs to site are large. The most obvious application is in long span bridges, further discussed in Session 2. However, the potential for high capacity pipelines in remote places and undersea should not be altogether overlooked. There appear to be few theoretical advantages of increasing yield strength in this area with respect to pipes made from welded plate, but it is tempting to speculate on the potential advantages of continuous manufacture of pipe at the point of laying, by winding high strength steel wire on to an extruded polymeric matrix.

The function of this introductory presentation is to extract and emphasize for further discussion those aspects in the papers which reflect the demands on the physical metallurgist arising from engineering requirements. They appear to be summarized as follows:

1. Further provision of understanding of the effects of alloying elements and impurities, and response to heat and deformation, such as will enable the steelmaker to produce products which are both economic and consistent in their mechanical properties, including notch toughness.
2. Further elucidation of the effects of welding thermal cycles and deformations, and their modification by heat treatments, where these are absolutely necessary.
3. Eludication of the surface and interfacial physical and chemical mechanisms involved with oxidation, corrosion and cracking, including environmental effects associated with fatigue, and possibly the surface characteristics for use with adhesive jointing and bonding with polymers.
4. Closer definition of all these properties aimed at securing agreement as to how they are to be measured, and the levels of improvements in magnitude and consistency which might realistically be sought.
5. Consideration should also be given to the propsects for new products, which should include higher strength steels and those such as wire suitable for composite constructions, and new or further developed joining processes for use on site and under water.

Discussion (Chairman: O. A. Kerensky, F.R.S. (Freeman Fox and Partners))

A. R. Flint (*Flint and Neil*)

The designer's primary aims are avoidance of collapse and of unserviceability, generally through excessive deformation. These aims require knowledge of load extension characteristics. Structural strength depends on tangent modulus and deflexion on secant modulus so that premature departures from elasticity and proportionality are of importance. Definition of yield stress by the drop of beam method, while satisfactory for mild steel, may not be so for high yield materials particularly when structural stability is critical. The present definition of yield may have to be replaced by 0.2 % proof stress as has been common for light alloys.

The influence of the limit of proportionality on structural strength has been observed in developing the design rules for plate girders for BS: 153, in the course of which tests undertaken on full-scale samples showed the proportional limit to be as low as 50 % of the nominal yield stress.

The T1 steels in the U.S.A. exhibit major reduction in tangent modulus at about 85 % of their yield stress in compression.

In components unaffected by instability the yield stress at which a plateau of extension occurs is of obvious significance. But advantage may only be taken of this if the structure is not so detailed as to permit local failure before the ductility may be mobilized. In such components, the ultimate tensile strength (u.t.s.) may still have relevance; thus the draft clauses for the new Steel Bridge Code requires that u.t.s. $\geqslant$ yield stress multiplied by 1.25 times the ratio of the gross to the net area at any bolted or rivetted connection. This provision will tend to govern the design of tension elements in materials having a yield stress in excess of that of grade 43. Bolted connections commonly have an efficiency $< 80\%$, and bolted rolled sections commonly suffer a reduction in gross area of about 30 % due to holes. Thus if the yield stress were to be fully utilized, it would be necessary to provide an ultimate stress of $37\frac{1}{2}\%$ greater than the yield stress. This is not achieved by the specifications for grades 50 and 55. The Draft Materials section of the Bridge Code, at present out for comment, permits ultimate strengths no greater than 10 % in excess of the yield stress!

These provisions need only apply to structures where redistribution of forces is to be allowed, and the lower limit for u.t.s. may prove adequate for bridge work where the connections are remote from points of maximum stress in primary members. The lower value of u.t.s. has been considered permissible given assured ductility. A minimum elongation of 16 % is suggested in the Draft Materials Code; this figure having been based on the work at Cambridge into the ductility requirements for plastic design. Furthermore, ductility of this order is desirable to relieve stress concentrations and to permit forming during fabrication.

What is ideally required is an elongaton of 5 % before either work hardening or necking begin, since the former inhibits stress redistribution and the latter is locally damaging.

The designer should ensure that the material properties at a particular part of a particular element have a known statistical probability distribution of achievement. The present so-called 'guaranteed minimum yield strength' represents neither a minimum nor a guarantee which has any significant worth from the viewpoint of providing compensation in the event of non-compliance. The variability of the quality within casts of steel has been shown to be at least as great as between casts and, as a result, there is poor correlation between mill test results and the actual component strength. It is noteworthy that our materials are still as variable as the extreme design loads.

The so-called characteristic value introduced under the cloak of limit-state design has not proved particularly helpful, other than by encouraging statistical control of material properties. The reliability of a structure, having a given characteristic strength may be shown to increase with variability in the material. Since it is known that rolled steels available in this country have a wide range for variability, it is clear that the characteristic values as at present defined are misleading. A more rational basis for design requires the definition of both mean and variance of product on a batch rather than a cast basis. A proposal for control procedures aimed to achieve this is included in the draft for comment of the Bridge Materials Standard. It is my view that the term 'characteristic strength' should be forthwith abolished or redefined as the mean and that all design rules should be based on a prescribed combination of mean and variance. In the absence of such a change, designers will continue to be landed with very varied products supplied to meet the same need.

The variability of steel quality seems to arise mainly not due to chemistry but due to metallurgical structure as influenced by rolling. It would seem desirable that materials be developed which are less sensitive to the brutal processes of forming structural components. It is too much to hope that a new family of economical steels will appear which have such properties and minimize residual stresses, perhaps in conjunction with new rolling techniques? If, at the same time, the advantages of improved strength due to rolling can be achieved the product will go far to satisfy the designer.

D. James (*British Aircraft Corporation*)

Flint (this discussion) referred to a new specification for structural steels calling for a minimum elongation of 16 %. In aeronautical structures high strength aluminium alloys with typical elongations of 7 or 8 % are commonly used. These may have short transverse elongations of 2 or 3 %. I am surprised at the very high figure which is being specified for a structural steel. The reasons for this are far from clear to me; it is likely that they are traditional rather than technical.

H. C. Cotton (*British Petroleum*)

Through thickness properties are of special importance when constructing large offshore platforms. This is evident from the many reports of lamellar tearing or what appears to be lamellar tearing experienced during the fabrication of large and thick weldments. It is not possible to simplify such joints as the location at which braces must meet the node are carefully calculated so there is little room for improvement in the geometry of such connections by which welding stresses could be minimized. The question of how to cater for punching shear and other effects presents significant problems and often gives rise to a can so interwoven by longitudinal and radial stiffners requiring full penetration welded attachments that the welder cannot even see the root of the weld. It seemed almost impossible to avoid a lamellar tearing problem under such circumstances especially as no opportunity presented itself under the prevailing conditions to practice special welding sequences. In association with certain steelmakers the cause of lamellar tearing was disclosed to be a basic anisotropic weakness in the steel attributable to inclusion morphology. This was cured and the lamellar tearing problem simply disappeared. In the future there will be many applications for such steel which may supplant the use of forgings for some applications. The use of steel plate with isotropic properties is bound to expand across a wide front of applications.

H. G. Wolfram (*West Gate Bridge Authority*)

Harper (1.1) has commented on the problems associated with distortion of thin plate structures due to application of heat.

This is one of the fields where close cooperation between metallurgist and engineer can help to overcome the problem. I am referring to heat straightening of stiffened plate structures which has been applied successfully during the fabrication of orthotropic deck panels up to 15 m × 3 m in size.

Before the use of heat straightening, metallurgical advice was obtained to ensure that the mechanical properties of the steel would not be affected adversely by the application of heat followed by air cooling.

However, although the problems of distortion of thin-plate structures can be overcome, I suspect that the future trend in bridge design will be away from thin plate structures towards heavier sections, largely because of:

(*a*) the increasng ratio of labour to material costs, but also because of

(*b*) the high costs associated wih the design of highly complex and sophisticated structures.

It is in this area that the engineer must make some predictions of his future requirements. I see the problems in two areas:

(i) the decline of toughness and weldability with increasing plate thickness,

(ii) the non-homogeneity of steel in the through-thickness direction as manifested by cases of lamellar tearing in cruciform and highly restrained weldments.

J. E. Roberts (*British Steel Corporation*)

Flint (this discussion) mentioned the statistical definition of strength, to recognize the inherent variability of steel plates and sections, and used the term 'characteristic strength'. Some of the problems arise because the metallurgist in the past has misled the engineer into believing that steel was uniform and homogeneous, when it is not. Nevertheless, conventional release test procedures have served adequately for quality control purposes for many years. The 'characteristic strength' concept produces considerable difficulties in application to steel production and clearance of material for dispatch, to say nothing of the question of how to deal with steel already supplied, and perhaps fabricated, should a drift in mean or standard deviation of strength be experienced. Certainly there is continual striving for a more consistent product, but the papers presented in this session all emphasize the need for economy, and so far the economic benefit of applying the characteristic strength concept has not been demonstrated. This is an area where collaboration between engineer and metallurgist is in action, a good deal of discussion having taken place between Flint and his colleagues, and steelmakers (including myself). The best description of the present state of play is 'a draw', but the game is still in progress.

J. H. Van der Veen (*Hoogovens*)

Ultimate strength is not a physical property of the steel but the point of instability in a particular test, the tensile test with parallel faces. As a steelmaker, I have no objection to a criterion on 'reserve ductility or strength beyond yield point' but please let it be a proven scientific criterion, such as, if justified, the strain hardening coefficient m.

W. L. Mercer (*British Gas Corporation*)

Could I comment on the need for a proper balance in development studies leading to higher strength materials. I think it is important that we look at all the required criteria together because only in this way can one be sure of covering any really critical factor important in a particular application. I think phenomena like corrosion, fatigue, etc. will be limiting in the majority of cases. For example, in the case of pipelines, one can say that there is obviously an ecnomic advantage in utilizing steels of 690 MPa (100000 lbf/in^2) yield strength. I have no doubt that it would be easy to provide steels with adequate levels of toughness to match these high levels of strength. However, I think the things which are going to trip us up in the end are corrosion behaviour, particularly stress-assisted corrosion processes and hydrogen embrittlement. Metallurgical development work needs to recognize these parallel potentially limiting factors and to incorporate them into the programme from the outset.

J. D. Harrison (*The Welding Institute*)

Is there any justification for requiring pipeline weldments to have sufficient toughness to arrest propagating cracks? This does not seem to be required by the British Gas Corporation but is still called for by some pipeline customers and is referred to as a requirement by Barr & Burdekin (2.2). The accident envisaged by Mercer (1.2) is breaking of the pipe by earthmoving equipment. A crack started in this way could not conceivably run for more than one pipe length in the weldment. The weldment should have sufficient toughness to prevent fracture initiation under normal service conditions. The specification by less enlightened customers of an arresting property in pipeline weldments can be a serious limitation on the materials and procedures used in construction.

R. R. Barr (*British Steel Corporation*)

I agree with Cotton (1.3) that the c.o.d. concept is not cut and dried, and in the paper by Burdekin and myself (2.2) we have discussed in detail some of the problems associated with the testing of weldments and the interpretation of the results. In these circumstances it is important to be able to rely on the results from wide plate tests and I would like to discuss further one or two points related to stored energy as referred to in Cotton's paper.

Cotton makes reference to the Robertson test and the question of stored energy and compares this to the wide plate test situation. I would like to point out that the essential difference between the wide plate and Robertson tests is that the former is a crack initiation test while the latter is a crack propagation test.

However, taking up the point about short crack initiation in the wide plate test, we consider that the stored energy in the B.S.C. test system is sufficiently high to drive a crack once initiated. I would like to quote some preliminary results from a test using a T-welded configuration where an attempt has been made to assess the dynamic response of the test frame. The notch was a surface notch sharpened by fatigue (55 mm long × 11 mm deep) and located in a m.m.a. weld metal (C–Mn) the steel being 25 mm thick, BS 4360:55E. Strain gauges were placed on a line parallel to the expected progress of the fracture and some 50 mm from the weld, running transverse to the plate as shown in figure 1. The gauges were placed at intervals of 75 mm from the centre line of the plate, but only one was placed on the opposite side from the weld. We recorded the response of these gauges during the course of the subsequent fracture at −30 °C, the fracture initiation stress being 528 N/mm^2.

The recording system proved to be capable of resolving the dynamic effects which occurred during fracture and this is illustrated in figure 1, where the signal voltages are illustrated over a time period of 8 ms. The gauge nearest to the plate centre line shows a sharp fall in strain as the fracture is initiated and propagates. The defect incidentally initiated and travelled through the transverse weld but arrested on the other side of the centre line, as a result of the increased propagation resistance of the parent plate. Moving out from the centre of the plate to the gauges nearest the edge of the plate, these show large transient rise of strain followed by a fall to the permanent plastic strain, indicating that there was sufficient stored energy to maintain the load on the remaining ligament giving a net increase in stress and hence the transient increased strain measurement.

The fracture arrested in the parent plate as mentioned previously and the gauge in this region showed a very large strain measurement being within a large plastic zone accompanying the arrest of the fracture.

In conclusion, these preliminary observations together with calculations on stored energy associated with the measured deflexion of the loading points, etc., should help to retain some confidence in the wide plate test as an initiation test in the circumstance described in Cotton's paper where there is a small zone of low toughness. This work will be continued and it is our intention to publish more details in the near future.

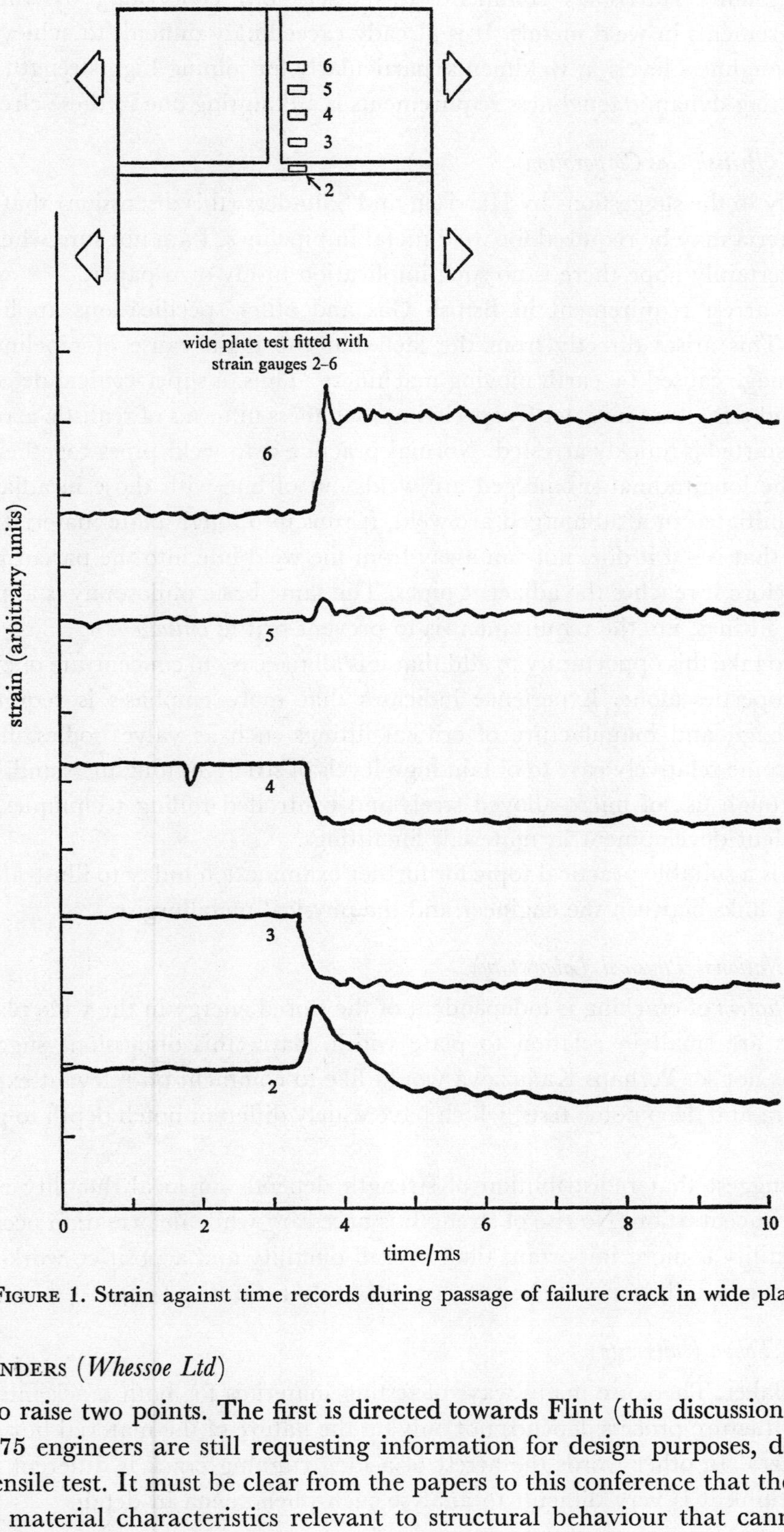

FIGURE 1. Strain against time records during passage of failure crack in wide plate test.

G. G. SAUNDERS (*Whessoe Ltd*)

I wish to raise two points. The first is directed towards Flint (this discussion). I am amazed that in 1975 engineers are still requesting information for design purposes, derived from the uniaxial tensile test. It must be clear from the papers to this conference that there are far more important material characteristics relevant to structural behaviour that cannot be obtained from such tests.

The second follows Harrison's comment (this discussion) concerning dynamic fracture toughness requirements in weld metals. It is already exceedingly difficult to achieve adequate static fracture toughness levels in weldments, particularly for joining high strength steels. The prospect of meeting dynamic toughness requirements is a daunting one in these circumstances.

W. L. Mercer (*British Gas Corporation*)

I wish to reply to the suggestions by Harrison and Saunders (this discussion) that unrealistic crack arrest criteria may be required for weld metal in pipelines. I am not sure where this idea originated – I certainly hope there is no such implication in my own paper.

The fracture arrest requirement in British Gas and other specifications applies to plate material only. This arises directly from the fact that the main cause of pipeline failure is mechanical damage caused by earth moving machinery. Thus, a super-critical defect can suddenly appear and a fracture initiate. By specifying toughness in terms of realistic arrest criteria, the fracture so started is quickly arrested. Normal practice is to weld pipes together to form a pipeline with the longitudinal submerged arc welds out of line with those in adjacent pipes. If a fracture is initiated in a submerged arc weld, it runs into tough plate material in the two adjacent pipes, that is – if it does not run away from the weld line into the parent plate of the initiator pipe before it reaches the adjacent pipes. The same basic philosophy is applied in the case of pipeline fittings, i.e. the requirement is to prevent brittle *initiation*.

I would like to take this opportunity to add that it is all too easy to concentrate one's attention on the pipe properties alone. Experience indicates that more emphasis is required on the specification, design and manufacture of critical fittings such as valve bodies, flanges, etc. While it has become relatively easy to obtain high levels of strength, toughness and weldability in line pipe through use of micro-alloyed steels and controlled rolling techniques, there has been no equivalent development in materials for fittings.

I suggest this is a suitable practical topic for further examination today to illustrate the value of close working links between the engineer and the physical metallurgist.

R. G. Baker (*National Physical Laboratory*)

Surely the *initiation* of cracking is independent of the stored energy in the wide plate test rig, provided cracks are small in relation to plate width. Barr (this discussion) suggested that perhaps this was not so. Perhaps Kanazawa would like to comment on relevant experience in the Wells–Kihara and deep notch tests, which have widely different notch depth to plate width ratios.

May I also suggest that redistribution of strength depends on local ductility adjacent to points of stress concentration. No rise of strength is necessary while deformation occurs. Surely then, notch ductility is more important than overall ductility and a positive work hardening coefficient, however small, is more relevant than y.s./u.t.s. ratio.

T. Kanazawa (*Tokyo University*)

I agree with Baker. There are many ways of testing materials for both crack initiation and arrest. Also the fracture process depends not only on the nature of the material but also on the test specimen sizes. In other words the arrest of a long running crack is different from short crack arrest. I think it is very difficult to analyse such phenomena in detail.

H. C. Cotton (*British Petroleum*)

Sometimes for light reading I keep in my bag a copy of *Welding in the World*; recently in volume 13 I read the final report of a working group on brittle fracture tests for weld metal. I am glad it was the final report because it seems from a cursory reading that confusion has been compounded as time has gone by. There is a tacit assumption I think by certain groups in Britain, that crack opening displacement is the only valid way to assess fracture initiation in yielding materials and no other methods are really worth considering yet, strangely enough, when one goes to the U.S.A. for example, one finds that almost nobody gives any thought to c.o.d., other techniques being used (K_{Ic}, d.w.t.t. etc.) to estimate low stress fracture risk.

Before we convince ourselves that we know everything that is to be known about crack initiation let us bear in mind that there are many other techniques in vogue which have received scant attention in this country. Furthermore the presently accepted criteria for c.o.d. assessment may be unduly conservative and hence costly. Because a weld shows transient brittleness in the c.o.d. test at a low applied stress it does not follow that the surrounding metal might not be able to arrest the small crack very quickly after initiation.

On the other hand in Wells cross welded wide plate tests, after a fracture has run for a short distance, it often arrests in more notch ductile material. Whether or not this is significant depends upon whether the load is still being applied as it would in a real structure where the actual load could be progressively rising rather than decaying as it does in many experimental arrangements.

In regard to the question of arresting a fracture which is propagating at some length in a large pipe line it is our conclusion that this is just not possible in the size of pipe and at the pressures and stresses that are commonly used for transmission pipelines. I believe that a Canadian Study Group has abandoned all ideas of arresting running cracks by means of metallurgical effects and are now resigned to using mechanical crack toppers, such as we use at B.P. We have some confidence in the wide plate test, we have some confidence in crack opening displacement, it might even be that we have a degree of confidence in the Nibbering test, but we think it prudent to keep all our options open at this time pending a clearer understanding of the mechanics of fracture initiation.

W. E. Duckworth (*Fulmer Research Inst.*)

May I comment on Baker's reference (this discussion) to tensile strength. Some time ago I would have supported the view of its irrelevance, but recent work at Fulmer suggests that the point of instability may be where the first microvoids, which ultimately coalesce into failure, may start. If this is confirmed then it may be an important measure of the ability of metals to withstand the onset of fracture in the regions of local stress mentioned by Baker. Thus, the instincts of the engineer in clinging to u.t.s. may be justified.

T. Gladman (*British Steel Corporation*)

May I endorse the view expressed by Van der Veen (this discussion) concerning the inadequacy of the u.t.s. as a design criterion. A high tensile strength is seen as a safety measure to guard against overstressing. The elongations associated with stresses above the yield are significant and would prove unacceptable in major structures. A safer approach might be to further increase the yield strength. If any plastic distortion is experienced, this is usually in some region of stress concentration where fracture, i.e. cracking, is far more likely to produce failure than any concept of plastic instability on a macroscale such as is inferred by tensile strength. Cavity development, although important in connection with ductile failure, has little to do with plastic instability. Experience points to the fact that cracking rather than plastic instability would cause failure, and the modern approaches to fracture should be followed through; this aspect will of course be dealt with in a later session.

A. R. Flint (*Flint and Neil*)

The strength of bolted or riveted connections in bridges are directly correlated with the ultimate tensile stress as obtained from the normal type of test, and if we are to make the best use of the yield stress that our steel produces, we do not want our connections to be weaker than the remainder of the structure, and we want to develop ductility in the structure remote from the connections, and it is for these reasons alone that I think we still need to consider the ultimate tensile stress.

G. P. Tilly (*Transport and Road Research Laboratory*)

One of the problems faced by designers of bridges and other steel structures is to assess collapse behaviour under compressive loading. It is now possible, by using advanced elasto-plastic analyses, to determine the post-buckling behaviour of steel plates having different degrees of initial imperfection, residual stress, and simple stiffening (Crisfield 1975). However, in making predictions it is necessary to use tensile stress–strain curves for the as-received material. In practice it is suspected that stress–strain curves, particularly in the heat affected zone, will be very different from the available data. In addition it is now known that compressive yield stresses are approximately 5 % higher than tensile yield stresses. It is therefore essential to have more accurate data to enable satisfactory comparisons to be made between analysis and experimental behaviour of structures. This is an area which requires research into material behaviour to aid the further development of analysis and improved design rules.

Another area in structural engineering requiring more information is that of connections made by high-strength friction-grip bolts. Such bolts are usually plated and can be susceptible to hydrogen embrittlement. There have been examples of delayed failures that have occurred up to four years after installation (Ito & Matsuyama 1970). In contrast laboratory testing has been conducted for much shorter durations and it has usually been considered that fractures due to hydrogen embrittlement occur relatively soon after tightening of the bolts; clearly this is not always the case and long term laboratory data are required to elucidate this situation. Current research is being conducted to determine behaviour of faying surfaces in joints being exposed for up to 10 years (T.R.R.L. 1974).

In steel bridges fatigue damage can occur and there have been several reported instances, such as those in the United States (Heins & Galambos 1974), of cracks particularly in designs having girders with cover plates welded to the flanges. One of the difficulties in assessing such fatigue is that bridges are designed for up to 120 years usage. During this service life, between 100 million and 1000 million fatigue cycles may be imposed by traffic and there may also be damage due to wind induced vibrations. Unfortunately laboratory data are usually for tests halted at endurances up to approximately ten million cycles and it is therefore necessary to extrapolate lives by approximately an order of magnitude in order to assess behaviour in design lifetimes. Fatigue cracking can also occur in welded regions subjected to pulsating compression.

An example of such behaviour occurred in the stiffener to cross beam welds of an orthotropic steel bridge panel subject to normal traffic at a site near Denham (Nunn 1974).

References

Crisfield, M. A. 1975 Full-range analysis of steel plates and stiffened plating under uniaxial compression. *Proc I.C.E.*, Part 2, December.

Heins, C. P. & Galambos, C. F. 1974 Bridge fatigue due to daily traffic. *Transportation Res. Rec.* No. 507. Washington: National Research Council.

Ito, A. & Matsuyama, S. 1970 Delayed failure of high strength bolts. *Railway Tech. Res. Inst.* (JNR), *Q. Rep.* **11**, (3) Sept.

Nunn, D. E. 1974 An investigation into the fatigue of welds in an experimental orthotropic bridge deck panel. *Report LR* 629. Crowthorne: Transport and Road Research Laboratory.

T.R.R.L. 1974 Long term exposure tests on high strength friction grip bolted connections. *Leaflet LF* 411, Crowthorne: Transport and Road Research Laboratory, October.

R. Dolby (*Welding Institute*)

I would like to refer to the use of quenched and tempered steels and in particular their low usage in this country even with the long experience in the United States and Japan. It seems from our position at the Welding Institute that one of the reasons for the low usage of the micro-alloyed quenched and tempered steels for tanks construction and other fabrications is one particular aspect of weldability. There are no cracking problems. Hydrogen induced cracking, hot cracking in the weld metals, lamellar tearing and so on, are no more likely to occur in this type of steel than in controlled rolled or normalized steels. The one stumbling block at the moment appears to be low weld metal toughness. There is a dilemma here because low carbon micro-alloyed steels are being used in Japan and are apparently satisfactory for these applications, but our experience is that the toughness of the weld metal regions can be very low. I think there would be wider usage of quenched and tempered steels in the U.K. if this particular problem could be resolved.

R. R. Barr (*British Steel Corporation*)

To refer to Dolby's remarks (this discussion) on quenched and tempered steels, low toughness at the root of a m.m.a. weld is apparently a problem when welding in the vertical-up position using a full weave. This situation is more widespread than may be realized and it is not confined to micro-alloyed quenched and tempered steels, nor is it associated only with British steels. It is a weld metal problem related to the thermal effects during welding and the fact that information has not been published elsewhere is probably due to the fact that it has not been detected in laboratory weld trials.

Referring to one of the points raised by Baker (this discussion), I did not mean to imply that the stored energy in a test rig would exert any control on the conditions for fracture initiation. I was commenting on the situation, referred to by Cotton, where a fracture might occur locally, perhaps within a region of low root toughness. I was trying to illustrate that, in the event of a short fracture being initiated, there is sufficient stored energy to maintain the load on the wide plate test and I hope that when we have additional data we can satisfy Cotton on this point.

J. Savas (*Republic Steel Corporation*)

At our Research Centre, we have found an interesting correlation between hardness of the weld metal and the heat affected zone area and toughness. We have found with some of the steels that if they are too highly alloyed you get exceptionally high hardness, and you have to resort to subsequent tempering of the heat affected weld zone. Thus, you can overcome the problem of using quenched and tempered steels in the welded condition. It all depends on the particular compositions you are looking at. We are not in a position to reveal our alloy developments in these areas, but it is suggested that it might be worth looking at these considerations if there is a problem with the use of quench and tempered steels in some applications.

CHAIRMAN – Is it a metallurgists' problem?

Yes, it is metallurgical due partly to the heat treatment that is required subsequent to welding to covercome the general problem of toughness. We have found that we can exactly double our toughness in experiments which use the implant test. If you are familiar with that test, you will appreciate that there is a thread in the sample that is welded over and that presents a sharp notch right in the heat affected zone. Thus, it is quite a critical test. Our results show that in most cases, depending on composition, you will require a post weld heat treatment.

CHAIRMAN – Is the cost of all this viable?

Well this has to be taken into consideration along with strength level.

One can work with steels having yield strengths of 690 MN/m^2 (100 000 lbf/in^2) or higher. I believe even 1000 MN/m^2 (150 000 lbf/in^2) yield strengths can be used in structures.

J. H. VAN DE VEEN (*Hoogovens*)

In talking about sufficient stored energy in tests on brittle crack propagation, I would like to draw attention to the question of whether sufficiently high stored energy is really 'available' at the right time and place for continuing crack propagation. Energy can travel no faster than the velocity of sound and this is of comparable magnitude in relation to brittle crack speed.

D. W. SMITH (*Mott Hay & Anderson and University of Dundee*)

Fatigue and jointing methods. Cotton (1.3) has rightly drawn attention to the fact that welding reduces all structural steels to the same fatigue strength, and this tends to inhibit the use of high strength steels. New methods of joining are consequently desirable.

Epoxy adhesives can join steel to steel and steel to concrete with a strength greater than the shear strength of concrete; and, with proper selection, they are remarkably tolerant of quite significant rust (but not grease) which may be on the surface before their application.

It would be of interest to know whether there is a metallurgical implication. Is the selection of suitable adhesive affected by the metallurgy of the steel? Is it necessary to repeat all the adhesive bond testing every time a change of trace element is made in the steel mix?

Another point is that adhesives may favour a return from thin plates to heavy steel sections. An adhesive joint is not as strong as the steel, and may therefore have to be wider than the steel part, as, for example, the flange of an I-section is wider than the web thickness.

J. C. CHAPMAN (*Wimpey*)

We are talking about the role of metallurgy in engineering practice. I work at the rough end of engineering, so perhaps you will not mind if I talk about metallurgical practitioners and their role in engineering practice. The practicing metallurgist has tremendous experience in diagnosing failures and recommending modifications. This experience, I think, equips him to contribute much more than perhaps he does at the moment to design, material specification, inspection, application and protection. He often has I think, an important role to play in the education of engineers, who need to know something about the practical behaviour of materials. In their training, to my mind, engineers are not taught enough about the practical behaviour of materials; they tend to get watered down courses on metallurgy – I think they want a more superficial but practical treatment of the actual behaviour of materials, and metallurgists have an important role to play in this. Where appropriate, practicing metallurgists should form part of the design team to advise on the matters I have mentioned; they should not simply be involved in the diagnosis of failures.

I think more attention needs to be given to the role of the metallurgist in bringing existing knowledge to bear on engineering problems. Naturally, we are talking mostly today about unusual problems, but most of the failures which occur are failures which could have been foreseen if enough previous attention had been given to them within the scope of existing knowledge. Equally in material development, high priority should be given to reliably and expeditiously producing materials to the existing specifications.

Perhaps I could just mention one or two examples where the practicing metallurgist has a most important role to play. Corrosion protection, at the design stage, is a very important area for the metallurgist. We talk a lot about the corrosion of steel but we think that when it is protected by concrete there is no problem. However, in my experience, when steel protected by concrete corrodes you are in even bigger trouble than when the steel is not so covered. Criteria need to be established for an adequate degree of protection in terms of aggregate purity, thickness of cover, crack width, exposure and environment.

In the choice of materials and design of details to prevent stress concentrations and fatigue, the metallurgist comes into his own because he is looking at these failures every day of his working life and we should be seeking his advice at an early stage. Ship structures are a very important area for the practicing metallurgist; failures may be induced by propeller induced vibration, wave forces and machinery vibration, and may occur in the machinery itself.

Could I just ask one question on development of new materials: is there a greater role for ductile iron castings? We have heard a lot about the diffiuclty of making connections, particularly welded connections in tubular members. Ductile iron has developed a lot in recent years; so far as I know there is a considerable difficulty in welding but is there any hope that it will be weldable in the future? Can we look forward to ductile irons being a tool at the disposal of both the engineer and the architect?

A. A. Wells (*Queens University*) (*Keynote Speaker*)

I simply want to put forward one thought in the remaining time. In the discussion this morning on two occasions the question arose as to the definition and development of a test. It came up in terms of Flint's point concerning the importance of the departure from linearity in the stress–strain curve, particularly with regard to compressive loading, and I think Tilly took that point up later. Certainly when Tilly was speaking, it came home to me that it was rather strange that this particular effect was not being qualified by a compressive test since one was dealing with something to prevent compressive failure. Then again, several speakers took up the question of the significance of the loading conditions in the wide plate test and in crack opening displacement tests, so it produced this thought which I think is important enough to put before you.

If you want to fine-down the actual point at which engineers and metallurgists react most with one another, it is at this point where tests have to be defined and developed. This job is characteristically done by task groups. When we have a meeting of the kind that we have this morning, we realize that there are so many raw topics that no one discussion can ever have sufficient time to go through to finality. This is the sort of thing that can be complemented by smaller groups of people working together in task groups when they do have a definite job to do; they are asked to come up with the specification of a test.

Some people will know exactly what I have in mind having seen the E 24 committee of the A.S.T.M. in action. In the United Kingdom, in years gone by, we had the Navy Department Advisory Committee on structural steel that provided the umbrella for quite a number of years for this kind of group of people to get together. May will recollect his committee, which was concerned with the high strength steel. It is my understanding at the moment that this kind of activity has suffered somewhat by the depression conditions which we met and hope we are going to get out of, and the thought I would like to leave is that we should look to the importance of task groups in this field.

The following two contributions were received after the meeting:

J. F. Lancaster (*Kellogg International Corporation*)

I think the embrittlement of steel in service deserves more attention than it receives at present. One is hampered in discussing this subject because of a lack of good quantitative information from the field. In particular, there is often a legal fog surrounding pressure vessel failures which makes it impossible to obtain other than under-the-counter information. Nevertheless, it is clear that catastrophic brittle failures of pressure vessels and other equipment continue to occur in process plants. This situation may be contrasted with that in shipping where brittle failure, which caused many losses at sea during the forties and early fifties, is no longer a serious problem. One possible reason for the relative lack of success in controlling brittle failure of pressure vessels is that modes of embrittlement in service are insufficiently understood. Two years ago there was a catastrophic brittle failure of a liquid ammonia storage tank in South Africa. There was a major escape of ammonia, as a result of which eleven men died. Investigation of the failure showed that the material of construction – BS 1501–151–28A – was chemically within the specification, but that one plate had a very high impact transition temperature and poor tensile ductility. There was no reason to doubt the as-delivered quality of the plate, and the investigators of the accident considered that a major contributory cause was strain-age embrittlement of the material. Clearly it is important to control susceptibility to this type of embrittlement. Nevertheless no such means of control exists in either British or American codes and standards. Moreover, there is very little information in the metallurgical literature on strain-ageing of pressure vessel steels, and virtually none on their ability to maintain notch-ductility over a period of time.

Strain-age embrittlement is only one of a number of possible embrittlement mechanisms that may afflict metallic equipment over a period of time. The risk of such embrittlement is well known in the hydrocarbon processing industry and not many operators would be happy to repeat a full hydrostatic test after a period of service. This is a problem that deserves study by the steel industry, bearing in mind that the quality of a material is best measured by how it behaves in service and not merely by its ability to pass the standard specification tests.

In presenting this viewpoint, I have been concerned to give examples of how the metallurgical profession can contribute to safety and reliability in process plants. In the future, both user requirements and government legislation will call for higher standards in this field, and it behoves metallurgists to anticipate such requirements and to be in a position to offer the improvement in both education and in materials that the development of our new-found resources will demand.

H. G. Wolfram (*West Gate Bridge Authority*)

The theme of the conference implies cooperative efforts by metallurgists and engineers. I think we should explore the avenues of cooperation in a little more detail. Just as there are metallurgists who are involved in research, or teaching or testing, or consulting, so there are engineers who are involved in various branches of their own discipline. I think it becomes necessary, therefore, to discuss also in more specific terms the stages at which metallurgists and engineers should be cooperatively involved and the scope of their association.

In my own branch of structural engineering, and particularly with reference to steel bridges, buildings and pipelines, experience has shown that for best results it is not only desirable, but also essential, for this cooperation to be established at the start of a project and to be continued through all its stages.

An early understanding by the metallurgist of the engineer's requirements for a particular structure with respect to strength, ductility, weldability and fatigue, will enable the metallurgist to advise the engineer on suitability and availability of steels for the particular purpose and to cooperate with him in the preparation of appropriate specifications for materials and workmanship, testing and standards of acceptance.

This early involvement of the metallurgist in the particular project provides him with the necessary background to enable him to act as adviser and consultant during later stages of fabrication, when inevitably metallurgical or welding problems arise which require expert attention.

Conversely, the anticipation of industry's future requirements enables metallurgists engaged in research and steel production to channel their efforts into productive fields.

I would suggest therefore, that as well as specialization and expertise in specific areas of metallurgy and engineering and cooperation between experts in these disciplines, we require broadly based cooperative efforts between metallurgists and engineers on a continuing basis for particular projects.

I am sure that this Rosenhain Centenary Conference will help greatly to achieve these objectives and would like to congratulate the organizers of the conference on their efforts.

1.4 Airframe material requirements

By D. James
British Aircraft Corporation Limited, Commercial Aircraft Division, Weybridge, Surrey

This short paper is restricted to a discussion on the characteristics of the principal light alloy materials which have been used in airframe construction over the last few decades. In particular, the response of the materials to the operating environment, as shown by problems of fatigue, notch sensitivity and stress corrosion, are discussed. These problems have been a major handicap to the achievement of maximum potential structural efficiencies. Reference is made to recent developments which offer the prospect of significant improvement.

The material in the paper is written in general terms and avoids reference to particular specifications, either national or proprietary.

Introduction

It is appropriate to introduce this paper by giving some indication as to the airframe designer's responsibility in the total field of the aeronautical industry, embracing design, manufacture and operation and subsequently to outline the manner in which the characteristic properties of the materials he selects contribute to the fulfilment of his responsibility.

Perhaps the best definition of a designer's responsibility is that agreed by the Fielden Committee on Engineering Design in the following terms:

> 'The designer's responsibility covers the whole process from conception to the issue of detailed instructions for production and his interest continues throughout the designed life of the product in service'.

Although this definition of a designer's responsibility was specifically agreed in the context of an enquiry into the mechanical engineering industry, few would dispute that it equally applies to any branch of engineering design. However, in today's environment in the aeronautical field, it could well be slightly amended to give recognition to the fact that the actual life in service is tending to increase significantly relative to the nominal life envisaged at the design stage. There are many aircraft in service today which are more than 20 years old and operators of commercial aircraft are looking for service lives of up to 30 years in the future. It is not suggested that these lives have been achieved, or will be expected to be achieved, without replacement of some parts, possibly more than once, during the relevant period. However, the main airframe is expected to last for such periods. This, then, is the duration of the period over which the designer's interest is expected to be maintained.

The ability to achieve these lives is very much influenced by the characteristics of the materials chosen for the airframe structure. Consideration of these is as vital a part of the design process as is consideration of the geometrical configuration.

Economic considerations, of course, play a significant part in the continuing demand for longer lives. It is becoming increasingly difficult for new designs to show major prospects of economic advantage over the already established designs. Structural developments using already established materials have tended to asymptote in terms of structural weight. Structural weight is a major parameter in the operating economics of an aircraft. If the airframe structural designer is to make his contribution to an advance in the operating economics, he can only do so through the advent of new materials offering the prospect of improved structural efficiency and reliability.

Paraphrasing slightly from the Fielden report, the designer's interest can be defined as ensuring, in the aeronautical context, that the aircraft will perform its pre-specified functions with the maximum economy and efficiency.

In the field of aeronautics, the pre-specified functions will be defined in a specification agreed between the designer and the customer. Generally, this will be in terms of the carriage of a given payload over a given range at a given speed. A minimum life may also be specified. Obviously, a pre-requisite of a satisfactory performance of the functions is that they shall be done safely throughout the expected operational life.

Safety requirements

Safety standards for commercial aircraft are defined in the form of general requirements, such as the British Civil Airworthiness Requirements in the U.K. or the Federal Aviation Requirements in the U.S.A. or, as is typically the case with military aircraft, in the aircraft specification itself. These will define the loading conditions which the aircraft structure must be designed to withstand in order to perform its specified functions. These will, in part, be in terms of conditions directly under the control of the pilot and, in part, determined by the operating environment. These requirements will be such that under reasonably foreseeable conditions, not only will the aircraft structure and the essential systems remain intact but also its flying characteristics will not be adversely affected.

Two conditions are normally defined, one a limit or proof load condition requiring that the loads be sustained without deformation of the structure such that its continued airworthiness is in doubt, and the other, an ultimate load condition requiring that the strength of the structure be such that collapse or critical rupture of the structure shall not occur before such loads have been sustained for at least a few seconds. The relation between these two conditions is normally such that the ultimate load is 1.5 times limit load or, in the case of military aircraft, sometimes 1.33 times proof load.

The two static strength conditions thus defined are expected to be met throughout the operational life of the aircraft. It is necessary, therefore, to ensure that the material used is such that the strength conditions can be satisfied, not only with the newly constructed aircraft but throughout the long service life during which it will be subjected to the environmental conditions typical of aircraft operations.

In discussing the requirements for airframe materials, it will be convenient to discuss the basic mechanical properties which traditionally have been regarded as providing a measure of the relative structural efficiencies of materials and then to discuss the impact of operational and environmental effects.

Basic mechanical properties – strength and stiffness

The normal basis of assessing the relative efficiencies of materials from the structural strength point of view is in terms of the ratio density: strength ρ/f. For components loaded in tension and robust components loaded in shear or compression f will be the ultimate tensile strength of the material if ultimate design strength is the criterion. If the limit or proof design strength is the criterion then f will be the relevant proof or offset yield strength. For less robust components loaded in shear or compression, instability may be the cause of collapse, in which case $\rho/E_t^{\frac{1}{2}}$ will be the measure of efficiency. E_t in this case is the tangent modulus of elasticity, which for failure at stresses below the elastic limit will equal E but for failures at stresses above the elastic limit will be a function of the stress at failure. The critical value of the stress in these cases is not only a function of the tangent modulus but is also a function of the geometry of the component.

From the materials point of view, the efficiency parameters indicate that the basic requirements are for high yield and ultimate tensile strengths, high modulus of elasticity and low density. This statement of requirements is indeed a platitude. In the context of airframe construction, its truth has been recognized from the earliest days. However, it is of interest in the context of this paper as a means of drawing attention to the fact that other characteristics of materials have tended to dominate research efforts in the materials field during the last two or three decades, as more experience of aircraft operation has been obtained. This is especially true of the aluminium alloys, the principal material for construction of the all metal aircraft. It is best illustrated by the history of the aluminium–zinc–magnesium series of alloys. In terms of the efficiency parameters quoted, there has been no improvement over nearly forty years. It is still the most efficient aluminium alloy currently used in aircraft construction.

The aircraft industry has a reputation, perhaps unjustified, for being a leader in the field of material development. What, then, are the factors which have mitigated against progress in the direction of improved efficiency of the aluminium alloys for so long? They are the factors which extended experience has taught us must often take precedence over simple structural efficiency, as measured in terms of basic mechanical properties, if the requirements for safety, reliability, and economic operation over a long service life are to be realized. They reflect the response of materials to their service environment and highlight the need for the designer to ensure that the design is such as to cater for the potential problems, either directly in the geometric detail of the design, or indirectly by providing for adequate preventative methods.

Service problems

The service problems, which have until very recently detracted from the attractions of the high structural efficiency potential of the Al–Zn–Mg series of aluminium alloys for airframe applications, have arisen mainly from three causes, which may often have occurred in combination. These are, fatigue, notch sensitivity and stress corrosion. These will be discussed separately.

Fatigue

The basic cause of fatigue is too well known to need further comment. Suffice it to say that, except when standing idle, the airframe is continuously exposed to fluctuating loads and hence to a fatigue inducing environment. There is ample evidence to show that the high structural efficiency potential offered by the basic mechanical property advances in the high strength aluminium alloys has not been matched by corresponding improvement in the basic fatigue endurance. In fact, there is some evidence to suggest that the reverse is true, not only in relative terms but also in absolute terms. The consequence is that in major parts of an airframe where fatigue considerations predominate, there is no advantage to be gained by the use of the high strength alloy. The stress limitations imposed by fatigue considerations are such that the ultimate strength potential of these alloys is not reached under normal ultimate design considerations. It is sometimes argued that there is still merit in having the excess static strength potential. However, this potential has often to be overridden by considerations of the likely behaviour of the structure in which a fatigue crack has developed. This will be determined by the notch sensitivity and fracture toughness of the material.

Notch sensitivity and fracture toughness

The term notch sensitivity is used to describe the deterioration in the behaviour under stress of a material in the presence of a stress raiser, as compared with its behaviour in the absence of a stress raiser. Premature failures have occurred under static stress conditions due to the presence of what, in general engineering terms, might be regarded as superficial imperfections. Such sensitivity to imperfections is obviously undesirable. The deterioration, be it in static strength or fatigue endurance, will be dependent on the sharpness of the notch and in most cases on its orientation relative to the local direction of grain flow. The Al–Zn–Mg alloys with the elongated grain structure have been shown to be particularly notch sensitive in the short transverse grain directions. In order to compare the notch sensitivity of materials, it is obviously necessary that the notched tensile strength be established on the basis of a standard notch.

An associated characteristic of materials is that which determines the rate of growth of cracks at stresses below the critical level, namely its fracture toughness.

The modern aircraft structure is designed to ensure that the risk of collapse of the structure due to failure of a single structural element is extremely remote. Structures designed on such principles are called 'fail-safe' structures. It is recognized that it is not always possible to detect defects, especially fatigue cracks, when they occur. It is highly desirable, therefore, that the materials used have the characteristic of toughness, i.e. an ability to resist the rapid development of a crack so as to permit of its detection by normal inspection procedures before a critical situation is reached. In addition to defining strength standards under normal conditions, airworthiness requirements also define strength standards for 'fail-safe' structures. These accept that defects cannot be completely avoided and, in consequence, some reduction of strength may have to be tolerated for periods between inspections. It is necessary then, for the designer to be able to assess the strength of the damaged structure. It is also necessary that he shall be able to assess the likely rate of crack growth so that acceptable inspection intervals may be defined.

The aluminium industry has made great efforts over the last decade to improve the fracture toughness of the Al–Zn–Mg alloys. Criteria have been developed which provide a reasonable basis for estimating the strength of the cracked structures and for predicting the rate of growth of fatigue cracks. The ability to do this is of the utmost importance in establishing safe inspection periods on airframe structures. In order to fix such intervals for the inspection of pressurized fuselages, for example, two questions need to be answered. These are: 'How long must a crack be before it can be detected?' and 'How long can it become before it leads to serious failure?' The design must be such that a crack which is approaching a detectable length at one inspection does not become catastrophic before the next. Preference will, therefore, be given to materials which satisfy the requirements of low notch sensitivity and high fracture toughness. In the past, this has led to the widespread use of the lower strength alloys in fatigue critical areas. The development of the improved versions of the higher strength alloys, together with techniques for predicting crack growth rates and the strength of cracked components, is therefore, a significant step forward.

In order that the designer can take full advantage of these advances, it will be necessary for the material manufacturer to provide the same standard of control over these qualities as has been the practice relative to basic mechanical properties. The determination of notched tensile strength is a straightforward matter. The direct determination of fracture toughness parameters is more demanding in terms of effort and material and will not be likely to be acceptable as a routine process. However, there is evidence of reasonable correlation between notch toughness parameters and the ratio of notched strength/tensile yield strength in the case of the aluminium alloys. There is, therefore, a good case for including a notch toughness standard in the material specification.

Stress corrosion

Stress corrosion failures have, perhaps, been the most difficult of the material related failure modes that the airframe designer has had to contend with in the last few decades. Every airframe manufacturer is able to quote examples of failures which have occurred, not only very early in the service life of an aircraft but sometimes before entry into service. Occasionally, failures have occurred on parts in store before assembly to components. Such failures have been established as being due to stress corrosion. The failure has been due to the presence of a sustained tensile stress in the component while it is exposed to a corrosive environment. The most disturbing aspect of such failures has been that the extent of corrosion present, as determined by normal methods of assessment, has been negligible. The sustained stress may arise from one or more of three causes, heat treatment quenching stresses, machining stresses or assembly stresses. Rarely does it arise from the normal service stress on the main airframe.

It must be accepted in the context of airframe manufacture and aircraft operation, that all environments are corrosive and that a sustained stress, therefore, means a stress corrosion hazard if the material is susceptible. It must also be accepted that it is not possible to design and build a structure to such closely controlled dimensions as to completely eliminate all assembly stresses. Stress corrosion potential is, therefore, unavoidable and the need exists for non-susceptible alloys. Material manufacturers have made great progress in this field too. Initially reduced susceptibility was achieved by modified heat treatment procedures which carried with them deterioration of the basic mechanical properties. Rival suppliers have developed proprietary procedures, the relative merits of which are difficult to establish. More recently, advances have been made in alloy development which have effectively restored the basic mechanical properties while maintaining good stress corrosion resistance. This has been achieved by a combination of changes in the balance of the major alloying elements and the introduction of new secondary alloying elements.

Intergranular corrosion

A major contributor to airframe maintenance costs is intergranular corrosion. In the context of stress corrosion, discussed in the preceding paragraph, the failures have generally been associated with intergranular corrosion of a very small order as measured by the amount of corrosion product present. The major problem in routine maintenance has been due to exfoliation corrosion, which reveals its presence very largely by the amount of corrosion product present.

The high degree of protection against intergranular corrosion offered by the aluminium cladding of sheet materials is not available on such products as extrusions and forgings, nor on the machined surfaces.

There appears to be no firm basis on which the structures designer can correlate simple laboratory corrosion test behaviour with service experience. He has, therefore, to provide adequate protection means. The operator must always be on the alert for evidence that intergranular corrosion is occurring and be ready to deal with it. A significant improvement in airline economics could be achieved if the resistance of the high strength aluminium alloys to this hazard was improved.

Creep and temperature effects

This paper would not be complete without some reference to the special problems associated with the use of the aluminium alloys in supersonic aircraft, such as Concorde. The previous sections have been primarily concerned with material characteristics which have been shown by experience to present difficulties, perhaps inadequately recognized at the design stage. The need for a long life at high temperature has resulted in very considerable effort being devoted to establishing the response of the material to prolonged exposure to high temperatures. Several aspects have to be considered in relation to continued structural integrity. The deterioration of mechanical properties of materials with rise in temperature has been known for many years, mainly as a result of the development work carried out in support of engine applications. Such work provided the basis on which the primary alloy used in the Concorde airframe was chosen. In addition to the decrease of strength at high temperature, there is a cumulative decrease which is time dependent. Finally, there is the non-recoverable strain 'creep' taking place during the period of stress at high temperature. On all three counts, the RR.58 material, developed initially for engine applications, was shown to be superior to all other available light alloys.

In the course of this development programme on the material, special effort has been devoted to examining the interaction between creep and fatigue behaviour. It has been shown that the effect of the combination is adverse on both counts. Another phenomenon which has been revealed is that under certain conditions subsurface creep deformations can give rise to the development of subsurface fatigue crack initiation and propagation.

Conclusion

The achieved efficiency of airframe structures for many years has fallen below that potentially achievable from simple considerations of the basic mechanical properties of the high strength aluminium alloys. The major inhibiting factors have been associated with:

(1) Failure to improve fatigue strengths of the alloys to match the improvements in basic mechanical properties.

(2) High notch sensitivity and poor toughness characteristics.

(3) Poor stress corrosion resistance.

Material producers have over the last decade made significant advances towards improving the toughness characteristics and stress corrosion resistance. These developments go some way to enabling the full potential efficiencies of the existing basic alloys to be realized.

Further developments in alloy technology have reached the stage when the prospect of a significant increase in modulus of elasticity, coupled with higher strength values, can be seen. It is important that adequate attention is paid in the pursuance of these objectives to the reliability aspects which have been discussed in this paper.

1.5 Undercarriage material requirements

By W. M. Imrie
Dowty Rotol Limited, Cheltenham Road, Gloucester, GL2 9QH

[Plate 6]

The main basic engineering requirements for undercarriage materials are described: while magnesium, aluminium and titanium alloys still continue to play a useful part, the ultra high tensile steels are now the preferred choice for many major structural items. The high standard of material quality and subsequent processing control necessary to ensure the integrity of components produced from these steels are described, with specific reference to the 300M alloy. Comparable data on fracture toughness and stress corrosion are given: it is explained why a fracture-mechanics approach to material selection has only limited application and that service reliability is dependent primarily on the steps taken to avoid crack initiation. Despite efforts to improve the intrinsic properties of these steels, with particular emphasis on reducing sub-critical crack growth of environmental origin, no worthwhile replacement for 300M has been found. It is concluded that the compromise between strength and brittleness has already been reached.

Introduction

The undercarriage and associated control equipment of an aircraft are subject to appreciable static loads while the aircraft is stationary on the ground, and to high, steady and dynamic stresses during taxiing, take-off and landing. Every effort is made to keep the weight down to an absolute minimum, which results in the structure being subjected to higher typical levels than most other major parts of the aircraft. At maximum design loads, some of the stresses may approach the yield strength, or indeed certain parts of the structure may be designed to accommodate limited permanent set.

General requirements

The main basic requirements sought by the designer are fourfold. First and foremost is that of reliability under service conditions, i.e. in addition to withstanding the calculated stresses, the undercarriage must also withstand normal handling by airport and maintenance personnel, stones, etc., thrown up by the wheels, and the corrosive environment of rain and salt-laden runways. This latter requirement includes adequate resistance to stress-corrosion cracking and a capability of being protected by electroplating, painting, etc. Additionally, the volume, weight and cost must be the minimum possible; these last two parameters are obviously largely opposed to one another and a compromise has usually to be made. Although minimum weight is highly desirable, it may not be cost-effective if it results in a significantly higher purchase price through the use of expensive materials or manufacturing techniques. Space limitations within the airframe structure often dictate the choice of materials used: where this factor is critical, steel becomes almost the only material choice, because its combination of strength and stiffness per unit volume greatly exceeds that of other conventional materials.

In simple terms, the weight of the undercarriage is determined primarily by the strength and density of the materials used. Modern stressing calculations take account of the shape of the actual section and the shape of the stress–strain curve, generally up to the 0.5% proof. Thus for parts subject primarily to bending, advantage can be taken of materials possessing a relatively high ratio of proof to tensile at a useable level of strain. Stiffness is also of importance on such parts and therefore a high value of the Young modulus is preferable.

A high resistance to impact is of little relevance: although emphasis is often placed by designers on this property in the case of steel, little concern has been given to the poor impact properties of the widely used aluminium alloys, which possess an Izod value in the order of 3–6J (2–4 ft lbf). A heavy landing is probably the closest approach to impact loading, but when transmitted to the undercarriage through the tyres and shock absorbers, the loads can hardly be considered impact in the sense of the Izod test. A good transverse ductility, particularly as measured by reduction of area, is of more importance as a material property for the avoidance of fracture by unforseen overload in those applications where the principal stresses are perpendicular to the grain direction.

TABLE 1. REQUIREMENTS OF MAJOR UNDERCARRIAGE MATERIALS

low cost	adequate toughness
low weight	good forgeability – heavy sections
low bulk	ease of heat treatment
high strength	good machinability
high stiffness	weldability
good resistance to:	
fatigue	hydrogen embrittlement
corrosion	crack initiation
stress corrosion	crack growth

In recent years, some attempt has been made to set fracture-toughness targets based on the capability of n.d.t. equipment to detect cracks of given lengths. However, because of the high strength materials used and the high applied stress levels, the critical crack lengths are very small. The basic design concept is one of safe-life rather than of fail-safe, i.e. little use can be made of residual strength in a cracked state. Resistance to crack-initiation and sub-critical crack-growth rate become more important criteria and ideally this rate should be slow enough to ensure that a crack will not grow to the critical size between two overhaul periods and could therefore be detected before complete failure took place.

Fatigue strength has become of more importance in recent years following the demand of long-life guarantees by the aircraft constructor. Experience has shown that when service failures do occur they originate at points of stress concentration: a typical K_t is 3.0. Undercarriages are often subjected to certain types of heavy loads applied a relatively small number of times and which therefore may contribute appreciably to the total fatigue damage. Low-cycle high-stress data (close to and exceeding the yield) under reverse-loading conditions are therefore required.

In addition to the above requirements, a solid-phase welding capability is sometimes necessary, which would normally limit the material choice to steel.

Table 1 summarizes the desirable characteristics of the major components of undercarriages.

MATERIALS

Four basic classes of alloy are possible contenders for the major structural items; these are, in order of increasing density (and stiffness), magnesium, aluminium, titanium and steel. The choice for any particular component depends on the size, shape, the stresses involved and the relative importance of the factors listed in table 1. In America, ultra-high tensile steel has taken over almost entirely for major items, and there is a general trend to follow this pattern in Europe. In view of this fact, and in order to limit the length and scope of this paper, only brief reference is made to the other three materials.

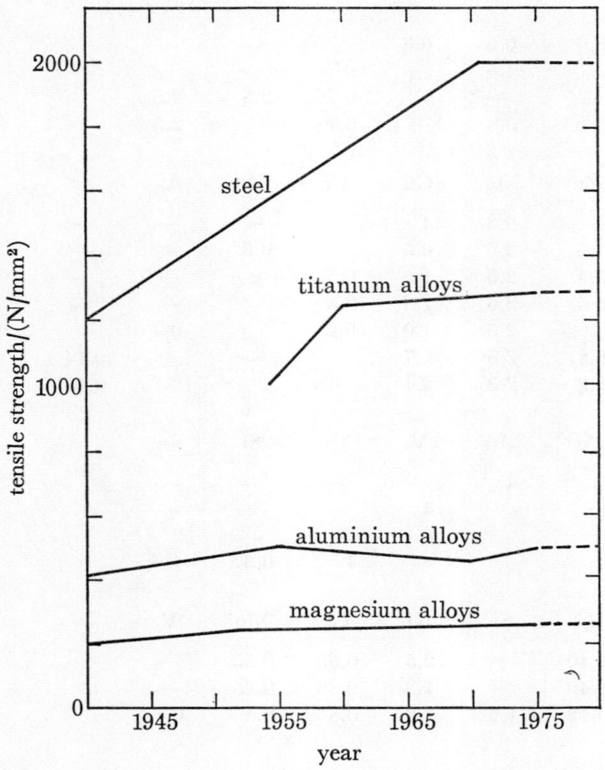

FIGURE 1. Increase in tensile strength of undercarriage materials from 1940.

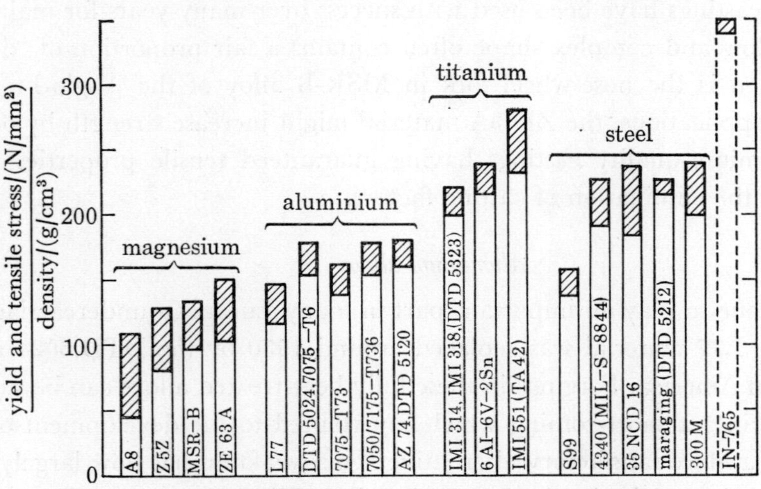

FIGURE 2. Yield and tensile stress to density ratios.

The increase in strength which has taken place in these four materials since around 1940 is illustrated in figure 1. Figure 2 compares some of the alloys on a specific-strength basis. Table 2 details the nominal chemical composition and tensile properties of the materials described below.

TABLE 2. NOMINAL COMPOSITION AND TENSILE PROPERTIES OF UNDERCARRIAGE MATERIALS

	nominal chemical composition							mechanical props.	
alloy								0.2 % p.s. N/mm²	t.s. N/mm²
Magnesium	Al	Zn	Mn	Zr	Ag	r.e.			
A8 (L121)	8.0	0.5	0.3	—	—	—	—	77	200
Z5Z (L127)	—	4.5	—	0.7	—	—	—	143	230
MSR-B (DTD.5035)	—	—	—	0.7	2.5	2.5		178	245
ZE63A (DTD.5045)	—	5.8	—	0.7	—	2.5		188	275
Aluminium	Zn	Mg	Cu	Cr	Mn	Ag	Zr		
L77	—	0.8	4	—	—	—	—	325	417
DTD.5024	5.5	2.5	0.5	—	0.5	—	—	420	495
7075-T6	5.5	2.5	1.5	0.3	—	—	—	430	510
-T73	5.5	2.5	1.5	0.3	—	—	—	385	455
AZ74	6.0	2.5	1.0	0.2	—	0.4	—	450	510
DTD.5120	6.2	2.5	1.7	—	—	—	0.14	450	520
7050-T736	6.2	2.3	2.4	—	—	—	0.12	420	490
Titanium	Al	Mn	V	Mo	Si	Sn			
IMI 314	4	4	—	—	—	—		927	1004
IMI 318	6	—	4	—	—	—		927	1004
6-6-2	6	—	6	—	—	2		980	1080
IMI 551	4	—	—	4	0.5	4		1130	1270
Steel	C	Si	Ni	Cr	Mo	V			
S99	0.40	—	2.5	0.65	0.55	—		1080	1240
4340	0.40	—	1.8	0.80	0.25	—		1500	1790
300M	0.42	1.6	1.8	0.8	0.4	0.1		1560	1900

Magnesium alloys

Magnesium sand castings have been used with success over many years for main-fittings, in which the basic hollow and complex shape often contains a fair proportion of 'dead' metal. A modern application is the nose-wheel fork in MSR-B alloy of the Anglo-French Jaguar aircraft. For future applications, the ZE63A material might increase strength by 5–10 %. The development of premium-quality castings having guaranteed tensile properties in specified areas should lead to the elimination of casting factors.

Aluminium alloys

These alloys continue to play an important part in many European undercarriages. For the highest strength, the L77 material was replaced around 1950 by the DTD.5024 alloy in the U.K. and 7075-T6 in America. Forgings in these fully heat-treated alloys can be susceptible to stress-corrosion cracking, a short-coming which contributed to the development of the ultra-high tensile steels as a more satisfactory alternative. This weakness has now largely been over-come by the introduction of duplex ageing to the alloy (DTD.5104, 7075-T73 etc.) but this does involve some sacrifice of strength (figure 2). The most commonly used alloy today is 7075-T73, but more modern alloys, e.g. DTD.5120, the German AZ74, 7050-T736 and 7175-T736, have been developed which have restored the general strength level of the T6 heat treatment while still possessing excellent resistance to stress-corrosion.

Titanium alloys

Despite their excellent corrosion resistance, only limited use has been made of the titanium alloys, probably because their commercial development coincided with the introduction of the less expensive ultra-high tensile steels. The strongest alloy in use today is less than 15% higher in specific-strength than 300M (figure 2) and there are few applications where this advantage warrants the considerable increase in cost of the raw material and of the subsequent manufacture. Some small forgings are made in the IMI 314 and 318 alloys: the largest forging is the pivot-bracket for the Harrier STOL aircraft in IMI 551. The 6–6–2 alloy has been tested successfully in America as a replacement for steel shock-strut cylinders.

Steel

Figure 1 indicates that the biggest advance in the strength of materials has been with the steels. Before around 1950, a tensile strength of 1250–1400 N/mm^2 was considered to be the strongest at which structural steel could be used, due to the claimed lack of sufficient toughness, fatigue life and tolerance to manufacturing defects at higher strength levels. However, evaluation of high-quality 4340 and similar materials heat treated to the higher strength level of 1800 N/mm^2, indicated that transverse ductility and toughness were not significantly reduced. These

TABLE 3. COMPOSITION, HEAT TREATMENT AND MECHANICAL PROPERTIES OF 300M

chemical composition		mechanical properties (transverse)	
carbon	0.39–0.44	0.2% proof stress	1560 N/mm^2
silicon	1.50–1.80		(101 tonf/in^2: 227 klbf/in^2)
manganese	0.60–0.90	tensile stress	1900–2100 N/mm^2
phosphorus†	0.015 max.		(123–136 tonf/in^2)
sulphur†	0.015 max.		(275–305 klbf/in^2)
chromium	0.70–0.95		
molybdenum	0.30–0.45	elongation	5%
nickel	1.65–2.00	reduction of area	25%
vanadium	0.05–0.10		
		heat treatment‡	
normalize	925 °C	1 h	
harden	870 °C	1 h; quench in warm oil (40 °C)	
temper	300 °C	Double temper; 2 h per temper; air cool between tempers	

† S+P 0.025 max. ‡ maximum ruling section: 100 mm.

findings have led to the use of several types of steel, both European and American, heat treated to strength levels of 1800 N/mm^2 and above (Imrie 1970, 1971). A vacuum-melted high-silicon variant of 4340, referred to as 300M, heat treated to a strength level of 1900–2100 N/mm^2 was introduced around 1965 and is now almost the universal choice in the U.S.A. and the U.K.: table 3 details the composition and other requirements. The manufacturing precautions necessary for the satisfactory and reliable use of this class of material, and with particular reference to 300M, have been described in detail elsewhere (Imrie 1971) and are summarized below, together with a brief review of the engineering properties.

Ultra-high tensile steel (300M)

Quality requirements for steel supplies

The steel is produced by the vacuum-arc or electro-slag remelting of air-melted ingots: this remelting lowers the hydrogen content, reducing susceptibility to flaking, and lowers the oxygen content, which minimizes the formation of oxide inclusions. The metallographic cleanness is tested to a high standard in accordance with ASTM Specification E45-69, ultrasonic examination of the billets has to meet Class A of MIL-I-8950B and finally the material must meet the stringent magnetic particle test requirements of AMS 2300.

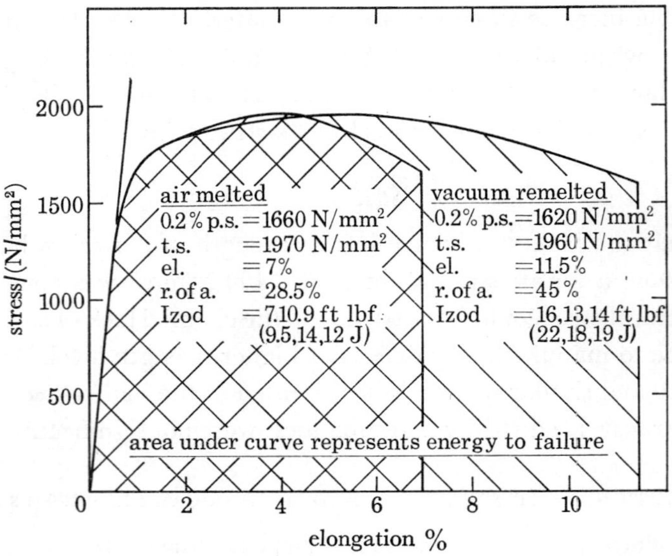

FIGURE 3. Energy absorbed in fracturing air and vacuum remelted 300M-transverse properties.

In addition to a higher degree of consistency in fracture-toughness and fatigue, the principal advantage gained from vacuum melting is an improvement in transverse ductility at these high strength levels. Figure 3 shows typical mechanical test data and stress-strain curves for air and vacuum-remelted 300M: the area under the curves represent the relative energy absorbed prior to fracture.

Processing control

Because of the hazards of machining the steel in the hardened condition, a large proportion of the surface area of parts is finish-machined before heat treatment. It is important to ensure that these surfaces are not functionally carburized or decarburized during heat treatment. Partial carburizing is particularly harmful as only a slight amount can lower transverse reduction of area from 30 to 8%: it also decreases stress corrosion resistance (see below). Decarburization will reduce fatigue life slightly but carburization has no effect. Following an assessment of the relation between micro-hardness and tensile strength, acceptable limits of hardness have been devised (figure 4). Parts in which distortion during conventional heat treatment can be a problem may be aus-bay hardened: minimum distortion is indicative of low residual tensile stresses and thus contributes to service reliability.

Grinding after heat treatment is rigidly controlled in order to ensure surface integrity, otherwise the temperature may become such that the surface layer is transformed to austenite.

Owing to the high rate of cooling from surface to core, brittle untempered martensite is then produced, followed, further in from the surface, by an over-tempered zone. The volume changes associated with these transformations can result in cracking or high residual surface stresses, sometimes as high as 700 N/mm². Particular care is taken to ensure that adequate and smooth radii, etc. are always present and that drill scores in bores, deep machining scratches, etc. are avoided. All parts are examined finally for grinding softening and re-hardening burns by etching in 4% nitric acid.

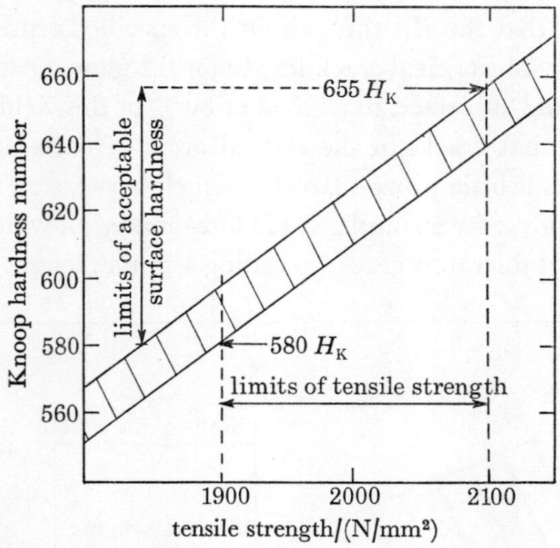

FIGURE 4. Limits of surface hardness indicating freedom from functional carburization or decarburization: 300M.

Following final machining and magnetic flaw detection, all surfaces are shot-peened. Corrosion protection is effected by high current-density cadmium plating, the embrittling effect of the plating bath being regularly monitored. Parts are thoroughly stress relieved before plating and de-embrittled after plating. All external surfaces are painted with epoxide primer and polyurethane finishing coat. Reasonable service maintenance is necessary to ensure that stone damage, etc. is blended and the effected area brush-plated with low-embrittlement cadmium solution in accordance with MIL-STD-865 (Rubinstein 1972).

Fatigue

There is a considerable amount of published data on the notched and unnotched fatigue strength of 300M (Imrie 1971; Aerospace Handbook), including some specifically related to undercarriage application (Harmsworth 1969), which enables the designer to predict service performance. In general, the steel appears to be as good as, if not better than, other comparable materials. The required high standard of cleanness and transverse reduction of area contribute to this satisfactory fatigue performance (Wells *et al.* 1972). The importance of avoiding surface decarburization has already been described, although the reduction in fatigue strength is less at higher stresses than at the fatigue limit. This is of advantage in the case of undercarriage parts in which the significant and potentially damaging fatigue stresses imposed during their design-life are not likely to exceed 10^5 cycles and in many applications are considerably less.

The specified shot-peening effects a significant improvement in notched fatigue life (Campbell 1971). Appreciable data are available concerning fatigue crack growth rate of high strength steels. In general, the propagation rate is directly proportional to the stress intensity range, and is largely independent of composition and strength level (Johnson & Paris 1968). A wet or humid environment contributes towards an even more rapid propagation.

Fracture toughness

Scatter bands of K_{Ic} results obtained from several steels (McKenna & Imrie 1969–72), together with reported data extracted from other sources, are illustrated in figure 5. It will be noted that 300M possesses a generally higher fracture toughness than the other low-alloy steels at comparable strength levels and that the K_{Ic} throughout the specified tensile strength range varies between 50 and 75 MN $m^{-\frac{3}{2}}$. The critical crack-length for the alloy, heat treated to its maximum permitted tensile strength and subjected to a stress of 80 % of the yield, is only in the order of 2.5 mm: thus it is obvious that cracks in the critical areas of parts just cannot be tolerated. However, this state of affairs is little worse than that which has existed for many years with the high strength aluminium alloys: for example, DTD.5024 material, which possesses a K_{Ic} in the order of 16 MN $m^{-\frac{3}{2}}$, will not tolerate a crack exceeding 4 mm in length when similarly stressed.

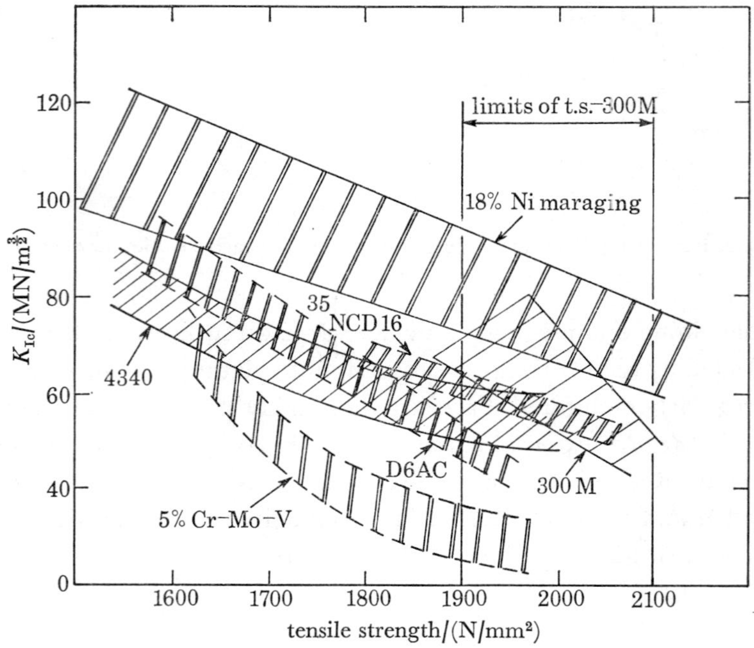

FIGURE 5. Fracture toughness plotted against tensile strength for ultra-high tensile steels.

There is little to choose between any of the available steels from a practical standpoint: unless crack growth rates can be significantly lowered, the magnitude of the differences in K_{Ic} will result in increased critical crack lengths which will take only marginally longer in time to develop. For this reason, the scatter in results, which may well be cause for concern in fail-safe designs, is not of much consequence in undercarriage applications. Even so, if damage tolerance analysis becomes a requirement of the aircraft designer, proposals to release steel with guaranteed K_{Ic} minima (I. E. Fogarty, personal communication) may also become of some importance. There are occasional applications where a relatively small increase in fracture toughness may be of advantage: for example, in the case of a thin-walled hydraulic shock-strut the applied stresses could well be such that the fracture toughness of one material may allow a crack to grow right through the wall resulting in nothing worse than an inconvenient oil leak, while another material of lower K_{Ic} would develop a crack which, after growing part-way through the wall, propagated rapidly to result in unstable shattering of the cylinder.

The bainitic 9Ni–4Co–0.45C alloy, originally claimed to possess a significantly higher fracture toughness, shows no improvement (Gardiner 1972).

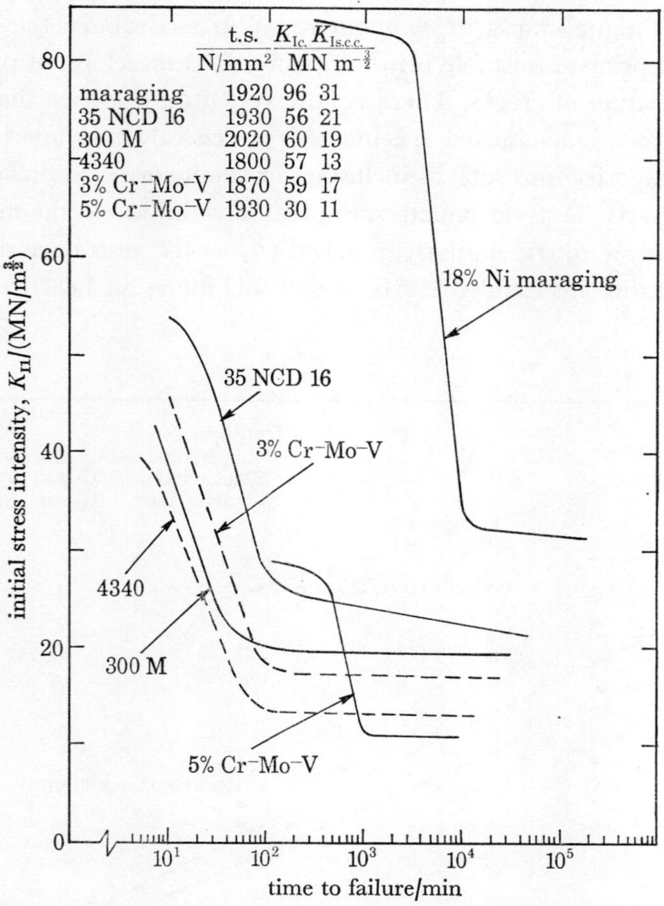

FIGURE 6. Stress corrosion cracking properties of ultra-high tensile steels.

Stress corrosion

Figure 6 indicates $K_{Is.c.c.}$ results obtained from production heats of six steels heat treated to the strengths shown and tested in a cantilever beam rig based on that devised by Brown (1966). All the low-alloy steel values fall within a band of 11–21 MN m$^{-3/2}$ and approximate to one third of the corresponding K_{Ic}: these values are far too low to serve as a basis for design related to a minimum acceptable crack length. Although the higher results obtained for maraging steel agree with those reported elsewhere (Dautovich & Floreen 1973; Benjamin & Steigerwald 1968), it must be borne in mind that some tests on this material, of both U.K. and American (Stavros & Paxton 1970; Carter 1970*a*) manufacture, have shown that it can be no better than the low-alloys, with reported values as low as 11–16 MN m$^{-3/2}$: it is perhaps of some significance that the U.K. test specimens were extracted from 110 mm diameter bar.

The time-to-failure of all the low alloy steels was substantially the same. The higher silicon in the 300M did not significantly retard failure when compared with 4340, as reported by

Carter (1969): similarly, although the 5% Cr–Mo–V material took slightly longer to fail, the increase was not of the magnitude (Steigerwald 1969) found on lower strength material of 1620 N/mm². The rates of crack propagation were not measured: such data for these steels are available (Carter 1969, 1970*b*) but would seem to be of little value because of the short failure times, which varied in these tests from two hours for the low alloys to less than seven days for the maraging steel. It follows that the velocities, combined with the small critical crack sizes, are such that similar fatigue cracks in components, if stressed above the $K_{Is.c.c.}$, are most unlikely to be detected prior to unstable brittle fracture. It is therefore, of paramount importance to avoid the initiation of cracks. There is however, little evidence that stress corrosion data obtained from pre-fatigue-cracked specimens is necessarily relevant to stress-corrosion crack-initiation from imperfections such as inclusions or small corrosion pits on the surfaces of typical undercarriage parts. Tests on smooth specimens have indicated the marked superiority of 300M over H 11 (Dreyer 1962): similarly, notched ($K_t = 4.7$) tests have shown 300M to be superior to 4340, maraging steel and the 9Ni–4Co–0.45C alloy, all heat treated to the 2000 N/mm² strength level.

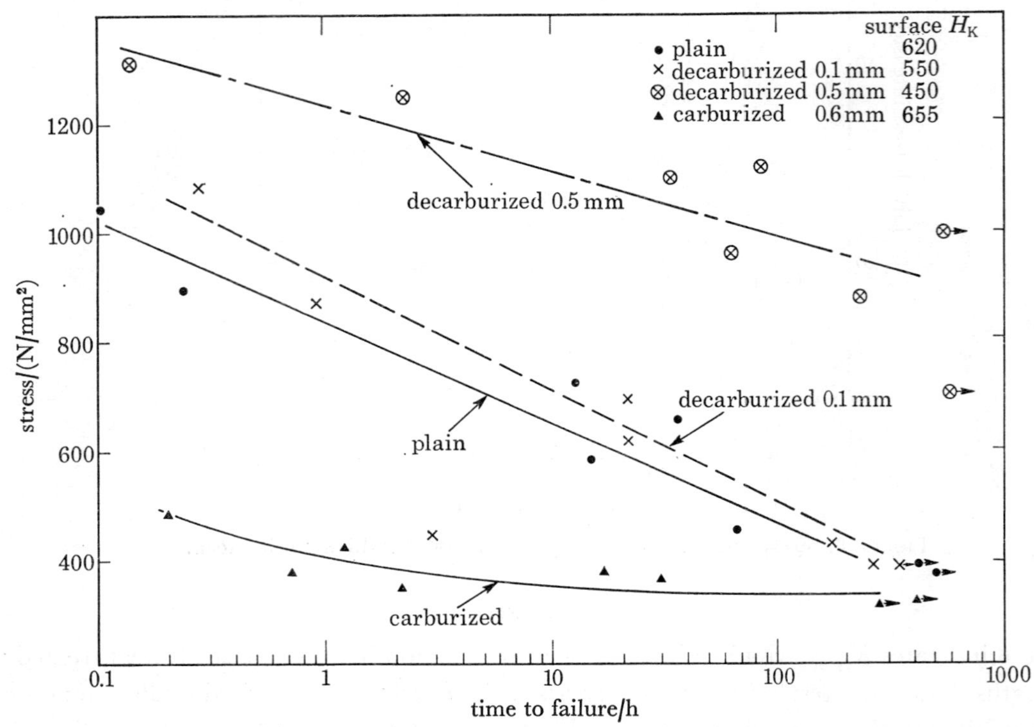

FIGURE 7. The effect of decarburization and carburization on resistance to stress corrosion cracking: 300M.

Slight surface carburization during heat treatment can reduce stress-corrosion resistance (McKenna 1973), figure 7. In these tests the cracks penetrated well beyond the carburized layer and as the crack stress corrosion propagation rate for the substrate material would be the same for all test specimens it is evident that it is the crack-initiation period which has been influenced by the surface condition.

The routine application of shot peening increases resistance to stress corrosion failure (Campbell 1971).

FIGURE 8. Nose undercarriage of the McDonnell Douglas DC-10 aircraft. 70 % of total weight in 300M steel.

Discussion

The general conclusion which could well be drawn from the fracture-mechanics data is that 300M and materials of similar strength are unsuitable because of the fast growth rates and small critical crack lengths. However, the facts are that the performance of these steels in actual service has in general been satisfactory and reliable. An American survey of failures of aircraft parts has indicated that only a small percentage were of high strength steel undercarriage components (Perlmutter & Carter 1973): most of these latter failures were associated with a stress-corrosion/hydrogen embrittlement mechanism, initiated by manufacturing defects. Although insufficient experience exists as yet in the U.K. on which to draw any definite conclusions, it is of interest to note that the only reported failures also originated from stress-corrosion cracking, in this case of maraging steel cylinders. These findings are indicative of the fact that the operating stresses were high and the $K_{\mathrm{Is.c.c.}}$ values of the steels were low. At high stress intensities, the rate of propagation of stress-corrosion cracks greatly exceeds that of fatigue crack growth: sub-critical flaw growth of environmental origin therefore becomes the major problem of very high strength steels, with fatigue probably being of less importance.

Since corrosion prevention would appear to be a major contributor to reliability, a high strength stainless steel is apparently technically desirable. However, evaluation of the developed Carpenter Custom 455 (12Cr–9Ni–2Cu) and the Crucible AFC.77 (15Cr–13Co–4Mo) stainless maraging steels has indicated that their stress-corrosion resistance in thicker section sizes is little better than that of 300M and thus corrosion protection would still need to be applied (Henthorne 1972; Caton & Carter 1973). Additional factors militating against the introduction of the stainless steels are the much higher costs, the generally lower mechanical properties and the practical problems involved in producing crack-free forgings.

It has been shown that improvements in fracture toughness values will not necessarily contribute very much towards reliability: it therefore follows that unless some new technique can be devised for significantly slowing down sub-critical crack-growth rates, the only approach to ensure service reliability is to take every possible precaution to prevent crack-initiation. This is achieved by the described close control of the manufacturing and processing procedures. The likelihood of inclusions acting as initiation sites is minimized by the high standard of quality control demanded during the various stages of manufacture of both the basic steel and the subsequent components. The high specified level of transverse reduction of area is further ensurance against crack initiation, as this property is inversely related to inclusion content (Austin & Goehler 1963). The features described to ensure surface-integrity (de-carburization control, grinding restrictions, stress relieving, shot peening, etc.) all contribute towards the avoidance of cracks. Finally, and possibly the most important in view of the results of the failure survey described above, corrosion pits, hydrogen embrittlement, etc. which could promote the initiation of stress corrosion cracking, are avoided by the carefully controlled plating, painting and maintenance procedures.

Figure 8, plate 6, shows the nose undercarriage of the DC-10 aircraft in which all the major structural components are of 300M steel.

Future developments

Developments of the very high strength steels in general are reviewed at this Conference by May. However, bearing in mind the generally satisfactory performance of 300M, any proposed improvement must take into consideration all the criteria defined in table 1 and not just one of them. For example, some of the developments directed towards increasing strength or fracture toughness by grain refinement, such as thermomechanical working (Zackay & Justusson 1962), high austenitizing temperatures (Zackay *et al.* 1972), or repeated cycles of rapid austenitizing and quenching (McDarmaid 1974), are unlikely to be very practical, economic or successful as applied to large or complex-shaped parts of varying cross-section. Similarly, alloy strengthening by direct transformation of austenite to an ultra-fine dispersion of carbides in ferrite (Honeycombe 1971) is a somewhat long-term possibility in so far as attaining the mechanical properties of table 3 are concerned: even so, this development is to be encouraged as representing a fundamentally different approach to attaining high strength with high toughness.

A modified 300M composition (0.4Mn–2.25Si–0.22V) has been proposed with a claimed minimum strength of 2140 N/mm^2 but the other properties offer no advantage (Ault 1971). A recently reported development is the Latrobe Multiphase MP.159 alloy (Cobalt Astracts) (25Ni–35Co–19Cr–9Fe–7Mo–3Ti–0.6Nb–0.2Al) which is claimed to possess a tensile strength exceeding 1830 N/mm^2 and a high resistance to stress-corrosion, but it is very unlikely that many of the other requirements, table 1, could be met.

It has been shown that worthwhile weight-savings in parts subjected primarily to bending could be made with maraging steel of 2750 N/mm^2 tensile strength (Webster 1971): the IN-765 high-purity steel (13Ni–15Co–10Mo–0.2Ti), figure 2, meets this requirement but the associated ductility is very low (Boesch & Cowan 1968). Even so, the alloy may have some limited future application for shock-absorber cylinders, etc.

In general, efforts to improve fracture toughness, including the evaluation of alloys of very high purity, have not so far been very successful: even in instances where some advantage in K_{Ic} has been obtained, the $K_{Is.c.c.}$ has remained substantially unaltered. In this latter connexion, it has been known for many years that small additions of copper increase the corrosion resistance of mild steel (Hudson 1948) and perhaps the effect on stress-corrosion resistance of similar additions to 300M is worthwhile investigating.

Several proposals have been made for improving on the standard corrosion protection of electrodeposited cadmium plate (Cole 1969), particularly in view of the evidence that hydrogen, liberated by the sacrificial removal of cadmium in a corrosive environment may, under conditions of high stress, enter the steel and accelerate failure by embrittlement (Dreyer 1962; Sink 1968). It has been shown that dispersed nickel particles in cadmium plating solutions reduce hydrogen embrittlement and confer increased protection (Sanderson *et al.* 1974). Aluminium, electro-deposited from non-aqueous solutions is now a proven process and offers superior corrosion and stress corrosion resistance combined with absence of hydrogen embrittlement (Lui & Rush 1972). Although likely to be much more expensive to install, this process may become worthwhile for applying to steels of greater strength than used currently or where solid-cadmium embrittlement (Fager & Spurr 1971) is a possible hazard.

Conclusions

The requirement for minimum weight and bulk in modern undercarriages is being met satisfactorily by the ultra high tensile steels: however, reliability is largely dependent on a very high standard of manufacturing and processing control necessary to ensure maximum resistance to crack initiation. Efforts must continue to improve the intrinsic properties of these steels, with particular emphasis on reducing crack growth rates and improving resistance to failure by stress corrosion.

There is not at the present time any sign of a worthwhile replacement for, or improvement on, the 300M alloy: indeed, the compromise between strength and brittleness of conventional alloys appears to have already been reached.

References (Imrie)

Aerospace Structural Metals Handbook. Vol. 1. Code 1217. Section 3.05. Traverse City, Michigan: Mechanical Properties Data Center.

Ault, R. T. *et al.* 1971 Development of an improved ultra-high strength steel for forged aircraft components. Technical Report AFML-TR-71-27, Wright-Patterson Air Force Base, 1971.

Austin, D. E. & Goehler, D. D. 1963 How inclusions affect ductility of steel forgings. *Metal Prog.* September 1963.

Benjamin, W. D. & Steigerwald, E. A. 1968 Environmentally induced delayed failures in martensitic high strength steels. Technical Report AFML-TR-68-80. Wright-Patterson Air Force Base, 1968.

Boesch, W. J. & Cowan, T. W. 1968 Evolution of a commercial 400 ksi grade maraging steel. Paper Presented at ASM Congress, Detroit, 1968.

Brown, B. F. 1966 New Stress-corrosion Cracking Test Procedure for High Strength Alloys. *Material Res. and Standards.* 6, no. 3, March 1966.

Campbell, J. E. 1971 Shot peening for improved fatigue properties and stress-corrosion resistance. Report MCIC-71-02. Columbus, Ohio: Metals and Ceramics Information Center.

Carter, C. S. 1969 The effect of silicon on the stress corrosion resistance of low alloy high strength steels. *Corrosion* **25**, 423.

Carter, C. S. 1970*a* Evaluation of a high purity maraging steel forging. AFML-TR-70-139, 1970.

Carter, C. S. 1970*b* Observations on the stress-corrosion crack propagation characteristics of high strength steels. Boeing Document D6-25274, p. 12, May 1970.

Caton, R. G. & Carter, C. S. 1973 Evaluation of AFC 77 martensitic stainless steel for airframe structural applications. Technical Report AFML-TR-73-182. Wright-Patterson Air Force Base.

Cobalt Abstracts 1974–1. London: Cobalt Information Centre.

Cole, H. G. 1969 The influence of surface and protective treatments on stress corrosion behaviour. AGARD Conference Proceedings No. 53, Istanbul 1969.

Dautovich, D. P. & Floreen, S. 1973 The stress corrosion and hydrogen embrittlement behaviour of maraging steels. National Association of Corrosion Engineers Conference, Unieux, France, June 1973.

Dreyer, G. A. 1962 Investigation of susceptibility of high strength martensitic steel alloys to stress corrosion. Report ASD-TDR-62-876, Wright-Patterson Air Force Base, September 1962.

Fager, D. N. & Spurr, W. F. 1971 Solid cadmium embrittlement: steel alloys. *Corrosion* **27**, no. 2.

Gardiner, R. W. 1972 Fracture Toughness of 9Ni–4Co–0.45C Steel in Bainitic and Martensitic Conditions. Royal Aircraft Establishment, Farnborough. Technical Memorandum MAT.149, 1972.

Harmsworth, C. L. 1969 Low cycle fatigue evaluation of titanium 6Al-6V-2Sn and 300M steel for landing gear applications. Technical Report AFML-TR-69-48. Wright-Patterson Air Force Base 1969.

Henthorne, M. 1972 Stress corrosion cracking of martensitic precipitation hardening stainless steels. AGARD Conference, Brussels. Proceedings No. 98, 1972.

Honeycombe, R. W. K. 1971 A new approach to the design of alloy steels. Iron and Steel Institute Special Report 145, 1971.

Hudson, J. C. 1948 Corrosion of iron and steel wires. *J. Iron and Steel Inst.* **160.**

Imrie, W. M. 1970 Maraging steel in the British Aerospace Industry. *Metal Forming.* January and February 1970.

Imrie, W. M. 1971 Ultra high tensile landing gear components – material selection and manufacture. *Jl R. aeronaut. Soc.* **75**, February.

Johnson, H. H. & Paris, P. C. 1968 Sub-critical crack growth. *Engng Fracture Mech.* **1.**

Lui, K. & Rush, P. 1972 Application of electroplated aluminium as a protective coating for fasteners. Presented at Advanced Technology Session, Westec Conference, Los Angeles, March 1972.

McDarmaid, D. S. 1974 Effect of austenite grain refinement on the plane strain fracture toughness and other mechanical properties of 300M steel. Royal Aircraft Establishment, Farnborough, Technical Report 74025. February 1974.

McKenna, J. 1973 Dowty Rotol Limited Materials Laboratory Report 73/9003.

McKenna, J. & Imrie, W. M. 1969–72 Dowty Rotol Limited Materials Laboratory Reports 69/9005, 70/9006, 72/9010, 72/9011 and 72/9014.

Perlmutter, I. & Carter, C. S. 1973 Material factors and reliability in the design of landing gear. TTCP Symposium, Cambridge. June 1973.

Rubinstein, M. 1972 Metal put-on tool solves complex aircraft maintenance problems. Presented at F.A.A. Maintenance Symposium. Oklahoma City. November 1972.

Sanderson, G. *et al.* 1974 Cadmium coatings containing dispersed nickel particles for the protection of high strength steels against corrosion. M.O.D. Report, D.Mat. No. 197, July 1974.

Sink, G. T. 1968 *Plating.* **55**, 449.

Stavros, A. J. & Paxton, H. W. 1970 Stress-corrosion cracking behaviour of 18 % Ni maraging steel. *Metall. Trans.* **1**, November 1970.

Steigerwald, E. A. 1969 Characteristics of environmentally-induced delayed failure in high strength steels. AGARD Conference Proceedings No. 53. Istanbul 1969.

Webster, D. 1971 Optimisation of material properties to achieve structural efficiency. *Metals and Materials.* August 1971, p. 257.

Wells, M. G. H. *et al.* 1972 The relationship of inclusions to toughness – ductility parameters in high strength steels. ASM Meeting. Cleveland 1972.

Zackay, V. F. *et al.* 1972 *Nature, Phys. Sci.* **236**, 108.

Zackay, V. F. & Justusson 1962 The properties of martensitic steels formed from strain-hardened austenite. High Strength Steels Conference, Harrogate 1962. Iron and Steel Institute. Special Report 76.

1.6 Gas turbine requirements

By R. J. E. Glenny† and B. E. Hopkins‡
† *Royal Aircraft Establishment, Farnborough, Hants*
‡ *National Physical Laboratory, Teddington, Middlesex*

The design of gas turbines varies widely according to the requirements of the application and the conditions of service and so, consequently, do the properties necessary in the materials of construction. The main variations on the basic cycle are discussed together with the features that may influence the choice of gas turbine. The importance of the two main operational variables, pressure ratio and turbine entry temperature, is emphasized. Aspects of the design of the main components are considered in relation to the properties required of the materials from which they are made. Higher temperatures and longer lives increase the demands on the metallurgical stability of the materials and their resistance to oxidation and hot corrosion.

Introduction

One of the outstanding features of the gas turbine is its versatility. This is apparent from the wide range of the applications during the 35 years or so of the existence of the gas turbine as a practical prime mover. The variety of forms and duties is such that it is not possible in a short presentation to do more than deal briefly with the main types of gas turbine available, indicate the importance of particular operational variables, and consider the design features and requirements of selected parts. The choice of components for discussion is based largely on their metallurgical or engineering interest, and some emphasis is placed on those operating at high stress and/or high temperature. A broad view is taken of the requirements for gas turbines in general but it is inevitable that aero gas turbines predominate because their state of development occupies a leading position.

General features of gas turbines

The versatility of the gas turbine arises both from the basic characteristics of the machine and the variations that can be introduced into the simple cycle to enhance particular features of the performance to suit the application. Figures 1*a–d* show some variations on the simple open cycle (Wigg 1955; Shepherd 1960; Simmons 1968) in which the output is in the form of shaft power which may be taken from either end of the machine. The splitting of the turbine duty into two parts that are not mechanically connected (figure 1*b*) introduces a degree of flexibility not obtainable with single shaft machines.

The use of a heat exchanger as shown in figure 1*c* to heat the working fluid by the exhaust gases before it enters the combustion system is an obvious way of increasing thermal efficiency which is essential if the gas turbine is to compete successfully in some applications, e.g. where power/weight ratio is a secondary consideration. Increased efficiency is obtained usually at the expense of pressure losses and substantial increase in size and cost, but considerable advances have been made with rotary ceramic heat exchangers for automotive turbines (Barnard 1974).

The work done by the compressor for a given pressure ratio may be reduced by cooling the air between two compression stages as shown in figure 1 *d*. The main effect is to increase the specific work output with some increase in efficiency at high pressure ratios. Intercooling is usually done with an external water supply and has the advantage of compactness.

Another possible variation is reheating of the partly expanded gases in the turbine. It is somewhat less effective than intercooling and, in view of the complications introduced, it has not as yet found much favour except for aero gas turbines.

The complex cycle, which incorporates intercooling, heat exchange and reheat, is capable

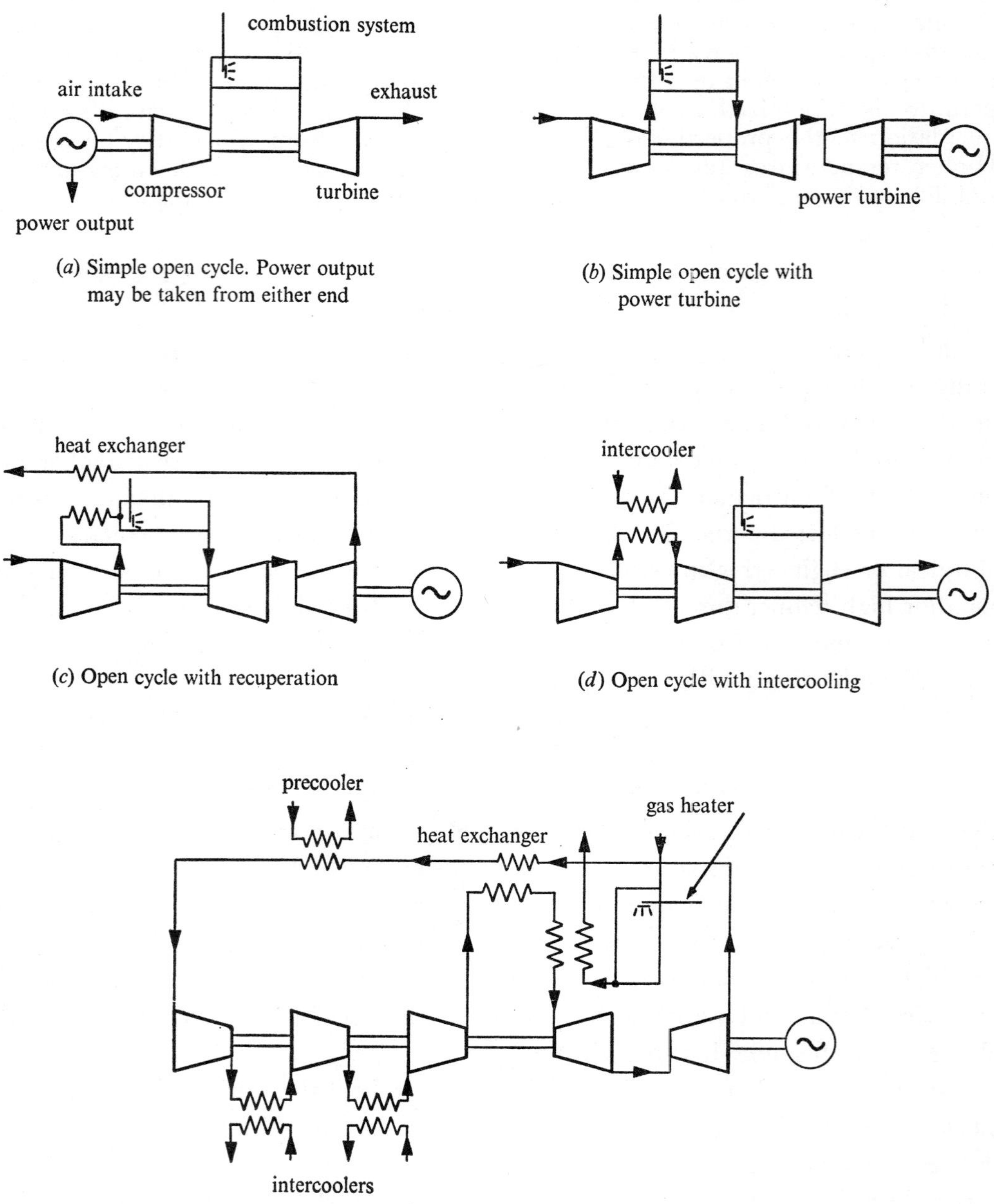

(*a*) Simple open cycle. Power output may be taken from either end

(*b*) Simple open cycle with power turbine

(*c*) Open cycle with recuperation

(*d*) Open cycle with intercooling

(*e*) Closed cycle

FIGURE 1. Some variations on open and closed cycle gas turbines.

of giving a high efficiency over a range of pressure ratios and a high work output. A number of arrangements are possible to suit the requirements.

The closed cycle, an example of which is shown in figure 1*e*, has a number of distinct features that affect the design of the plant. The working fluid, which need not be air, is recirculated without change in composition and has to be cooled before entering the compressor. The combustion system of the open cycle is replaced by external heating which incurs practical problems that limit the maximum temperature of the gas to levels below those usually obtainable with an open cycle. The method of controlling the output differs from that in the open cycle where both the speed of the compressor (and hence the pressure ratio of the gas) and the turbine inlet temperature are varied so that part-load efficiencies are lower than at the design point. With the closed cycle, the output can be controlled by the datum pressure level of the system by means of a reservoir. The mass flow rate is thus varied but the pressure ratios and temperatures are constant so that it is possible to operate at the design point irrespective of load, and maximum efficiency is thus maintained. The minimum pressure can be above atmospheric so that high outputs can be obtained with relatively small units. The closed cycle has considerable potential for nuclear power plants (Worley & Campbell 1970) where helium could be used as the coolant in the reactor and the working fluid in the turbine. The design of the components of the gas turbine, such as the compressor and the turbine, is much influenced by the physical properties of the working fluid so that sets operating on a closed cycle may differ considerably from those operating on an open cycle.

Jet engines represent a distinct category in that the output is a high velocity gas stream rather than shaft power, and so the expansion is carried out in two stages, the first through a turbine to provide just sufficient work to drive the compressor, the second through a nozzle to provide a high velocity jet. Of the variations to the basic cycle for shaft power engines outlined above, namely heat exchange, intercooling and reheat, heat exchange is not used for jet engines; short-term cooling (and increased mass flow rate) may be used in the compressor by spraying water or a water-methanol mixture into the intake to give improved thrust on take-off; and reheat may be used, for military aircraft in particular, to give increased thrust (at low efficiency) for take-off, climb and in combat situations. The design of the exhaust system is influenced by reheat, which requires variable propelling nozzles, and by other factors such as thrust reversal, vectored thrust and noise suppression.

There is also some variety in the type of component available, e.g. the compressor, and in the type of fuel that can be burnt, although in the latter respect the choice is more limited than was originally anticipated because of the fouling and corrosion that are apt to occur with some fuels.

The choice of plant is naturally influenced by features of the gas turbine of most relevance to any particular application. Basically, the gas turbine has outstandingly low weight and volume for a given output when compared with other prime movers except when high efficiency is required and complex cycles or heat exchange of high effectiveness are resorted to. Low weight is important in all aircraft applications but particularly so in lift engines. The range of plants available continues to increase from small units having outputs of tens of horsepower (several kilowatts) to land-based installations of around 100000 hp (75 MW) and jet engines of over 40000 lbf (180 kN) thrust.

The infrequency and low cost of maintenance, low installation costs and relative independence of auxiliary services are factors that may have an over-riding influence on the choice of gas turbine. Comparative ease of starting in cold conditions or by remote control, coupled with a short warm-up time, may be important in some applications. The generally high self-running conditions, which require the supply of considerable power for starting, have to be catered for but can be influenced by the design of the components.

Importance of pressure ratio and temperature

The degree of compression of the working fluid achieved in the compressor and the turbine entry temperature are two basic variables in all gas turbines (Bragg 1952; Shepherd 1960). Figures 2*a* and *b* show the variation of the heat added in combustion and the work quantities for a simple open cycle for two turbine entry temperatures. The work output is seen to be a maximum at an intermediate pressure ratio, and the thermal efficiency is a maximum at a higher pressure ratio, so that in any particular application the relative importance of work output and efficiency have to be considered. Comparison of figures 2*a* and *b* shows that the work output is increased considerably and the efficiency to a smaller extent by increasing the ratio of the maximum to the minimum temperature of the cycle. This emphasizes the importance of a high turbine entry temperature for a given compressor entry temperature. The importance of high efficiencies in the compressor and the turbine can also be deduced from

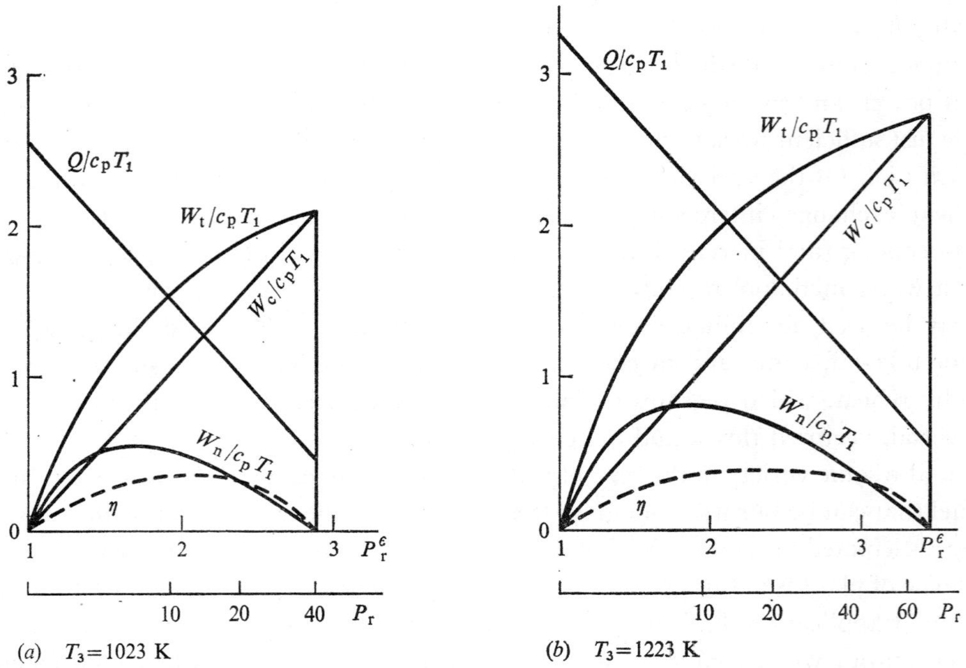

FIGURE 2. Heat and work quantities for a simple open cycle showing the effects of pressure ratio and turbine entry temperature.

W_c, work done in compressor; W_t, work done by turbine; W_n, work output $= W_t - W_c$; Q, heat added in combustion; η, thermal efficiency; ϵ, $1-1/K$ where

$$K = \frac{\text{specific heat at constant pressure}}{\text{specific heat at constant volume}};$$

C_p, specific heat at constant pressure; T_1, compressor entry temp. $= 288$ K; T_3, turbine entry temperature; P_r, pressure ratio in compressor.

Small losses neglected and constant values assumed for compressor and turbine efficiencies and for C_p and K.

figure 2 because the work output is the difference between two large quantities and is much influenced by the component efficiencies. The variation of efficiency and output with design point temperature and pressure ratio for a simple open cycle is illustrated in figure 3.

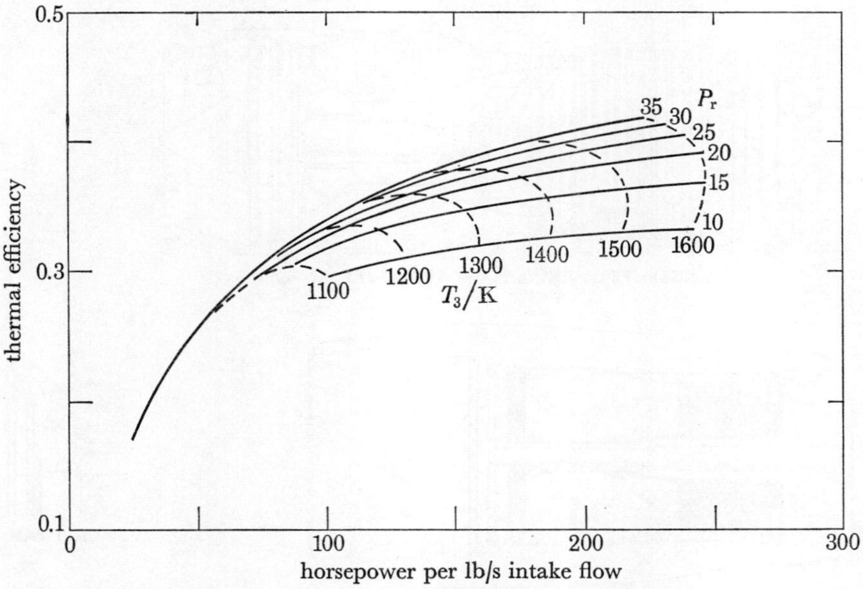

FIGURE 3. Variation of efficiency and output with design point temperature and pressure for a simple open cycle.

COMPRESSORS AND TURBINES

In compressors and turbines the effectiveness of the transfer of energy to and from the gas respectively is governed by the fluid dynamics of the system so that their design is largely the province of the aerodynamicist.

Radial flow or centrifugal compressors are used mainly in small shaft power units of comparatively low pressure ratios where their simplicity, ruggedness and light weight are advantageous, the overall diameter can be accommodated and maximum efficiency is not an overriding criterion. There are some combined axial-radial flow compressors in use. However, axial flow compressors are the most common and highly developed. They are relatively compact, have a small frontal area for a given thrust which is important in aero engines, and are capable of high mass flows and pressure ratios. There is a limit to the pressure ratio usefully achieved on a single shaft, and the design of aero engines has progressed from single-spool (e.g. the Avon) to two-spool (e.g. the Conway and Olympus) and three-spool machines (e.g. the RB 211), each shaft being driven by a turbine (see figure 4).

The moving blades in an axial compressor are carried on disks or a drum or a combination of both. The intervening stator blades are carried by the outer casing or by retaining rings which are attached to the outer casing, and they may be of variable incidence at the intake. The longer stator blades of the early stages may be shrouded to reduce vibration due to variations in the air flow.

In modern aero engines, part of the air from the low pressure compressor of a twin spool assembly, or from a fan on single, twin or triple spool assemblies, by-passes the combustion system and turbine and makes its own contribution to the thrust. This increases the propulsive efficiency of an aircraft and, in particular, narrows the gap in efficiency between a pure jet and a turbo-prop aircraft at speeds below about 720 km/h (Rolls Royce (1971) Ltd, 1973). The by-pass ratio may be 6:1 or higher in the most recent aero engines (Sawyer's Gas Turbine Catalogue 1974).

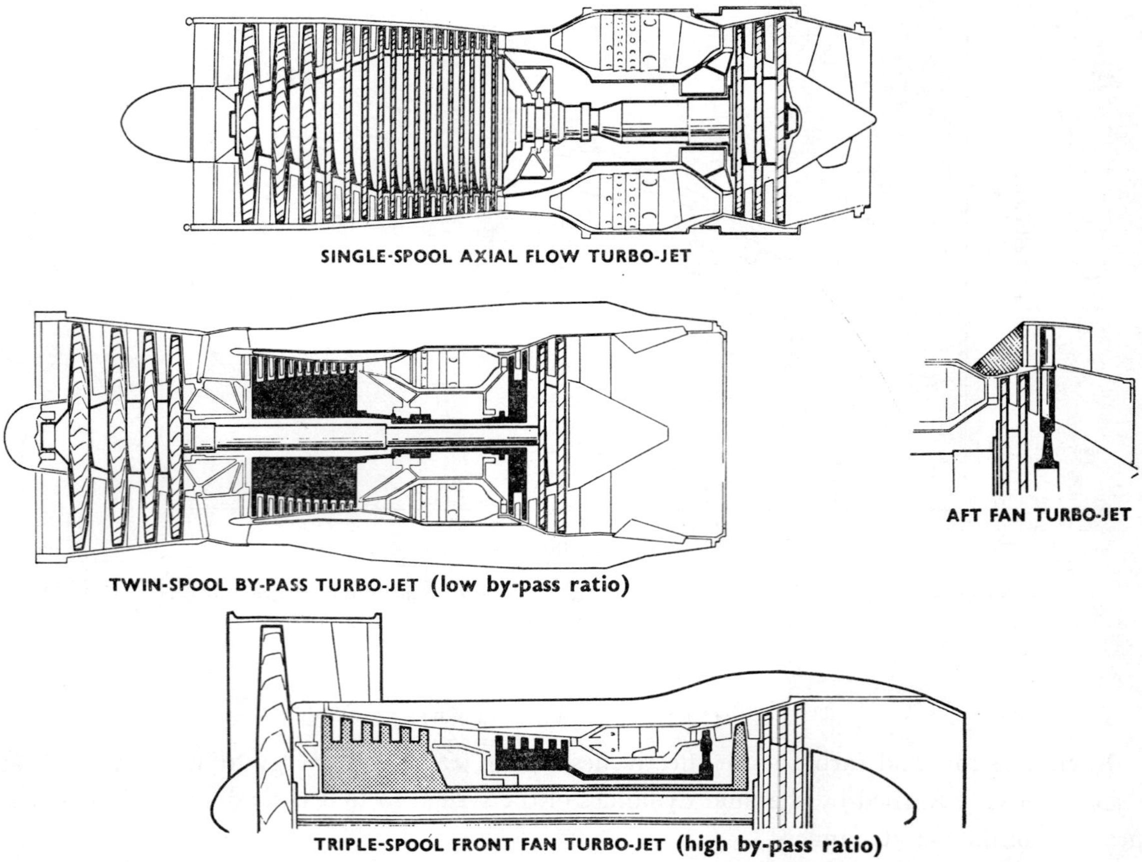

FIGURE 4. Mechanical arrangement of some turbo-jet engines (Rolls-Royce (1971) Ltd, 1973).

The turbine and the compressor have a number of features in common, and similar remarks apply to the relative advantages of radial and axial flow in turbines as in compressors. The design of axial turbines in particular is much influenced by air cooling. In addition to pressurizing oil seals and cooling the lubricating oil, some of the air from the compressor is used to cool externally various parts such as the turbine disk faces and the turbine blade roots. It is common practice for the turbine blades to be provided with platforms that interlock from one stage to the next and so give a continuous cooling annulus between the hot gas and the rotor. Internal cooling of moving blades and nozzle guide vanes (stator blades) in the higher pressure stages of the turbine enables the turbine entry temperature of the gas to be increased very considerably. The technique has been used in production aero engines for civil use in the U.K. since about 1960. Cooling technology has developed progressively so that in modern aero engines the turbine entry temperature of the gas may be up to 300 °C higher than metal temperatures. Turbine entry temperatures are currently around 1550 K in the RB 211 series of engines while temperatures in the range 1650–1800 K are used in development work (Fulton 1975).

Cooling is especially important with supersonic aircraft where the compressor entry temperature of the air may be 150 °C at the supersonic cruise condition so that the turbine entry temperature has to be increased correspondingly to maintain the same performance as in subsonic flight (Hooker & Garnier 1964).

Figure 5 gives an example of the variation of the cooling effectiveness with the proportion of cooling air. Defining cooling effectiveness as $(T_g - T_b)/(T_g - T_c)$, where T_g, T_b and T_c are the mean gas, blade and cooling air temperatures, an effectiveness of 0.5 enables an increase in T_b from alloy development to provide about twice this increment in gas temperature (Glenny, Northwood & Burwood-Smith 1975).

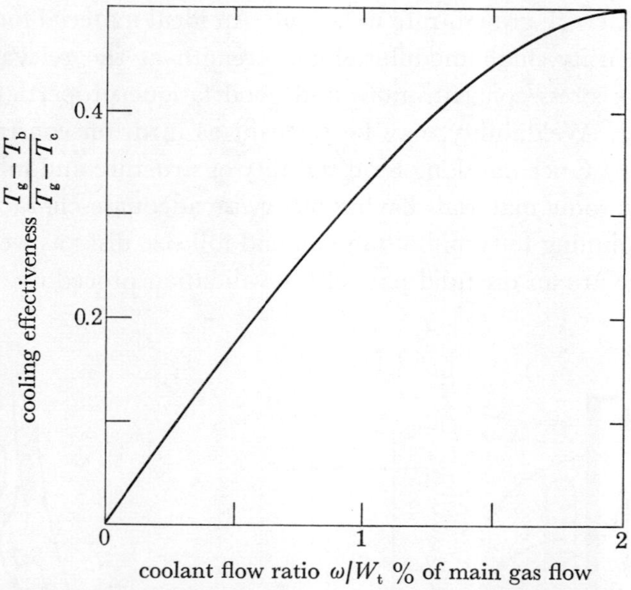

FIGURE 5. Variation of cooling effectiveness of air-cooled turbine blade (Hooker & Garnier 1964). T_b, blade metal temperature; T_c, cooling air temperature; T_g, turbine entry temperature; ω, weight of cooling air; W_t, weight of hot gas.

DISKS

The basis of the design of disks has advanced through a number of stages. Initially the burst speed was the major criterion. The values of the stresses calculated assuming completely elastic behaviour, with the usual allowances for stress concentrations at holes and blade-fixing grooves, are unrealistically high because stress relaxation occurs by plastic deformation. Reasonably accurate values for bursting speed are obtained when the redistributed peak tangential stress is equated to the true stress at the ultimate in a tensile test (Islip 1967). It was formerly the practice to set the design speed to give a proportion of the bursting stress so calculated, say 50%, to allow for other factors such as fatigue.

Although burst speed is still considered, more emphasis is now placed on performance of the material in low cycle fatigue for the stresses arising from a start-up/shut-down cycle, which are both mechanical and thermal in origin (Hall 1967; Waters & Norris 1974). There may still be some uncertainties in the values of the stresses, their variation with time and the importance of strain cycling so that a factor of safety is incorporated. It is the onset of a detectable crack that is important. A life of 30000 cycles is a typical requirement for civil aircraft. The rate of propagation of the crack must be such that growth to a critical size cannot occur within one inspection period.

Other factors that have to be considered include whether creep and distortion due to blade loading are likely to impose limitations on disk performance. High frequency stresses arising from blade vibration or other causes have to be effectively designed-out because of the very large number of cycles involved, so that the behaviour of the material in high cycle fatigue is not usually a problem.

In all the rotating parts low density and high elastic modulus are desirable characteristics so that materials are assessed on the basis of specific strength and modulus over the whole range of working temperature. There is also a general connexion between elastic moduli and certain mechanical properties, e.g. crack growth rate in fatigue. An ideal material for disks (Cockcroft 1973) would have low density, high modulus, high strength at the relevant temperatures, sufficient ductility to relax stress concentrations and good fatigue properties including crack propagation characteristics. Weldability may be relevant as in drum constructions for compressor disks. Oxidation resistance and long term stability of structure and mechanical properties may be problems with some materials having otherwise adequate characteristics.

Overspeed and cyclic spinning tests on both model and full size disks as well as the building up of experience in service are an essential part of the validation procedure.

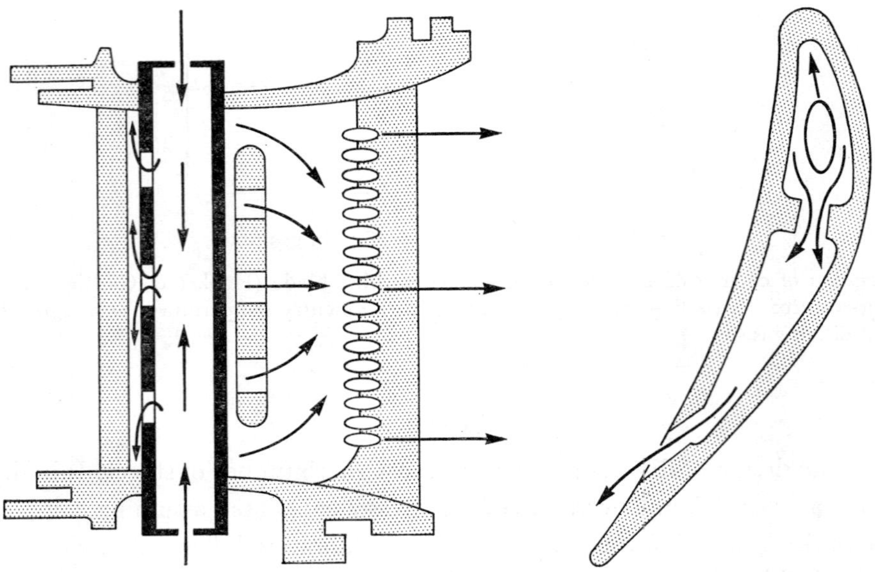

FIGURE 6. Cooled nozzle guide vane for the Conway engine (Hare & Malley 1966).

TURBINE VANES

The first stage nozzle guide vanes are normally the hottest components in the turbine. The peak temperature may vary from one vane to another, e.g. because of non-uniformity in the temperature pattern at the exit from a multiple combustion chamber (see p. 63). It may also be at a different position along the span in different vanes. The problem in the design of cooled nozzle guide vanes is thus one of providing adequate cooling for the worst condition (Hare & Malley 1966). The first stage vanes are normally supported at both ends and are subjected to a relatively low bending stress by the gas pressure. The primary purpose of the cooling is to keep the highest metal temperature below that at which oxidation would be excessive; the material then usually has adequate strength to resist the mechanical loads.

Although temperatures are lower at the second stage vanes, the stresses may be considerably higher as in the Spey engine where the vanes are cantilevers with the outer ends fixed and the inner ends carrying an interstage seal subject to the drop in gas pressure across the vanes (Hare & Malley 1966). The primary object of the cooling in this case is to lower the temperature to a level at which the creep strength is adequate with resistance to oxidation an important secondary consideration. Figure 6 shows a well developed cooled nozzle guide vane for the Conway engine.

Similar considerations apply to the requirements for uncooled vanes.

Thermal stresses can result in distortion and thermal fatigue cracking at the leading and trailing edges of guide vanes, but as the steady mechanical stresses are lower than in moving blades, they can be tolerated to a greater degree before a vane becomes unserviceable. Thermal fatigue problems tend to be alleviated by air cooling because there is less constraint on the surface layers during the heating cycle and compressive stresses are developed in leading and trailing edges but the chordwise temperature variation may be much larger than in uncooled vanes.

Failure in nozzle guide vanes (Hare & Malley 1966; Sullivan, Varin & Donachie 1969; Restall 1975) may take the form of distortion or cracking due to repeated thermal shocks, excessive bowing due to the gas bending stress, oxidation or erosion of the leading edge or, sometimes, local melting at hot spots arising from irregularity of combustion. Although internal cooling and external protection reduce the incidence of local melting and corrosion respectively, more refractory alloys are always under consideration.

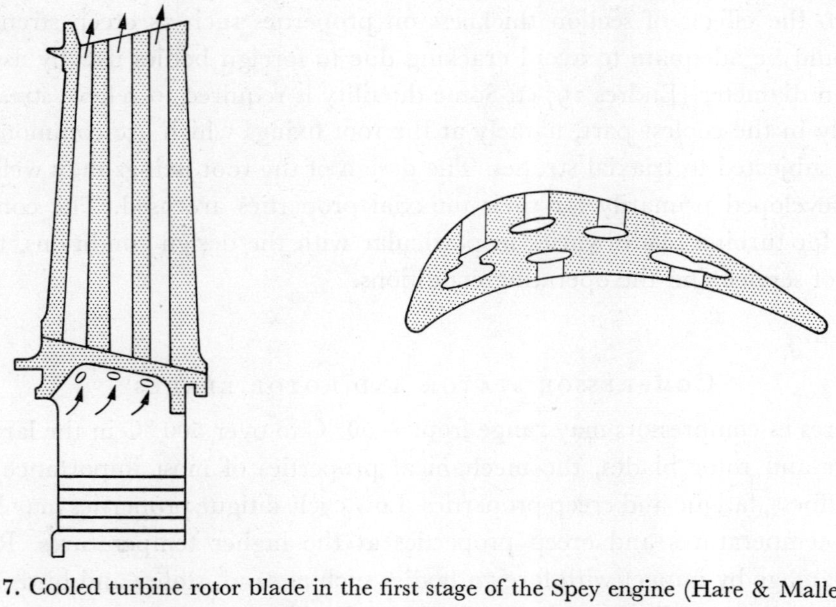

FIGURE 7. Cooled turbine rotor blade in the first stage of the Spey engine (Hare & Malley 1966).

TURBINE BLADES

The high pressure turbine blades operate in the most stringent conditions. Unlike nozzle guide vanes, all the blades of a particular stage experience the same temperature distribution because of their rotation. The peak temperatures are lower than for the vanes by up to 130 °C (Hare & Malley 1966) in the first stage but the rotating blades have to withstand a high centrifugal stress. While the average chordwise temperature may be a maximum near the middle of the span, the centrifugal stress increases from the tip to the root of a blade and the section of minimum creep life may be at a quarter of the length of the blade from the root (Hare & Malley 1966). Typically the stress in this region varies from 120–200 N/mm^2 (8–13 tonf/in^2) for an h.p. blade of density 8 g/cm^3 depending on the stage and the design of the blade (Glenny *et al.* 1975).

Figure 7 shows the design of cooling passages in the first stage rotating blade of the Spey engine. The creep life of an uncooled blade can be estimated simply from the stress, temperature and creep rupture strength of the material. For a cooled blade there are usually large differences in temperature across a chord section that result in thermal stresses the magnitude of which may be up to three times that of the centrifugal stress. Large tensile stresses are developed in the cooler core and low tensile or compressive stresses in the hotter leading and trailing edges. These stresses relax with time due to creep (Ainley 1957; Barnes 1967). The problem of estimating creep life is thus much more complicated for a cooled blade, and a finite element method of doing this has been described (Halls & Baker 1969; Hare & Malley 1966). Incorporation of thermal stresses arising from heating/cooling cycles further complicates the estimation of the life in view of the changes with time in the chordwise temperature pattern and of effects on the behaviour of the material.

Creep properties, low cycle (including thermal) fatigue and oxidation resistance are factors of primary concern for the aerofoils of rotating blades. Cooled blades are complex in shape and are expensive to produce so that ability to manufacture at an acceptable cost is particularly important. Wall thicknesses are less in cooled than in uncooled blades, and so consideration has to be given to the effects of section thickness on properties such as creep strength. Impact resistance should be adequate to avoid cracking due to foreign bodies usually assumed to be about 1 mm in diameter (Endres 1974). Some ductility is required to relieve stress concentrations especially in the coolest part, namely at the root fixings which are commonly of fir tree form and are subjected to triaxial stresses. The design of the root fixings may well be affected if materials developed primarily for their uniaxial properties are used. The combination of requirements for turbine blades varies in particular with the design conditions, the expected life, the type of service and the operating conditions.

Compressor stator and rotor blades

Temperatures in compressors may range from −50 °C to over 500 °C in the largest engines. For the stator and rotor blades, the mechanical properties of most importance are specific tensile and stiffness, fatigue and creep properties. Low cycle fatigue properties may be dominant at the lower temperatures and creep properties at the higher temperatures. Resistance to erosion and damage by impact with foreign bodies such as sand, stones and birds are required in varying degrees depending on the application. Resistance to corrosion is required; the conditions may be severe as in applications where some ingestion of salt is hardly avoidable. Joining may require specialized techniques (Meleka 1968). When combustion of the material (Ti alloys) is a possibility, means have to be employed to reduce the risk to a low level and to contain a fire within the engine.

Combustion systems

The function of the combustion system in open cycle gas turbines is to heat the air from the compressor to the design turbine entry temperature by internal combustion of fuel which is normally gaseous or liquid. The requirements (Shepherd 1960) include minimum loss of pressure, as complete combustion as possible, reliability of ignition, acceptable temperature distribution at the exit which is not always easy to achieve, reasonable velocity distribution which is easier to achieve in most designs, minimum volume and weight (particularly important in aero engines) and durability/reliability. The last is obviously important because of damage to the turbine should parts of the combustion system become detached.

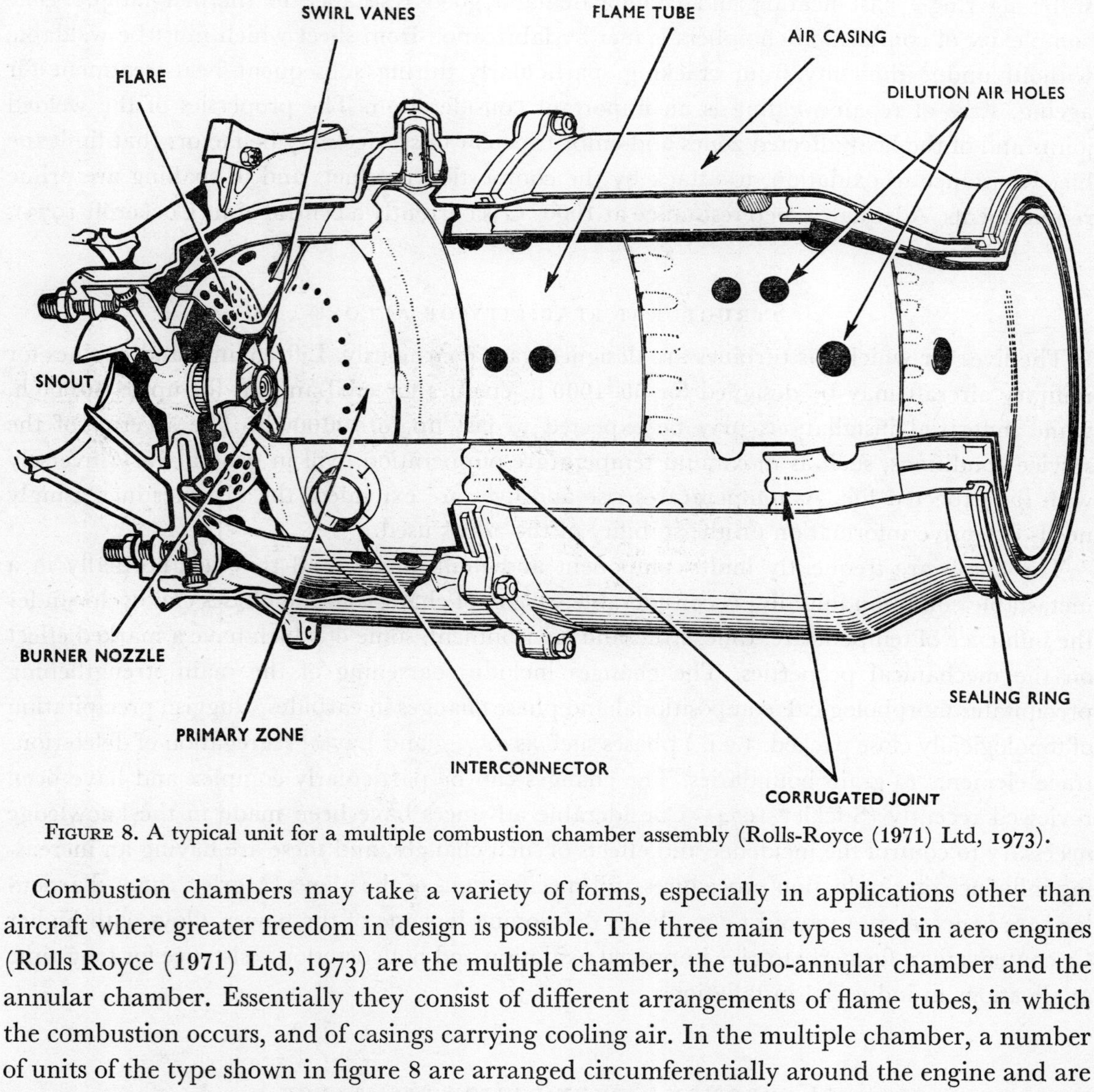

Figure 8. A typical unit for a multiple combustion chamber assembly (Rolls-Royce (1971) Ltd, 1973).

Combustion chambers may take a variety of forms, especially in applications other than aircraft where greater freedom in design is possible. The three main types used in aero engines (Rolls Royce (1971) Ltd, 1973) are the multiple chamber, the tubo-annular chamber and the annular chamber. Essentially they consist of different arrangements of flame tubes, in which the combustion occurs, and of casings carrying cooling air. In the multiple chamber, a number of units of the type shown in figure 8 are arranged circumferentially around the engine and are interconnected. In the tubo-annular chamber, the flame tubes have common inner and outer air casings of annular form. In the annular chamber there is a single annular flame tube with inner and outer air casings. The annular chamber has a number of advantages including the best combustion efficiency and compactness thus reducing weight and cost.

Typically, with paraffin as the fuel, about 30 % of the air from the compressor is supplied for combustion which should take place at as high a temperature as possible (1800–2000 °C). The remainder of the air is used to cool the flame tube walls and to dilute and cool the combustion products to the required turbine entry temperature. The temperatures quoted for the flame tube walls for aero gas turbines are an average of 800 °C with hot spots at higher than 1000 °C (Cockcroft 1973). The flame tubes are usually entirely metallic but ceramic liners are used successfully in some industrial units (Simmons 1968; Endres 1974).

The pressure of the air outside the flame tube is greater than that of the gases inside so that creep strength over a range of temperatures is required to avoid buckling with the consequent development of hot spots, which can occur in spite of good design involving devices such as stiffening rings. Fast heating and cooling demand good resistance to thermal fatigue. The complexity of combustion chambers is met by fabrication from sheet which must be weldable without undue difficulty from cracking, particularly during subsequent heat-treatment for ageing. Ease of repair welding is an important consideration. The properties of the welded joints and of the heat affected zones and embrittlement in service may be factors that limit the life. Resistance to oxidation, to attack by the combustion products and to spalling are prime requirements. Adequate creep resistance at 1000 °C is currently a central aim (Cockcroft 1973).

Structural stability of alloys

The lives for which gas turbines are designed vary enormously. Lift engines and engines for military aircraft may be designed for 50–1000 h, engines for civil aircraft for up to 30000 h, while industrial installations may be expected to last up to 100000 h. The severity of the service conditions, such as maximum temperature of operation, will in general vary inversely with the expected life. As temperatures rise and lives are extended, the designer increasingly needs extensive information on the stability of the alloys used.

The alloys are frequently multi-component and multi-phase, and they are normally in a metastable condition initially. A considerable variety of changes in the phases can occur under the influence of temperature, time, stress and environment, some of which have a marked effect on the mechanical properties. The changes include coarsening of the main strengthening precipitate; morphological, compositional and phase changes in carbides; sluggish precipitation of topologically close packed (t.c.p.) phases such as σ, χ, μ and Laves; segregation of deleterious trace elements to grain boundaries. The changes can be particularly complex and have been reviewed recently (Stickler 1974). Considerable advances have been made in the knowledge necessary to control the incidence and effects of such changes, and these are having an increasing influence on production procedures and specifications of the alloys (Morley 1974). Particular care is necessary to avoid t.c.p. phases developing in some of the newer alloys with higher Cr contents than formerly to give improved oxidation and hot corrosion resistance for long-time applications in industrial installations.

Corrosion and oxidation resistance

The corrosive conditions in open cycle gas turbines depend on the quality of the air and the fuel used for combustion. In the compressor the main problems arise from salt in the air, and they are met in particular in marine and hovercraft engines. Effective designs of filters are necessary to reduce the salt content of the air to low levels, regular washing or grit blasting procedures are used to remove deposited salt, and protective coatings of various types depending on the material and the temperature of operation have been developed (Nolte 1966; Stoeckley 1966; Hardy 1966; Spiers & Weinstein 1969). Pitting corrosion can have a serious effect on the fatigue properties as in steel compressor blades (Speirs & Weinstein 1969) but hot salt corrosion of Ti alloys does not appear to have emerged as a problem in service even though it is encountered in the laboratory (Glenny *et al.* 1975).

The main impurities in the fuel that can cause corrosion by the combustion gases are sulphur and vanadium compounds while sea water may be a contaminant of the fuel or the air. The air is also a possible source of sulphur compounds.

Distillate fuels as used for aero engines have a low ash content so that corrosion by vanadium pentoxide or sodium vanadates is avoided. The view has been expressed (Felix 1974) that the importance of corrosion by vanadium is decreasing in gas turbines in general because of the greater availability of fuel with low vanadium contents.

On the other hand, combined attack by chlorine and sulphur is a serious problem to which much attention is being devoted (Stringer 1972; Stringer & Whittle 1974). Improvements are being sought in the corrosion resistance of the basic alloys and in the protective coatings required for oxidation resistance at the highest temperatures of nozzle guide vanes and rotor blades of the h.p. turbine in advanced engines. Coatings so far developed are of the diffusion type and, therefore, have a limited life-temperature capability. The effects of coatings on the mechanical properties of the basis alloy have to be considered; as for corrosion and oxidation resistance the alloy/coating combination is best considered as a unit. In particular, the ductility of a coating needs to be sufficient to avoid unduly deleterious effects on thermal fatigue and creep properties and resistance to damage by impact of foreign bodies.

In many stationary installations the life of the hottest components is determined by corrosion resistance rather than mechanical properties (Felix 1974), and this may also hold in other applications including some aircraft duties. There is thus no doubt that the problem of resistance to hot corrosion and oxidation is important, and the development of gas turbines may ultimately be limited by it.

Trends and prospects

As designs become more sophisticated and life cycle costs are considered *ab initio*, the need for radically new materials is tempered by the desire to use proven materials at minimum cost. In the immediate future more emphasis will be given to the development of processing techniques that lead to a lower cost or technically superior component. The area of uncertainty about the efficacy of any new material or process must be reduced to the minimum before commitment. Advanced materials such as boron or carbon fibre composites, silicon nitride or directionally solidified eutectics may well be used only in special applications that exploit their outstanding properties and tolerate their limitations, e.g. low toughness or high cost. Reappraisal of overall requirements will be more evident and involve even closer collaboration between the designer and the materials scientist/engineer.

References (Glenny & Hopkins)

Ainley, D. G. 1957 Report and Memorandum 3013 H.M.S.O.
Barnard, M. C. S. 1974 *J. Instn Metall.* **6**, 62.
Barnes, J. F. 1967 Bulleid Memorial Lectures, University of Nottingham.
Bragg, S. L. 1952 *Iron Steel Inst., Spec. Rep.* **43**, 23.
Cockcroft, M. G. 1973 Conference on Selection of Materials in Machine Design, p. 69. London: Inst. Mech. Engrs.
Endres, W. 1974 *High-temperature materials in gas turbines*, p. 1. Amsterdam: Elsevier.
Felix, W. 1974 *High-temperature materials in gas turbines*, p. 15. Amsterdam: Elsevier.
Fulton, K. 1975 *Flight International*, 2 January, 1.
Glenny, R. J. E., Northwood, J. E. & Burwood-Smith, A. 1975 *Int. Metall. Rev.* 1975, **20**, 1.
Hall, G. L. B. 1967 International Conference on Thermal and High-Strain Cycle Fatigue, p. 28. London: Inst. of Metals.
Halls, G. A. & Baker, S. G. 1969 9th International Aeronautical Congress, Paris.
Hardy, G. 1966 ASME Paper 66-GT/M-31.
Hare, A. & Malley, H. H. 1966 SAE Paper 660053.
Hooker, S. G. & Garnier, M. 1964 *Jl R. aeronaut. Soc.* **68**, 454.
Islip, L. 1967 Bulleid Memorial Lectures, University of Nottingham.
Meleka, A. H. 1968 *International conference on science, technology and application of titanium*, p. 1117. London: Pergamon.
Morley, M. 1974 *The Nimonic alloys*, 2nd edn (ed. W. Betteridge & J. Heslop), p. 23. London: Edward Arnold.
Nolte, G. H. 1966 ASME Paper 66-GT/M-35.
Restall, J. E. 1975 To be published.
Rolls Royce (1971) Ltd 1973 *The jet engine*, 3rd edn, revised, pp. 226 and 33.
Sawyer's Gas Turbine Catalogue, 1974 edition, **12**, 6–41.
Shepherd, D. G. 1960 *Introduction to the gas turbine*, 2nd edn. London: Constable.
Simmons, C. R. 1968 *Gas turbine manual*. London: Temple Press.
Speirs, K. K. & Weinstein, M. 1969 Chromalloy A12 Protection Systems. Chromalloy Corp.
Stickler, R. 1974 *High-temperature materials in gas turbines*, p. 115. Amsterdam: Elsevier.
Stoeckley, E. E. 1966 ASME Paper 66-GT/M-22.
Stringer, J. 1972 MCIC Report 72008. NTIS, U.S. Department of Commerce.
Stringer, J. & Whittle, D. F. 1974 *High-temperature materials in gas turbines*, p. 283. Amsterdam: Elsevier.
Sullivan, C. P., Varin, J. D. & Donachie, M. J. 1969 *ASM Metals Engng Q.* **9**, no. 2, p. 16.
Waters, C. J. & Norris, G. M. 1974 Report R335 National Gas Turbine Establishment.
Wigg, R. E. 1955 *Gas turbine principles and practice* (ed. H. Roxbee Cox). London: Newnes.
Worley, N. G. & Campbell, J. W. E. 1970 *Nuclear gas turbines*, Proc. Int. Conference, p. 1. London: British Nuclear Energy Soc.

KEYNOTE SPEECH

J. Fielding (*Hawker Siddeley Aviation Ltd*)

During this session we are still considering engineering requirements but concentrating on aircraft – airframes, undercarriages, and engines.

The aims of this conference were to bring engineers and metallurgists together, and to discuss the design requirements and how the various requirements lead to different priorities. There are unfortunately no simple cases. It is important to note that aircraft designers wish to increase the strength of materials as far as safety and reliability will allow.

This is precisely what Rosenhain did in the past and I cannot overemphasize the importance of very close associations between engineers and metallurgists in the aviation business. We will therefore consider the designers requirements, how the metallurgist has been able to help in the past, and what is required in the future.

First airframe material requirements, and we have the paper by James of B.A.C., Weybridge. He details the designers responsibility, of producing an efficient design to meet requirements with maximum economy, and emphasizes that the airframe must now last for at least 20 years.

Design for tension, shear, and compression cases are required and formerly 'proof stress'. T.s., and E were the main concern (and with the materials used these were almost sufficient). However, over the past 20 years problems of fatigue, corrosion, stress corrosion, and notch sensitivity, have required considerable attention and later the effects of elevated temperatures.

The paper details some of the troubles which have developed with airframe materials and the highlights are perhaps:

1. The failure to improve fatigue properties to match the improvements in basic mechanical properties.
2. The problem of notch sensitivity, poor fracture toughness and rapid rates of crack propagation. These factors all make their contribution to low fatigue life – early crack initiation, fast propagation and shorter critical crack lengths.
3. Poor resistance to all types of corrosion – stress corrosion, exfoliation, and general corrosion.

The paper also indicates the results of the various researches by material manufacturers, government establishments, N.P.L., and research associations. For aluminium alloys these are considerable and are dealt with in detail by Hunsicker of the Aluminium Co. of America in Session 3. This combined metallurgical and engineering research has:

1. Provided aluminium alloys some with greatly improved resistance to SCC and some with 'virtual immunity'.
2. Provided aluminium alloys free from exfoliation corrosion worries.
3. Reduced notch sensitivity and improved fracture toughness.

At first these improvements were at the expense of significant and sometimes unacceptable strength reductions, but recent developments are recovering the lost strength features. These represent considerable achievements by physical metallurgy, with help from the engineers, the result giving an acceptable product improvement at some extra cost. However, we note that apart from slight improvements in rate of crack propagation, and better toughness, the fatigue position has *not* been improved. This is perhaps James' main requirement for the future.

The aim must be a combined metallurgical and engineering attack on the following fronts:

(1) *Fatigue.* To improve basic fatigue properties – the period of initiation described later by Forsythe.
To slow down the rate of crack propagation.
To further improve and reduce the scatter in toughness.

The prospects are not too good at present but the requirement and challenge to the metallurgist is still there. Improvements are possible with 'surface effects', working, peening, and preventing fretting, and also by increased protection both to delay crack initiation and retard propagation. The designer is of course helping himself with fail safe approaches, crack stoppers, weld bonding, and fastener bonding.

We should concentrate some of our discussion on the fatigue aspects. What is required? Is it a matter of metallurgical structure features, or is it just explained by a fracture mechanics approach of cracking from insoluble intermetallic constituents – failure from the defect nearest to the peak stress?

(2) After crack initiation we have a period of rate of crack propagation. The lower strength naturally aged alloys show the slowest rates but all the stronger alloys show very similar rates of crack propagation, and so like the basic fatigue position we cannot effectively use the

stronger materials without the rates escalating in a serious way. The designer really does need slower rates of propagation from the stronger alloys and if possible increased toughness.

(3) The third requirement is for better resistance to stress corrosion cracking. The designer does of course, help himself as far as possible by using engineering technology to mitigate the effects of fatigue. It would also be helpful to establish acceptable criteria for exfoliation corrosion, stress corrosion, and corrosion fatigue.

James concentrated most of his comments on aluminium alloys, quite understandable as they are the main structural material at present and for many years to come. This does not mean that there is not considerable interest in other materials, steels, titanium alloys, carbon fibre reinforced epoxy resins, and non metallic materials. However, Imrie has covered high strength steels in the second paper we are considering, and Morton will describe both airframe and engine uses of titanium alloys in his paper (3.9).

We should therefore pass on to Imrie's paper on undercarriage material requirements.

Many of the requirements are similar to those for airframes, i.e. resistance to corrosion, stress corrosion, handling, and general reliability. There are however, as Imrie shows, some important differences, they have sustained loads under hydraulic pressure, severe taxiing and landing loads, and a more hazardous corrosive environment. The larger undercarriages are required to fit into a restricted space and so ultra high strength steels are often the preferred choice for many structural items.

As for airframes, the design is primarily on proof, maximum tensile stress, and elastic modulus. The izod values normally taken on steels are not really relevant even for heavy landings as the loads are hardly of the impact test rates after transmission through shock absorbers and tyres. The transverse grain ductility as measured by percentage reduction in area is, however, very relevant and important.

Imrie then considers the possibilities of the fracture mechanics approach and we note how the very high applied stresses combine with the toughness of the steel to give very short critical crack lengths, which are too small for detection by the present capabilities of N.D.T. equipment. This means that very high strength steel undercarriages are not 'fail safe'. Resistance to crack initiation and subcritical crack growth rate become more important than fracture toughness.

The paper briefly describes suitable aluminium and titanium alloys but as steel has taken over for the larger items he concentrates on steel, and the almost universally used 300 M, so called as it has a tensile stress of 300 000 lbf/in^2 (130 tonf/in^2; 2000 MN/m^2). The complete process of steel melting, heat treatment and processing to avoid undesirable surface carburizing or decarburizing, grinding and machining, must be carefully controlled to ensure that parts will be satisfactory in all respects. This is followed by shot peening to improve fatigue life, very careful plating and final painting.

The paper then considers fatigue and indicates that the significant and potentialy damaging stresses are at the high load – low endurance end of the endurance curve, and it is notable that the required standard of cleanness, indicated by the high transverse reduction of area, is effective in this respect. The controlled shot peening extends the fatigue life to the formation of significant cracks.

The rate of crack progagation is mainly dependent on E and hence there is little difference for similar type steels over a wide strength range.

When considering fracture toughness, Imrie's figure 5 indicates a satisfactory level when compared to other steels and while not sufficient to provide a 'fail safe' condition, a high toughness can be advantageous in some cases. The point to remember is that the permitted design loads are high (using the high material strength), and at these high operating stresses the material toughness will only allow critical crack lengths of 2.5 mm.

The rates of crack propagation for both fatigue and stress corrosion propagation, are similar to those for aluminium alloys at the applied stresses, and there is insufficient time between 'initiation' and failure to allow any cracking in critical areas. Doubling the fracture toughness would only marginally improve crack length, and to approach a 'fail safe' condition, a toughness of around 200 MN $m^{-\frac{3}{2}}$ (180 klbf/in^2) would be required. Increases in toughness are however, always welcome, they can and have prevented failure in one or two specific cases.

Stress corrosion has caused serious cracking and failure of undercarriage forgings in past years, mostly with aluminium alloys but in some cases with high strength steels. The incidence of all stress corrosion cracking on airframes and undercarriages was at a maximum in 1968–70 and is now fortunately considerably reduced. Designers have provided better protection, used the more resistant materials, and processing improvements have avoided surface troubles due to carburization or decarburization. The surface compression stresses introduced by peening have also been helpful. However, the improvements have not removed the stress corrosion hazards and the paper comments on the threshold values for pre-cracked speciments in wet salt conditions. The table shown in figure 6 indicates that the K_{Iscc} values for the steels considered are only approximately one third of the values for the fracture toughness K_{Ic}. This means that it is again of paramount importance to avoid the initiation of cracks.

Imrie has noted that in spite of the apparent weakness shown by the fracture mechanics approach the service performance of undercarriage parts made from ultra high strength steels has been generally satisfactory. His main hazards are still stress corrosion/hydrogen embrittlement cracking, and fatigue cracking, and he therefore emphasizes all the precautions necessary to prevent crack initiation, and to maintain surface integrity. These include the close control of all aspects of heat treatment, avoiding any machining or grinding defects, careful peening, special plating and the best paint schemes, The steel used is either vacuum remelted or electroslag refined to obtain cleanness and the high reduction in area values. The existing reliability is only possible with this very high standard of process control.

The aims for the future must be to improve resistance to crack initiation by any mechanism, and also to reduce the rates of crack growth. These improvements must not be at the expense of the practicability of the manufacturing processes and cost aspects must not be forgotten.

It is interesting to note that Imrie considers that the compromise between strength and brittleness may already have been reached with conventional steels, an interesting point for discussion. The problems with undercarriages are similar in many ways to those in James' paper, the requirement for high strength coupled with good fatigue and toughness, and resistance to all forms of corrosion.

The paper by Glenny & Hopkins describes the wide applications over 35 years of the gas turbine as a practical prime mover. The authors have considered the main types of turbine available and have chosen several components for discussion, the choice being to cover both metallurgical or engineering interest. The paper broadly covers gas turbines in general but concentrates on those for aircraft.

Gas turbines are versatile and appear in the various forms indicated in figure 1 of the paper. The output can be in the form of shaft power, with the small units of 37 kW (50 hp) to the large land based units of around 75 MW (100000 h.p.). The jet engines for aircraft have the power output as a high velocity gas stream, and can generate more than 178 kN (40000 lbf) thrust. Figure 4 in the paper indicates the mechanical arrangement of some R.R. turbojets and shows the disks, blades, and turbine vanes highlighted in the paper.

Disks. Early design was mainly based on the burst speed and the practice was to set the design speed to give say, 50 % of the bursting stresses. This practice usually led to satisfactory fatigue performance. Although burst speed is still considered, more emphasis is now placed on the low cycle fatigue stresses developing from the mechanical and thermal stresses of the start up–shut down cycle, and civil aricraft require a life of 30000 cycles.

The main requirements are:

Low density, High elastic modulus, High strength } throughout the required temperature range

Satisfactory ductility and fatigue

Slow rate of crack propagation

Turbine vanes. The first stage nozzle guide vanes are normally the hottest components in the turbine, and the problem is to provide adequate cooling and prevent oxidation. Failure may take the form of distortion, fatigue caused by thermal shock, oxidation, erosion or even local melting. There is always a requirement for more refactory alloys.

Turbine blades. The peak temperatures are lower than those for the vanes but the blades must also withstand high centrifugal stresses. The primary concern is therefore with creep, fatigue, and resistance to oxidation. In the compressor the stator and rotor blades require good specific tensile strength and stiffness, resistance to fatigue and creep, with the additional requirements of resistance to erosion by sand, and the impact of stones or birds. Corrosion due to the ingestion of salt is also an important requirement.

In highlighting points for discussion three aspects may be noted:

1. To consider and discuss the engineering and operational requirements of land based turbines, e.g. for the Central Electricity Generating Board, hovercraft, and other applications which could use the convenience (automatic switch on and off), the good power to weight ratio, and easy mounting on light foundations. However, the low efficiency must be noted.
2. The most important problems, i.e. corrosion due to salt, sulphur, and fuel additives, the surface stability, and the long life necessary between overhauls.
3. For aircraft turbines the requirements are for still higher operating temperatures, and Nicholson's paper to be given later is encouraging in this respect. The resistance to hot corrosion and oxidation often determines life and may well limit the development of gas turbines.

Finally the importance of cost aspects should not be underestimated.

Discussion (Chairman: J. Fielding (*Hawker Siddeley Aviation Ltd*))

M. G. Gemmill (*Central Electricity Generating Board*)

The C.E.G.B. interest in gas turbines has developed steadily during the past decade from auxiliary service functions in large fossil-fuelled power stations to small power stations, entirely of gas turbine plant, whose principal purpose is to meet peak load demands. Here the ability of the gas turbine to be started up very rapidly is an important attribute.

The great majority of these gas turbine units have been derived from the use of established aero engines, such as the Avon and Olympus, as gas generators to drive a power turbine. These units are subject to planned maintenance after, at the most, 2000 h of operation when burning distillate fuel.

There have been instances of blade corrosion problems due to sulphidation attack and related high sodium levels in the fuel; the solution to this problem has been to control the fuel quality. Two prototype industrial gas turbines, each of *ca.* 55 MW output, are due to be commissioned at one of the Board's power stations in the near future. Here the aimed-for operational life before undertaking planned maintenance will be *ca.* 20 000 h. This places greater emphasis on the need to appreciate any time-dependent process affecting engineering performance. From a materials standpoint these are corrosion resistance, thermal and high strain fatigue and creep-rupture.

Specific problems under study in blade materials are the consequences of corrosion-resistant coatings upon the mechanical properties and the limits of acceptability of defects. The latter involves crack growth monitoring under conditions of creep, high strain and high cycle fatigue.

As the future emphasis should be directed towards gaining a better understanding of material behaviour in the projected engineering situations the physical metallurgist has to think beyond the metals themselves and consider, for example, the interactions that occur between metals and coatings.

Mr H. E. Gresham (*Rolls Royce* (1971) *Ltd*)

Another important use of the gas turbine is as a gas pumping engine where modified Avon engines have run without any maintenance for over 30 000 h and are now approaching 40 000 h. This is a state of reliability far exceeding any other high duty mechanical contrivance and shows the effect of environment and starting and stopping on the life of an engine. In an ideal environment design factors come into play and give the right answers, but as soon as we start putting salt in the fuel we get trouble with the hot parts of our engine and we have such effects as thermal fatigue, stress corrosion, corrosion fatigue; so the story goes on.

It is necessary to design the material for the application so that an Admiralty turbine operating in the sea is of a different material from one used in the air, where the cleaner environment enables higher stressed materials to be used.

A. D. Hall (*Westland Helicopters Ltd*)

When considering factors that inhibit the achievement of maximum material properties, I do not think that we can forget fretting which is often an overriding phenomenon. Now it may be that this has no place in a conference of physical metallurgy but I would be most interested to hear other people's views.

There are materials problems that seem to be peculiar to helicopters. On a conventional helicopter, the main rotor blades are attached to the hub through hinges which allow freedom in the flapping, lag and pitch sense. On the Lynx helicopter, partly to reduce maintenance and partly to improve control power, we have substituted the flat and lag hinges by elements in bending. Here the requirements are high fatigue strength with low stiffness, coupled with the inevitable requirements of low weight, low notch sensitivity and good fracture toughness properties. Our solution to this problem was the 'workhorse' titanium alloy, Ti 6A1 4V, in the annealed condition. We found, however, that the effect of high mean stready stress on fatigue strength was more critical than we had first predicted and we had to design our way out of this problem with the penalty of higher rotor control power than was strictly necessary. Another area of interest is the main rotor blades. The current generation of helicopters has, on the whole, blades made of hollow aluminium alloy extrusions. On the Lynx, however, the 'matched stiffness' concept has led to a blade requiring appreciable stiffness taper in flap and lag. Our solution has been to fabricate the blade in stainless steel sheet section bonded together. Requirements here were formability, good fatigue characteristics and integrity and high erosion resistance.

We have still a long way to go with the full adoption of fail-safe characteristics for the rotor and transmission systems of helicopters. This requirement, together with the need to produce complex shapes for blades and hub, is leading us to look at other means of construction. Here the ability to tailor stiffness to design concepts, the 'fail-slow' and easy formability characteristics make reinforced plastics, both glass and/or carbon future contenders in this field.

Finally a word on gear box casings where high rigidity, stability and stiffness and, of course, low weight are needed. We have used thick magnesium alloy castings but one of the problems here is to provide satisfactory corrosion protection. I believe that we are overdue for new thinking in this area.

H. Tyrer (*British Aircraft Corporation*)

The commonness of problems between the 'lightweights' and the 'heavyweights' has already emphasized the necessity of discussion between metallurgists and engineers in each particular activity. I think it is now obvious that the metallurgists and engineers from the different activities should get together. Referring to the significance in heavy engineering of tensile strength, I would like to suggest that a proof stress plus total energy to fracture might be a more valid approach to what is being sought. I would like to ask James, considering that the fatigue resistance of light alloys has not improved in the last 20 or 30 years, what can be done by direct design to improve the general fatigue resistance of structures? I would also like to ask Imrie why maraging steels have been pushed to one side. Considering that they are readily weldable, the increase in design efficiency so obtained could probably offset the slightly lower mechanical properties.

D. James (*British Aircraft Corporation*)

Hall (this discussion) made a reference to the problem of fretting, suggesting that this conference was the wrong place in which to discuss it as it was not a materials characteristic. I suggest that it is and that it should be dealt with by metallurgists and materials engineers. It is essentially a surface stress problem which should respond to surface treatment developments.

My company has been very conscious of the problem of surface fretting as a cause of fatigue in pseudo-static joints in airframes for many years and we have made extensive use of interference fit pins or bolts as a means of overcoming the problem. This solution does not lend itself to the type of problem Hall probably has in mind. Hence the suggestion that surface treatment developments should be explored.

Tyrer (this discussion) asked if we can improve conditions by recognizing the limitations of the material and adapt our design to get the maximum out of it. This, of course, we do in various ways. The use of interference fits in joints, which has just been mentioned, is one example. Another is the attention paid to providing a standard of corrosion protection adequate to deal with the anticipated environment.

One speaker this morning was critical of the use of the ultimate tensile stress as a measure of material quality on the grounds that what was really being demonstrated in the tensile test was an instability. To me this is an academic controversy. What better method can be found for demonstrating the point at which such instability occurs than the simple tensile test?

I. L. G. Baillie (*British Aircraft Corporation*)

Although James (1.4) and Imrie (1.5) both recorded a lack of progress for aluminium alloys I would express some optimism that the peripheral behaviour problems of fracture mechanics, stress corrosion, etc. are now being brought under control. In this event, aluminium may be similar to the balanced set of properties which Irvine reports for 300 M.

Considering the future balance of properties I would ask whether Mr James envisages going through a learning curve phase of increasing working stresses with materials at present strength levels, or whether he sees a direct improvement of static properties as being of immediate importance. As a corollary to this, I would ask whether James considers that absolute freedom from exfoliation and stress corrosion troubles is essential. This can be associated with attempting to seek more representative methods of checking corrosion performance.

I would also refer to the future development of acceptable aluminium alloys. Imrie, in figure 1 of his paper, shows 2 % per year increase in strength for steels from 1940 to 1970, and I wonder what rate of improvement of *working* stresses should be regarded as a target. For example, for successive generations for aircraft of a given type (e.g. subsonic medium range airliner, supersonic fighter, etc.), hopefully say each 5 years, this 2 % per year rate would give 10 % improvement for each new aircraft. Would James regard this as a reasonable target for future development work?

Finally, I would refer to Imrie's paper. He shows that, from 1970, the u.t.s. of *all* undercarriage materials has remained constant and he seems to be generally satisfied with this state of affairs. I find this rather surprising, for users of materials normally make very different noises even in the presence of material suppliers and developers.

Does Imrie imply that there are other considerations which prevent further developments? For example

(*a*) is undercarriage stiffness now an overriding constraint on higher working stresses.

(*b*) are thicknesses now so reduced that they cannot be further reduced without introducing unacceptable handling problems?

(*c*) is the chance of success so low that it is not justifiable to use the increasily limited development funds to search for them?

G. W. Meetham (*Rolls Royce* (1971) *Ltd*)

I would like to comment on the compromise situations in materials selection for gas turbine engines, with particular reference to the paper by Glenny & Hopkins (1.6). Unfortunately there is a broad inverse relationship between the strength and corrosion resistance of the superalloys used for the highest temperature components in the engine. As a family cobalt alloys are more corrosion resistant in contaminated environments than nickel alloys, but as Nicholson (3.8) points out in his paper, the problem of finding an adequate strengthening precipitate for the former remains unsolved. Without wishing to minimize the corrosion problems experienced by industrial engines it must be pointed out that aero gas turbines in, for example, helicopter applications experience at least equally severe environments. These problems may well have common solutions.

High strength and low density tend to be incompatible characteristics in superalloys. The latest alloys contain high density elements such as Mo, W and Ta, with beneficial effects on strength but adverse effects on alloy density. Finally, in considering the conference title there is an example of a very practical contribution that I should like to highlight. This concerns the design of cooled turbine rotor blades. The designer needs to know the total temperature distribution in his blade designs. Conventional temperature measuring techniques such as thermal paints and thermocouples provide valuable but incomplete information. It is much more informative to section such turbine blades and carry out a detailed examination, using optical and electron microscopy, of the microstructure.

L. P. Pook (*National Engineering Laboratory*)

Basic design is done before a fracture analysis. To do a fracture analysis we must first decide whether we are dealing with a cracked or an uncracked situation. James (1.4) deals with aluminium airframe structures in which cracks occur early during life – therefore a cracked situation. By contrast Imrie (1.5) deals with high strength undercarriage materials and attempts to delay crack initiation as long as possible – essentially an uncracked situation. The resistance to crack growth (by any mechanism) is perhaps the best definition of the rather elusive concept of toughness.

In a cracked situation the fracture mechanism becomes important, and life is largely dependent on fatigue crack growth. Unfortunately physical metallurgy can do very little about fatigue crack growth rates, whereas it has been able to help in the uncracked situation. Confusion over the significance of the values obtained in various mechanical tests, for example what value of K_{Ic} should be specified, arises when considering which tests are to be done. Some tests appear in material specifications simply to ensure that the material is of consistant quality, i.e. they are quality control tests, whereas others are used to provide numbers which are used in design, i.e. they are design information tests. The aircraft industry does not seem to believe its own calculations and proceeds to expensive service simulation tests.

I conclude with two points referring to James's paper. First, creep damage can only offset fatigue behaviour when pores develop, i.e. when tertiary creep is reached. The possible detrimental effect is likely to be swamped by larger cracks already present in the structure. Secondly, increasing efficiency of aircraft structures offers prospects of improvement in fatigue crack growth resistance as this is more dependent on the Young modulus than strength.

R. Wedge (*Rolls Royce* (1971) *Ltd*)

Several speakers have referred to the safety of structures. In the aircraft engine field we are concerned with the integrity of the rotating parts. The energy we have to contain, when a crack propagates through a turbine disk through to its final burst, is equivalent to arresting a motor car of 1 tonne travelling at 100 km/h. It is obvious that the aircraft manufacturers are not able to contain those missiles, so we must address ourselves to the crack propagation and fracture toughness behaviour and considerable effort is put in on the physical metallurgy side. Full scale testing is demanded – a very expensive and time consuming activity.

R. G. Baker (*National Physical Laboratory*)

Is full advantage taken of the improved weldability of maraging steels as compared with 300 M for undercarriages? Perhaps some of the problems of making large forgings can be avoided by joining sub-components. Could this in principle offset the greater cost of maraging steels? Do we need to heat them as much if we can join them more easily?

I would like to question the use of the Charpy or Izod test either as a definitive or quality assurance test for high strength materials. Little of the Charpy/Izod energy is really concerned with a relevant measure of fracture toughness. Should we perhaps think of pre-cracked Charpy specimens or do we need more sophisticated tests even for quality assurance?

H. K. Farmery (*Ministry of Defence* (*Procurement Executive*))

In the aircraft industry environment-assisted-failure of aircraft materials is generally referred to cases which occur under static loading conditions such as stress-corrosion or hydrogen embrittlement or under zero load such as exfoliation attack. It is rare to refer to corrosion fatigue which is appropriate to the dynamic loading conditions prevailing in aircraft, and it is even rarer to see test specimens or structures being fatigue tested in an environment more aggressive than that of the shop floor – hardly typical of service conditions. Yet there is ample evidence stretching back from the work of Gough & Sopwith on simple S/N curves at the N.P.L. to the fracture mechanics approach on pre-cracked specimens of today to suggest that for many high strength materials of aircraft interest, corrosive environments as innocuous as plain water can have an appreciable effect in reducing fatigue resistance – especially at low stress concentrations.

Do Imrie and James rate corrosion fatigue a real hazard to designers in the growth of sub-critical cracks which must be taken into account when assessing the relative merits of materials for aircraft structures?

M. J. May (*British Steel Corporation*)

It is important to remember that, in utilizing a criterion such as the impact value in a Charpy or Izod test as a basis for material selection, the total energy involved in fracture is made up of two parts. One relates to the initiation of a crack from a specific notch configuration while the other relates to the propagation of the crack through the specimen. Considering the Charpy impact properties for a typical ultra high strength steel tempered over a range of temperatures there are several tempering conditions that could be used to achieve an acceptable level of Charpy energy to fracture, however, if one were to conduct identical tests on specimens containing a crack rather than a machined notch the energy to fracture would be markedly reduced, the magnitude of the reduction being dependent on the tempering temperature (as can be seen in figure 1). In the highest strength condition (T = 200–300 °C) only 2 J of a total fracture energy of 20 J is associated with crack propagation the other 18 J being associated with initiation from the machined notch. Unfortunately, in the real situation cracks and discontinuities can be such as to eliminate or markedly reduce the large energy requirement for initiation. I think this reinforces the point made previously by Baker (this discussion).

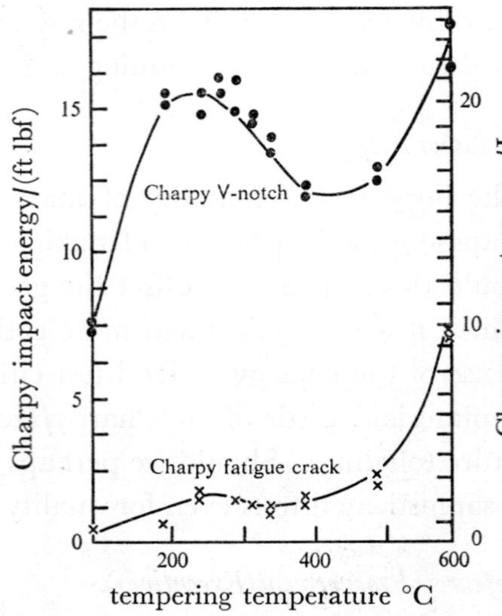

FIGURE 1. Longitudinal fracture properties of the high silicon Ni-Cr-Mo-V steel (cast FZ 0115).

Nevertheless, such tests can be used and will continue to be used for quality assurance but have considerable shortcomings in relation to material selection for design requirements.

I would agree with Imrie (1.5) that of the ultra-high-strength low alloy steels the 300 M grade has one of the highest combinations of strength and toughness at the 1650 N/mm^2 proof strength level (2000 N/mm^2 tensile strength). However, in terms of fracture resistance considerable benefit is gained by reducing the strength level to 1450–1500 N/mm^2 in terms of the initial crack size for fracture at a given stress level. Perhaps even greater importance is the marked increase in resistance to stress corrosion cracking that can be obtained in the lower strength condition.

It is argued by the undercarriage designer that any reduction in strength level imposes an additional weight penalty and if adopted would still result in critical crack sizes that are borderline from a non-destructive inspection point of view. These are, of course, valid points and emphasize the need to adopt the attention to detail and quality control outlined in Imrie's paper. It is perhaps salutory to reflect, followng the theme suggested by Kelly (centenary lecture), that when a scientific analysis raises doubts on material selection the requirements and needs of trade necessitate the development of an appropriate art. Such I believe is the situation which has been developed successfully over the years by the undercarriage designer and manufacturer.

D. JAMES (*British Aircraft Corporation*)

In reply to Baillie (this discussion) I would remind him of some discussion I had with him some 15 years ago over a proposal to use a stainless steel FV 520 in the welded and precipitation hardened condition at high stress levels on an aircraft landing gear. He warned me of the hazards of stress corrosion in such a condition on the basis of the published data. I suggested that in looking at the published data in the context of the proposed application, we should look very carefully at the comparative environments of the proposed application and that relevant to the published data. In practice, the application has been completely satisfactory over a long period of commercial airline operation.

Baillie raised a question as to the benefits to be obtained from higher strengths in new materials. Strength levels should not be considered in isolation. In many cases modulus of elasticity as affecting the stiffness of a structure is of equal, if not greater, importance.

Pook (this discussion) raised some doubts about the feasibility of getting improvements in modulus of elasticity. I am assured by my metallurgical friends that these improvements are possible. One particular experimental alloy with which we had some experience a number of years ago, was the lithium bearing alloy X2020 produced by Alcoa. Its higher strength and modulus of elasticity compared with the aluminium-zinc-magnesium alloys was unfortunately associated with very poor manipulative qualities, which made fabrication difficult. I understand that refinements of composition have been developed which made this basic type of alloy a reasonable prospect for the future.

It was suggested in discussion that airframes are designed for cracked conditions and perhaps landing gears for uncracked conditions. We do not design the airframe to have cracks but we do accept that accidental damage does occur and we have to take account of these possibilities as well as those of fatigue crack development in our use of materials. It was also suggested that there should be no need to carry out full scale testing in view of the very detailed and elaborate design calculations carried out. I would like to think that this is a correct assumption and I try to press it in the interests of economy but it is very difficult to persuade the airworthiness authorities of the validity of the case.

In reply to Farmery (this discussion), I agree we do not do very much testing which is positively identified with corrosion fatigue. It would be extremely difficult to draw meaningful quantitative conclusions from such tests in view of the uncertainties of the relationship between test time histories and real time histories under these conditions. However, much full scale fatigue testing is done under corrosive conditions, e.g. long term pressure cabin tests in water tanks.

W. M. Imrie (*Dowty Rotol Ltd*)

Pook (this discussion) indicated the difficulty of quantifying fracture-toughness as a design criterion. While agreeing I would in fact give consideration to the K_{Ic} of a potential material, but only as a preliminary screening evaluation. Improved toughness could be significant from the standpoint of greater reliability and is a desirable factor in raising the confidence level in the mind of the user.

Farmery (this discussion) asked about corrosion fatigue. Dr Forsyth (2.3) explained that crack growth rate is dependent upon the test frequency. I do not think that the published laboratory data are very relevant to undercarriage design, where the accent is on avoiding the onset of corrosion and of crack-initiation.

Tyrer (this discussion) thought that energy-to-fracture was an important criterion in design. I support this view and refer to figure 3 of my paper (1.5), which illustrates how the introduction of vacuum-remelted steel, with its resultant improved transverse ductility, leads to an increase in energy-to-failure.

Tyrer also asked why we had chosen 300 M in preference to other low alloy steels and additionally, would not maraging steel be a superior material? With regard to 300 M, the choice was made because of the vast data available from American sources, particularly in large sections. Additionally, our own work has indicated that it is certainly as good as, if not better than, all other comparable materials. Even so I believe that, for example, a Ni-Co-Mo-V alloy could be developed in the U.K. to perform satisfactorily but it is unlikely that the funds would be available on which the necessary confidence level could be acquired.

With regard to maraging steel, I would say we are very satisfied with its technical performance on the Harrier undercarriage, but that the easier manufacturing route was insufficient to offset the higher cost of the alloy in heavy sections, when compared with 300 M. Other factors militating against its use are:

(*a*) the capital tied-up in purchasing economic batches of forgings which have to be stored for many months awaiting machining,

(*b*) the financial loss which occurs if, for example a machinist makes a false cut in a large component,

(*c*) the lack of consistency in transverse ductility in large forgings, and

(*d*) the high degree of temperature control required by the forger in order to avoid thermal embrittlement.

I would also point out that the application of both 300 M and maraging steel in undercarriages is dealt with in detail in the references at the end of my paper.

Baker (this discussion) referred to the advantages of the weldability of maraging steel but we have shown that 300 M is capable of being solid-phase welded with almost 100 % parent-metal properties at the weld.

May referred (this discussion) to work at B.S.C. which has shown that some of the materials possessed a much higher degree of toughness than 300 M, although the strengths were lower. Although it has been generally accepted for a very long time that lower strengths are normally associated with higher toughness in the case of undercarriages such increases cannot be accomplished at the expense of strength.

2. DESIGN IMPLICATIONS FOR MATERIALS SELECTION

2.1 Strength as a design parameter

By A. A. Wells
David Keir Building, Stranmillis Road, Belfast, BT7 1NN, Ireland

The paper starts with a review of structural dead and live loading which shows that the latter will in some cases increase indefinitely as service experience is accumulated, so that the ideal of complete safety is unattainable and gives place to alternative probabilistic concepts. Brief reviews of the functions and limitations of structural and stress analysis and of the origins of stress concentrations are followed by considerations both historical and contemporary, of individual mechanical properties from the viewpoint of design. Attention is mainly given to yielding in tension and compression, and fracture. Residual stress effects on buckling are considered both as design and material development parameters. Considerable attention is given to scatter in notch ductility tests, and an elementary analysis of probabilities of failures is presented which both correlates with world wide experience of storage tank and ship failures and suggests that the diminished failure probabilities now being demanded in pressure vessels are realistic and attainable in terms of practical quality control. The review concludes with speculations on the potential for high strength steels in future civil structures, and observations on methods of metal production as they may influence structural design.

Introduction

The name of Rosenhain is commonly associated with recognition of intergranular fracture in polycrystalline metals as a dominant mode at elevated temperatures, and with the prescription of a heat treatable aluminium alloy with favourable elevated temperature properties much used for the cast pistons and cylinder heads of internal combustion engines. These complementary contributions symbolically embrace the activities which link physical metallurgy with engineering practice, and thereby provide a background to this paper. Since design is a topic with many interpretations, and application to numerous products, it is obviously necessary to accept a restricted scope, which conveniently relates to those structures and components where strength is a principal limitation, and to the more common modes of failure considered in design. If generalizations are possible, they are most likely to belong to those structures which are repetitively designed (rather than mass-produced), where design codes occupy an important place. The paper traces some of the factors which influence the choice of design loads for civil structures such as bridges, pressure vessels and pipelines, and the hulls and airframes of ships and aircraft, in the manner directed towards the control of collapse. It goes on to consider the principal mechanical properties that are invoked, and methods of dimensioning. Other important modes of failure are recognized in relation to their control by design, such as fatigue, stress corrosion and creep, but the treatments will be found in other papers. The objective throughout is to identify those mechanical properties which can best be augmented or controlled in the metallurgical domain.

The loading of structures

All the structures listed above are considered to sustain so-called dead and live loads. Dead loads embrace those due to gravity and distributed pressures, which may be hydrostatic or even steady state hydro- or aerodynamic. Live loads may be defined as those from all sources which vary during the life of the structure; they may be considered as static or dynamic dependent upon the rate of change in an inertial sense. It is less important for the present purpose to seek rigorous definitions for these two types of load than to emphasize the characteristics of each. Thus, the stresses arising from dead loads increase with the size of the structure, as may be seen by considering the self weight stresses in a suspended tie bar, and impose definite upper limits on the dimensions of some structures, such as the spans of large bridges. However, dead loads are usually clearly defined in magnitude; being steady they do not normally introduce the risk of fatigue cracking. On the contrary, live loads are often indeterminate, or dependent upon how much punishment the structure is intended to accept without collapse. They may be repeated few or many times and can give rise to fatigue damage. As will later appear, the optimal choices of material to withstand dead and live loads can be different so that the need for compromise is evident at the outset.

It is quite important to recognize that there must often be an element of formalism in the prescription of live loads for design purposes. The loading of floors in buildings is among the simplest and U.K. codes of practice recognize classes of occupation from 30 to 200, being the design loads in pounds force per square foot (i.e. from 1.5 to 10 kN/m^2). For single family houses the lowest class may correspond with 3 tonnes of furniture in a living room; the heaviest class is for book and stationery stores. It is totally impracticable to consider floor loadings as they are from one hour to the next, since the quantitative effects of change of use cannot be anticipated, although deliberate allowance in design might be made for specific heavy fixed items for which the effects could exceed these limits. Bridge loadings in many countries distinguish between individual and groups of axle loads in such a way that the assumed loading density decreases continuously with increase of span. This allows for the inevitability of sustaining at least one very heavy vehicle or group, but rejects the necessity to accept a hypothetical queue of such vehicles.

The statistical nature of live loads has been recognized in some cases for many years. The limiting design data tabulated below were determined from many field observations and published in summary form relative to ship (Bull 1954) and airframe (Taylor 1955) conditions in 1954 and 1955 respectively.

wavelength/ft (m)	100 (30)	200 (61)	400 (122)	800 (244)	Bull (1954)
wave height/ft (m)	14 (4.3)	22 (6.7)	35 (10.7)	52 (15.9)	
miles (1.6 km) per vertical gust	10	10^2	10^3	10^4	Taylor (1955)
gust velocity/(ft/s) (m/s)	10 (3)	18 (5.5)	26 (8)	34 (10.3)	

For waves with a length approximately equal to that of a ship hull the wave height determines the change of bending moment arising from differential buoyancy together with a hydrodynamic correction. The gust velocity, with allowances made for duration and aero-elastic effects, can determine the acceleration loading on an airframe, but is only one of the sources of such loading for design purposes. The recognition of loadings such as these implies (1) that the observed intensities would increase with the accumulation of operating experience and (2) that there must be a limit to the severity of a loading event to be accepted as legitimate for design purposes. In the Building Research Station document of 1970 (Anon 1970) on the assessment of wind loads on buildings in the U.K., the information is presented on a probabilistic basis so that the designer may opt for severities corresponding with building life expectations from 5 to 50 years. At the other extreme, the allowances of acceptable pressure excursions in pressure and storage vessels and pipelines are rigidly limited in almost all cases. On the one hand, failures in such vessels can be dangerous at radii well beyond those of the structure. On the other hand, effective pressure relief devices are available in almost all cases. These few examples are sufficient to illustrate the heterogeneity of the design problem with respect to assumed loadings.

Structural and stress analysis

It is important to distinguish these choices or specifications of real or formalized, dead or live loading from the stress analyses which follow in the design procedure, since the latter involve further rationalizations of a separate character. A few simple stress or bending moment calculations can be made which are dependent only upon the requirements of equilibrium. The most obvious examples are the tie rod, the cylindrical (or axisymmetric) 'thin' pressure vessel, and the cantilever or simply supported beam; they represent the statically determinate situations. For statically indeterminate counterparts it is necessary at the outset to define a stress–strain relation, and to recognize a precondition of the compatibility of strains, one of the most elementary examples being the condition of plane cross sections remaining plane as in a tie rod or simple beam. In fact, the very concept of a statically determinate structure is a non-existent ideal, justified by crude but useful assumptions, such as that the stress is uniformly distributed in a tie rod of uniform cross-section, or in the wall of a pressure vessel whose thickness is small compared with its radius of curvature. The latter assumption is clearly untrue in an aluminium overhead electrical conductor having a steel core, where two different moduli of elasticity are involved. A better assumption arising directly from the principle of compatibility of strains is that any extension of the cable leads to equal tensile strains in its constituent strands. If the law of elasticity is then introduced the twin principles of statical indeterminacy have been fully accepted. The stress–strain relation involves two separate notions, namely the dependence of one upon the other, and the degree of reversibility of the relation during unloading and reloading. The common design assumption is that the relation is linear-elastic, involving constant elastic moduli and complete reversibility. However, no designer can afford to forget the departures from this ideal which occur in practice, leading to non-linearity, plasticity, hysteresis, etc.

The significance of stress concentration

The occurrence of stress non-uniformity and concentration effects is a direct consequence of the intervention of the stress–strain compatibility restrictions, and the satisfactory control of stress distributions is an objective of structural and stress analysis. It is more accurate to conclude that stress concentrations are ameliorated by the procedure of dimensioning aided by stress analysis, than that they are removed. Complete removal is simply not feasible in many cases, such as those involving concentrated loadings, or sequential loadings of different character, such as those due independently to internal pressure and transient heat flow.

Because of these limitations it is necessary for almost all structures to exhibit a rugged resistance to or tolerance of stress concentrations, and the contribution of the metallurgist in this regard has been of far-reaching importance. Resistance to stress concentration, at its extreme form in the region at the tip of a crack, is conferred by mobilizing the properties of plasticity, and ductility.

Since an important aspect of design is contained in the creative process of synthesizing geometric forms, complemented by quantitative stress analyses applied to each, it is appropriate to seek contrasting examples in this way, rather than to confine to analytic comparisons. The pressurized cylinder provided with a peripheral opening is a valuable fundamental example. Such could be made in practice from metal plate, with a cut-out, or it could consist of a double helical winding of wire or strong fibre, held together in a suitable cast matrix and having the fibres locally drawn aside to provide a diamond shaped opening. Two important features in the first case are a stress distribution in which the hoop is double the axial component, and a stress concentration at the opening, part of which is due to the disturbance in direction of the principal stresses there, and another part is due to the local out-of-plane forces exerted by internal pressure on any closure gap. Some amelioration of the stress concentration, which may be represented by a factor of 3 or more, can be produced by local thickening at the edge of the opening, but it would be difficult to reduce the factor below 2. In contrast with the isotropic metal form, the wound cylinder as described exhibits practically no stress concentration, because the fibres can be bunched at the edges of the opening; by design, they also follow the principal stress directions. Nevertheless, the fibre composite construction starts with one disadvantage compared with the isotropic metal, which is that each fibre can only sustain stress in one direction; the existence from considerations of equilibrium of a complex stress distribution in the cylinder calls for a dual winding, which is actually more voluminous at a given primary stress level than for an isotropic metal. It is the function of the matrix to adjust the stresses in the superimposed windings to the applied loads, which it does by accepting interfacial shear and normal stresses. The matrix in the wound pressurized cylinder case may also, for instance, have to accept extraneous loadings from transient heat flow through the thickness. This adaptation might require almost incompatible physical and mechanical properties of the matrix, and the calculation of the resulting thermal stress distributions would not be easy. It is perhaps a remarkable tribute to the favourable mechanical properties of some strong fibres that composite constructions in which they are used have achieved such outstanding successes.

The concept of redundancy has at times been confused with statical indeterminacy, since the number of members in a loaded plane frame relative to the number of joints must not exceed a certain limit if the forces in the members are to be determined solely by equations of force equilibrium. A redundant structure for the purposes of design is one in which there are alternative load paths if components are removed or destroyed. The wound cylinder is a good example of a redundant structure, provided that the bond between fibres and matrix is sound (fit for its intended purpose), and the concept of compatibility of strains is at least partially upheld. Such a structure may not need to use materials of superior ductility, because damage can be localized through redundancy, but this topic will require further discussion.

The basic mechanical properties of structural material

Since few structures are entirely statically determinate it follows that the elastic moduli are important as determinants of stress distributions as well as deflexions. The extensional and shear moduli are coupled through the Poisson ratio. These properties are sensibly isotropic for most metals. Important physical properties which may be categorized with the elastic moduli are the density, which influences elastic wave velocities, vibration characteristics, and such features as the aeroelastic response to transient loading, as well as the obvious feature of structure mass. The list is extended to thermal expansion coefficient in regard to thermal stresses, and the thermal conductivity and specific heat which contribute to thermal diffusivity which is the property through which thermal stresses are most influenced. The designers' view (which may be a drastic oversimplification), is that these properties are intrinsic to a material, and not much susceptible to metallurgical manipulation.

The stress limit at which plastic flow occurs is so central to structural design that it is designated by several variables. The simplest is the yield stress in uniaxial tension or compression; it corresponds at the microscopic scale with the shear stress at which there is ordered relative displacement at slip bands, and at the crystallographic scale with a corresponding movement of dislocations. It is necessary to add that design is dominated by an alternative continuum view of yield, especially as it is affected by complex stresses. The simplistic view, that yield and the orientation of the resulting irreversible components of strain is determined by the maximum shear stress, is supplemented by an alternative criterion of shear strain energy. Although the two concepts lead to broadly similar results, there is firm evidence from many experiments on metal specimens selected for isotropy that the shear strain energy criterion of yielding is preferred. While it is not easy to explain the anomaly, it may be associated with the change from the single to the polycrystalline mode of deformation, whereby a strongly statistical element is introduced. In spite of the absence of an easily perceived mechanism of yield associated with a shear strain energy criterion, the latter possesses an overwhelming advantage for purposes of calculation of elastic-plastic deformations of non-uniformly shaped and loaded bodies, which is that of the continuous slip surface in stress space. The shear strain energy criterion, unlike that of maximum shear, is independent of the choice of the two critically interacting principal stresses in the presence of a third principal stress.

Certain metals clearly show the phenomena of slip and a recognizable yield stress, which may even require the distinction to be made between upper and lower yield stresses in laboratory specimens which are carefully set up and axially loaded. These phenomena are rare in the modern technological metals, which alternatively require the arbitrary definition of a limit of proportionality. This is more coarse than it first appears to be, since it involves a departure from linearity between stress and strain associated with a non-linear component of strain of the same order as the linear component. The metallurgist, in consideration of a wide selection of metals and alloys, examined as small very carefully loaded specimens, could identify many plausible sources for such non-linear behaviour, associated with heterogeneities of microstructure. The influence of micro-creep deformations might also be included. Such non-linearities would be associated with irreversibility or permanent set. The polymer technologist would additionally have to consider non-linear elastic deformations arising both from large strains and the influences of long chain molecules, as in rubber.

It is quite important in this interfacial area between metallurgist and engineer to recognize that residual stress effects are the most important sources of irreversible non-linearity at low applied strains, as seen by the engineering designer. This may be illustrated by the fortuitous result of an experiment, designed to verify the newly calculated elastic buckling characteristic for metal rods potted in foamed polymer, in the manner that steel bars are introduced into reinforced concrete. The degree of support from an elastic matrix even as light as expanded polystyrene is such that there should be large buckling strengths for steel rods of unlimited lengths (Wells 1972). Although this may be verified, certain axially loaded and matrix supported mild steel bars failed unexpectedly by plastic hingeing at well below yield strengths. That this was due almost entirely to the ordinary residual stresses of manufacture was proved both by prestretching and normalizing of replacement bars. These bars all buckled at yield point stresses as originally anticipated. The significance of this observation is that the degree of refinement in the specified or accepted limit of proportionality is governed by care in manufacture. 0.1% offset strain might be reasonable for carefully wrought and heat treated strong alloys for aircraft applications, where 0.5% offset would be realistic for mass-produced structural steels. The engineering designer should recognize the penalty that he may have to pay in steel structures extensively loaded in compression, of effective elastic modulus significantly lower than the recognized value, or the alternative of a restricted design stress. It is fortunate that the proportion of compression structures in the total population of structures is relatively small, and each is easily recognizable so that it may receive due attention, without the need for applying precautionary measures to the material supply over the whole field.

A historical perspective must be encompassed in a review of the alternative mechanical property of ultimate strength with its associated strain measures of elongation and reduction of area. Any physical or mechanical property when it is first measured becomes a legitimate design parameter. The need to distinguish between tests for the extraction of design data and alternatives for quality control inevitably follows the growth of experience with testing methods and results. The two types of tests diverge because the one must be realistic, even if costly, whereas the other must be inexpensively repeatable. The crude measurement of ultimate strength became important in the 19th century with the transition from cast to wrought iron and rolled steel structures, and the progression was driven by instant if ill-defined recognition of the advantage of ductility.† Yet is is sometimes forgotten, for want of attention to the literature of the day, that the mechanical test for ultimate strength was frequently supplemented by realistic large scale tests for the accumulation of design data. Those conducted by Fairbairn over a century ago on full scale riveted iron bridge girders are an excellent example, from which it was concluded that up to 10^5 or 10^6 applications of load could be sustained if the calculated stresses were confined in design to 60 N/mm^2 (4 tonf/in^2). The ultimate strength test was therefore conducted for the principal purpose of quality control almost from its inception, and the difference between the strength that it indicated and the value usable in design was inevitably contained within a large factor of safety. The retention of ultimate strength for quality control is therefore more defensible than as a design parameter.

† This recognition is identified in the account by Todhunter & Pearson (*History of the theory of elasticity*, vol. 1, 1886, p. 519) of the British Association meeting of 1837–8, illustrating the characteristically aggressive style of 19th century discussions. It relates to Fairbairn's comparison tests of full sized beams of different types of cast iron in which the yielding phenomena appeared. The full quotation reads,

'Obviously little is determined by Fairbairn's experiments. He suggests that Hodgkinson or Barlow should investigate these matters mathematically, thus showing how unconscious he was of the infinite difficulties which beset the subject of after-strain.'

There is a considerable difference in many underdeveloped metals between the yield strength or limit of proportionality and the ultimate strength, and much of this is due to strain hardening capacity. However, it is recognized with most structures that the maintenance of dimensional integrity is of paramount importance. Under these circumstances the ultimate strength is out of reach as a design parameter, as was particularly well demonstrated in the long series of tests initiated by Sir John Baker on welded steel frames (Baker, Horne & Heyman 1956), conducted for the purpose of exploring and developing the plastic design method. When many of these structures reached their ultimate load carrying capacity they had changed from structures into link bar type mechanisms. A proportion of others, although first exhibiting plastic hinges, failed by local buckling or cracking before the ultimate load carrying capacities were reached.

During the now significant period during which the limit of proportionality has been pursued as a quality control as well as a design parameter with regard to structural steels the former has gradually increased through metallurgical development to match the ultimate strength. Many problems have been solved during this time relating to extraneous fracture modes associated for instance with strain ageing and brittleness, and the need to make welded joints. It has sometimes been questioned with laudable caution whether this closing of the gap between yield and ultimate strengths would lead to situations in which structural stability would be lost for want of strainhardening capacity. Such situations could perhaps be met in uniformly loaded tie bars, but are not because of intervention of the principle of compatible strains. Perhaps the only example worth recording relating to a similar event, with reference to appearance rather than integrity concerns the formation of stretcher strains in deep pressings. It may be concluded that deformation instabilities in high yield metals in structural service are more dependent upon the shape of the stress–strain curve at the limit of proportionality and the presence of elastic residual stresses than the absence of strain hardening capacity.

It is of interest to note in passing that one of the last sources of confusion between yield and ultimate strength design disappeared in recent years in the complex field of boiler design, where structural integrity must be guaranteed over a remarkably wide temperature range. The anomaly during a period of assimilation of yield point factored design for pressure vessels used near ambient atmospheric temperatures was that the occurrence of a discrete yield limit vanished reversibly in the material of any one vessel as it approached steam temperatures. With regard to the important question of hydrostatically testing the completed boiler at room temperature an elaborate conceptual fiction could have developed whereby the boiler would be tested at a pressure factored to the required steam pressure according to the ratios of arbitrarily defined yield strengths at the extremes of temperature. More earthy considerations prevail, in which there is an agreed design stress limit at each temperature interval for each steel type, determined with reference to each of the several failure modes which can intervene at least on a notional basis, such as excessive deformation, reversed yield at stress concentrations (or non-shakedown), stable or unstable cracking, creep, fatigue, corrosion. This is a case where both limit of proportionality and ultimate strength may legitimately remain as material quality control variables, although neither has exclusive significance in design. The numerical factors that might have related them to design stresses have diminished significance, because credence is given to a wider variety of mechanical tests.

The gas transmission pipe situation described in another paper at this symposium lends itself ideally to yield point design and quality control, since it relates to large areas of membrane material at uniform hoop tensile stress, and variations of operating temperature are of less importance. Hydrostatic testing to yield of completed pipeline sections can then contribute significantly to quality assurance by artificially inducing local failures in sub-standard section at the stage before gas pressure is introduced.

If it is argued that plastic strains as large as those associated with ultimate strength cannot be mobilized in structural design, it follows that the measures of elongation and reduction of area are also of significance only in terms of material quality control. In the case of structural steels this assertion is powerfully supported by the observation that brittle cleavage fractures are seldom exhibited in plain tensile tests (although the same cannot always be said concerning steel bars for concrete reinforcement, where they are of obscure origin). In order to assess the resistance of a material to brittle cracking, as is very well known, it is necessary to employ a notched test specimen such that crack extension is clearly reproduced. The result may then be expressed as a fracture toughness measurement. In response to an urgent need, but slowly and painfully over the past fifty years, the development of fracture mechanics has begun to permit the translation of fracture toughness measurements into estimates of the limiting stress and dimensional conditions for extension of initial crack-like defects in structures. This step, to the designer, has a potential significance comparable with that of the introduction of stress analysis as an aid to interpreting the significance of yield strength measurements. Nevertheless, the long time scale in the development of stress analysis should be noted, since Galileo (1564–1642) is known to have stated the hypothetical relation between strength and stress as it is now universally recognized. It is not the function of this paper to place a disproportionate emphasis on the function of fracture mechanics. In pursuing the main aim, of delineating the mutual impingement of material development and engineering design, it may be appropriate to set down certain propositions for discussion, and suggest lines of development.

(1) The fracture toughness property assesses the tolerance of a material to stress concentrations and notch effects, and the resistance to extension by cracking of unwanted but inbuilt defects.

(2) A fracture toughness test must at least reproduce in the material the mode of failure against which the structure is to be protected.

(3) If the mode of failure is reproduced in both tests a fracture toughness measurement can be correlated with such independent measurements of ductility as elongation and reduction of area, but usually only within very narrow limits, or in the qualitative sense.

(4) In order to be both quantitative and reproducible, bearing in mind that it is dependent upon specimen or component size and shape as well as material strength, a fracture toughness measurement must conform to well defined criteria of validity.

These propositions, reduced to a severe brevity, illustrate once again that design and material development have at least some different objectives. Despite long sustained efforts to produce a fracture toughness test that could also be used repetitively for quality control, two sources of incompatibility remain. These are, firstly, that the test for producing design data requires a specimen whose thickness is similar to that of the component section being assessed and, secondly, that instrument transducers mounted on the specimen cannot easily be obviated. Such tests do not meet the need for inexpensive reproducibility at the quality control stage.

There is active development of fracture toughness testing throughout the U.S.A., Europe and Japan, and the above described incompatibility can be expected to diminish in importance with the growth of experience. An interesting recent lead from U.S. work addressed to the proof of heavy section steels has been the recognition, in tests of geometrically similar compact tension fracture toughness specimens, prepared from uniform sections of pedigreed material, that the size or thickness effect on the fracture toughness measure is reduced almost to the point of elimination if the test is concluded at the stage of first recognition of flat crack growth. This more closely defines the problem rather than solves it, because two important practical objections remain, firstly, that the measured fracture toughness is severely reduced in many cases, through leaving out of account the toughness component associated with formation of a full crack to the point of maximum load in the specimen and, secondly, because the required instrumentation attached to the specimen is even more sensitive and demanding than before. It is recognized, in the relatively few full scale components that have been tested to failure with full instrumentation applied to crack growth, that the margin between the first signs of crack extension and full instability is even more pronounced than in any laboratory test specimen. Thus, a measurement of fracture toughness at the first recognition of crack extension is often very conservative and will become more so as the sensitivity of detection is increased. While it is the function of mechanical testing to identify the limits of safe operating conditions in structures, it is no advantage in the final synthesis to overemphasize one property, particularly when this is at the expense of other properties. Such a dilemma is by no means unknown in material development.

The significance of scatter of results in mechanical testing

An encounter with scatter of results is inevitable in physical and mechanical testing. Its recognition leads initially to selection of output, so that the best material is allocated to the most demanding structures or customers. A more insidious result is that inferior material is thereby diverted to unwary or undemanding users, so that the second stage is to accumulate experience on the sources of scatter, so that they may be more successfully avoided at the outset.

The presence of scatter in the conventional strength properties of metals is mainly seen in the literature of the last century. Kirkaldy's published results from a test house (1862) show, not only that the common structural steels of the time were perhaps on average half as strong as the mild steels of today, but that the range of scatter was very high, and certainly several times higher in proportion than is customary today from particular sources.

The complaint concerning scatter is nowadays addressed more to fracture toughness or notch ductility than to yield or ultimate strength. In the case of notch ductility of ferritic steels measured in the quality control Charpy V test this scatter is mainly centred on the properties measured near the transition temperature, because the vagaries of a change of fracture mode are fully represented. The perspective view is well illustrated in a world-wide collection of results obtained by Lloyds Register of Shipping, quoted by Boyd (1970) and reproduced in figure 1*a*. When reduced to a histogram (figure 1*b*) with respect to energy absorption there is negligible doubt that the distribution is normal or Gaussian (99+% probability by the χ^2 test). The scatter, expressed as standard deviation divided by the mean, is 35% (the conventional term 'standard error' is misleading). This situation has probably improved significantly in the intervening 17 years, but it is considered likely that the counterpart, as seen in more restricted contemporary samples, would still reveal a scatter of 15%, even if restricted to the upper shelf, at which most pressure vessels would be expected to operate.

Many of the reasons for this scatter are well understood by metallurgists, and have to do with the consequences of segregation both on a microscopic scale, and as a result of the complex process of cooling and solidification of ingots. There are further sources in mechanical working. Substantial positional and orientation heterogeneities can be consistent in the production of plate; they may be seen as inconsistent when plate products, components and even whole structures are sampled, because of random dispositions of material. The closer that metal production approaches a continuous process, the smaller these inconsistencies can be expected to be, but continuous processes can be notably inflexible with respect to the marketplace, so that the problem acquires a strong economic element. The information that it is believed that the metallurgist needs from the engineering designer concerns the consequences of scatter of properties, so that it may be known how much control is suited to various outlets.

This question cannot be fully answered without recourse to each in the range of different structures that are made. It cannot be answered without considering the consequences of

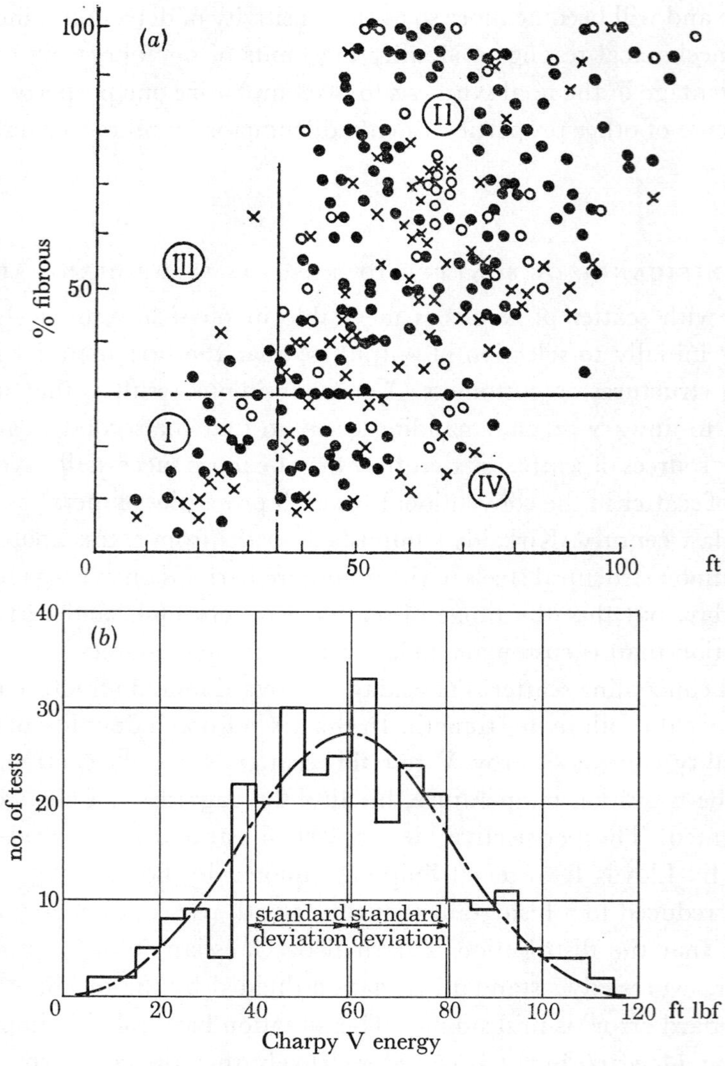

FIGURE 1. (*a*) Plot of results of Charpy V-notch tests at 0 °C on a large number of plates, taken at random from the (then) current production, varying in thickness from $\frac{3}{4}$ to $1\frac{7}{8}$ in of many steelworks all over the world (by courtesy of Royal Institution of Naval Architects). ○, Up to and including 0.80 in; ●, over 0.80 in to 1.20 in; ×, over 1.20 in.

(*b*) World-wide distribution of structural steel notch toughness values at 0 °C prior to 1958 after Boyd (1970). 1 in = 25.4 mm; 1 ft lbf = 1.35582 J.

redundancy of members in a structure. However, in the case of notch ductility and crack extension, if facts are fully faced, it may be accepted that there is negligible redundancy, and this proposition is endorsed even if the best properties of high strength filament wound components are alternatively considered. The tests of Wright & Iannuzzi (1973) at University of Tennessee on carbon fibre epoxy specimens show that critical crack lengths in planes normal to the fibres are of the same order as would be expected in high strength metal specimens.

The methods of fracture mechanics permit a translation from scatter in fracture toughness to scatter in critical crack length in a structure. The statistical consideration of defect populations would then permit the estimation of structural failure probabilities, in non-redundant structures where there is no crack arrest.

It is now widely recognized at national levels and in the international scene that the failure probability approach to structural safety is helpful, whereas the alternative demand for completely proven safety is unattainable. The probability approach is helpful because the statistical data on failures can be produced, and even show a remarkable consistency, and because the approach identifies areas where improvements can be made and monitors whether the steps have been successful. The original analysis of ship failures by Williams & Ellinger (Williams 1954) is well known. Boyd (1970) also quotes actuarial and economic data derived for the U.S. population of about 6000 large oil storage tanks up to 1954, which exhibited one brittle failure per 3400 tank-years in service, or a probability of 3×10^{-4} per tank-year. The corresponding ship-year probability (Boyd 1970) for breaking in two was 1.2×10^{-4} between 1949 and 1963. Similar assessments more recently made for land pressure vessels within that period (Phillips & Warwick 1968; Kellermann & Tietzle 1969; O'Neil & Jordan 1972), considering bursting failures almost all of which occurred at the time of hydrostatic test immediately following manufacture, point to a comparable probability of about 10^{-5}. The compromise that obtains, between what can be achieved in terms of safety and what is socially acceptable seems to take into account, at least qualitatively, the likely severity of the resulting accident, and demands a detectable improvement rate through each decade. For the purposes of comparison the U.K. government statistics (Anon 1974) show that road accident death probabilities were 7.5×10^{-4} in 1957 and 4.4×10^{-4} in 1973, per vehicle-year. The recorded injury rates per vehicle-year were 48 and 47 times greater, respectively, in those years.

The existence of such failure probability data justified analyses such as those of Becher & Pedersen (1973) and O'Neil & Jordan (1972), whereby the cumulative frequency of interaction of a distribution of defects and a distribution of critical crack lengths is determined. In the elementary version given below the assumption of a normal distribution of critical crack lengths is justified by the data of figure 1, together with the observation that critical crack length is always dimensionally proportional to the toughness measure of G_c or J_c, as it is to the square of the corresponding critical stress intensity K_c. In so far as it is a measure of fracture toughness, energy absorption in the Charpy V test is proportional to J_c, rather than K_c. There is fortunately no concern for the present purposes with the slope of the correlation between critical crack length and Charpy V energy, bearing in mind that this also involves stress level, specimen and component geometry, and the degree to which fracture in the test is representative of that in the structure.

Unfortunately, there are at present very few available statistical data on the distribution and densities of crack defect sizes or lengths in structures, which are mainly those fabricated by welding. In the absence of such information it is plausible to examine a normal distribution of lengths in the thickness direction of surface cracks. The population will contain a majority of short cracks, but not an infinite number of those approaching zero length. The normal distribution of defect lengths may then be conveniently described in terms of zero mean length and a standard deviation, as in figure 2. If x is crack length, the probability of interaction P of an actual crack length and a critical crack length, is

$$P = \int_0^\infty P_x N_x \, dx^*, \tag{1}$$

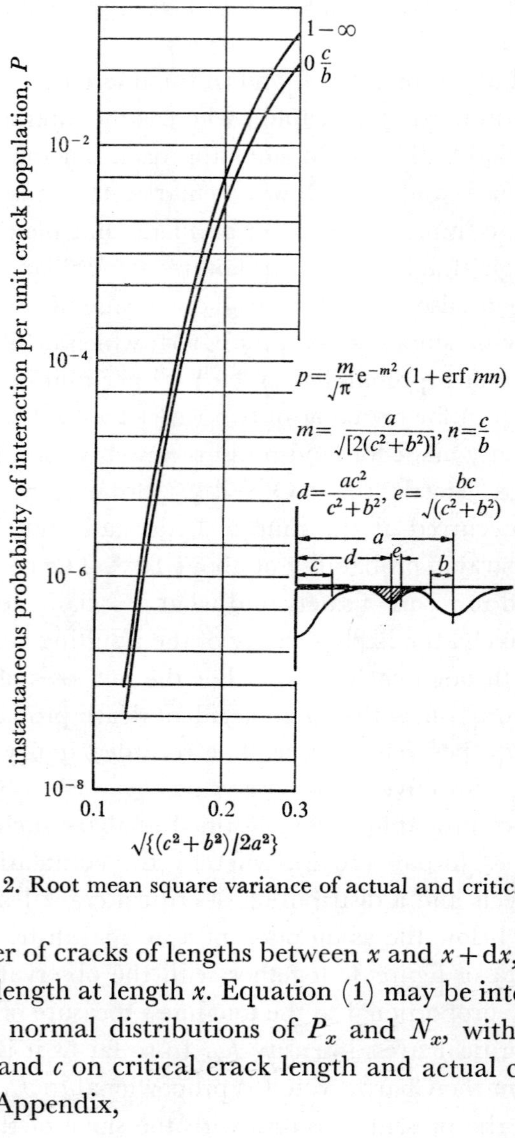

FIGURE 2. Root mean square variance of actual and critical crack lengths.

where N_x is the number of cracks of lengths between x and $x+dx$, and P_x is the probability of a crack being of critical length at length x. Equation (1) may be integrated in closed form for unit crack population and normal distributions of P_x and N_x, with mean critical length a, and standard deviations b and c on critical crack length and actual crack length respectively. The result, derived in the Appendix,

$$P = \frac{m}{\sqrt{\pi}} e^{-m^2}(1+\operatorname{erf} mn), \quad m = \frac{a}{\sqrt{[2(b^2+c^2)]}}, \quad n = \frac{c}{b} \tag{2}$$

depends strongly upon the term m, which is proportional to the reciprocal of an r.m.s. specimen of actual crack and critical crack lengths, referred to mean critical crack length. The bracketed term can vary between 1 and 2, and represents the cut-off at zero crack length. P is plotted

* See also p. 198.

against the r.m.s. scatter $\sqrt{\{(b^2+c^2)/2a^2\}}$ for various values of c/b in figure 2. From the viewpoint of material development figure 2 usefully summarizes the relative effect on failure probability that increasing the consistency of the fracture toughness property can have.

For instance, in the cases reported by Boyd (1970) for oil storage tanks and ships, where the materials could be considered to conform to the relative fracture toughness distribution of figure 1, the ratio b/a would be 0.35. Such structures are habitually operated at low stresses, and critical crack (semi-) lengths are large, of the order a = 300 mm. The standard deviation actual crack length c would be known from observation to be negligible in comparison, so that r.m.s. scatter can be put at $0.35/\sqrt{2} = 0.25$, for which the calculated total probability of failure throughout life for unit crack population, from figure 2, is 1.8×10^{-2}. But the observed failure probability per structure-year is 3×10^{-4} for oil storage tanks, so that for an average life of 20 years this result is consistent with an average defect population of $\frac{3 \times 10^{-4} \times 20}{1.8 \times 10^{-2}} = \frac{1}{3}$ defect per structure. This is quite plausible, and would also be compatible with an observed relative insensitivity to defect sizes, because most of the r.m.s. scatter is due to material variability. At another extreme, perhaps representative of more highly stressed pressure vessels operated above a fracture appearance transition temperature, the standard deviations on critical crack and actual crack lengths would be comparable. Under such conditions a failure probability of 10^{-5} per vessel year with a 20 year life ($P = 2 \times 10^{-4}$) and an average of $\frac{1}{3}$ defect per vessel would require a scatter on fracture toughness not exceeding 17%. The corresponding figure at 10^{-6} would be 15%. These are readily attainable values with the use of standard quality control methods. The suggested limits are also attainable with respect to defect sizes, and encourage improvements in both cases.

There has been much complaint concerning the scatter observed in toughness measurements using the c.o.d. criterion, and there will be similar observations with J_c measurements under elastic-plastic conditions as experience is accumulated, since there is firm evidence to indicate that the two criteria are closely related. Probability of failure considerations should indicate that the improvements in consistency being aimed at are not only realistic in terms of material supply, but that the industries concerned, with specification bodies, could usefully consider the abandonment of material acceptance rules based upon arbitrary minimum values, in favour of more realistic standard deviations and means.

Future prospects for strong materials in civil structures

The metallurgist's crowning technological achievement during the past century has been to raise appreciably the limit of proportionality of structural metals, and to cope successfully with many of the modes of failure, the risk of which is thereby reduced. Most of the opportunities created by these developments have already been exploited in aerospace applications. At the present juncture, manufacturing industry at large stands at the threshold in using some of these possibilities, and the substitution of quenched and tempered ferritic steels for as-worked and normalized versions is a good example. This substitution in bridges and pressure vessels, which has gone further in the U.S. than elsewhere, has led to a few additional collapse problems, but very many more successes, including the heaviest load bearing components in long span bridges. For perspective, it may be noted that the steel employed in the Forth rail bridge was high strength in its time, being at about the same level as in the recently completed Rio-Niteroi long span bridge.

The factors which influence the decision to make greater use of quenched and tempered steel in civil structures are delicately balanced, and worthy of an attempt at definition. The problem of maintaining fracture toughness is an obvious one, and is by no means insoluble, including the fusion welded configuration. The counterproblem is susceptibility to compressive buckling instability at increased operating stresses, without benefit of a comparable increase of elastic modulus. The designer's path between these extremes may be compared with progress along a narrow corridor in which, for the sake of better access, it is necessary to prise the walls apart. Thus, the attractive way in which to avoid notch ductility and weldability problems is to use thin sections, and to avoid buckling problems is to use thick ones. The designer has immense scope on the creative front in which to compromise with these extremes, but the very existence of this scope may delay the emergence and recognition of the best configurations. A few of the characteristic approaches may be listed, in order that their relevance may be observed. The medium-to-long span bridge is uppermost in consideration in this exercise, because strength with economy is here most challenging.

The redundancy approach is well established, and typified in the interleaved thin plate and stranded wire constructions respectively of the two Forth bridges. The fracture problem is largely offset by use of thin sections, but the disadvantage is aggregation of surface. Every British schoolboy knows that it takes 7 years to paint the first Forth bridge and that, as soon as it is finished it is recommenced. The history of corrosion protection of stranded wire suspension cables is also protracted, but the problems are soluble.

Despite the Quebec bridge compression failures, the logic is sound of using truss constructions to avoid buckling failures, because the individual members are shorter, while being of compact cross-section. The section thicknesses of these members reach the limit for easily achieved fracture resistance. Joints between truss members may require a rare skill to design in detail at the larger sizes. In spite of such difficulties, there are good modern examples of welded and bolted truss bridges of medium spans in Japan and the U.S. The former are seen from the high speed Tokaido railway line, and demonstrate that the truss type of design can be aesthetically attractive.

The thin walled welded box girder bridge favoured in Europe is optimal for manufacture and erection. It has a good appearance, and has established a revolution in reducing bridge building costs. It minimizes the fracture problem by using relatively thin butt welded plates. Unfortunately, recent experiences have made evident a susceptibility to buckling problems during erection. This disadvantage, and another due to the fatigue cracking propensity of fillet welded connexions associated with wheel loads on the deck may be traced to the need for internal stiffening diaphragms, stringers and transverse beams, in order to stiffen against concentrated loads and buckling.

The first mentioned two types of bridge structure are compatible with high strength steel construction, although the advantage is diminished by the high labour content of these constructional forms. Few engineers would now envisage scope for building box girder bridges with quenched and tempered steels or would see any reward for doing so. This course of action would entail higher working stresses, thinner sections, closer stiffener spacings and dimensional tolerances to obviate the risk of buckling during erection, and therefore a significantly higher labour content and cost. The fatigue strength of fillet welded attachments is not increased when yield strength is increased, so that higher working stresses would entail more susceptibility to fatigue cracking, as well as higher live load deflexions.

The prospect of using higher strength steels in box girders is increased if more use is made of the composite principle. This has been successfully applied in the combination of concrete decks and welded steel underframes. There have recently been encouraging trials at Queen's University, Belfast of laboratory scale box girders stiffened with expanded polymeric cores (Wells 1972; Montgomery 1974). These cores are lighter than diaphragms and stringers and obviate fillet welded joints. The advantages of the continuous core are firstly, that the buckling strength conferred on the skin by the continuous core depends upon the relative elastic moduli of skin and core and not at all upon the thickness of the skin, and secondly that the inner surface of the box is protected from atmospheric corrosion. A long standing parallel in the aircraft industry is seen in the use of honeycomb construction, and the present generation of hovercraft makes a substantial structural use of this form. The generality of core supported construction should also have long term potential for the hulls of ships.

Although this discussion, for the sake of brevity, has centred on bridge constructions, it should be emphasized that the steel successfully developed for heavy walled welded pressure vessels for light water reactors is quenched and tempered, and is of substantial yield strength. These considerations all suggest that there is much remaining potential for the further utilization of high strength steels, and it is likely that the same is true for other metals.

The influence of production methods on mechanical properties used in design

It is widely recognized that it is unsuitable for material users to specify methods of production or to impose positive limits on chemical compositions of materials. The most effective interface resides with the mechanical and metallurgical properties through which suitability for service is determined. Nevertheless, it is inevitable that consumers should take an intelligent interest in production methods, since they bear on the important questions of specification and tolerances. Changes of production methods may also introduce changes of performance which are detected after the event, and which must be defined with the object of modifying specifications. It may even be necessary to introduce new forms of test.

The need to sustain and enhance weldability has demanded much attention at the specification interface in recent years. One example for consideration could be the incidence of lamellar tearing due to restraint of weld cooling contractions at T joints. The plastic strains that are induced in cooling are considerable. Whereas the restraint of longitudinal deformations at fusion welded butt joints could account for values of about 1% plastic strain, the shear components near the molten weld pool have been demonstrated by Moiré measurements, such as those due to Johnson (1972) and at N.C.R.E. (Cargill 1968) to be as large as 5%. Lamellar tears are formed where the ductility of the material is insufficient in the through-thickness direction to sustain such strains. Through-thickness ductility may be impaired where mechanical working has been used to good effect to maximize in-plane ductility. It was shown in 1952 by Soete and by Matton-Sjöberg that steel in which a low brittle transition temperature has been induced with respect to the longitudinal and transverse directions, by adopting a low roll finishing temperature, may in some circumstances owe its high Charpy V energy absorption to incipient lamellar tears, which are clearly evident in the fracture surfaces, and which appear to diminish triaxiality by relaxing through-thickness stresses. This may be one reason why lamellar tears at T weld joints seem to have caused few, if any, brittle cleavage fracture casualties.

With the perspective of several years since the buckling failures of box girder bridge beams during erection it is still possible that there are further lessons to be learned. Careful studies both of dimensional irregularities and of weld residual stress distributions have been made, and the likely contributions of these to inadequate buckling resistances have been assessed. They do not quite completely explain the below-yield failures in some of the stiffened panel tests that were conducted. A factor that was recorded in at least one case was the incidence in skin plates, at the time of secondary damage, of locally propagated lamellar tears at stiffener fillet welds. Being secondary impact damage these tears had no perceptible influence on the incidence of the primary failures. The very existence of these tears also confirmed the resistance to brittle cleavage of the materials. Controlled rolling was used in the manufacture of these materials. In the light of the results of the compression tests of expanded polymer supported steel bars briefly reported earlier in this paper, the question is prompted as to the residual stress effects within the plate thickness on compressive stress–strain curves of virgin plate material (as contrasted with weldments), in the sense that they might be affected by low finishing temperatures in controlled rolling. Such tests do not appear to have been conducted. The purpose of this question is not to diminish the standing of the controlled rolling production method, which is both valuable and a definite commitment, but to suggest a limited investigation. If the effect were found to be significant, the substitution of stretched, normalized or higher yield steels (without change of design stress) in certain compression structures or components might be found to be a better optimized solution to the problem than the imposition of barely attainable dimensional tolerances in welded fabrication.

Discussion and conclusions

This paper ends as it began, with an acknowledgement that a memorial symposium is not just an abstract but respectful gesture. That the paper is partly concerned with earlier references is a reminder that effective retrieval of information is an important adjunct to new and original work. A perspective view is intrinsic to the type of review that has been attempted.

Firm conclusions are not easily extracted from such a broad topic and brief treatment, but two appear to justify emphasis. Firstly, in material testing at the interface between metallurgist and design engineer, there should be a conscious recognition of the different objectives of design data accumulation and quality control. The same tests will suffice for both in some but not in all cases. This situation is compatible with a total synthesis if accompanied by an adequate and sustained exchange of information. Secondly, at the stage of refinement following a full development of mechanical properties in a material, it may be concluded that considerable attention should be given in design as well as material supply to the consistency with which mechanical properties are achieved. Both social and economic advantages may be gained thereby.

Without committing them to the views expressed here the author wishes to thank Professor P. B. Hirsch, Dr R. O'Neil and Dr R. Nichols for their active stimulus and help towards acquiring an appreciation of the probabilistic treatment of scatter in notch toughness tests.

References (Wells)

Anon 1970 The assessment of wind loads, *Building Research Station Digest* 119, July 1970.

Anon 1974 *Basic road statistics* 1974. British Road Federation.

Baker, J. F., Horne, M. R. & Heyman, J. 1956 *The steel skeleton*, vol. 2. Cambridge University Press.

Becher, P. E. & Pedersen, A. 1973 2nd International Conference on Structural Mechanics in Reactor Technology, Berlin, September 1973, Paper M6/4.

Boyd, G. M. 1970 *Brittle fracture in steel structures*, pp. 5–11, Fig. 6.4. Navy Department Advisory Committee on Structural Steel. London: Butterworth.

Bull, F. B. 1954 *Structural trials on Ocean Vulcan and Clan Alpine*; *Discussion of results*, *Report R*12, p. 47. H.M.S.O.: Admiralty Ship Welding Committee.

Cargill, J. 1968 Measuring strains under welds using the Moiré fringe technique, MOD, NCRE Report 551, December.

Johnson, L. D. 1972 Moiré fringe analysis of welding strains. Ph.D. Thesis, University of Strathclyde, 1972.

Kellermann, O. & Tietzle, A. 1969 I.A.E.A. *Tech. Rep.* No. 99.

Kirkaldy, D. 1862 Experiments on wrought iron and steel, 4 Corunna Street, Glasgow, 15 July 1862.

Matton-Sjöberg, P. 1952–53 *West of Scotland Iron and Steel Inst.* **60**, 1952–53.

Montgomery, S. J. 1974 Core supported structures. M.Sc. Thesis, Queen's University, Belfast, 1974.

O'Neil, R. & Jordan, G. M. 1972 I.Mech.E., App. Mech. Group Conf., Periodic inspection of pressure vessels, C48/72, p. 140.

Phillips, C. A. G. & Warwick, R. G. 1968 A survey of defects in pressure vessels. ANSB(S)R162, 1968.

Soete, W. 1952–53 *J. West of Scotland Iron and Steel Inst.* **60**, 1952–53.

Taylor, J. 1955 *Symposium on fatigue of metals*. Nottingham University, September 1955.

Wells, A. A. 1972 *Composites*, May 1972, p. 112.

Williams, M. L. 1954 A.S.T.M., S.T.P. 158, September, 1954, p. 11.

Wright, M. A. & Iannuzzi, F. A. 1973 *J. Composite Materials* **7**, 430.

Appendix. Probability of interaction of actual and critical crack lengths

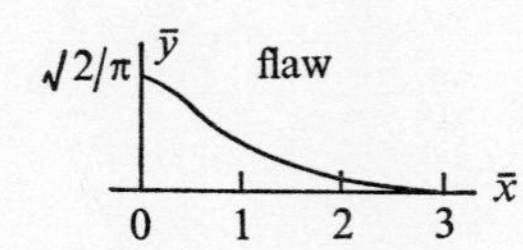

$$y = 1\sqrt{(2/\pi)}\, e^{-\frac{1}{2}\bar{x}^2}$$

$$\int_0^\infty \bar{y}\, d\bar{x} = 1, \quad \int_0^\infty \bar{x}^2\bar{y}\, d\bar{x} = 1$$

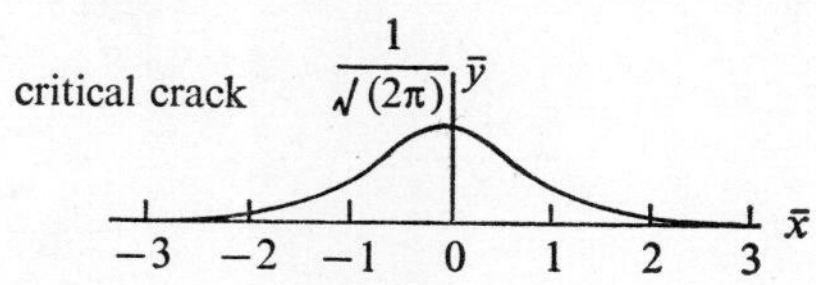

$$\bar{y} = 1/\sqrt{(2\pi)}\, e^{-\frac{1}{2}\bar{x}^2}$$

$$\int_{-\infty}^\infty \bar{y}\, d\bar{x} = 1, \quad \int_{-\infty}^\infty \bar{x}^2\bar{y}\, d\bar{x} = 1$$

Let mean flaw depth = 0, with standard deviation c, and critical crack depth = a, with standard deviation b. The distance x is measured from mid-plane.

The required probability of interaction is given by

$$P = \frac{a^2}{bc}\sqrt{\frac{2}{\pi}}\sqrt{\frac{1}{2\pi}}\int_0^\infty \exp\left(-\frac{x^2}{2c^2}\right)\left\{-\frac{1}{2}\frac{(a-x)^2}{b^2}\right\} d\left(\frac{x}{a}\right) = \frac{a}{\pi bc}\int_0^\infty \exp\left[-\frac{1}{2}\left\{\frac{x^2}{c^2}+\frac{(a-x)^2}{b^2}\right\}\right] dx. \tag{1}$$

Put
$$\frac{1}{2}\left[\frac{x^2}{c^2}+\frac{(a-x)^2}{b^2}\right] = A^2(x+B)^2+C;$$

then
$$A^2 = \frac{b^2+c^2}{2b^2c^2}, \quad B = -\frac{ac^2}{b^2+c^2}, \quad C = \frac{a^2}{2(b^2+c^2)},$$

and
$$\int_0^\infty e^{-A^2(x+B)^2-C}\, dx = e^{-C}\int_0^\infty e^{-A^2(x+B)^2}\, dx = e^{-C}\int_B^\infty e^{-A^2u^2}\, du = \frac{e^{-C}}{A}\int_{AB}^\infty e^{-v^2}\, dv$$
$$= \frac{\sqrt{\pi}\, e^{-C}}{2A}(1-\operatorname{erf} AB).$$

Then $$P = \frac{1}{\sqrt{\pi}} \frac{a}{\sqrt{\{2(b^2+c^2)\}}} \mathrm{e}^{-a^2/2(b^2+c^2)} \left(1+\operatorname{erf}\frac{c}{b}\frac{a}{\sqrt{\{2(b^2+c^2)\}}}\right) = \frac{m}{\sqrt{\pi}} \mathrm{e}^{-m^2}(1+\operatorname{erf} mn). \tag{2}$$

Maximum interaction occurs at depth $$-B = d = \frac{ac^2}{b^2+c^2}, \tag{3}$$

the corresponding standard deviation $$\frac{1}{\sqrt{2}\,A} = e = \frac{bc}{\sqrt{b^2+c^2}}. \tag{4}$$

2.2 Design against brittle fracture

By R. R. Barr† and F. M. Burdekin‡
† *British Steel Corporation General Steels Division, Research Centre (Scotland), Meadow Road, Motherwell ML1 1QB, Lanarkshire*
‡ *Messrs Sandberg, Consulting, Inspecting and Testing Engineers 40 Grosvenor Gardens, London SW 1*

[Plate 7]

The paper assesses the current position and future trends in a complete definition of the problem of brittle fracture. It is concluded that there is a need for a more uniform approach to the specification of toughness requirements in present design standards, and ways of achieving this are indicated. The acceptance of the presence of defects leads to the use of fracture mechanics techniques in defining conditions for avoiding fracture initiation. A new treatment for a toe crack in a fillet weld is illustrated, as an example of this approach.

Because of interpretation and experimental difficulties in assessing the fracture toughness of weld metals and heat affected zones, it is concluded that it would be inappropriate to use the c.o.d. test, on its own, to specify material toughness levels for National Standards. Some of these difficulties are illustrated and areas for future research are discussed, including the role of the physical metallurgist.

Introduction

In all major steel producing countries of the world, extensive research funds have been committed towards seeking an all-embracing solution to the problem of brittle fracture. It is not the intention in this paper to dwell on a historical review of past efforts, but to turn instead to an assessment of the immediate current position and the likely future trends in a complete definition of the problem. Major catastrophic brittle fractures are relatively rare today, despite a move to more onerous design conditions, and this is entirely due to research already carried out to gain an understanding of the many parameters which have to be controlled to eliminate the possibility of fracture.

The physical metallurgist has contributed greatly to an understanding of how various factors such as yield strength, grain size, basic microstructure and the existence of second phase particles such as carbides and inclusions influence the conditions for unstable fracture. In engineering terms, the greatest recent advance in seeking to control the problem has undoubtedly been developments in fracture mechanics as a means of defining the critical parameters for fracture initiation. There are, however, a number of aspects which require further research to achieve an agreed rational basis for design against brittle fracture which recognizes the existence of defects and which can define them as either tolerable or non-tolerable. Some of these factors are dealt with in the present paper.

Although the incidence of brittle fracture has decreased it is still feared as a possible cause of catastrophic failure and it is natural, in these circumstances, that the introduction of very large fabrications as offshore installations in the North Sea should lead to an upsurge of activity in brittle fracture research. The more onerous conditions of this particular application and the uncertainty in estimating operating stresses in complex designs, together with the probability that defects may grow due to severe fatigue loading conditions, have encouraged efforts to define brittle fracture in terms of acceptable defect limits using fracture mechanics principles. Oil and gas finds have also led to a considerable increase in the tonnage of large pipelines and in the case of natural gas in particular a study of the principles of design against unstable fibrous fracture is being extensively examined.

The present paper examines some of the additional refinements which may be necessary in an engineering definition of tolerable defect sizes and discusses ways in which physical metallurgy can assist in solving existing problems.

Present bases for design

In any structure tension stresses inevitably occur in some regions of the structure and, in most welded fabrications, externally applied operating stresses are magnified locally at geometric stress concentrations. Additionally, there may be a further contribution in the form of residual tensile stresses. In theory, brittle failures can be avoided by elimination of defects from tension stress areas. Unfortunately, if the material toughness is too low, due perhaps to metallurgical embrittlement, and the stress level is high enough, fracture initiation may occur from defects which are too small to be detected, even allowing for full non-destructive testing and inspection.

In the United Kingdom and Europe, the selection of steels for containment and support structures is based on the general assumption that undetected weld defects may be present and that fracture initiation from these defects should be avoided by selection of materials of a high enough toughness level. A second, and more conservative basis for design, is stated to be employed in the United States, which accepts that it may be impossible to guarantee completely that fracture initiation will not occur from a brittle area of a weld. Thus a base metal with sufficient toughness to arrest a crack emerging from such a region is required. A third option is applied to 'high risk' structures, especially where the stored energy is high, as encountered in natural gas pipelines. In this situation the design is based on the possibility that fracture initiation might occur as a result of some unexpected external source, and that all regions of the weld and parent plate should be capable of resistance to fracture propagation.

However, the avoidance of fracture initiation is, by far, the most commonly applied basis for design and there is, as yet, no uniformly agreed technical basis being practiced to define the various toughness requirements applicable over a range of different structures. The necessary toughness will change for various structure depending upon a variety of factors, but designs standards are frequently unrelated in terms of 'clear-cut' adjustments to toughness requirements, according to conditions of service and the complexity of the welded joints in the structure. This situation may be illustrated by reference to some current British Standards.

BS 1515 Appendix C for pressure vessels and BS 2654 and BS 4741 for storage tanks, use the results of welded wide plate tests to assess fitness for service, the results of these tests being related to the Charpy impact test energy values for parent plate in the form of test temperatures to achieve certain minimum impact energies.

Other general structural standards such as BS 153 for bridges, BS 449 for steel buildings, quote a maximum thickness for a particular grade of steel, the level of thickness being mainly related to practical experience and Charpy impact levels in parent plate. Thus the basis for selecting a steel with adequate fracture toughness is different and as a result the structural standards permit plate steels to be used up to 75 and 100 mm thick in the as-welded condition, whereas the containment standards recommend stress relief above 38 mm. A comparison between these various standards including more recent proposals under the Merrison Rules for box girder bridges, and proposals being considered for offshore structures is shown in figure 1, for a specific commonly used design temperature (−10 °C). It is assumed tacitly that the differences between standards are related to variations in design stresses and stress concentration features. However, modern bridges can have areas of stress concentration as high as, for example, nozzles in a pressure vessel and some caution is warranted in using steels to the maximum thickness given in current BS 153 and BS 449.

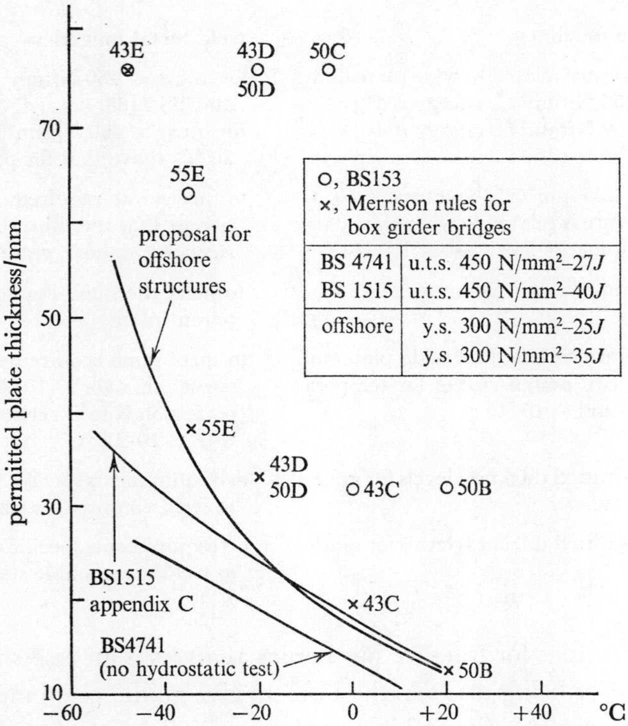

O, BS153 ×, Merrison rules for box girder bridges	
BS 4741	u.t.s. 450 N/mm²–27*J*
BS 1515	u.t.s. 450 N/mm²–40*J*
offshore	y.s. 300 N/mm²–25*J*
	y.s. 300 N/mm²–35*J*

FIGURE 1. Comparison of toughness for various British Standards. Design temperature −10 °C.

Standards also differ in the way they fix or recommend weld metal and heat affected zone toughness, as summarized in table 1, and there is a need to clarify the approach to the specification of these toughness levels, and the way in which they are measured, so that differences between standards can be clearly reconciled.

FRACTURE MECHANICS ANALYSES

The applications of the concept of 'plane strain fracture toughness' K_{Ic}, are limited to materials which fracture without serious deviation from linear elastic conditions (ASTM 1967). However, for many lower strength structural materials, such as the steels used for bridges, offshore structures and pipelines, it is found that valid plane strain fracture toughness measurements cannot be made for the thicknesses, temperatures and rates of loading of practical interest. To compare toughness measurements in regions beyond the validity of linear elastic conditions, and to enable some assessment to be made of effects at geometric stress concentrations where local yielding occurs, the general yielding fracture mechanics concept of 'crack opening displacement' (c.o.d.) was introduced by Wells (1961, 1971). Developments in the U.S.A. in this area have tended to concentrate on the 'crack tip contour integral' J introduced by Rice (1968), Rice *et al.* (1973). A review of these two methods of studying crack tip conditions and assessing toughness outside linear elastic conditions was prepared recently by Turner & Burdekin (1974).

TABLE 1. COMPARISON OF WELD METAL TOUGHNESS REQUIREMENTS IN VARIOUS STANDARDS

standard	basis for plate toughness	weld metal toughness
BS 1515C	C_v test temperature related to wide plate data: for u.t.s. $\leqslant$ 450 N/mm² C_v energy is 27J. for u.t.s. > 450 N/mm² C_v energy is 40J.	for u.t.s. $\leqslant$ 450 N/mm² – conform to BS 639 and BS 4165. for u.t.s. > 450 N/mm² C_v test temperature + 20 °C above that for parent plate
BS 2654	for plates > 12.5 mm or above grade 43. C_v test temperature is related to wide plate data; same impact energy levels as for BS 1515	to meet same requirements as for parent plate except that the min. C_v energy is 27J for all steels even above grade 43
BS 4741	based on a similar requirement to BS 2654	to meet the same impact requirements as for parent plate
offshore proposal	C_v test temperatures based on wide plate and c.o.d. data; two design curves for temperatures +4 °C and −10 °C	to meet same requirements for +4 °C design curve; thus for −10 °C design temperature corresponds to a relaxation in test temperature of 10–12 °C
BS 153	maximum permitted thickness levels for grades of BS 4360	no requirements specific to this standard, refers to weld consumable standards
BS 449	maximum permitted thickness levels for grades of BS 4360	no requirements specific to this standard, refers to weld consumable standards

One of the major applications for fracture mechanics analyses is an assessment of the significance of defects. Attempts are being made in the U.K. to give guidance on methods for deriving acceptance levels for flaws in welds (Burdekin 1975*b*) and similar steps have been made at the International Institute of Welding (Burdekin 1975*a*). As a conservative assumption, the present use of c.o.d. method at stress concentrations requires that the flaw be considered to lie entirely with the maximum strain field as a uniform strain up to a specified distance from the peak position. Recent finite element studies of stress intensity factors and crack tip contour integrals in such regions have suggested that this is over conservative, where yielding at the stress concentration region is entirely contained within elastic material. The magnitude of the peak strain may be greater than purely elastic strains but the strain gradients are such that defects of any size are not contained entirely within a uniform field.

Figure 2 shows the finite element predictions of stress intensity factor magnification for defects at the toe of 45° fillet welds of different sizes relative to the plate thickness, as compared to the basic case for the same defect without the fillet weld. These have been deduced from original computations by Hayes & Maddox (1972) Maddox (1972). The results show that the magnification effect at the toe of the fillet dies away very rapidly, and even for a fillet of leg length equal to the plate thickness the effect has largely died away at a flaw size/thickness(t/e) ratio of 0.1.

The IIW and the provisional of draft BS rules require the flaw to be considered as in a uniform field of three times the applied stress, for flaws up to $t/e = 0.15$. As a simplified treatment a flaw size parameter $\bar{a}$ is given by:

$$\bar{a} = C(K_{Ic}/\sigma_y)^2 \quad \text{or} \quad \bar{a} = C(\delta_c/\epsilon_y),$$

where σ_y is yield stress, δ_c critical c.o.d., ϵ_y yield strain and C a constant dependent on applied (+residual) stress conditions (Burdekin & Dawes 1971).

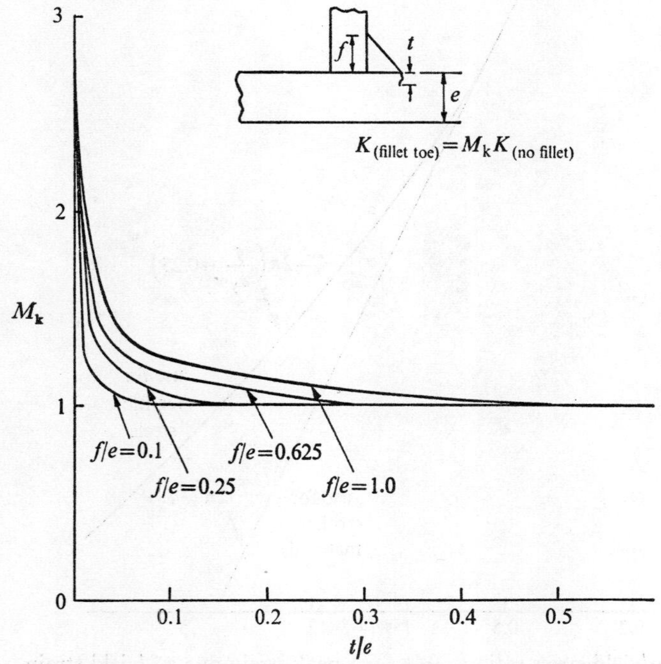

FIGURE 2. Stress intensity magnification factors due to fillet weld (45°).

Figure 3 shows the curve for values of the constant C against stress or strain level, as adopted in the IIW and BS documents. The effects of defects at the toe of different fillet weld sizes may be found by adjusting the stress/strain axis in figure 3 using the appropriate value of Mk (stress intensity magnification) from figure 2. The basic curve of figure 3 is used to derive the defect tolerance parameter $\bar{a}$, which is converted to actual flaw dimensions using the curves of figure 4, which take account of thickness effects, etc. The curves of figure 3 incorporate a basic safety factor of 2 on defect size. The application of the above treatment is illustrated for a particular size of fillet and T-butt weld in figure 5. This graph illustrates the toughness level in terms of c.o.d. which would be required to prevent initiation of a toe crack from within a particular t/e ratio and compares the discontinuous steps involved in the assumption that the flaw be considered to be within a uniform M_k field for t/e ratio up to 0.15. From this example it can be seen that very low levels of toughness would be necessary to permit fracture initiation from a small toe crack. Thus, there would be little chance of fracture from a small undetected defect unless the h.a.z. toughness was very low. The principles involved in this treatment can be extended to include stress concentration features of T-butt welds (see figure 5) and other small zones of low toughness due to metallurgical features. If the critical defect size calculated is greater than the size of the zone of low toughness or the zone of influence of the stress

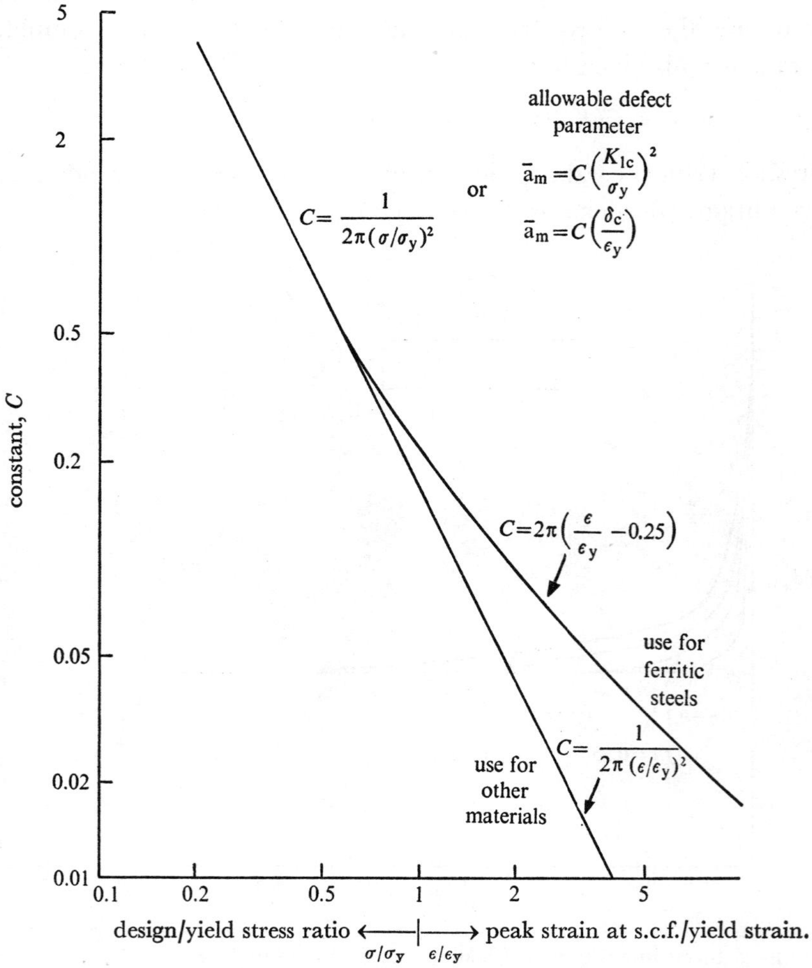

FIGURE 3. Values of constant C for different loading conditions.

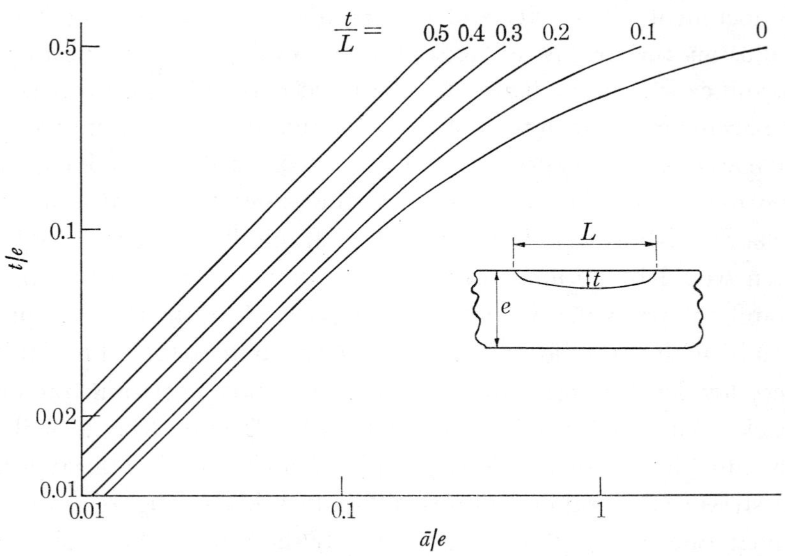

FIGURE 4. Relation between actual defect dimensions and the parameter $\bar{a}$; surface defects.

concentration, it cannot be considered to be possible for initiation to occur within the influence of either of these features. The metallurgical features are considered in more detail in the next section.

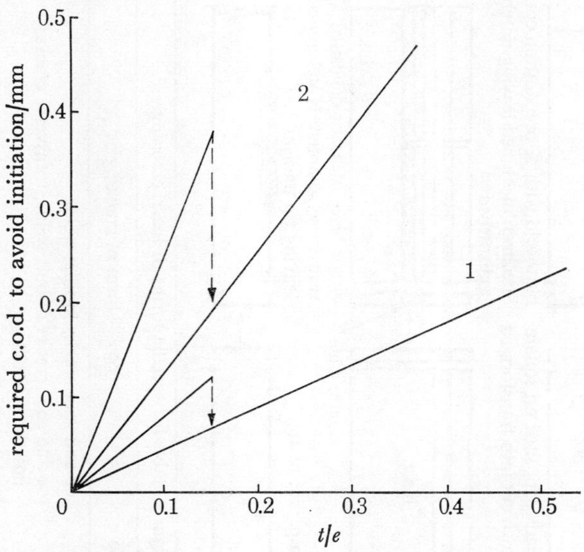

FIGURE 5. Comparison of c.o.d. levels to avoid initiation from a defect within a given t/e ratio. 1, fillet weld, $e = 25$ mm, $f/e = 0.25$ (figure 2); 2, T-butt weld, $e = 75$ mm, $f/e = 0.25$.

METALLURGICAL IMPLICATIONS

The previous section has shown how fracture mechanics theory, using laboratory test parameters, K_{Ic} or c.o.d., may be applied to the calculation of acceptable defect levels. Defects almost invariably occur in or close to a welded joint and the regions of lowest fracture toughness are generally associated with weld metal or heat affected zones, especially in high strength low alloy ferritic steels currently used in large tonnages for many fabrications. When considering a fracture initiation criterion, the toughness of the uniform plate material remote from stress concentration regions and residual stress effects is largely irrelevant and it is the response of the microstructure of these steels to welding variables, which is important. For non-grain controlled steels damage due to strain ageing effects especially in the presence of a pre-existing defect has been extensively examined (Woodley, Burdekin & Wells 1964). Present trends are to use higher yield strength and tougher grain controlled steels which are not susceptible to strain ageing and, as a result, considerable attention has been focused, more recently, on the fracture toughness of weld metals and heat affected zones (Baker *et al.* 1973).

(*a*) *Test piece requirements*

Certain difficulties are associated with the design of welded laboratory fracture toughness tests to simulate welded joints in practical structures and the interpretation of subsequent results.

A comprehensive survey by Dolby & Archer (1971) demonstrated how fracture toughness tests could be made to examine h.a.z. toughness with the notch orientation (preferably a sharp fatigue crack) simulating that of the more common defects. Test designs were sub-divided into those with the notch positioned after welding, and those with the notch made at an

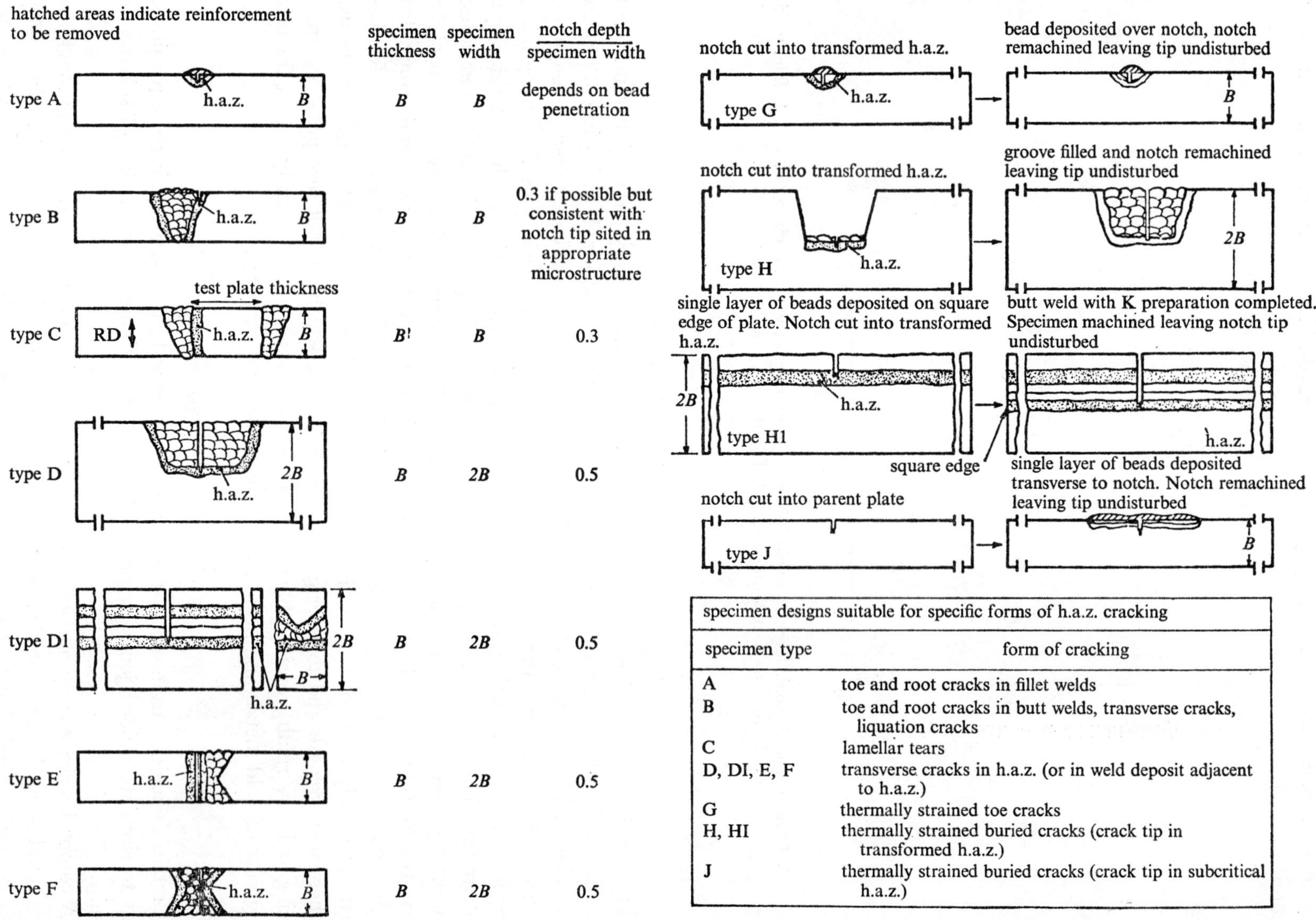

specimen designs suitable for specific forms of h.a.z. cracking	
specimen type	form of cracking
A	toe and root cracks in fillet welds
B	toe and root cracks in butt welds, transverse cracks, liquation cracks
C	lamellar tears
D, DI, E, F	transverse cracks in h.a.z. (or in weld deposit adjacent to h.a.z.)
G	thermally strained toe cracks
H, HI	thermally strained buried cracks (crack tip in transformed h.a.z.)
J	thermally strained buried cracks (crack tip in subcritical h.a.z.)

FIGURE 6. C.o.d. test piece designs for the assessment of welded joints.

intermediate stage to measure thermal straining effects. Dolby (1974) subsequently illustrated the influence of notch orientation on c.o.d. test results proposing minor modifications to the test piece designs. A full summary of the test piece designs proposed are reproduced in figure 6.

For an assessment of weld metal toughness the test specimen shown in figure 7*a* is mainly used, although surface notched specimens can also be tested. However, for general assessment the former test is favoured, although it can be criticized on the basis that the notch orientation is unlike a practical weld metal defect.

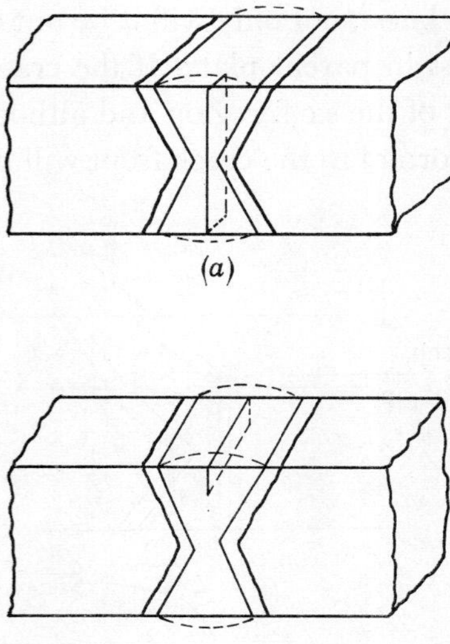

(*a*)

(*b*)

FIGURE 7. Test specimens for assessment of weld metals.

(*b*) *Measurement of heat affected zone toughness*

There are a number of difficulties in the interpretation of h.a.z. fracture toughness which involve both the physical metallurgist and the engineering fracture mechanics expert. In many cases these factors have been overlooked in the promotion of the fracture mechanics techniques, but they are of fundamental importance and are mainly concerned with the changes in metallurgical structures involved. This is an area where opinions about the best parameter to use, J or c.o.d., recede in terms of practical importance.

Dolby (1974) has concluded that significant differences in results could be obtained using different notch orientations. In addition, the toughness measurements could be markedly influenced by the adjacent weld metal toughness and yield strength, since the shape of the plastic zone formed at the notch tip prior to fracture could be affected. Another problem is the requirement to position the notch tip very accurately due to the steep microstructural gradients and variations in siting the notch are generally the reason for the observed scatter in results, as obtained by Archer (1974) and George (1975). However, provided sufficient care is taken to site the notch and subsequently check its position after testing acceptable results can be obtained for fatigue notched specimens, as shown in figure 8. Furthermore, where a small amount of fibrous tearing precedes fracture the notch tip can move out of the region being investigated. This was noted by Widgery (1974) when studying the influence of deoxidation practice on the toughness of mild steel weld metal.

Heat affected zone fracture toughness can also be studied using simulated structures, although care must be taken to match the required weld h.a.z. in terms of austenitic grain size and constituent structures. One drawback in using this technique is that the size of specimen which can be subjected to the appropriate thermal cycle is limited and the effects of triaxiality related to the use of thick plates cannot be reproduced. In addition, the test will not show the interaction effects with weld metal or parent plate referred to above.

A fillet weld situation was discussed in the previous section in terms of the influence of stress concentration on the tolerance for defects. The metallurgical implications are also illustrated in figure 9, plate 7. If the crack is very small it may well be within the s.c.f. region. However, if it lies vertically through the thickness, not only will it be out of the s.c.f. region, but it will be in a region of improved toughness in parent plate. If the crack, as is more likely, follows the fusion line it will have moved out of the s.c.f. region and although the basic fracture toughness may not alter, the applied stress normal to the crack front will have been considerably reduced.

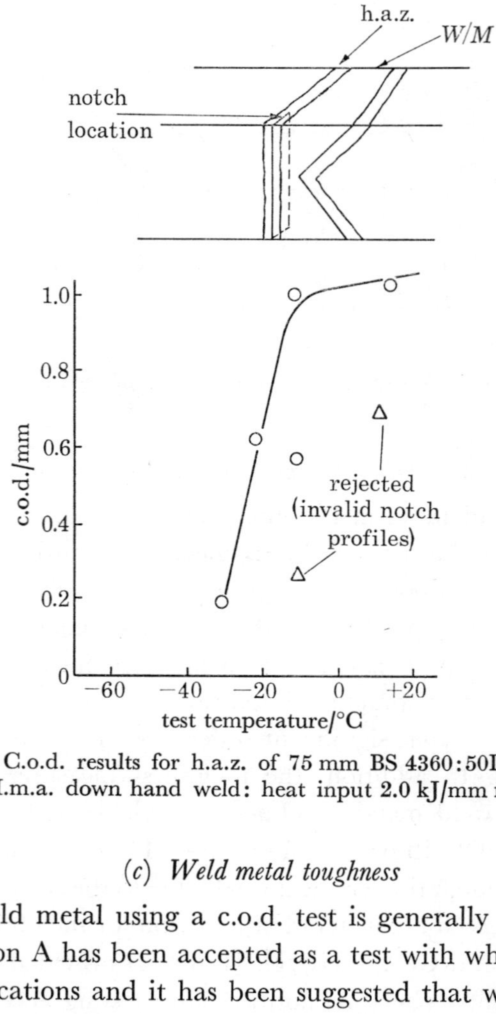

FIGURE 8. C.o.d. results for h.a.z. of 75 mm BS 4360:50D (modified). (M.m.a. down hand weld: heat input 2.0 kJ/mm max.)

(*c*) *Weld metal toughness*

The toughness of a weld metal using a c.o.d. test is generally carried out as indicated in figure 7. Test configuration A has been accepted as a test with which to identify consumables suitable for specific applications and it has been suggested that weld consumables should be required to pass a minimum c.o.d. value using this test. However, there are a number of factors associated with the metallurgical features of a weld metal which need to be taken into account and the following example illustrates how variations in toughness within the weld metal can influence fracture toughness measurements. An example is shown in figure 10 of the Charpy V-notch impact strength measured on a m.m.a. vertical up-weld using a 3 % Ni electrode. The

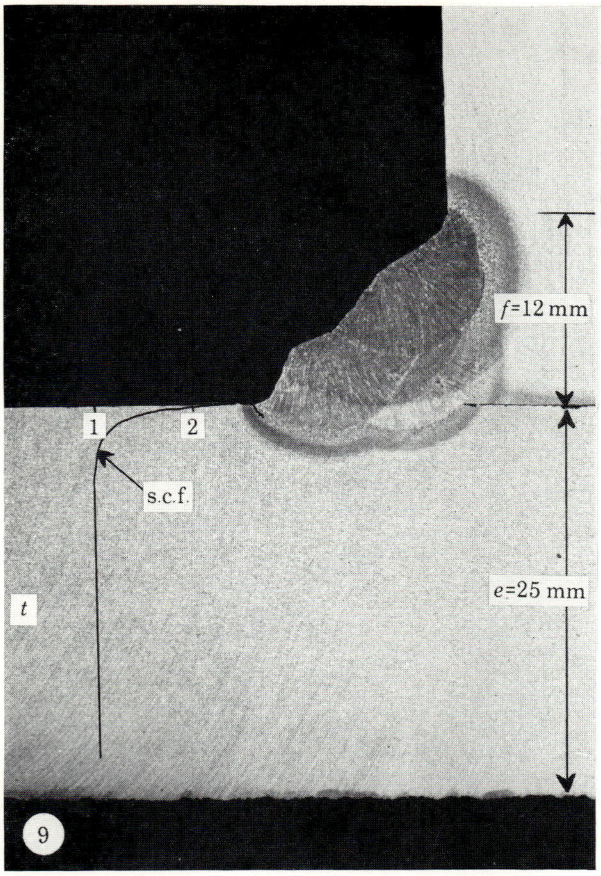

FIGURE 9. Toe crack in a fillet weld. (Magn. × 2.)

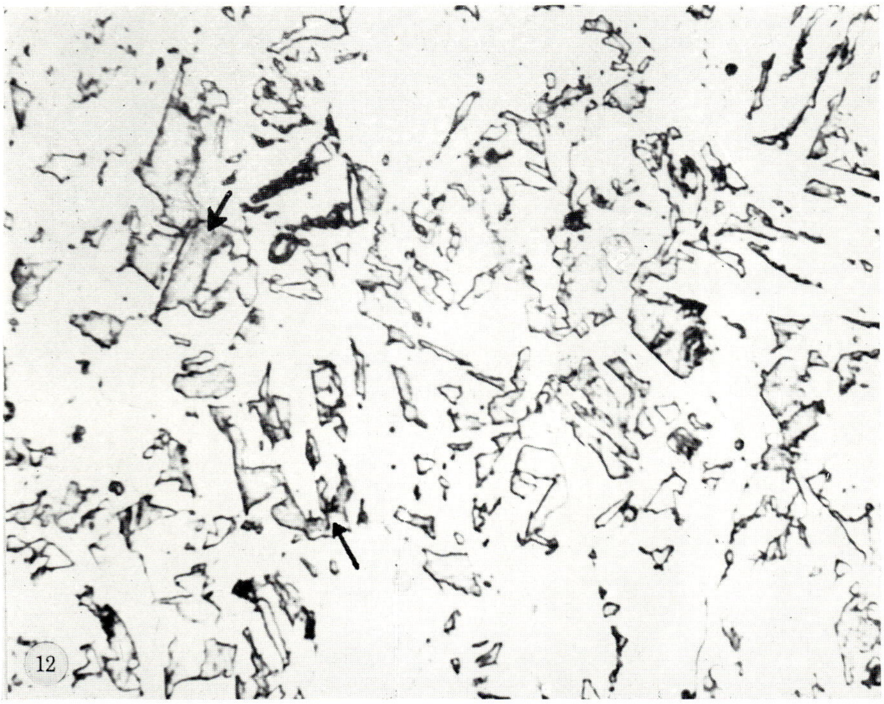

FIGURE 12. Areas of blocky lath martensite microphase in weld metal. Etched in 4% picral. (Magn. × 2000.)

(*Facing p.* 158)

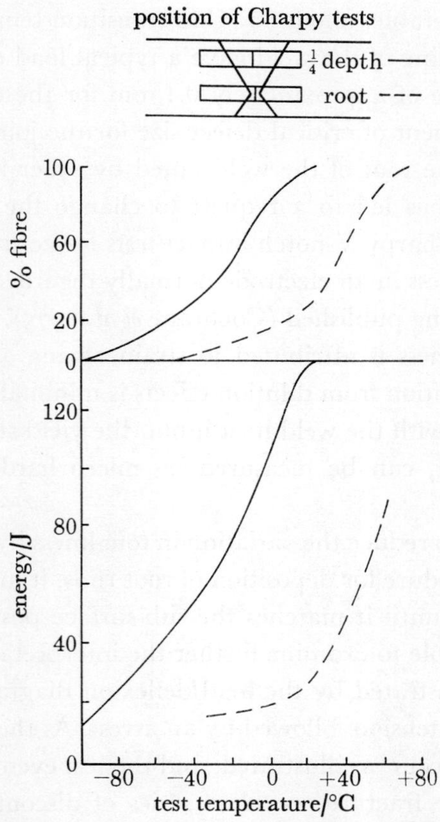

FIGURE 10. Charpy V-notch impact results for a m.m.a. – 3 % Ni electrode vertical-up weld. (Steel sample: yield strength 520 N/mm², plate thickness 40 mm.)

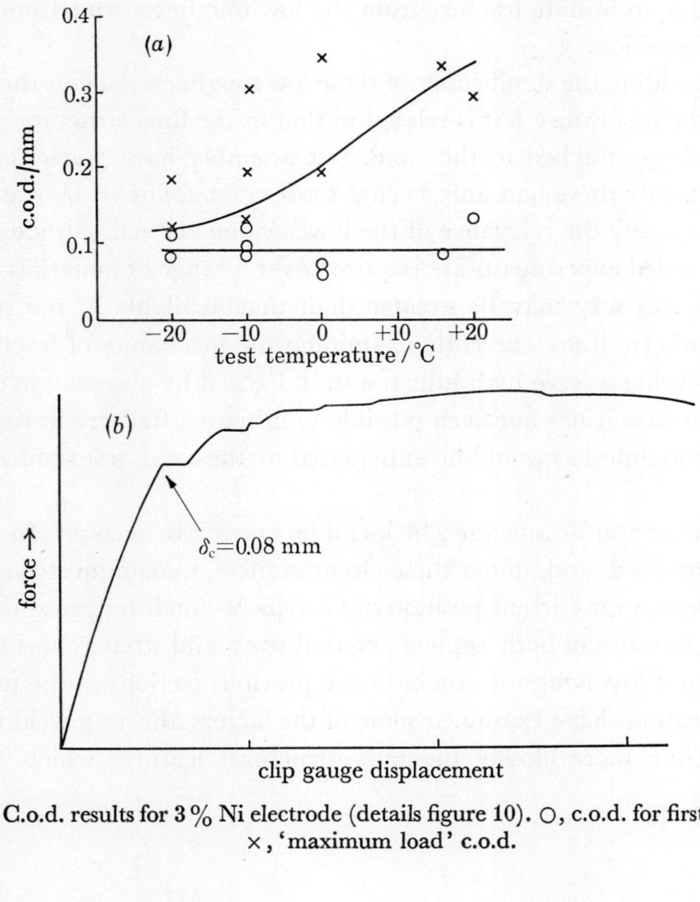

FIGURE 11. C.o.d. results for 3 % Ni electrode (details figure 10). ○, c.o.d. for first crack extension; ×, 'maximum load' c.o.d.

variation in toughness is considerable, about 55 °C in transition temperature. C.o.d. tests with the notch cut along the centre line of the weld gave a typical load displacement graph shown in figure 11, with a c.o.d. value of approximately 0.1 mm for the first short crack extension. This leads to a very low assessment of critical defect size for the joint as a whole.

Lower toughness values in the root of the weld noted by other investigators (Nicholson & Rogers 1974; Harrison 1975) has led to a request to change the method of specifying the position of the quality control Charpy V-notch impact tests for certain applications (BS 1515). The reasons for low root toughness in an electrode normally regarded as giving a tough deposit will be dealt with in a paper being published (Cochrane *et al.* 1975). At this stage it is sufficient to indicate that the low toughness is attributed to strain ageing and that the effect of steel composition in terms of contribution from dilution effects is minimal. The extent of the zone of low toughness, which will vary with the weld heat input, the yield strength of the steel and the degree of restraint at the joint, can be measured by micro hardness surveys and detailed metallographic examination.

Although it may be possible to reduce the variation in toughness by modifications to electrode composition or changes in procedure for deposition of root runs, it may be difficult or uneconomic, to improve the toughness until it matches the sub-surface position or even parent plate properties. It is therefore desirable to examine further the interpretation with respect to actual structures. For the example illustrated by the load/deflexion diagram in figure 11, the c.o.d. test gives a short initial crack extension followed by an arrest. As the load rises, a further series of short unstable fractures can occur, as illustrated, and the test eventually reaches a maximum load value. Examination of the fracture reveals a series of discontinuous central 'cleavage' fracture areas with extensive shear lips typical of a very ductile outer region. It may be possible to accept that, if the dynamic toughness of the weld metal in these outer regions were adequate, it would not be possible to initiate fracture from the low toughness zone from a chance sharp defect at that precise position.

Another factor in deciding the significance of these low toughness zones is the extent to which the stored energy of the laboratory test is related to that in the final structure. Harrison (1975) has reported that springs attached to the c.o.d. test assembly have given complete unstable fractures, where previously there had only been a short crack followed by stable growth. This observation is used to justify the relevance of the lower value of c.o.d. Although this feature is important, further detailed experiments are required over a range of materials, since the stored energy introduced in this way may be greater than that available in the structure. B.S.C. Research Centre, Motherwell are currently examining the mechanics of fracture over a wide range of loading rates using a servo hydraulic test unit, backed by observation of the behaviour in wide plate tests. To date it has not been possible to initiate a fracture in the wide plate test from the zone of low toughness as would be anticipated by the c.o.d. test results shown in figure 11 (see table 2).

Stress distribution may also be a factor which can be taken into account. In many structures bending stresses are involved, and, under these circumstances, maximum stresses are located in the outer surface (a reason for current position of Charpy V-notch test) and the stresses at the weld root will be low in terms of both applied external stress and stress concentration. Thus an argument similar to that involving toe cracks in the previous section can be used.

Many recent publications have examined some of the factors affecting weld metal toughness in an attempt to define more closely the microstructural features which control fracture toughness. This is an area of work where there is scope for a considerable research effort aimed at understanding the effect of welding parameters and the composition of the consumables.

Colvin (1974) has surveyed recently the influence of various factors governing weld metal toughness including variations in the composition of fluxes, since development in the offshore application in particular had demonstrated that a high proportion of weld consumables available were not capable of consistently achieving the necessary 'standard' of toughness. In

Table 2. Wide plate results (single weld transverse to direction of loading) on weldment with Charpy and c.o.d. properties given in figures 10 and 11

test temperature	notch detail	maximum stress N/mm²	maximum strain and other details
−10 °C	surface notch 9 mm deep × 56 mm long	610	unbroken overall strain 1.0 % local strain 2.5 %
−10 °C	through-thickness notch 20 mm long	627	unbroken overall strain 2.0 % local strain 3.0 % c.o.d. − 1.32 mm by feeler gauge

particular, the author suggested that there was a need to apply principles from metallurgy and other branches of science to fluxes for consumables and to examine methods for obtaining very low hydrogen weld metals. Deoxidation practice and its effect on weld metal toughness has been examined by Widgery (1974). He considered that two factors were of importance in determining the toughness levels, the extent on non-metallics and the nature of the transformation products including the effects of alloying elements in solution. Steel (1971) has demonstrated that *S* reduced the toughness of m.m.a. mild steel weld deposits by reducing the upper shelf of the transition curve for Charpy and c.o.d. tests. To study the basic weld metal toughness Widgery deposited test welds in a grooved plate, the groove being previously buttered to minimize dilution effects. One of the problems encountered was that ductile fracture by micro void coalescence occurred at a c.o.d. level of about 0.15 mm, and the maximum load c.o.d. was sometimes reached after an initial fibrous tear had moved out of the weld region being examined. This fact illustrates that the c.o.d. measured in this situation is not applicable to a defect tolerance calculation.

A detailed examination of the microstructure of submerged arc weld metals has been carried out by Garland & Kirkwood (1975), to explain some of the current anomalies in weld metal toughness behaviour, such as erratic and occasional low toughness results and the influence of flux combinations. Comparatively large areas of coarse blocky lath martensite (see figure 12, plate 7), together with varying proportions of bainitic carbides were identified in a range of weld metals and this structure exerted a deleterious influence upon as-welded impact properties. A new weld metal toughness model was developed to explain the comprehensive set of results obtained.

Uniformity in approach to design

The above sections have shown some aspects of the engineering approach to fracture safe design and the metallurgical features associated with fracture toughness measurements in welded zones. Although fracture mechanics theory is fairly advanced, there is no easy way to a completely rational basis for design over a range of structural applications. One of the main problems in rationalizing these standards is that laboratory toughness data are not available covering a sufficient range of materials (including weld metal and h.a.z.) to permit a uniform treatment. Thus there is considerable scope for more authoritative data using a variety of welding parameters and for work in defining the metallurgical factors governing weld metal and h.a.z. toughness.

Until the interpretation difficulties associated with the c.o.d. test and described above have been clarified, it would be inappropriate to use the c.o.d. test, on its own, to specify material toughness levels for national standards. At this stage, however, major users and fabricators after consideration of design philosophy in terms of operating stresses and identifiable defects which can be rectified economically, can use the c.o.d. test as a basis for a toughness specification. In the present circumstances it would also be advisable to carry out wide plate tests designed to simulate practical welds thus bridging the gap between the small scale laboratory test and the final structure. If this step is not taken it could lead to an overconservative prediction of the toughness requirements with subsequent difficulties in material procurement or the economic penalties of prolonged fabrication times (Gulvin & George 1974).

In the wide plate test a notch is cut to a given size at the required position in a weld and the test determines the stress and strain conditions to give failure at a particular test temperature. This test has the advantage of being sufficiently large to represent a piece of structure and can demonstrate the true brittle fracture problem of failure at low applied stresses. Considerable modifications are being made to the design of the test and the measurements made. A new design of loading ram has increased the displacement available in the loading system and sharp crack-like defects are being studied using pre-cut surface and embedded notches, extended by fatigue prior to testing, in place of the original chevron slit-notch which made a true assessment of the test difficult in terms of fracture mechanics. Today attempts are being made to analyse the large scale test results, not only in terms of overall strain to fracture related to a strain condition at, for example, nozzle connexion ($4 \times$ yield strain), but in relation to crack opening displacement and other local strain measurements to seek a more definitive relationship with the small scale c.o.d. toughness test. The wide plate test can be used to examine the true significance of zones of low toughness in weld metal and h.a.z. to clarify the various points of view now being put forward.

Since the wide plate test is expensive to perform, it is evident that many of the measurements required on the factors influencing toughness in various zones of the weld must come from smaller scale c.o.d. or Charpy tests. The c.o.d. test is suitable for this type of assessment since the Charpy test cannot produce reliable information on the relation between stress level and defect size and material resistance. Some attempts to relate the Charpy test direct to service experience has been made. One of these methods, involving arbitrary adjustments to the Charpy energy absorption level and testing temperature for different service duties, stress levels, thickness, etc., was prepared by a Committee of N.D.A.C.S.S. This method is the basis of the present requirements of the Merrison Committee – Box Girder Bridge Rules – notch ductility. This leads to requirements which are more stringent than existing U.K. Bridge Standard requirements in BS 153. This may, however, be due to differences in assumed minimum design temperature selected.

The present authors consider that for many constructions, including bridges, it is sensible to consider a dual approach. For any high stress concentration, or highly restrained regions in a structure, the fracture mechanics and strain analyses show that much higher toughness is necessary to tolerate a flaw of a given (detectable) size, than is the case for regions stressed only to normal design stresses (plus residual stresses if present). Thus it is suggested that consideration be given to adopting two different levels of notch ductility in tension regions, a lower level for general tension areas, and a higher level for severe stress concentration regions.

For non-stress concentration regions (excluding proof loading) the most severe stress conditions which should be applied are a maximum design stress of $\frac{2}{3}$ yield stress plus maximum residual stress of yield level (total $\epsilon/\epsilon_y = 1.67$). For severe stress concentrations, successful material selection based on wide plate tests has adopted a maximum plastic strain condition of $4\epsilon_y$ (total $\epsilon/\epsilon_y = 5$). Thus it is suggested that the notch ductility requirements for as-welded structures could perhaps be explored in terms of two cases:

	case 1	case 2
stress/strain level	$\epsilon/\epsilon_y = 1.67$	$\epsilon/\epsilon_y = 5$
value of constant C (from figure 3)	0.11	0.033
flaw parameter/thickness ratio $\bar{a}/e$ (assumed as an example)	0.2	0.2
toughness required (at minimum service temperature) from		
from $\bar{a} = c(\delta_c/\epsilon_y)$	$\delta_c = 2\epsilon_y e$	$\delta_c = 6\epsilon_y e$ ($\doteqdot 2\pi\epsilon_y e$)

These two toughness levels may be considered in the light of the levels necessary for transition behaviour away from plane strain conditions to occur. The toughness for case 1 is consistent with the condition at which deviation from plane strain first occurs, while that for case 2 is very close to the screening level assumed by Wells (1965) for the fully established plane stress condition.

Figure 2 shows that at these toughness levels and flaw size/thickness ratios, the additional effects of simple fillet welds can be accepted without the need to allow for further stress concentrations.

It is considered that if such an approach proved feasible in respect of bridge construction, category 1, would be applicable to welded tension areas of all plate girder and box girder steel bridges, with the exception of high global stress concentration regions. Any such stress concentration areas should be treated as requiring toughness up to category 2 level. In particular tension flanges in continuous structures with load bearing diaphragms or stiffeners attached by heavy welding (T-butt welds or fillets with f/e greater than 1), should be treated as category 2. It would be necessary to achieve these toughness levels in all areas of welded joints, and material with flame cut or sheared edges, or punched holes in tension regions should be treated in the same way as welded tension regions. This dual approach to toughness levels could, of course, be adopted in a number of other standards.

In practice the quality control of structural steel notch ductility is achieved by the Charpy V-notch impact test. Although there is no reliable theoretical correlation between Charpy energy absorption and fracture toughness, measured as K_{Ic} or δ_c, it is still essential to retain the Charpy test for quality control purposes, especially for parent steel properties. It will still be necessary to retain the Charpy test for weld procedural checks and it is therefore important to relate the quality control acceptance level and position of the Charpy test to the fracture toughness test by empirical relationships, which can, assist within certain limits.

However, to implement a philosophy as outlined above for design standards it would be necessary for a body of standardized data on fracture mechanics tests for different grades of structural steels with different welding conditions and consumables to be available. It is also implied that butt welds are tested to exclude defects of greater size than one fifth of the thickness as assumed previously.

In producing this data it is essential to examine the toughness requirements for weld metal and h.a.z. in relation to the particular application and not merely to assume that the minimum requirements for these regions should be the same as for parent plate. In many cases minimum toughness levels for weld metal may be lower than those applied to the parent steel.

Conclusions

The paper has discussed some of the existing alternative basis for design against brittle fracture, and has outlined a possible future approach which would allow rational judgements to be made on material toughness requirements for particular applications.

This philosophy could not be implemented without the joint efforts of the physical metallurgist and the fracture mechanics and design specialists. It also requires the clarification of certain problem areas in fracture toughness testing, and the accumulation of authoritative data on different grades of structural steels with different welding conditions and consumables.

References (Barr & Burdekin)

Archer, G. L. 1974 Fracture toughness in the HAZ of a C-Mn steel and a low alloy steel. *Met. Const. & B. Weld J.* December, pp. 369–371.

ASTM, STP 410 1967 Plane strain crack toughness testing of high strength metallic materials.

Baker, R. G., Barr, R. R., Gulvin, T. F. & Terry, P. 1973 The use of wide plate test data for design against brittle fracture. 2nd Int. Conf., Pressure Vessel Technology. Part II. Materials, Fabrication and Inspection, 11–75, pp. 1049–1058.

BS 1515: Discussions and proposal to amend BS.1515 Appendix C.

Burdekin, F. M. 1975*a* IIW Doc X-749-74; Report of Working Group on significance of defects. Published in *Welding in the World*, January.

Burdekin, F. M. 1975*b* Comments on the draft British Standard for the derivation of acceptance levels for defects in welds. Welding Institute Autumn Meeting Conference, November.

Burdekin, F. M. & Dawes, M. G. 1971 Practical use of yielding and linear elastic fracture mechanics with particular reference to pressure vessels. *Proc. I. Mech. E. Conf.*, London, May 1971, C5, pp. 28–37.

Cochrane, R. C., Garland, J. & Terry, P. 1975 Paper to be published.

Colvin, P. 1974 Weld metal toughness, basic fluxes – developments – applications. *Weld and Met. Fabr.* November, pp. 362–371.

Dolby, M. A. 1974 Influence of defect orientation on HAZ fracture toughness measurements. *Met. Const. & B. Weld J.* July, pp. 228–233.

Dolby, M. A. & Archer, G. L. 1971 The assessment of HAZ fracture toughness. *Proc. I. Mech. E., Conf.*, London, May 1971, C53, pp. 190–199.

Garland, J. G. & Kirkwood, P. R. 1975 Towards improved weld metal toughness. W.I. Spring Conf., London, April.

George, M. J. 1975 HAZ results on BS.4360: Grade 50 steels. To be published.

Gulvin, T. F. & George, M. J. 1974 The influence of weld metal toughness on the brittle fracture resistance of welded joints. B.S.C., Report PROD.149/5/74.

Harrison, J. D. 1975 Discussion at W.I. Seminar. Weld control in offshore construction. Newcastle, February.

Hayes, D. J. & Maddox, S. J. 1972 The stress intensity factor of a crack at the toe of a fillet weld. *Welding Inst. Res. Bull.* **13**, No. 1.

Maddox, S. J. 1972 An analysis of fatigue cracks in fillet welded joints. Welding Institute Report E/49/72.

Nicolson, S. & Rogers, P. F. 1974 *Proc. W.I. Conf.* Welding in offshore constructions. Newcastle.

Rice, J. R. 1968 Mathematical analysis in the mechanics of fracture. *Fracture; an advanced treatise*, Vol. 2 (ed. Liebowitz). London: Academic Press.

Rice, J. R., Paris, P. C. & Merkle, J. C. 1973 Some further results of *J* integral analysis and estimates. ASTM, STP.536, pp. 231–245.

Steel, A. C. 1971 The effects of sulphur and phosphorus on the toughness of mild steel weld metal. W.I. Report M/64/71.
Turner, C. E. & Burdekin, F. M. 1974 Review of the current status of yielding fracture mechanics. *Atomic Energy Rev.* **12**, No. 3.
Wells, A. A. 1961 Proceedings crack propagation symposium. Cranfield **1**, 210–230.
Wells, A. A. 1965 I.I.W. Houdremont Lecture on Notched bar impact testing. *Br. Weld. J.* **13**, No. 2.
Wells, A. A. 1971 The status of COD in fracture mechanics. Canadian Congress of Applied Mechanics.
Widgery, D. J. 1974 Deoxidation practice and toughness of mild steel weld metal. *Weld. Res.* IWT, Vol. **4**, Vol. **2**, pp. 54–80.
Woodley, C. C., Burdekin, F. M. & Wells, A. A. 1964 Mild steel for pressure vessel equipment at sub-zero temperatures. *Br. Weld. J.* **11**, 123–136.

2.3 Fatigue as a design limitation

By P. J. E. Forsyth
Materials Department, Royal Aircraft Establishment, Farnborough, Hants

[Plate 8]

Fatigue in metal parts can result from the repeated application of loads considerably below those required for general yielding, that is at levels where stress raisers have their greatest effect. The design limitations thus imposed can often be considerably reduced by introducing some local pre-loading.

A corrosive environment can greatly reduce fatigue strength and the fatigue test data used for a design should, as far as possible, have been obtained under conditions that simulate the anticipated service environment of the part.

Another limitation is imposed by material variability, particularly the chance appearance of defects in fatigue sensitive regions. This must be considered as the greatest hazard and emphasizes the importance of effective non-destructive inspection.

The design penalty imposed by the presence of fatigue loads arises from the fact that in order to maintain complete integrity of a part subjected to a large number of uniform load applications, that repeated load, certainly for aluminium alloys, must not induce local stresses greater than about one third of the yield strength of the alloy. Mean stress and corrosive environment reduce the level of this safe cyclic stress even further and thereby limit the load that can be indefinitely repeated without failure. The local stress in the bore of a hole, or at any other form of stress raiser, may easily be three times greater than the net section stress, and although such a geometric irregularity may have little effect in reducing the tensile rupture strength it will most likely halve the fatigue strength of the part. From these considerations it is clear that the repeated load to cause fatigue failure would be about one sixth of the steady load that would cause yielding. As an example of this the notched fatigue strength of the newly developed aluminium alloy 7050-T73651, which might be taken as representative of the most technologically advanced material in the aircraft structures field, has been reported as 75.8 MN m^{-2} compared with a yield strength of 441.3 MN m^{-2}.

This strength reduction factor is based on materials test data that can never fully represent the component in service with respect to size, surface finish and environment, and will usually give an optimistic view of fatigue strength. It is, therefore, only a guide to the designer who will have to consider, in detail, these various deleterious aspects and decide how to minimize the fatigue stresses and maximize the local resistance to fatigue cracking. Needless to say that if weight was not an important criterion the problem would be greatly diminished.

It is important to consider how far the engineer has been successful in designing against fatigue and what the factors are that cause a continuing incidence of fatigue failures.

One important one is that the designer may not be fully acquainted with the spectrum of repeated loads that the part will receive, perhaps, in some cases, because its service function is imperfectly understood. This is a fundamental shortcoming whereas any lack of knowledge of the local stresses developed by recognized loads can be covered by component testing. The next important factor, which is the one of most concern to metallurgists, is the departure of the material used to make the part from the standards of that on which the data was obtained. By this I mean, not only the variations in the metallurgical quality of the material, but the way it behaves in the final shape and in its service environment.

If we assume that, for a particular part, the loads were known accurately, including the cumulative damage effects from mixed level loads, and the design was geometrically correct, the question is, what are the other factors that might still cause a premature service fatigue failure? Several of these factors can be readily stated:

(1) Surface roughness, bad machining.

(2) The service environment might be more corrosive than that of the tests conducted to obtain the fatigue strength data.

(3) Fretting might occur.

(4) Residual stresses might be higher than anticipated.

(5) Manufacturing defects that might occur in some parts will not be discovered by the methods of inspection used.

Taking (1), (2) and (3), assuming that the designer uses the most appropriate fatigue strength data available to him we must ask, How far can the conditions of the test and the test material be allowed to depart from reality? Both corrosion and fretting can be considerably reduced if not entirely eliminated by attention to design detail, but it must be recognized that they would, if allowed to occur, considerably reduce fatigue life. (1), (4) and (5) are related to manufacturing procedures. The residual stresses may in some cases, be reduced to negligible proportions and the control of surface finish is an obvious requirement. However, manufacturing defects remain as a particular problem in which, although their incidence may be reduced by careful manufacturing procedures, non destructive testing must be used as the ultimate safeguard against their existence in service.

All of these factors will reduce the fatigue strength of the part below that which might have been first estimated, and it is the purpose of this paper to consider their importance in correct relation to the purely mechanical factors that may control the local fatigue stresses.

In many branches of engineering bolted joints represent the most important fatigue sensitive points. Whereas in an unjointed component, or even in a welded structure if we ignore the presence of defects, the local stresses are governed by load, shape and dimensions, in a bolted joint there are the added complications of bolt fit, bolt tension and interface friction. These engineering factors which may be independently varied can all considerably affect the local bolt hole stresses and therefore the fatigue life of the joint.

Compensatory factors operate, e.g. filling a hole with a fitted pin or bolt reduces the geometric stress concentration of the hole and may even double the fatigue strength, although this benefit will certainly be lost if fretting is allowed to occur. Furthermore, the stress range experienced in the holes can be reduced by the frictional resistance at the flange faces which is derived from bolt tension. These features are all part of the mechanical arrangement of any bolted joint, but special measures may be adopted to enhance the beneficial effects by:

(1) Using interference fit bolts which introduce residual tension stresses in the surrounding metal thereby shielding it from the full stress range. The fatigue strength gain with this method using different interferences is shown in figure 1 (Smith 1965).

(2) Deforming and thereby cold working the bores of holes by a variety of methods such as

ballizing, rolling or stress coining. The improvements to be gained by the use of stress coining are shown in figure 2 (Speakman 1970).

(3) The use of peening with shot or glass beads which may produce beneficial residual stresses, cold work the microstructure, and by abrasion and peening removing surface blemishes. In this should be included the attention to surface finish as a general measure. Figure 3 shows the effects of a variety of surface finishes (Manson 1971).

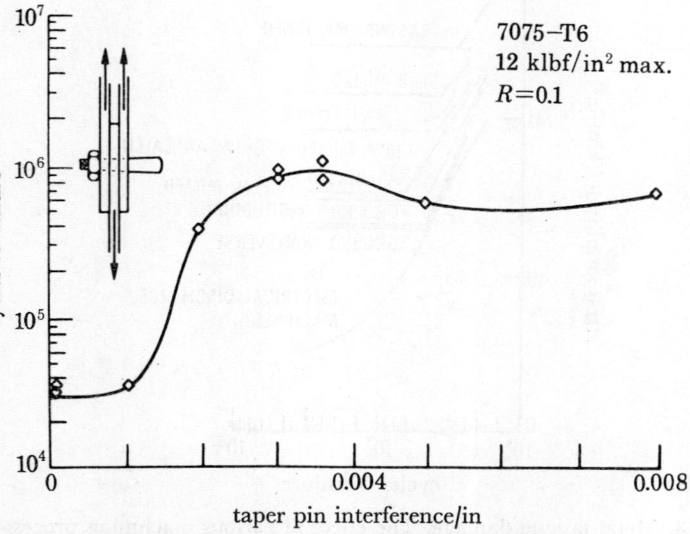

FIGURE 1. Effect of taper pin interference in fatigue life of small lugs.

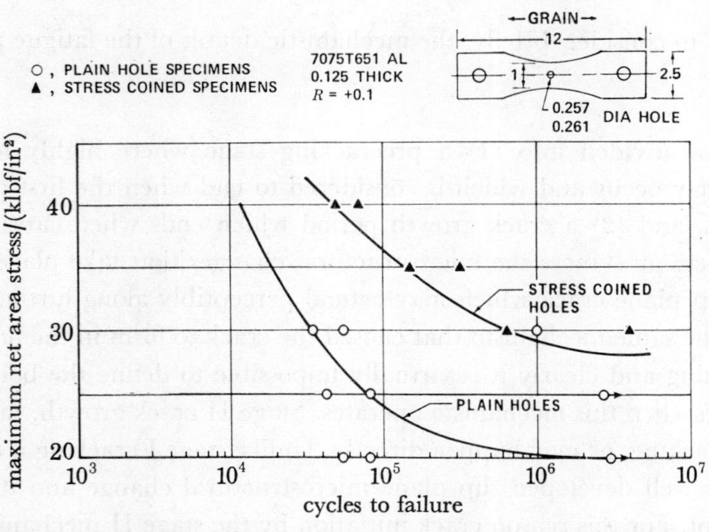

FIGURE 2. High fatigue resistance in metals and alloys. Comparison test of specimens with a conventional, plain hole and a stress-coined hole.

To this list, which is specifically concerned with joints, we could add such ways of introducing beneficial residual stresses such as by prestretching notched parts. Again, the fatigue strength can be increased by a factor of 3 by such treatments.

All of these practices must be applied with care and it is usually necessary to institute developmental work to optimize the fatigue strength of the particular part to which the treatment is applied, nevertheless it can be seen that mechanical methods are extremely effective in improving fatigue strength, and in some cases the fatigue stress concentration factor can be completely removed.

It is now proposed to show how the metallurgical contribution compares with these mechanical aspects in its effect on component fatigue strength, and how and where environment will play a predominant role.

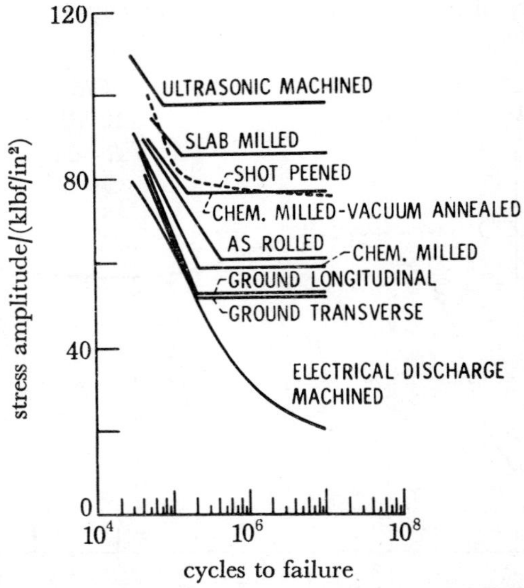

FIGURE 3. Metal fatigue damage. The effect of various machining processes on the bending fatigue strength of Ti-5Al-2.5Sn alloy.

It is first necessary to consider, briefly, the mechanistic details of the fatigue phenomenon.

(i) *Fatigue life*

Fatigue life can be divided into (1) a precracking stage where highly localized microstructural changes may occur and which is considered to end when the first quite arbitrarily defined crack occurs, and (2) a crack growth period which ends when catastrophic rupture takes over. In some circumstances the microstructural changes that take place on certain slip planes produce a slip plane crack which may extend perceptibly along further slip planes by what appears to be the same mechanism that caused the crack to form in the first place. This is termed stage I cracking and clearly it is virtually impossible to define the boundary between initiation and growth when this mechanism operates. Stage II crack growth, may either follow a period of stage I cracking, or may happen directly. Unlike stage I cracking it does not appear to develop from any well developed slip plane microstructural change and its appearance is therefore more abrupt. For this reason crack initiation by the stage II mechanism, or even the stage I to stage II changeover can be better defined than the start of stage I cracking. With the realization that fracture mechanics principles are inapplicable at very short crack lengths, which would nearly always include these indeterminate changes, engineers have arbitrarily chosen the crack propagation period to start with the appearance of a crack visible to a hand lens, and this for convenience, is taken by some workers to be 0.005 in (0.127 mm).

The service conditions of stressing, stress concentration, the nature of the surface and the environment can greatly affect the proportion of the fatigue life occupied in forming the first crack. The effects of stress level and stress concentration are shown in figure 4 (Pearson 1971).

For any potential fatigue situation the designer must decide whether fatigue life will be governed mainly by initiation or mainly by propagation of a crack, and furthermore whether the major part of the growth period (even on his arbitrary definition of crack initiation) occurs while the crack tip is still within the stress gradient of the stress raiser, or whether there will be a long growth period in what may be considered, in fracture mechanics terms, as approaching infinite plate conditions.

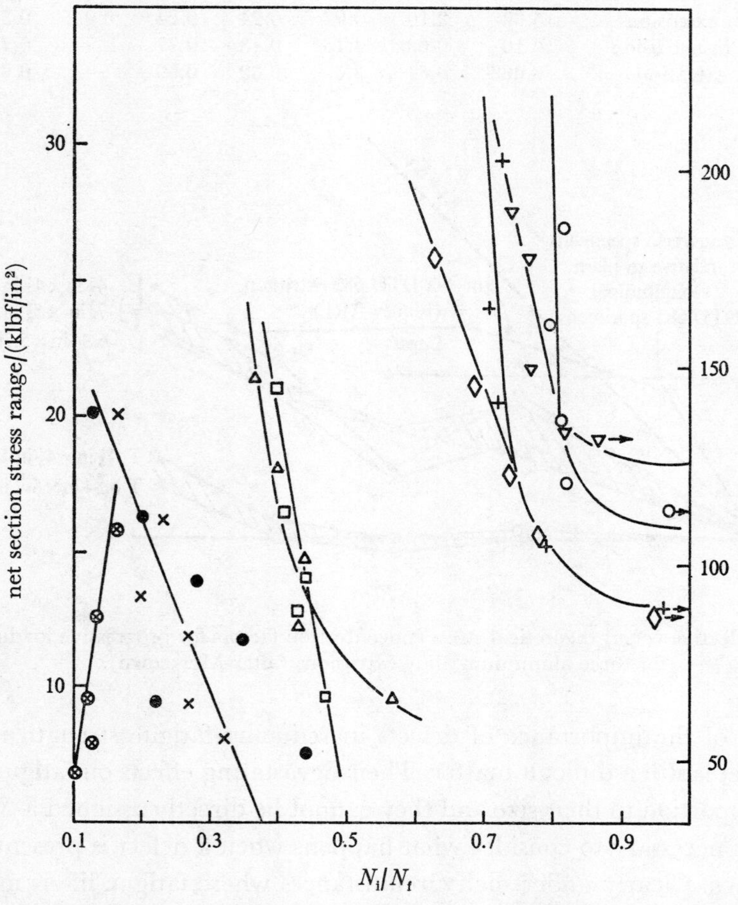

FIGURE 4. Plot of N_i/N_f against net section stress range. Notch depth 0.05 in (0.127 mm): □, $r = 0.005$ in (0.127 mm); ▽, $r = 0.025$ in (0.635 mm). Notch depth 0.125 in (3.17 mm): ●, $r = 0.002$ in (0.051 mm); ×, $r = 0.005$ in (0.127 mm); △, $r = 0.01$ in (0.254 mm); ○, $r = 0.025$ in (0.635 mm). Notch depth 0.25 in (6.35 mm): ⊗, $r = 0.002$ in (0.127 mm); ◇, $r = 0.015$ in (0.385 mm); +, $r = 0.025$ in (0.635 mm).

Where the crack growth period is the major part of the life, and particularly where the crack is relatively long, the fracture mechanics approach will be adequate in defining crack life, but in the circumstances where fatigue is initiation controlled, then the metallurgist must advise on the anticipated crack behaviour and so on the protective measures best suited to resist cracking.

This means that he must know what microstructure to expect in the fatigue sensitive regions of the design with regard to the methods of manufacture and surface treatment of the chosen alloy. Yield strength (either at room or at other temperatures), availability, and cost rather than fatigue strength may have been overruling factors in the choice of material, and this will make it even more important to optimize fatigue resistance.

(ii) *Microstructural features*

Because fatigue is very sensitive to certain microstructural features a primary requirement will be to set up the inspection standards with regard to allowable defects, surface finish and treatment (including permissible decarburization), grain size, flow direction of the structure, etc.

	Zn	Mg	Cu	Fe	Si	Cr	Mn	Ti	Ni
DTD 683 4¼ in × 4¼ in extrusion	5.69	3.10	0.92	0.24	0.24	0.05	0.33	—	—
BS L. 65 6.84 in × 5.7 in extrusion	0.10	0.68	4.15	0.43	0.77	—	0.72	0.15	0.03
BS L. 65 7⅞ in × 5⅝ in extrusion	0.068	0.64	3.6	0.32	0.69	—	0.92	0.09	—

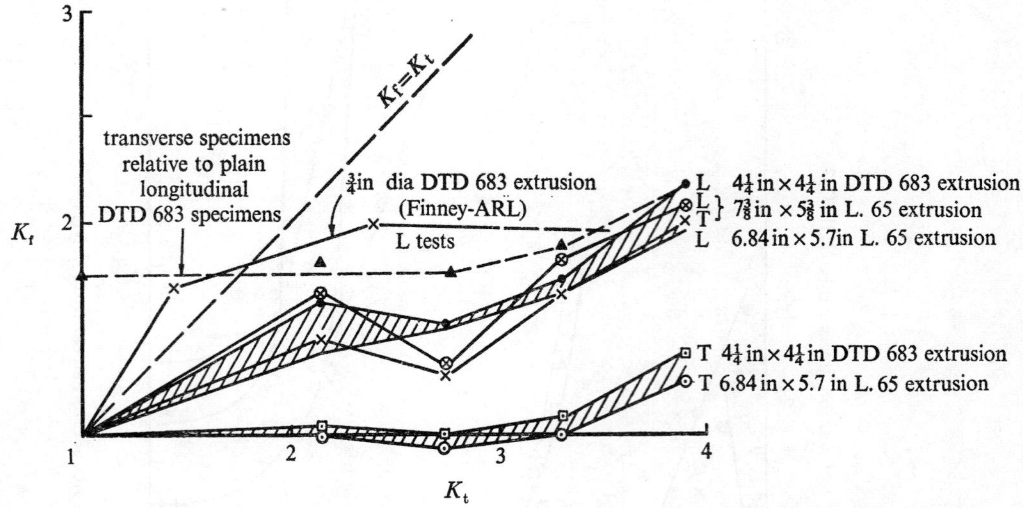

FIGURE 5. Effective versus theoretical stress concentration factors for progressive loading fatigue for three aluminium alloy extrusions (after McKeown).

The assessment of the importance of defects in reducing fatigue strength is of paramount importance and yet is still a difficult matter. Their devastating effects on fatigue strength seems to be out of all proportion to their size and they cannot be directly assigned a K_t value that has any meaning. It is necessary to consider what happens when a defect is present in terms of the fatigue life sequence. Clearly under such circumstances where fatigue life is mainly composed of a crack growth period their importance in initiating cracks will be minimal except where massive inclusions are concerned that might subtract most of the low rate growth period. In other circumstances where the life is mainly taken up with forming the first crack then a defect may replace this crack and thus eliminate the initiation period. This could easily reduce the component life to a tenth of that expected. Now consider what increase in alternating stress would have been necessary in the absence of the defect to produce this reduced life. We can, by this indirect method, assign an equivalent K_f value to the defect. This equivalent K_f can be observed at work when considering directionality effects. This is shown in figure 5 where it can be seen that the material, when stressed in the short transverse direction has an inherent structure contribution where the $K_f = 1.75$ which is not swamped until the K_t is increased to about 3. These K_f values resulting from microscopic defects may thus be as effective as a bolt hole in reducing fatigue strength and it is this incongruity that has confused the matter.

As would be expected, large defects have a greater fatigue strength reducing effect than small ones, and the effect of any particular size is increased with increasing matrix yield strength.

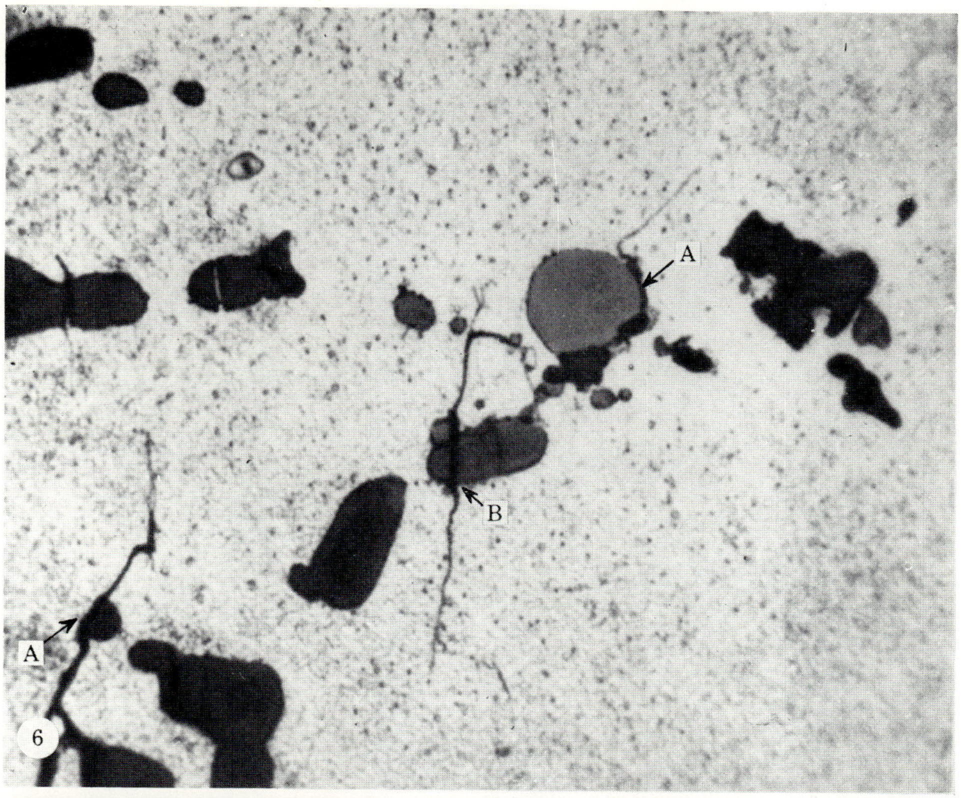

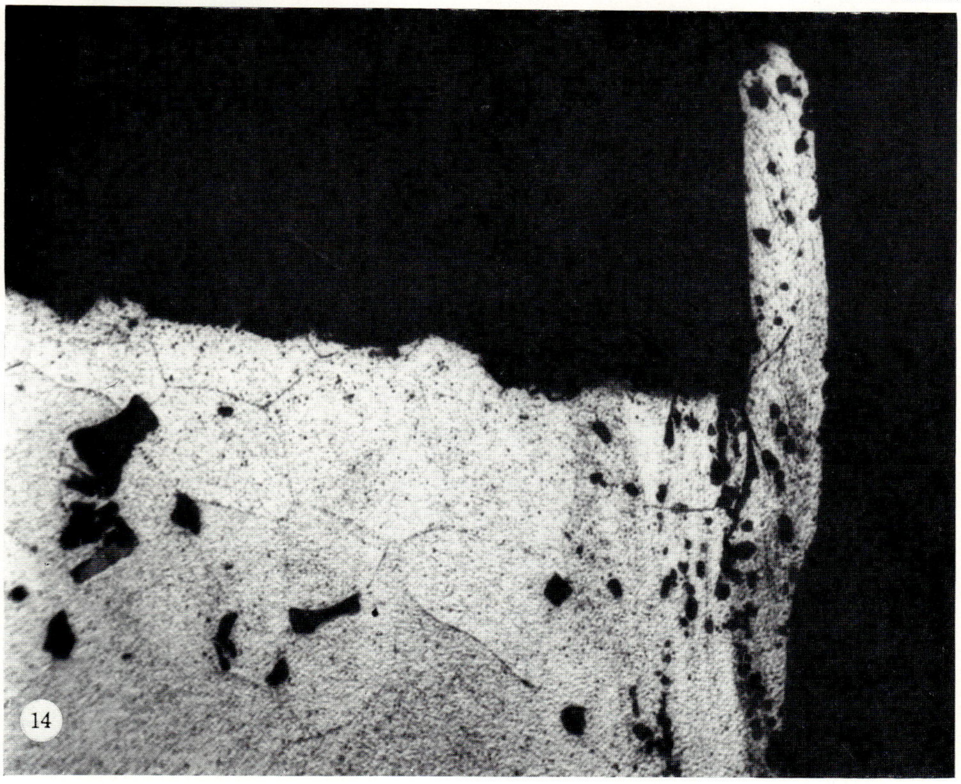

FIGURE 6. Fatigue crack origins at intermetallic particles in an aluminium alloy. (Magn. × 1000.)

FIGURE 14. Subsurface origin of fatigue fracture where the aluminium alloy surface has been previously cold-worked by drilling. (Magn. × 500.)

(*Facing p.* 172)

There also appears to be some limiting size below which defects have no appreciable effect on strength although they may still eventually be the features that initiate the first cracks.

One aspect of the problem is related to the unlikely but finite chance appearance of a large defect in the sensitive region. Its effect may be missed by coupon testing but experienced in parts where a considerably larger volume of material is being subjected to fatigue stress. This would seem to be only a question of the different probabilities of one of these defects existing in a significant position in the two sets of circumstances.

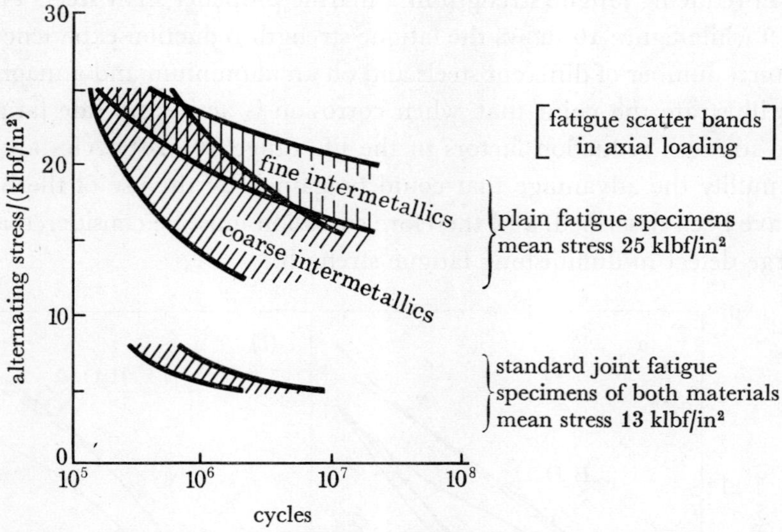

FIGURE 7. *S–N* curves for an aluminium alloy DTD 5024 made to two different compositions (after Faulkner & Fielding 1967).

Fatigue cracks often initiate at particles as can be seen in figure 6, plate 8, either at the particle matrix interface (A) or if the particle is cracked, as an extension of this crack into the matrix (B). The effect of recognizable metallurgical stress raisers such as intermetallic particles in reducing fatigue life has been demonstrated for plain fatigue specimens as shown in figure 7. But there is no doubt that any advantage that may accrue from reducing particle size or any other stress raising defects can be largely lost in the presence of the usual stress concentrations of any working part such as a joint, as is also indicated in figure 7. In this respect fretting cracks or corrosion pits must be equally recognized as stress concentrations and in fact the failures in these joints had originated by fretting.

(iii) *Corrosion fatigue*

Corrosion can reduce fatigue strength by providing stress raising pits, but the term corrosion fatigue implies something more. The generally accepted definition of stress corrosion can be modified to describe corrosion fatigue as 'the greater deterioration in the strength of a material through the conjoint action of fatigue stress and exposure to a corrosive environment than would occur by the separate but additive action of these agencies'. Not all environments that would cause corrosion accelerate fatigue and a more specific requirement has been stated as 'environments of an oxidizing nature', although it has been pointed out that corrosion fatigue could occur in the absence of oxygen where the cathodic reaction is evolution of hydrogen. As would be expected, where a time dependent phenomenon such as corrosion is operative, the fatigue frequency has a considerable effect on fatigue strength.

Air itself must be considered as a corrosive environment and it can be demonstrated that water vapour has a particularly detrimental effect on the fatigue crack growth rate of aluminium alloys, as shown in figure 8 (Bradshaw & Wheeler 1969).

Sea water is a classic corrodent giving trouble in marine applications such as ships and harbour installations, etc. The presence of chloride is particularly detrimental to most metals used for engineering purposes both from the general corrosion and from corrosion fatigue aspects. As an example sea water has been shown to be considerably more effective than fresh water in reducing fatigue strength in a marine propeller shaft steel. These results are shown in figure 9 while figure 10 shows the fatigue strength reduction experienced with sea water or sea spray on a number of different steels and on an aluminium and a magnesium alloy. All of these curves illustrate the point that, when corrosion is active, no safe fatigue limit exists, and the fatigue strength reduction factors in the life range 10^7–10^8 cycles are about 3. Such a factor could nullify the advantage that could be gained by the use of the best mechanical methods that have been described, and therefore corrosion must be considered as powerful an agency as the large defect in diminishing fatigue strength.

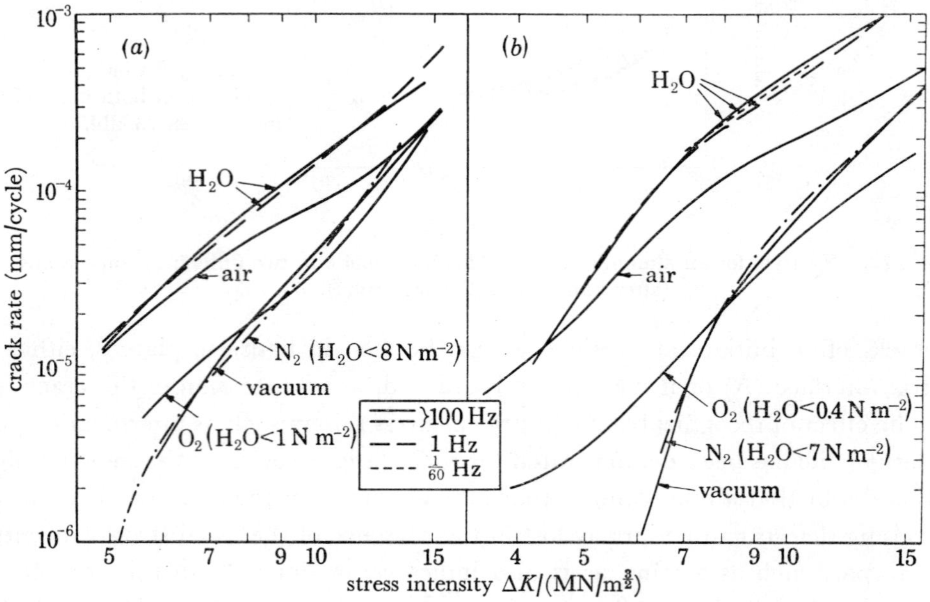

FIGURE 8. Crack rates in laboratory air, water vapour (2 kN/m². 15 Torr), oxygen (1 atm) nitrogen (1 atm) and vacuum at 100 Hz (*a*) 5070A alloy; (*b*) 683 alloy with rates in water vapour at 1 Hz and $\frac{1}{60}$ Hz.

Mean tensile stress reduces the air fatigue limit and increases the growth rate of cracks. When a corrodent is present the position is considerably more complicated. The alloy may suffer from stress corrosion under this steady stress and this form of cracking differs in nature from true corrosion fatigue. In some cases, such as with the strong aluminium alloys, stress corrosion cracking is essentially intergranular whereas true corrosion fatigue cracking will most likely be transgranular. One must be extremely careful in defining these situations because low frequency cyclic loading can cause an intergranular crack that behaves in all respects like a stress corrosion crack, and in the author's opinion should be classified as such, whereas in the same material a higher frequency will produce a true corrosion fatigue crack.

It is clear that if mixed modes of failure can occur both mean stress or R value (stress ratio) and frequency will not only influence strength but also mode of failure. There is no doubt that

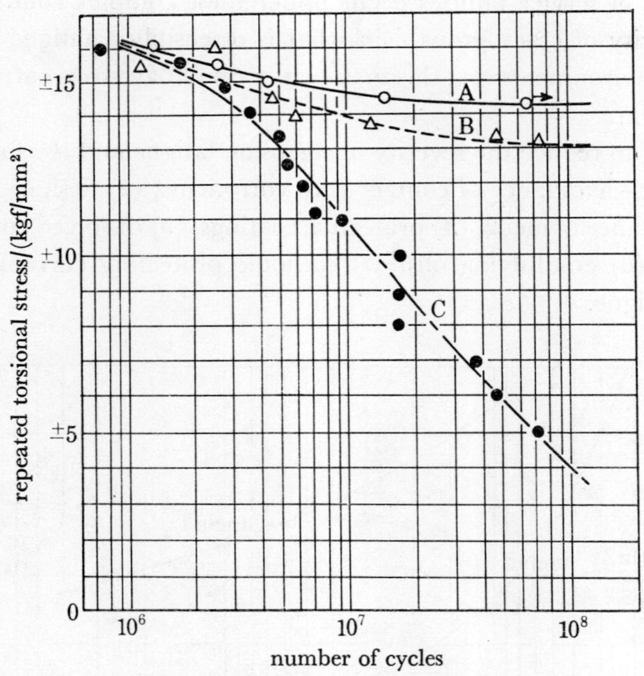

FIGURE 9. *S–N* curves: A, in air; B, with fresh-water corrosion; C, with sea-water corrosion (after Hara 1956).

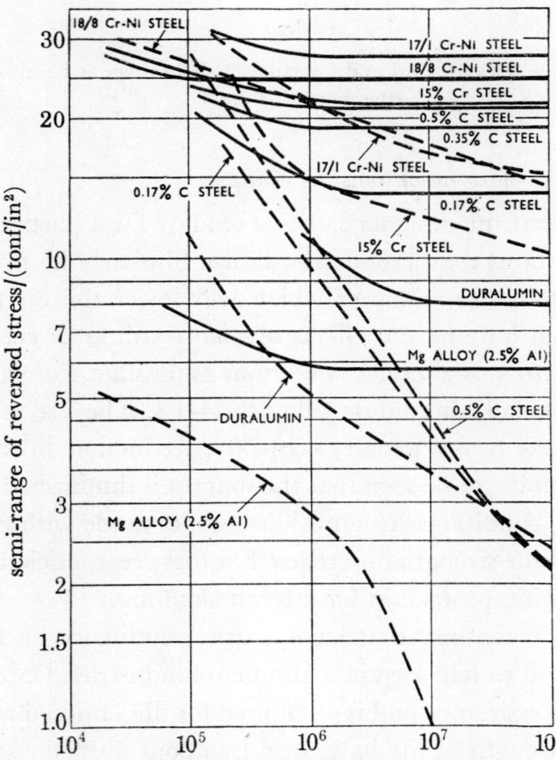

FIGURE 10. Fatigue and corrosion-fatigue tests on aircraft materials (Gough & Sopwith 1933). *S–N* curves. ——, in air; – – –, in sea-water or sea-spray.

a considerable number of service failures occur under these complex conditions and there is evidence that the severity of crack growth, in what is ostensibly a fatigue situation, may be different for different environments and this may be the result of a stress corrosion contribution if a mean stress is present.

Methods can be used to reduce the severity of corrosion fatigue and the following have been listed: (1) arrangement – avoidance of contact with corrodents, (2) design – as for air fatigue, (3) basic corrosion resistance of metal, (4) protective coatings, (5) dissolved corrosion inhibitors† forming surface films, (6) emulsifying oils, (7) cathodic protecting currents, (8) mechanical methods – as for air fatigue.

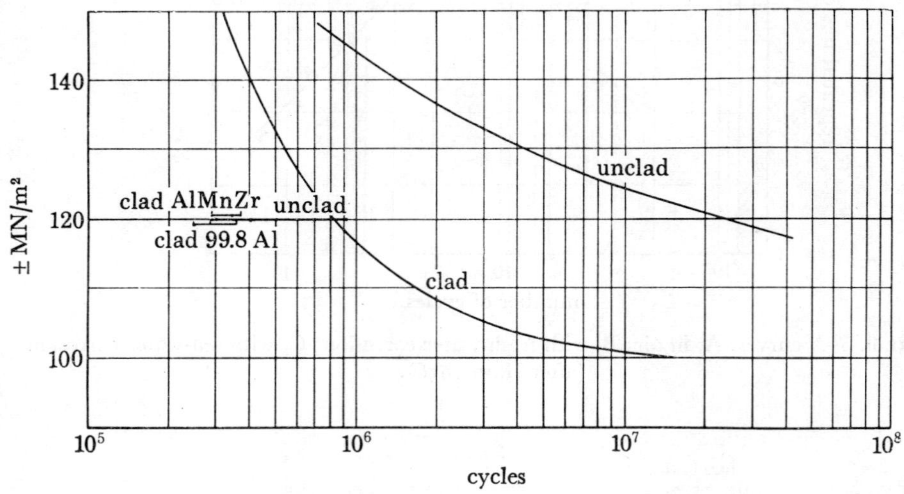

FIGURE 11. Corrosion fatigue lives for clad and unclad 2024-T3 shown as bands on air fatigue curves (after Forsyth 1974).

(iv) *Fatigue strength dependence on corrosion resisting coatings*

It was indicated earlier that the designer's choice of alloy for a particular part is most likely dictated by other considerations than fatigue resistance. Similarly he may have to use surface treatments for corrosion resistance, some of which will lower the air fatigue strength. It is particularly important when judging the effects of such coatings to ensure that the data has been obtained under as near realistic service conditions as possible. For example figure 11 shows *S*/*N* curves for clad and unclad aluminium alloy 2024-T3. The use of cladding for general corrosion resistance carries a heavy penalty – 40–50% reduction in fatigue strength unless corrosion is introduced when it can be seen that the margin is diminished. Similarly it is known that, even in laboratory air conditions structural assemblies made with clad and unclad alloys do not show this marked fatigue strength difference. For these reasons cladding is still considered a valuable method of corrosion protection for aircraft skinning.

Other corrosion protective coatings that have a strong influence on fatigue are chromium and cadmium. These are both widely used in a number of industries. Chromium may be chosen both for corrosion and wear resistance and is often used for dimensional recovery of worn parts. It can reduce the fatigue strength of the basis steel by about 50% for some medium strength steels. The percentage loss in fatigue strength is generally greater for stronger steels and for

† One must emphasize the danger of situations where the inhibitor almost, but not quite, keeps the corrosion inhibiting film in repair thus providing a few anodic places that receive the whole intensity of attack.

thicker coatings. The loss of fatigue strength with this coating can be attributed to several factors including:

(*a*) residual tension stresses developed in the deposit as indicated in figure 12 (Williams & Hammond 1955).

(*b*) The low effective fatigue strength of the chromium which most likely results from the presence of numerous cracks.

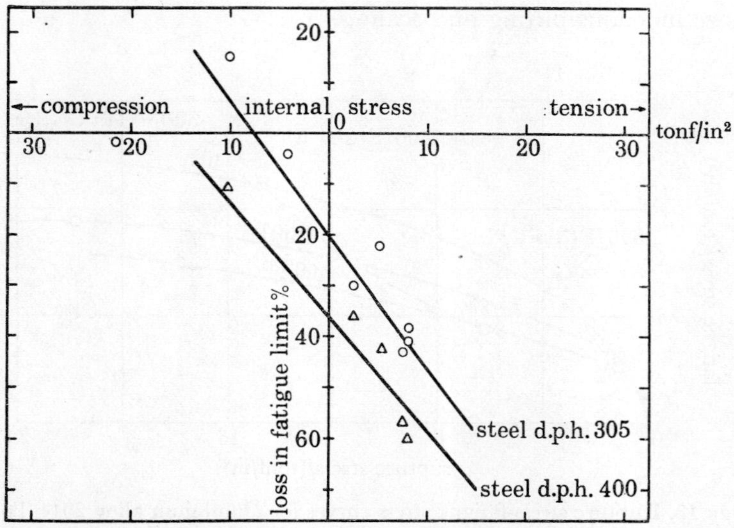

FIGURE 12. Relationship between residual internal stress of chromium and percentage loss in fatigue strength for En 25 steel.

Co-deposited hydrogen may be present in both the chromium and the basis metal and this may, if allowed to remain, result in an embrittling effect, although its direct effect on the fatigue mechanism is not clear. Post plating heat treatment, for those steels where this would be acceptable can restore some of this lost fatigue strength. Deposited cadmium can also cause some loss of fatigue strength although generally far less than observed with chromium. Again, co-deposited hydrogen can be troublesome. The detrimental effects of plating can be nullified by prior shot peening and even grit blasting, both of which will induce beneficial residual stresses into the basis steel surface. It is now common practice, particularly for high tensile steels to shot peen before plating where fatigue strength is important.

(v) *Temperature*

The detailed mechanism of fatigue crack formation and also the crack growth path may be different for different operating temperatures and, as a general rule, fatigue strength is reduced with increasing temperature although in some steels the strain ageing effect may cause some fatigue strength improvement by going to moderately elevated temperatures. There is also some evidence that crack growth resistance in some alloys may be reduced at low temperatures. This somewhat conflicting experience indicates the complexity of the crack initiation and growth behaviour particularly where creep may interact with fatigue. Where mean stress is present elevated temperature fatigue is complicated by creep stress relaxation, and additionally, cyclic stress increases creep rate. Such interactions must be considered at the design stage of any component that will be subjected to elevated temperatures in service. The relation between independently applied fatigue stress and creep stress required to produce failure at equal times and temperatures for 2014 aluminium alloy is shown in figure 13, but the simultaneous action of the two would be expected to reduce life considerably.

A considerable body of data has been amassed in which these interactions have been brought into play. With increasing temperature the environmental aspects can also become particularly troublesome; oxidation, carburization, decarburization and sulphidation may all play a part in degrading the surface in a variety of ways including grain boundary penetration and cracking or even the more inocuous pitting and scaling.

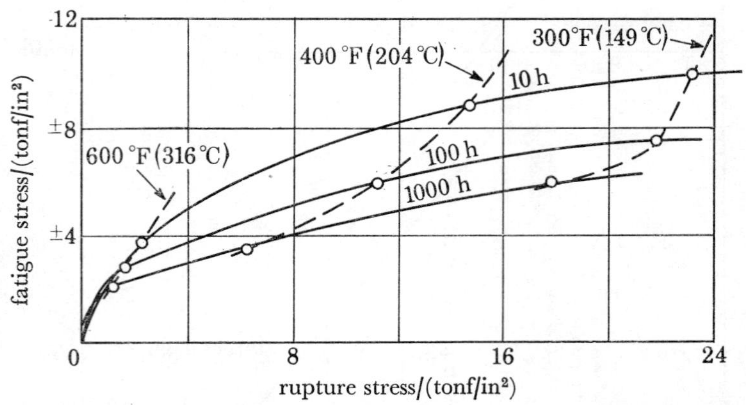

FIGURE 13. Rupture stress-fatigue stress curves for aluminium alloy 2014-T6 for rupture in equal times at equal temperatures (after Allen & Forrest 1956).

It is clear that increasing temperature, by accelerating corrosion and diffusion and by relaxing residual stresses must throw the emphasis from mechanical methods of fatigue inhibition to metallurgical methods. Other factors such as rest periods, strain ageing, etc. now play a part and, as with normal temperature fatigue, corrosion resisting coatings must be considered as playing an important part.

Even moderate temperatures, e.g. 100–150 °C, can affect the fatigue strength of some aluminium alloys by softening the cold worked layer that can result from machining (Forsyth 1972; Heath-Smith & Aplin 1971). When the cold worked layer is present sub-surface crack initiation is observed, as shown in figure 14, but prior exposure of the specimen to some hundreds of hours at 150 °C can soften the worked layer resulting in surface initiation of the cracks and reduced fatigue life.

(vi) *The fatigue behaviour of weldments*

There are a large variety of metal joining methods available to the designer and the appropriate choice for a particular situation will depend on the two metals involved, the size and shape of the parts, and service requirements. Modern methods have been evolved that result in minimum general heating from very limited fusion zones, e.g. electron beam welding, but the fatigue strength is still basically dependent on (1) geometry, (2) weld defects, and (3) local microstructural changes.

Thus the behaviour of the weld is very dependent on the occurrence of defects in the most highly locally stressed regions. Local plasticity in these regions is generally experienced on the first service load application. This arises from the residual stresses induced during the thermal cycle of the welding process, and to a large measure, this relaxation must benefit the subsequent fatigue resistance of the weldment. This stress relief can be obtained in some circumstances by purposely post weld straining.

Because defects are so important in fatigue, welding demands particular flaw detection care and for the same reason no sensible data on fatigue strengths can be given, except perhaps to indicate the fatigue strength reducing effect of dressed butt welds where there is no section change and therefore where any weld defects are operating in plain fatigue conditions. Under these circumstances it is possible to obtain parent metal failures usually in the heat affected zones, but the real controlling factor on fatigue strength is the presence of weld defects. The various defects stem from several causes and it would seem impossible to assign precise K_t or K_f values, however, it has been stated that, from experience, a $K_t = 3.0$ and a $K_f = 1.5$ would be typical for the case of incomplete weld penetration.

(vii) *Conclusion*

The design limitations that would be imposed by fatigue loads can be reduced considerably by engineering methods aimed at minimizing the severity of stress raisers. This can be done by attention to notch geometry and by various methods of prestressing. In the most favourable circumstances the effects of the stress raiser may be practically eliminated, even to the point of suppressing the detrimental effects of local flaws. In this last respect surface deformation methods such as peening and rolling can be particularly beneficial. In components where a plain fatigue situation is encountered then the problem of the chance defect is at its most acute. It is in these same circumstances where a long uncracked life is expected that any metallurgical defect, including corrosion pitting, becomes most important as it seriously reduces this part of the life by presenting an early crack.

Improvements in design against fatigue are being actively sought by engineers, but a pressing problem for the metallurgist is to obtain a better understanding of the influence of the various types of defect, and perhaps most important of all, reliable non-destructive inspection methods must be developed that can detect the important flaws.

With a full appreciation of the importance of the various types of defect in reducing fatigue strength and with the availability of adequate n.d.t. a large reduction in the incidence of fatigue failures would be achieved. On our present understanding of the nature of fatigue this approach must be far more profitable than that of alloy development.

References (Forsyth)

Allen, N. P. & Forrest, P. G. 1956 The influence of temperature on the fatigue of metals. International Conference on Fatigue of Metals. Institution of Mechanical Engineers. London.

Bradshaw, F. J. & Wheeler, C. 1969 The influence of gaseous environment and fatigue frequency on the growth of fatigue cracks in some aluminium alloys. *Int. J. Fracture Mech.* **5**, 255–268.

Faulkner, R. A. & Fielding, J. 1967 Comprehensive fatigue tests on a high strength aluminium alloy containing fine and coarse insoluble intermetallic constituents. Hawker Siddeley Aviation (Manchester) Report No. Eng/Res/MIS/MAN/408 (MOD supported work).

Forsyth, P. J. E. 1972 Microstructural changes that drilling and reaming can cause in the bore holes in DTD 5014 (RR 58 extrusions). *Aircraft Engng.*

Forsyth, P. J. E. 1974 The fatigue behaviour of aluminium alloy 2024-T3 clad with 99.8% aluminium and with aluminium 1% manganese 0.1% zirconium alloy. *R.A.E. Tech. Report* 74170.

Gough, H. J. & Sopwith, D. G. 1933 *J. Iron & Steel Inst.* **127**, 301.

Hara, Saburo 1956 An investigation on the corrosion fatigue of marine propeller shafts. International Conference on Fatigue of Metals. Institution of Mechanical Engineers. London.

Heath-Smith, J. & Aplin, J. 1971 The effect of an application of heat on the fatigue performance under random loading of a notched specimen of DTD 5014 (RR 58) material. *R.A.E. Tech. Rep.* 71209.

McKeown, J. 1959 Notch sensitivity and directionality in fatigue in light alloy extrusions as shown by endurance and progressive loading tests. *Metallurgia* **59**, 31–38.

Manson, S. S. 1971 Avoidance, control and repair of fatigue damage. Metal Fatigue Damage. ASTM, STP 495.

Pearson, S. 1971 Fatigue crack initiation and propagation in half inch thick specimens of an aluminium alloy. *R.A.E. Tech. Rep.* 71109.

Smith, C. R. 1965 Interference fasteners for fatigue life improvement. *Exptl Mech.* **5**, 19A–23A.

Speakman, E. R. 1970 Fatigue life improvement through stress coining methods. Achievement of high fatigue resistance in metals and alloys. ASTM, STP 467.

Williams, C. & Hammond, R. A. F. 1955 The effect of chromium plating on the fatigue strength of steel. Armament Research & Development Establishment Report (M) 6/55, September.

2.4 High temperature mechanical properties as design parameters

By G. D. Branch

C.E.G.B., Chelham House, Bath Road, Cheltenham, Gloucestershire GL53 7LS

[Plates 9–11]

The contribution of physical metallurgy to high temperature design is discussed under the headings of research, development, data collection, structural analysis, failure analysis, material selection, plant operation and failure investigation. Particular attention is drawn to the problems of transferring our available technology into our operating plant and it is suggested that further development of the status of the materials engineer is required. The paper highlights the significant contributions made by the metallurgist in microstructural studies, materials development, uniaxial data collection and failure investigations. For the future, it is suggested that much work remains to be done in the areas of crack growth, structural analysis and plant monitoring.

Introduction

I intend to discuss the contribution which physical metallurgy has made to the subject of high temperature design. There are several ways in which to do this but, recognizing the aims and objectives of this centenary Conference, I have decided on a personal, subjective view based on an involvement over a period approaching twenty years. I therefore apologize at the outset to those friends, colleagues and associates who have made worthy contributions which are unrecognized in this paper. I trust that the decision of not attempting a comprehensive, authorative review will be justified by provoking discussion or at least stimulating thought.

I will begin by defining my subject – high temperature design. Design may be defined as a determined purpose, a pictorial representation or a cunning contrivance. I will say no more than remark that all three of these definitions are recognizable to engineers and metallurgists who are involved in the design of high temperature components. I will define temperature as high if it is sufficient to allow time dependent deformation to occur, on either a macro- or micro-scale, under the loadings and environment imposed by the working conditions. Physical metallurgy allows us to equate the observed behaviour of metals with the established laws of nature.

The physical metallurgist has contributed to high temperature design in various areas; these being research, development, data collection, structural analysis, failure analysis, material selection, plant operation and failure investigation. I define the first six of these as 'front end' activities and the last two as 'back end' activities. In an ideal world, there would be no need for the practising metallurgist to be involved in 'back end' activities as the successful operation of plant from a materials point of view would be guaranteed by attention to the 'front end'. The world is, of course, far from ideal and failures do occur from time to time. Nevertheless, it is my personal view that the overall economy of our high temperature plant would benefit from increased attention to the 'front end'. In purely financial terms, there is no argument in view of the high and increasing cost of plant failure. The problem is one of resource utilization and we need to devote increased attention to transferring our technology into our plant. This technology is gained not only from our research activities but also from our operational experience and failure investigations. We therefore need metallurgists in the design function who are capable of assimilating, and are motivated to assimilate, the available information and injecting it into what I have called the 'front end'. I make these general points in concluding my introduction, as I am sure that they would have been in keeping with Walter Rosenhain's devotion to understanding the behaviour of metals and to utilizing this understanding to the benefit of future generations including our own.

Research

Around the 1920s it was recognized that creep might present a real threat to the development of high temperature plant and intense activity took place to further the understanding of the phenomenon which had been demonstrated by Andrade (1910, 1914) before World War I. This research has yielded sufficient information to guide the development of commercial alloys, by applying our knowledge of the qualitative effects on deformation behaviour and fracture ductility of factors such as grain size, prior cold work, soluble alloying elements, coherent precipitation, dispersion hardening, insoluble impurities and heat treatment.

The extensive research information is insufficient to enable us to predict creep behaviour, in useful quantitative terms, from first principles and, in view of the complexities of commercial alloys, there would appear to be little merit in devoting further resource to the study of fundamentals for this purpose. Let us learn to use what we know before we struggle to know more!

Over more recent years, the research worker has tended to divert his attention away from the constant load, constant temperature studies of smooth specimens to considering creep under variable testing conditions, creep in complex stress situations and creep crack growth. Progress has been slow as a result of the complexities of experimental testing and the number of variables involved. Also, most resource has gone into attempts to simulate practical design problems. Such work may produce useful data for a particular problem, but the experiments by their very nature are of little help in solving the next problem. However, we are now beginning to break the problem down into areas which may allow planned research to proceed in the near future. Three basic areas are structural analysis, crack initiation and crack propagation.

A man before his time in the area of structural analysis was A. E. Johnson whose lifetime devotion was commemorated recently by the publication of an authoritative volume on creep design (Smith & Nicholson 1971). The problem is to apply our simple laboratory data to predicting the changes of stress and strain distribution with time in complex engineering structures. The physical metallurgist will need to develop improved methods for summing the cumulative effects of prior history on creep deformation behaviour.

Crack initiation is of little importance in most high temperature engineering situations, as future designs will be forced to assume the presence of defects of a size which can not be detected by non-destructive testing.

Crack propagation, on the other hand, is vitally important and fortunately there is the prospect of significant progress in this area. We know that the rate of crack propagation is both material and structure dependent (Neate & Siverns 1973–74) and the physical metallurgist must define the material factors controlling the relation between crack extension and displacement.

Development

The most widely used materials in the creep range are the ferritic steels and they present a wide range of creep strengths as a result of their great sensitivity to structural change caused by relatively minor changes in composition or heat treatment. These steels include the C/Mn and other structural steels originally developed for room temperature use, the low alloy steels specifically developed for creep applications and the higher alloy steels with other advantages such as corrosion resistance or high proof strength.

The austenitic steels have also been widely used in the creep range and many complex proprietary alloys have been developed for special applications, particularly in the aero-engine and gas turbine industry. Some of these alloys contain so much alloying element that licence is required to call them steels at all. I therefore have little hesitation in considering the nickel- and cobalt-base alloys together with the austenitic steels.

Creep resistance is also required of many other non-ferrous metals and suitable alloys have been developed. Probably the most important are aluminium-, magnesium- and titanium-base alloys, but alloys of zirconium, copper, lead, tin and zinc have also been subjected to detailed study.

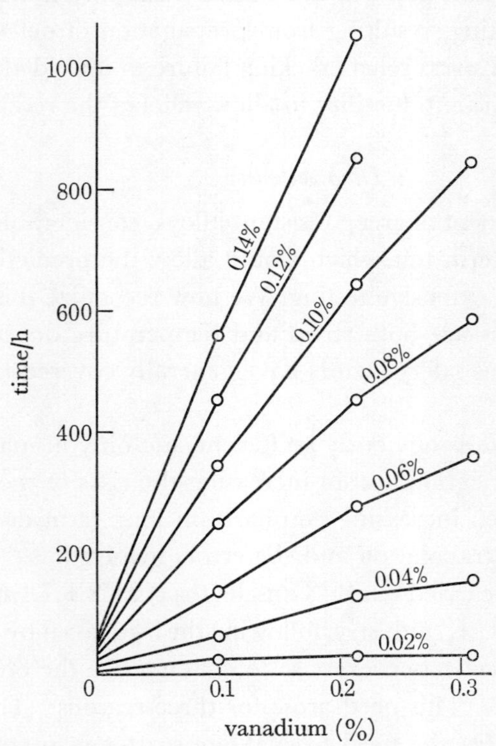

FIGURE 1. Effect of vanadium on the time to reach various creep strains at 4 tonf/in² (60 MN/m²) and 600 °C (Glen 1948). 0.6 % molybdenum steel – normalized 900 °C, tempered 690 °C 24 h.

The physical metallurgist's contribution to the development of creep resistant alloys has been of paramount importance and I would highlight the work of Tapsell (Tapsell & Ridley 1945) and his successors at the N.P.L., Glen (1948) whose classical work led to the identification of latent creep resistance resulting from precipitation and strain ageing during creep and by Bardgett (1948) and his successors at Swinden laboratories. These workers were primarily concerned with the development of low alloy steels containing Cr, Mo and V for application in steam power plant and in the oil industry. A typical example of the thorough nature of this early development work is illustrated in figure 1, taken from Glen's work which shows the effects of variations in vanadium content on the creep strength of a 0.6% molybdenum steel. The development of creep resistant austenitic steels and nickel- and cobalt-based alloys can be traced to the advent of the jet engine, although applications have been found more recently in steam power plant and in the nuclear industry. Various strengthening mechanisms have been used including carbides, nitrides, intermetallic precipitates and warm working and the majority of the work has been carried out by the leading manufacturers of such materials, who have marketed their own proprietary alloys.

The metallurgical profession has much to be proud of in making possible the development of more efficient power plant, process plant and aero-engines. If any criticism can be made, it must be on the grounds of insufficient attention being paid to important factors apart from creep strength. The practical use of creep resistant alloys has been limited by shortcomings in ductility, fabricability and corrosion resistance rather than a lack of strength. Occasional failures have been reported and figures 2–4, plates 9–10, illustrate three examples involving Cr Mo V steel. Figure 2 shows a high temperature bolt failure resulting from insufficient ductility to absorb creep strain at a point of stress concentration. Figure 3 shows a high strain fatigue failure in a turbine casting, resulting from propagation of defects in a region of high thermal stress. Figure 4 shows a stress relief cracking failure in a welded joint, resulting from a heat affected zone having insufficient ductility to allow relief of the residual welding stresses.

Data collection

In the early years of development of creep resistant alloys, considerable effort was devoted to the search for a suitable short term test which would allow the prediction of long term creep behaviour without the need for extensive testing. We now recognize this to have been a pious hope because creep mechanisms are both stress and temperature dependent but fortune has smiled on the designer in that his safety factors have generally covered the errors in extrapolation from short term data.

Conservatism in design stresses is obviously inefficient, not only in material costs but also in adding to the difficulties of fabrication and in increasing the risks of thermal fatigue or brittle failure. There has therefore been increasing emphasis on long term data collection in recent years to reduce the extent of extrapolation and the errors involved.

The first long term tests were carried out by Tapsell (1944) at N.P.L. in the early 1940s, but it was not until the late 1950s that U.K. industry, following the lead given by West Germany, recognized the need for a vast data collection exercise to characterize the creep properties of commonly used creep resistant steels. This need arose for three reasons: (1) the above mentioned uncertainties in extrapolation, (2) the need to evaluate scatter in material properties arising from unavoidable chemistry and process variations, and (3) the intended publication by the International Standards Organization (I.S.O.) of a Boiler Code (1968), which specified 100000 h properties as design criteria.

The main U.K. contributors to long term data collection have been the Electrical Research Association (E.R.A.) with its industrial sponsors, the National Engineering Laboratory and the British Steelmakers' Creep Committee. The work of these organizations has been of great value to the designers of high temperature plant in the U.K. and, more recently, throughout the world because of the lead taken in deriving agreed I.S.O. properties for incorporation in design codes (ISO/DIS 2604–2607). The collected International creep-rupture data, of which a typical

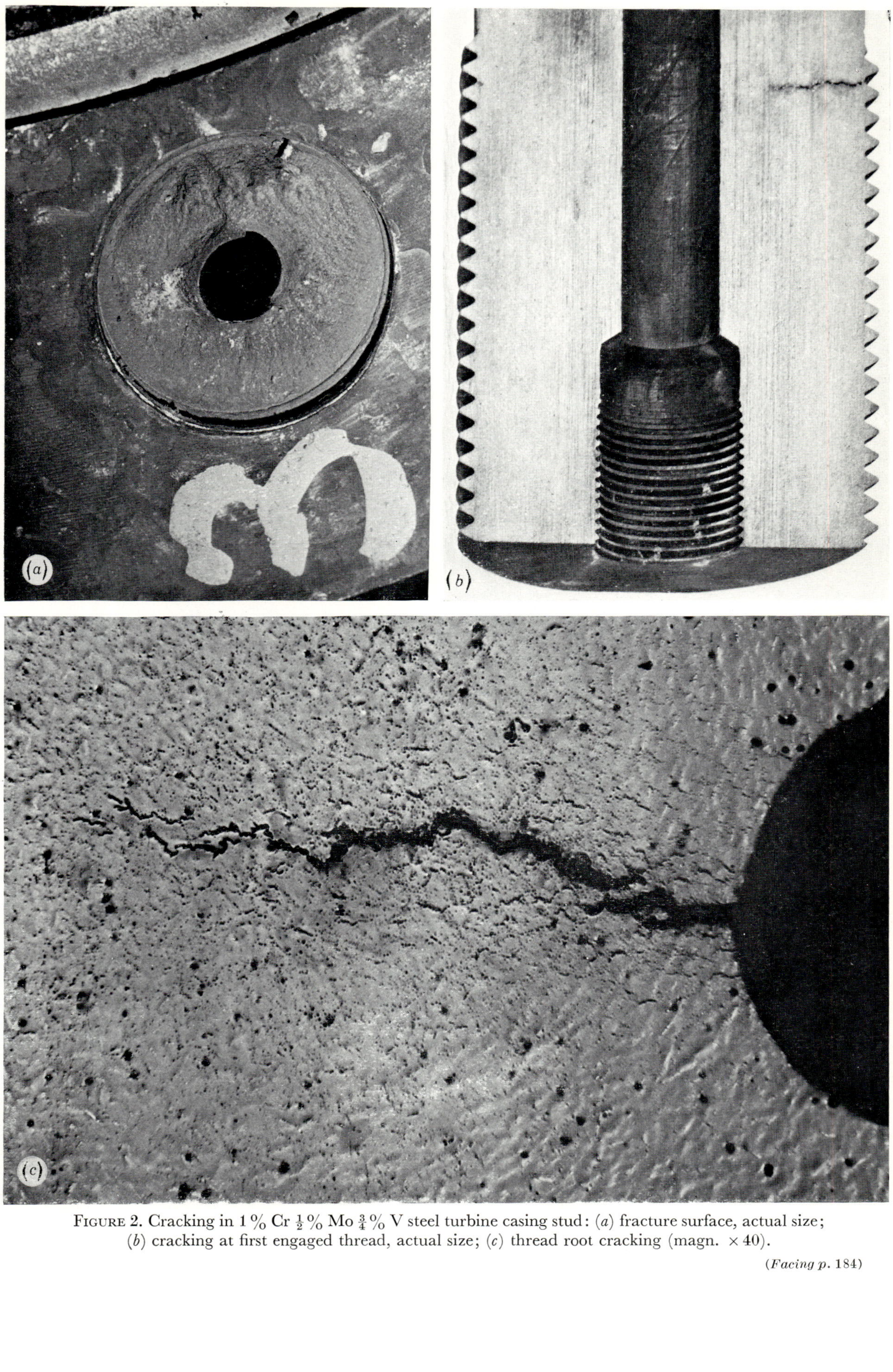

FIGURE 2. Cracking in 1 % Cr $\frac{1}{2}$ % Mo $\frac{3}{4}$ % V steel turbine casing stud: (*a*) fracture surface, actual size; (*b*) cracking at first engaged thread, actual size; (*c*) thread root cracking (magn. $\times 40$).

(*Facing p.* 184)

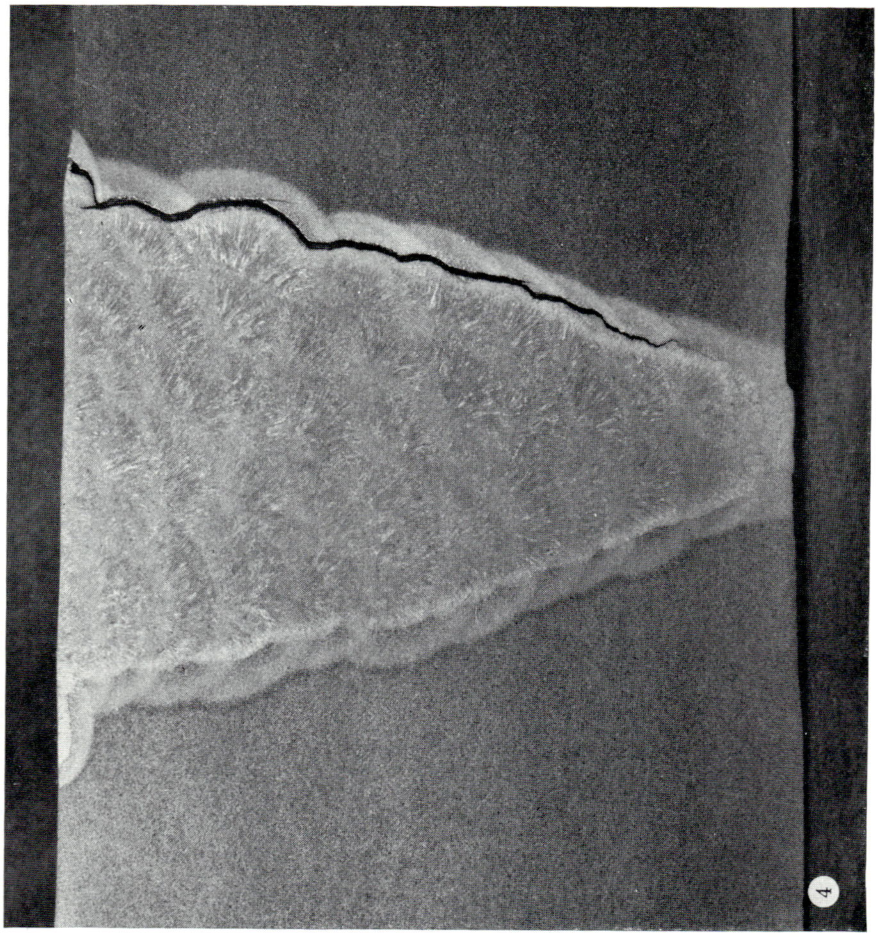

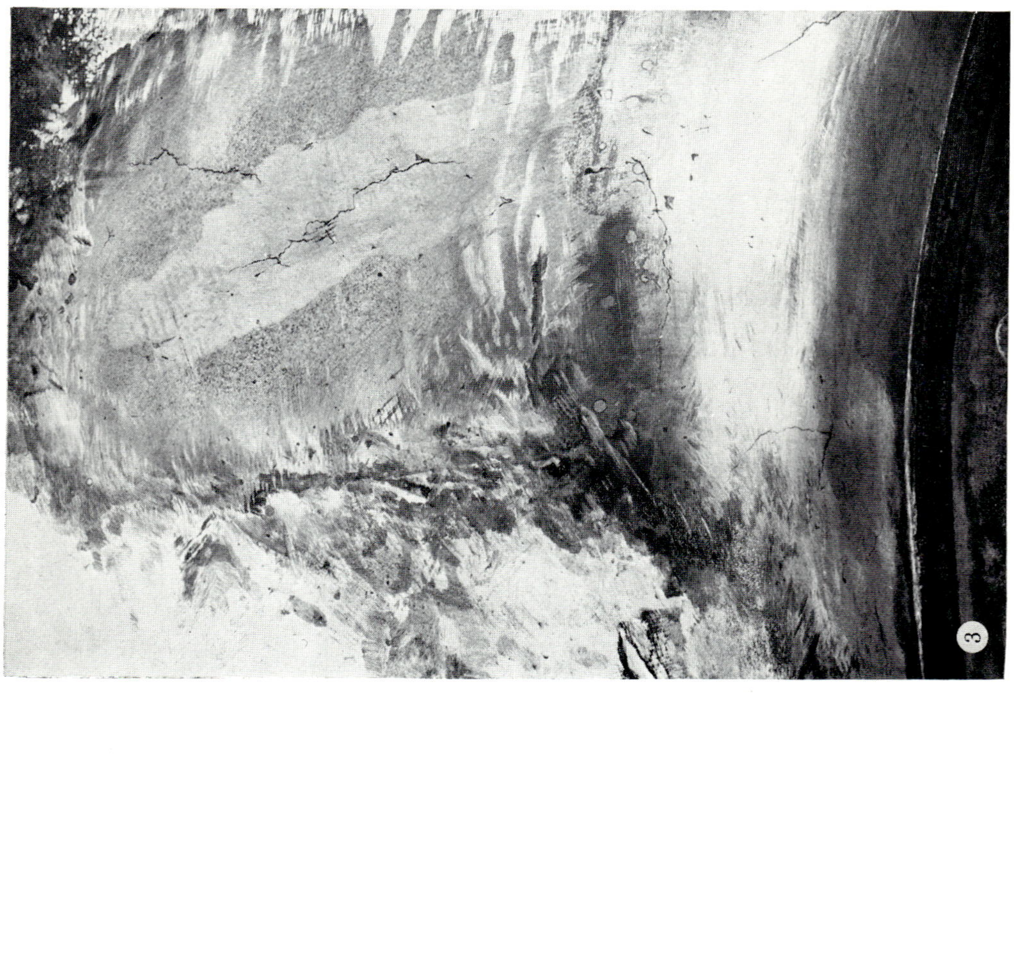

FIGURE 3. Cracking developing during service from weld repair and internal shrinkage in a turbine casting.

FIGURE 4. Reheat cracking in the heat affected zone of a Cr Mo V butt weld.

example for $\frac{1}{2}$Cr $\frac{1}{2}$Mo $\frac{1}{4}$V steel at 575 °C is shown in figure 5, have been analysed to produce design information of the type shown in figure 6. Experience gained from this long term data collection has enabled extrapolation to be put on a more secure footing and internationally agreed procedures are now available to provide a rational basis for comparison between steels. The sponsored work at the E.R.A. laboratories has been particularly valuable in that the planned experimental programmes have yielded valuable knowledge of the metallurgical factors controlling long term creep behaviour, e.g. the effect of bainite content in Cr Mo V castings (Murphy & Branch 1969) (figure 7). It is a pity that not all other data collection programmes have been so well planned or the test materials so well characterized. Foresight in this area would have increased our knowledge still further.

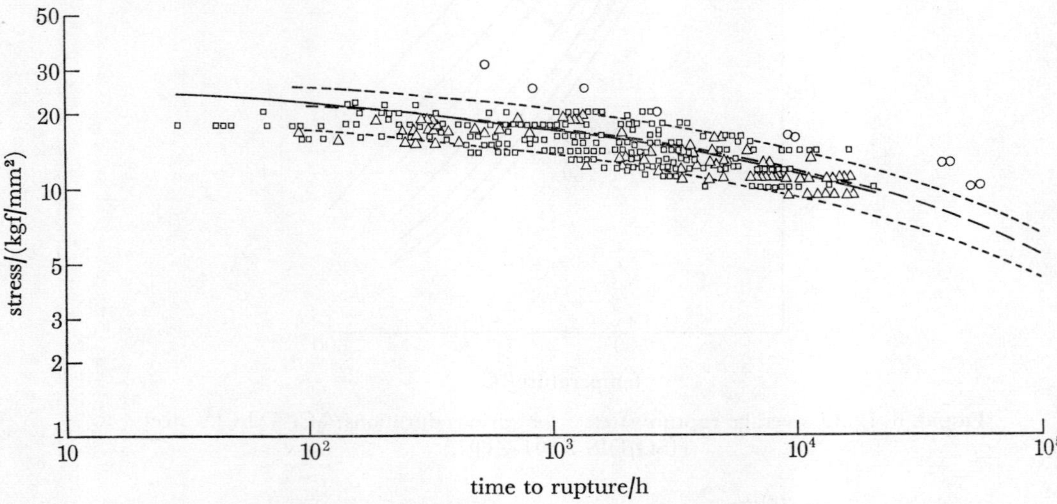

FIGURE 5. $\frac{1}{2}$Cr $\frac{1}{2}$Mo $\frac{1}{4}$V steel – I.S.O. stress-rupture data at 575 °C. (1 kgf/mm² ≈ 9.8 MN/m².) (ISO/TC 17/WG 10/ETP–SG 9.)

Structural analysis

A discussion of structural analysis must begin by considering the initial design phase of determining basic scantlings. Codes are available throughout the world which relate the design stress in the creep range to either the creep-rupture strength or the creep strength at 100000 h or longer (Findley 1973–74).

The latter criterion, as a result of the lack of relevant data, is usually replaced by an equally questionable criterion based on the creep rate at 1000 h (Tuppeny 1973–74). The long term data collection programmes, discussed earlier, are clearly a relevant contribution by the metallurgist to high temperature design.

The real problems occur in the verification of a design generated by the codified approach. The economic incentive to get the most out of our plant and the materials of construction demands an even closer look at areas of structural discontinuity under the complex loading conditions imposed by the plant operator. The consequences of failure, in both economic and safety terms, are becoming of increasing concern to the designer.

A prerequisite of assessing failure risk is the need to compute the stress and strain history at points of structural discontinuity. As early as 1935, Bailey (1935) demonstrated the ability to perform creep calculations but, in spite of the intensive efforts of Johnson *et al.* (1962), it is only

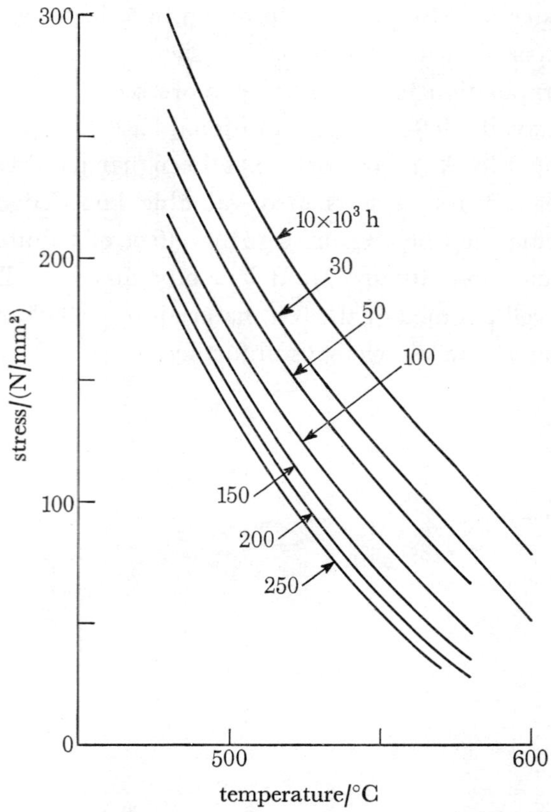

FIGURE 6. I.S.O. average rupture stresses for various durations: $\frac{1}{2}$Cr $\frac{1}{2}$Mo $\frac{1}{4}$V steel. (ISO/DIS Z 604–Z 607.)

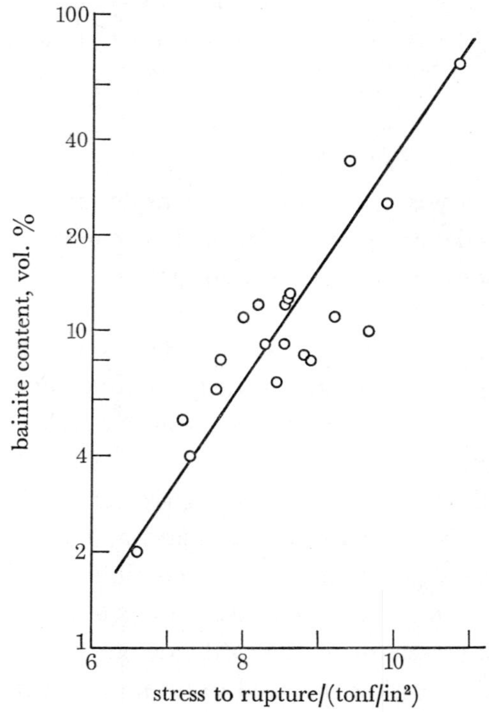

FIGURE 7. The relation between creep-rupture strength and bainite content of Cr Mo V castings; 10000 h at 575 °C (1 tonf/in² ≈ 15 MN/m²) (Murphy & Branch 1969).

recently that substantial progress has been made. Surprisingly, this progress has been made in two separate directions.

Approximate design methods, which rely on the identification of a reference point or points within a structure, the behaviour of which determines the response of the structure as a whole, are useful in that a single uniaxial test can be made following the loading conditions applicable at the reference point (Hayhurst *et al.* 1973). More exact analyses can however be made with the use of finite element methods, made possible by the advent of large digital computers. Given sufficient time and money, it is a feasible proposition to consider the analysis of any structure, no matter how complex its shape and the loading conditions imposed. However, there are snags resulting from our lack of knowledge on materials behaviour, which is generally limited to strain/time data from constant load, constant temperature tests such as those shown in figure 8 for a Cr Mo V rotor forging.

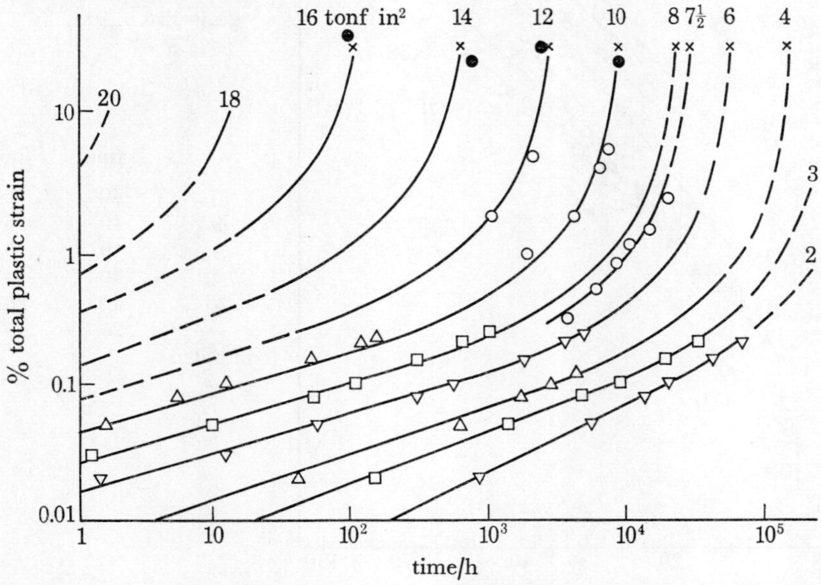

FIGURE 8. Cr Mo V rotor forging: estimated and actual creep curves, 575 °C (Branch 1972). △, □, ▽, creep tests; creep rupture: ×, predicted fracture; ○ interruption measurement; ●, fracture.

The analysis is less time consuming if constitutive equations are available to describe materials behaviour. Unfortunately, such equations are of very doubtful validity outside the range of parameter values covered in the experimental testing. The metallurgist is presented with a daunting task in providing materials data which recognizes the cumulative effects of a prior history of variable loadings. The enormity of the task is compounded by variations in material properties and residual stresses resulting from fabrication, particularly welding. This latter difficulty can however be eased by removing welds from areas of stress concentration.

A further problem in structural analysis is that of applying uniaxial data to the practical multiaxial stress situation. There are considerable experimental data to back the use of criteria such as Von Mises condition for plasticity or high stress creep situations, but further experimental work is desirable in the low stress, grain boundary deformation dominated, situation. Independent of the analytical predictions of deformation in complex situations, there is also a need to know the strain capacity of materials. This is one area where the conventional uniaxial test gives no quantitative guidance and it seems better to use the crack propagation approach discussed below.

Failure analysis

The useful life of many components is limited by failure resulting from crack growth, rather than by deformation. Furthermore, in view of the fact that freedom from cracks cannot be guaranteed in engineering structures, there is always an implicit need for the designer to assure himself that a crack at the limit of detection will not lead to failure in service. The safe behaviour of an engineering structure containing a crack is highly dependent on the ability of the material at the crack tip to deform plastically without crack extension, thus relieving the stress concentration at the crack tip.

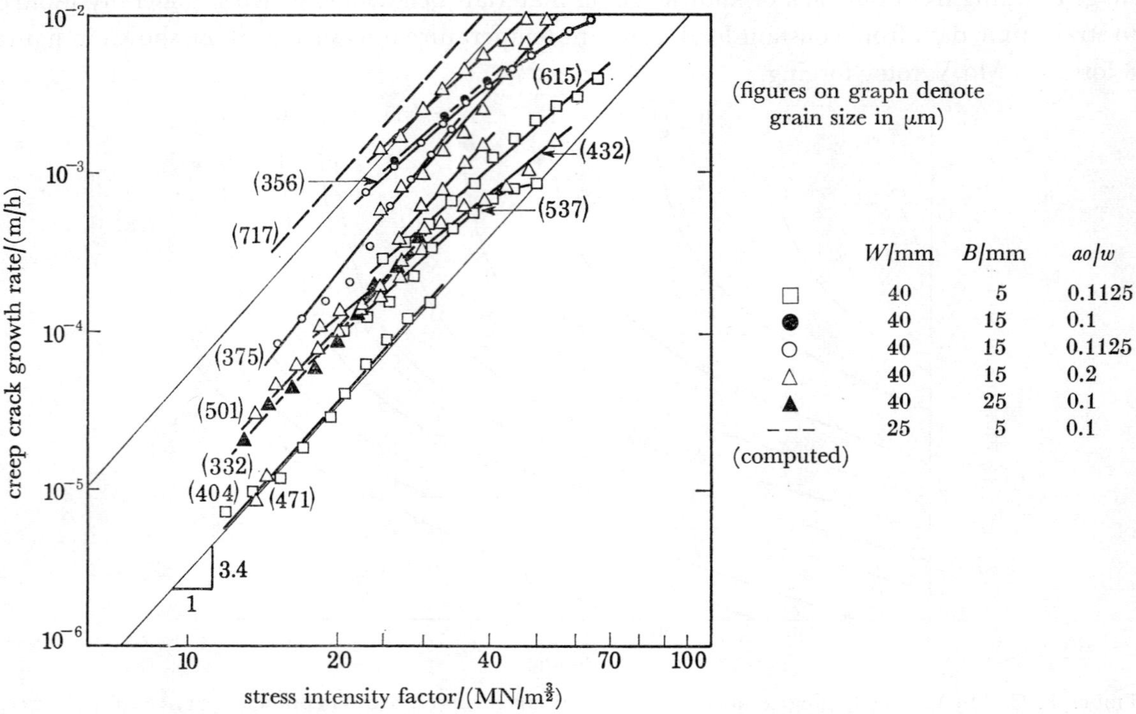

FIGURE 9. Crack growth rate as a function of stress intensity factor $\frac{1}{2}$Cr $\frac{1}{2}$Mo $\frac{1}{4}$V quenched (Neate & Siverns 1973).

For those circumstances where no crack extension occurs, life can be predicted from conventional creep-rupture data by limit analysis (Goodall 1974). At the other extreme where deformations remote from the crack tip are negligible, failure times can be computed from crack growth data obtained from laboratory tests of the fracture mechanics type using compact tension testpieces (McEvily & Wells 1973–74). Typical data of this type for Cr Mo V pipe steels oil quenched from 1300 °C are shown in figure 9. The analytical requirements are a knowledge of stress intensities at the crack tip and laboratory data at the relevant stress intensities, paying due recognition to those factors controlling crack growth – environment, rate of change of stress, temperature and microstructure. The intermediate region where crack growth requires significant general deformation is imperfectly understood at present but methods of analysis based on crack opening displacement have been suggested (Wells 1963).

The failure mode of structures tends to become more brittle not only as the material becomes more brittle but also as the size of the structure increases and as the ratio of crack size/section size increases from 0 or decreases from 1 (I. W. Goodall & C. H. A. T. Townley, personal

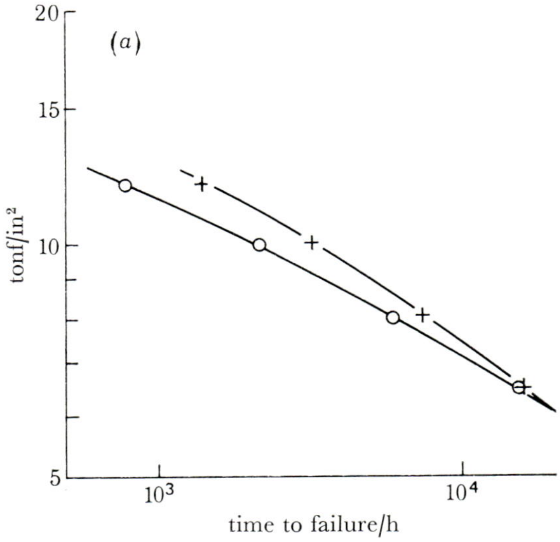

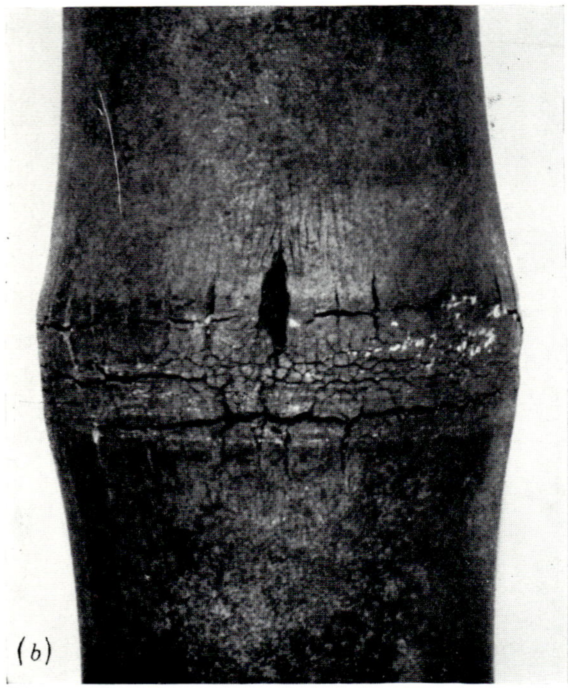

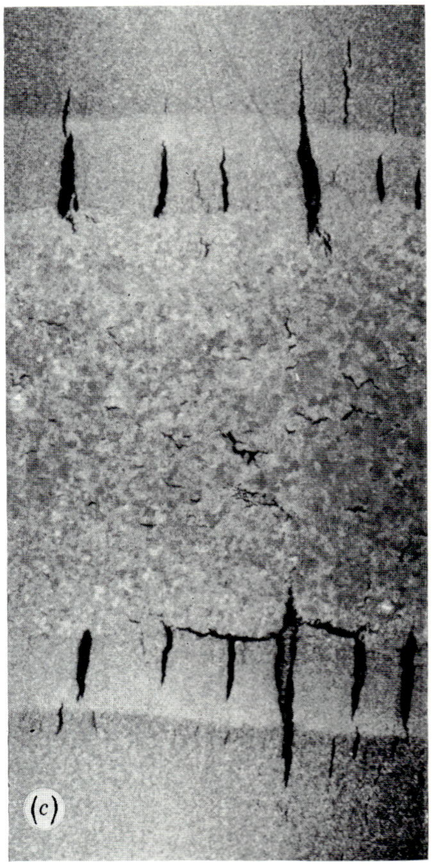

FIGURE 10. Internal pressure creep-rupture properties of welded Cr Mo V tubes at 575 °C (Branch & Williamson 1972). (*a*) Time to failure as a function of mean diameter stress: +, parent tube; ○, welded with 1Cr $\frac{1}{2}$Mo weld metal; (*b*) general view of 1Cr $\frac{1}{2}$Mo weld metal failure ($\frac{3}{4}$ size); (*c*) h.a.z. cracking associated with 1Cr $\frac{1}{2}$Mo weld metal.

(*Facing p.* 189)

communication). These factors are understood qualitatively but, unfortunately, not quantitatively. This leaves much interesting work on crack growth for the metallurgist, but his work will need to be planned very carefully to isolate the effects of geometry, stress, stress range, frequency, rate of change of stress, hold times, temperature, prior history and environment. He will also need to instrument his tests to measure displacements. Quite a challenge, but I suspect a worth while one in view of the potential benefits.

Materials selection

The careful reader will have noted that I have chosen to illustrate this paper with information relating to the family of low alloy Cr Mo V steels which are widely used in steam power plant for high temperature pipework, boiler headers, valve bodies, turbine casings, rotor forgings and bolts. It will also have been noted that the steels have been developed through carefully planned research and development programmes and that extensive data have been collected and processed to produce design information. In more recent years, creep strain data and crack growth data have been produced to permit detailed stress analyses and failure analyses. These latter data requirements have become necessary as a result of isolated failures and a somewhat larger number of cracks which can be found with the help of modern non-destructive testing techniques. When we add our plant experience, to be discussed later, it can be seen that the present day materials engineer has a vast body of metallurgical knowledge at his disposal to aid the crucial 'front end' process of materials selection.

Apart from this knowledge and an equal knowledge of other factors relevant to material selection – cost, delivery, fabrication, weldability, corrosion resistance, fracture toughness at ambient temperatures and tolerance to abuse – the materials engineer must have sufficient status in his organization to play an active role in determining company policy. In return for being given this status, the materials engineer must of course be willing to accept full responsibility for the decisions taken as a result of the contribution which he makes.

I believe that the metallurgical profession as a whole can make significant progress, both for its members and for the manufacturing and process industries, by improving the status of the materials engineer in general to the levels obtaining in our more progressive companies. The process is one of education, training and publicity. The quality of the metallurgist is not in doubt, as witnessed by his contributions to research and development and to failure investigations. Why should he not be capable of accepting a more positive role in the engineering function?

Plant operation

Having highlighted some of the uncertainties in our knowledge of high temperature behaviour, I would now like to suggest that some of these gaps could be filled by increased studies during and after plant operation. This information requirement has only been partially met by work on experimental engineering structures, such as the internal pressure creep-rupture testing of welded tubes (figure 10, plate 11). Such experimental work is difficult to promote in view of its high cost and long time scales, together with its uncertain relevance to real structures. As a result, it can only be justified as part of the development of high capital cost projects in industries such as nuclear power and aero-space.

Studies of operating plant can only be effectively pursued within an integrated organization, as they require very careful planning in the design phase. Sufficient spare material must be procured, not only to fully characterize the material in terms of the conventional design properties, including the effects of fabrication, but also to study the change of properties resulting from service exposure. Extensive instrumentation must be installed to monitor the service environment and its influence on the structure during operation. Instrumentation is required to monitor temperatures, strains, pressures, chemistry of the environment and crack growth, both during operation and during plant shut-downs.

I do not underestimate the practical difficulties of such work and the costs in design, construction and operation would not be negligible. However, the potential future benefit is so great that I believe that increased effort in this area could be justified.

Independent of any plant oriented studies of mechanical behaviour at high temperatures, the metallurgist might perhaps play a more significant role in the control of plant operations. There is no doubt that some of our plant problems could have been prevented with a greater appreciation by the operator of material limitations. I am thinking particularly of those failures resulting from thermal stresses, in excess of those postulated at the design stage, arising from excessive heating and cooling rates during the start-up and shut-down of plant (Wyatt & Gemmill 1963).

Failure investigations

A significant proportion of our total knowledge on the high temperature behaviour of materials has been obtained from carefully planned investigations into plant failures. One important observation is that only a small minority of creep failures can be attributed to a lack of creep strength and these few have been due, either to using a material with properties below those assumed by the designer or to operating under conditions more severe than the design conditions (Wyatt & Gemmill 1963).

The majority of creep failures have resulted from an inability to accommodate strain in local regions of engineering structures without cracking (figures 2–4). These strains arise generally in a multiaxial stress situation and result from a variety of sources: (1) mechanical loading, (2) thermal stresses, (3) system stresses, (4) local stress concentrations, (5) relief of fabrication stresses, (6) relief of thermal stresses.

It must also be recognized that these strain requirements can arise in areas where the microstructure of the material has been changed during the process of fabrication.

There is an obvious obligation on designers and plant operators to do what they can to minimize these strain requirements but there is an overriding obligation on the materials engineer, using the information available from the physical metallurgist, to provide materials with an optimum combination of strength and ductility. The lesson to be learnt from our failure investigations is that this combination must be available, not only in the virgin state but also in the as-fabricated state. This leads to a need to understand and control the effects of fabrication, particularly in those cases where we are forced for other reasons to use materials which are not particularly tolerant to abuse during fabrication.

Concluding remarks

This paper has been prepared to initiate some discussion on the conference theme of examining the impact of physical metallurgy on engineering practice in the area of design. The subject is much too wide to provide the hard evidence within a short paper for many of the views expressed but I trust I have conveyed some impression of the extent of the metallurgist's contributions.

Areas of significant and generally successful impact include:

(1) A qualitative understanding of the microstructural features controlling high temperature strength and ductility.

(2) The development of materials of increased creep strength.

(3) The collection and collation of long term uniaxial data, although in retrospect this might have been better organized.

(4) An encouraging start to studies of high temperature crack growth.

(5) Technically enlightening studies of plant failures.

Areas where the metallurgist has not, in the past, been quite as successful as he might have been include:

(6) Giving sufficient emphasis to creep ductility.

(7) Examining the consequences of successful, as well as unsuccessful, plant operation.

(8) Transferring his knowledge to the benefit of engineering plant.

For the future, much work remains for the metallurgist to assist in improving the overall economy of high temperature plant. My predictions are that we will make significant progress in the area of high temperature crack growth and that we will continue to struggle with the problems of the development of constitutive equations to cope with the analysis of variable loading in complex stress situations. I would hope for extensive monitoring of selected operational plant as a valuable aid towards solving the latter problem but, above all, I would look towards further development of the status and role of the materials engineer.

There is an urgent need for a shift from classical physical metallurgy to a greater involvement of the metallurgist in the design function. This is the only way to transfer effectively our knowledge into our plant. I wonder if Walter Rosenhain would have agreed with me, and whether the educators of future metallurgists will accept the challenge.

This paper is published by permission of the Central Electricity Generating Board.

References (Branch)

Andrade, E. N. da C. 1910 *Proc. R. Soc. Lond.* A **84**, 1.

Andrade, E. N. da C. 1914 *Proc. R. Soc. Lond.* A **90**, 329.

Bailey, R. W. 1935 *Proc. Instn mech. Engrs* **131**.

Bardgett, W. E. 1948 *J. Iron Steel Inst.* **160**, 143.

Branch, G. D. 1972 *The presentation of creep strain data.* British Steel Corporation.

Branch, G. D. & Williamson, J. 1972 *Creep of welded structures.* International Institute of Welding.

Findley, G. E. 1973–74 International conference on creep and fatigue in elevated temperature applications. Institution of Mechanical Engineers.

Glen, J. 1948 *J. Iron Steel Inst.* **158**, 37.

Goodall, I. W., Cockroft, R. D. H. & Chubb, E. J. 1974 CEGB Report No. RD/B/N.3059.

Hayhurst, D. R. *et al.* 1973–74 International conference on creep and fatigue in elevated temperature applications. Institution of Mechanical Engineers.

ISO/DIS 2604-2607 1972 *Steel products for pressure purposes. Quality requirements.*

ISO/TC17/WG10/ETP-SG9 Elevated temperature tensile and stress rupture properties of ½Cr ½Mo ¼V steel.

I.S.O. Recommendation R831 1968 *Rules for construction of stationary boilers.*

Johnson, A. E., Henderson, J. & Khan, B. 1962 *Complex stress creep, relaxation and fracture of metallic alloys.* Edinburgh: H.M.S.O.

McEvily, A. J. & Wells, C. H. 1973–74 International conference on creep and fatigue in elevated temperature applications. Institution of Mechanical Engineers.

Murphy, M. C. & Branch, G. D. 1969 *J. Iron Steel Inst.* **207**, 1347.

Neate, G. J. & Siverns, M. J. 1973–74 International conference on creep and fatigue in elevated temperature applications. Institution of Mechanical Engineers.

Smith, A. I. & Nicolson, A. M. 1971 *Advances in creep design*. London: Applied Science Publishers.
Tapsell, H. J. 1944 *Proc. Instn mech. Engrs* **151**, 54.
Tapsell, H. J. & Ridley, R. W. 1945 *Proc. Instn mech. Engrs* **153**, 181.
Tupenny, W. H. Jr. 1973–74 International conference on creep and fatigue in elevated temperature applications. Institution of Mechanical Engineers.
Wells, A. A. 1963 B.W.R.A. Research Report M13/63.
Wyatt, L. M. & Gemmill, M. G. 1963 Joint international conference on creep. Institution of Mechanical Engineers.

KEYNOTE SPEECH

By R. G. Baker (*National Physical Laboratory, Teddington, Middlesex*)

Introduction

The requirements of the engineer must be expressed as recognizable goals for those involved in material research and development if technological resources are to be used efficiently. This requires the development of reliable physical models together with the establishment of standard test procedures and the accumulation of relevant data on the important metal characteristics.

It may be helpful to think of an analogy in considering the demands made on the materials scientist. A designer seeking to innovate in engineering may be compared to a farmer in the land of the blind who unknowingly buys a farm situated on the edge of a steep and very irregular precipice, which is initially unfenced. When his stock progressively disappear, he suspects the existence of a precipice and, securely roped, probes for the position of its edge. Having established this at a number of points he proceeds to build fences between those points. In that way he cuts down most of his stock losses, just as, in the absence of any reliable basis for design, an engineer would have to work by trial and error towards a system of design which worked within the experience available to him.

We can carry the analogy still further, since land-slips could bring the cliff edge within the fence and there would be further stock losses until its position was corrected. In the engineering case, unexpectedly severe conditions or perhaps variability of material could lead to problems until the specification or codes were amended. The farmer's problem by now has attracted the attention of a wise man, who advises him to take back his fence at least a quarter of a mile from the nearest point where he has lost a cow. He adopts this completely safe solution, but unfortunately his grazing area is now so severely restricted, that his cows begin to die of malnutrition and he considers going out of farming altogether. In engineering, safe solutions are often easy to prescribe but, if not realistic in cost, they may lead to abandonment of the project or the adoption of a less safe alternative which only gives no cause for concern because of ignorance. The ideal solution is naturally an economic system of tests and prediction which enables the farmer to map the edge of the cliff precisely, and build his fence far enough away from it to reduce the risk of cattle loss through landslip to an acceptable level. In engineering, this corresponds to the physical modelling of the potential problem and establishing a safety factor consistent with the required reliability.

To summarize: potential problems must be identified and their cause established in qualitative terms. Assessment methods must then be defined which at least reproduce their essential featues. Quantitative engineering requirements must then be related to material characteristics such as flow stress or toughness which can be measured in relatively cheap and simple tests. Safety factors must then be established which are consistent with the required reliability. All these things are the concern of the metallurgist and it is the concern of the session to see how far he has succeeded.

Strength

Wells argues that, whatever the aspect of strength in engineering structures, provided no time dependent deformation occurs and variable loading can be neglected, flow stress is the relevant material characteristic. Neither u.t.s. nor y.s./u.t.s. are of importance in design. On the other hand, there must be sufficient ductility to accommodate the plastic strains which occur at stress concentrations, when a structure is loaded. This is a significant conclusion because the optimum combination of high strength with high toughness and good weldability is found in steels with a high yield to ultimate ratio. It is thus counterproductive to cling to an unnecessary regard for the u.t.s., if it is really unnecessary. Nevertheless u.t.s. is a desirable material characteristic to quote in specifications, since it is very sensitive to composition and is thus useful as a check on quality assurance where strength is to be combined with some characteristic such as weldability.

Wells emphasizes that an intrinsically high modulus of elasticity is not always realized in the structure. The presence of residual stresses can reduce the effective modulus, and Wells illustrates this with reference to the case of rods surrounded by foamed plastic material, which failed by buckling well below yield stress. If there is general agreement with his interpretation, then perhaps there is scope for reduced design factors in compression members to be offset against stress relief. There may also be scope for altering production processes to produce stress free products in some cases if it proves economically worthwhile. Incidentally, does the over stressing practiced in proof stressing line pipe reduce the risk of buckling during laying in the North Sea?

Wells draws attention to the importance of structural reliability and cites fracture toughness to exemplify the application of probability theory to estimate reliability. Presumably, the risk of failure by plastic collapse depends on the probability that flow stress may fall below the design stress. Failures of this type are exceedingly rare except when an unexpected factor is encountered such as the use, by mistake, of a metal with flow stress below the design stress. This suggests that design factors are generally too high but, whether or not that is so, design stress is certainly based on a measurement of flow stress taken at a standard position in plate, strip or section, and a design factor is used which, on the basis of experience, takes account of variability due to segregation, variation of temperature deformation history; within-cast as well as cast to cast variation. This raises two points: 'Could we reduce design factors by using a reliability approach to the definition of yield stress?' and 'Is there benefit in reducing the scatter of properties'. The latter question should be considered with care. 'People get the parliaments they deserve', it is said, and customers certainly get the materials they deserve. In general, any improvement in quality means an increase in production cost, and thus an increase in price. This must be carefully balanced against the benefits.

Fracture toughness

Great advances in understanding of fracture toughness have been made over the last 20 years. The production of Liberty ships during the last war corresponds very closely with my analogy. More than 50 were lost 'over the precipice' by brittle fracture before a rule of thumb method for material selection was evolved. We can now identify two general bases of design;

against initation and progagation respectively. There is widespread recognition that criteria for material selection should depend on the severity of application. This is well illustrated in Barr & Burdekin's figure 1 (paper 2.2).

Fracture mechanics enables reliability theory to be applied as outlined in Wells's paper. Thus, in principle, it should be possible to define critical fracture toughness parameters appropriate for the required reliability in the light of scatter in material properties, and of the NDT methods used. This seems to be very considerable progress, but there are a number of implications.

Wells emphasized the advantages of reduced scatter in material properties. Figure 1 of Wells' paper (2.1) shows the considerable scatter which apparently exists in fracture toughness. If we have to live with that sort of scatter, we can still design rationally against brittle fracture. Potentially, material could be used much more economically if it were more consistent and figure 1 suggests there is quite a lot to go for! Perhaps recent improvements in material quality and processing techniques have reduced scatter significantly. The physical metallurgist has achieved a good qualitative understanding of the factors involved in brittle fracture. Whether or not a more precise relationship between microstructural parameters and the behaviour of definitive details would be an advantage, is something which requires further debate. However, there is an apparent lack, at present, of a reliable small scale standard initiation test for general use with the lower strength, more ductile materials. The c.o.d. approach is undoubtedly safe, although non-rigorous. However, particularly in very ductile materials with high yield to ultimate ratio, the criteria given by the c.o.d. test can be very over-conservative. The results of the test also may be unreliable where one is trying to predict the behaviour of a structure from tests on a small region of material. Barr & Burdekin draw attention to both of these points and they merit further discussion. However, whatever the qualifications on the c.o.d. test it seems clear that, used in conjunction with a more definitive test such as the wide plate test, it can provide valuable supplementary information. The advantage over the more precise linear elastic fracture mechanics as a criterion for design is illustrated by Barr & Burdekin in their figure 3.

One further point must be mentioned and that is the role of the Charpy test. The Charpy energy is a complex mixture of that required for initiation from the existing notch and propagation. As such, it is completely useless as a definitive test. Local correlations, however, can be made with more definitive tests, and this makes it extremely valuable for quality assurance purposes, and in order to characterize materials falling within certain broad types for material selection. The Charpy test then is here to stay and there is no need to replace it providing no one is deceived about the nature of the information it gives.

Fatigue

Forsyth (2.3) highlights the need to define precisely initiation and propagation respectively. He adopts an arbitrary definition for convenience; that initiation refers to the development of a crack to 0.127 mm. This concept is useful in planning an engineering design and inspection strategy but useless as a criterion for the materials scientist, whose concern lies in modelling the fatigue process. Are we always dealing with progagation from the most potent pre-existing defect in practice or must we invoke models for homogeneous initiation? Forsyth is clearly unsure, he notes that there '...appears to be some limiting size below which defects have no appreciable effect on strength although they may still eventually be the features that initiate the first cracks'. If fatigue always started from the most potent defects, as the work at B.I.S.R.A., during the last decade seemed to suggest, one would expect that the "intrinsic fatigue strength' would be some function of the defect size population. In the B.I.S.R.A. work, it was found that both defect size and defect type were important.

In work on the fatigue of welded joints it was found that defect acuity as well as size was significant. Could the lack of this concept inhibit an explanation of the relatively simpler case where there is no serious stress concentration? It might perhaps resolve Forsyth's difficulty; 'the devastating effects on fatigue strength of defects appears to be out of all proportion to their size and they cannot be correctly assigned a K_t value which has any meaning'.

Whatever the truth of the matter, it appears essential to consider the effect of surface treatment and environment as separate, though most important factors. One can readily see why compressive surface stresses would have a beneficial effect on the fatigue life. To what extent are surface treatments reliable throughout life? How can protection be lost, and is there potential for further improvement?

The more complicated question of corrosion fatigue seems at present enormously important in a number of areas, and also very intractable. One feels instinctively that the measurement of crack growth rates, as shown by Forsyth in figure 8, is the correct method of approach; provided that the methods used can be related uniquely to conditions in practice. It is perhaps both cowardly and at the same time realistic to suggest that at the present time there is no reliable basis for a unique relationship. Two questions, I think, are of great importance:

1. To what extent and how reliably can one live with corrosion fatigue by relying on protective methods such as coating, cathodic protection, inhibition, etc?
2. To what extent is the factor of three suggested by Forsyth either reliable or of general application in practice? If it is a limit value, then this is perhaps an important point in relation to design.

High temperature design

In this field there is a good qualitative understanding and some quantitative relations between microstructure and the behaviour of specimens tested uniaxially. This has been of value both in the development of materials and in the analysis of behaviour in practice. At the same time, perhaps because of the intractable nature of the problem, there has been little progress in modelling material characteristics in terms of true design needs and establishing realistic design criteria on a reliability basis. Branch (2.4) emphasizes that failure in creep, as a result of simple underdesign, is virtually unknown. Creep failures certainly occur, but these are almost invariably because of temperature excursions above those designed, or through the use by mistake of a material with inadequate creep strength.

Branch also refers to an extremely important group of failures, which occur early in plant life at stress concentrations; perhaps with the added complication of thermal stresses, system stresses and the relief of fabrication stresses. These have led to 'fracture mechanics' type tests in which precracked specimens have been tested and crack growth rate plotted as a function of stress intensity, as shown in Branch's figure 9. Instinctively one feels that this is the logical way to approach the problem, but it does give rise to the question of how far fracture mechanics criteria, such as stress intensity, can be applicable to the viscoelastic behaviour at high temperatures

Evidently, ductility in the vicinity of stress concentrations is of immense importance, and should perhaps form one of the design criteria. It seems to me unlikely that there will be a simple relation between notch ductility under these circumstances, and ductility in a parallel sided uniaxial test. Perhaps the many uniaxial creep tests on parallel sided specimens should be supplemented by more on specimens with sharp notches in them.

Another point brought out by Branch is the importance of taking the effect of biaxiality and triaxiality into account in design. It is unrealistic to expect that one could ever, or would ever wish to replace all uniaxial creep data with triaxial creep data, but we should at least know how the one relates to the other, and whether there is a unique relation. Can we get such information on properly designed test specimens, waisted or notched to give biaxial or triaxial stresses, or must we undertake the complex and costly testing of structures?

In conclusion, physical metallurgy appears to provide an invaluable basis for conceptual understanding of behaviour, and goals have been defined for materials development. Two factors emerge:

In certain areas, the traditional goals should be modified. Those now defined as important include: improved performance and consistency in yield stress, definition of small scale initiation and propagation tests for brittle fracture, propagation behaviour in fatigue and the improvement of long term and short term notched ductility in creep in biaxial and triaxial systems. In most cases this in no way renders redundant the traditional tests, which will remain economic and invaluable methods of implementing quality assurance.

There is a clear challenge to the material scientist to work with the engineer in evolving constituitive relationships to explain the behaviour of engineering details. Finally, if we are to make the best use of our resources without prejudicing safety, reliability concepts should be introduced to provide a logical basis for design factors, through relating them directly to the reliability required by society for the structures concerned.

Discussion (Chairman – D. McLean (*National Physical Laboratory*))

F. A. Leckie (*University of Leicester*)

I share Baker's opinion (Keynote) that properly formulated constitutive equations form the essential link between material behaviour on the one hand and the laws of mechanics on the other. In this way it is possible to relate material behaviour to the performance of components occurring in engineering practice. The procedure has, in general, been limited to the material mechanical properties and little attempt has been made to include in the constitutive equations, suitable descriptions of the physical mechanisms which take place within the material. There is evidence however, that attempts are now being made, especially in the field of fracture and creep mechanics, to relate in a quantitative way, the physical processes occurring within the material to the behaviour of components with notches.

If, in this way, the physical mechanisms can be related to component performance by quantitative means then it should be possible to design materials to meet engineering requirements in a systematic manner.

N. Hansen (*Danish Atomic Energy Commission*)

A nuclear problem relevant to the theme of this afternoon's session is the design and performance of fuel elements for water reactors. The fuel was designed in the 1950s and materials were developed to fit the design requirements. The fuel elements behaved satisfactorily in-pile except for some difficulties due to lack of control in the manufacturing process. By the end of the 1960s it was necessary for reactors not only to work under base load but also to follow the requirements from the grid. This drastic change in operating conditions led to overstraining of the cladding. The problem has been circumvented by restrictions on operating conditions, an expensive solution costing perhaps U.S. \$5–10 × 10^6 per year for a single station.

The problem was to analyse the behaviour of fuel elements under irradiation and to predict as closely as possible the resistance of the fuel to power cycling. To solve this complex problem mathematicians, computer experts, engineers and metallurgists worked together to develop a fuel element model consisting of a generalized finite element model, including a series of physical sub-models. Parallel to this mathematical and physical modelling, engineering programmes were carried out to test the model predictions.

This model work has had a number of positive results: (*a*) An improved understanding of the relevant problems to be solved in this field, (*b*) A closure of the gap between laboratory tests and what is required in engineering practice and (*c*) With reference to Branch (2.4), it had helped in a rapid transfer of physical metallurgy results into engineering application.

J. T. Barnby (*University of Aston*)

In relation to Forsyth's paper (2.3), I should like to comment on the suggested method of design against fatigue crack initiation by choosing an arbitrary size for the nucleus and counting the number of fatigue cycles to produce such a nucleus. If one does this, there is a practical basis which should be usable for design since number of cycles for initiation, N_1, is related to the stress intensity factor ΔK according to:

$$N_1 \propto (1/\Delta K)^m.$$

An initiating notch can also be allowed for. However, I should like to make a distinction between fatigue crack initiation from a large stress concentration, such as some shape factor, and fatigue crack initiation from small surface or internal defects. In the former case, so long as the nucleus size is much less than a critical defect size, then design against 'initiation' is practicable.

In the second case, two problems arise. As we choose a smaller and smaller nucleus size we find that the law above changes in terms of the material properties, namely the coefficient and the power m; though the *form* of the law does not change. This may arise because we are dealing with an extra geometrical factor such as the growth of many small and differently shaped nuclei. The second problem is that, at smaller nucleus sizes, we reach down to the level of inhomogeneity of the structure. If both of these factors are allowed for perhaps we shall have a true capability for design against fatigue crack initiation and perhaps also we shall see that initiation behaviour is directly related to the propagation behaviour which is now quite well known.

Comment by G. M. Boyd (presented by J. van der Veen)

Figure 1 of Wells's paper (2.1), originally appeared in a paper by Hodgson & Boyd (*Trans. Inst. R. Naval Architects* **100**, 1958) and its object was simply to show that a large proportion (75 %) of the steels then available in the world would comply with a certain criterion, i.e., minimum energy 47 J (35 ft lbf) and 30 % fibrous fracture at 0 °C.

Since the steels surveyed were taken virtually at random from all over the world, at a time when there were no specification requirements for Charpy values, it is not surprirsing that they conform to a Gaussian distribution with a fairly wide dispersion. This is even more probable considering that the values plotted were at a particular, arbitrarily chosen, temperature, i.e., 0 °C.

When the values were plotted separately for each individual steelworks (as stated in the original paper) the individual scatters were much smaller. The points tended to form 'clouds' in the field of the diagram, and the positions of these 'clouds' clearly reflected the steelmaking practices of the individual steelmakers.

Since then (1955–58), specifications have included Charpy requirements, and the 'clouds' now tend to be confined to the region II of the diagram.

J. van der Veen (*Hoogovens, Netherlands*)

In Wells's paper (2.1) reference is made to design habits for bridges, in which loading density increases with increase of span. The necessity for long bridges to accept a hypothetical queue of (heavy) vehicles is rejected. I think this can sometimes be a dangerous method since in heavily populated areas, such as Holland, and especially around Rotterdam, almost every day, queues of heavy traffic occur during peak hours on rather long span bridges.

A. A. Wells (*The Queens' University of Belfast*)

In a reappraisal of the synthesis of equation (1) in my paper (2.1), with regard to the probability of interaction of an actual crack length and a critical crack length, it is necessary to draw attention to two objections. Firstly, the interaction will also occur if the critical crack length is less than the actual defect length, as well as when the two are equal. This is a question of the most suitable modelling for the physical event. However, there is also a second objection, which is fundamental, that the probability P_x should have been expressed as $P_x \, dx$. The preferred alternative expression, due to Sir Peter Hirsch (unpublished communication) is

$$P = \int_0^\infty \left(\int_0^x P_\alpha \, d\alpha \right) N_x \, dx. \tag{1a}$$

It is quite fortuitous, but nonetheless fortunate that the evaluation of the original equation, detailed in the Appendix, is the means of showing that equation (1*a*) may also be calculated to a good approximation in closed form, leading to a result that differs little from equation (2). Use is made (*a*) of the series expansion,

$$\int_x^\infty e^{-\xi^2} \, d\xi = \frac{e^{-x^2}}{2x} \left(1 - \frac{1}{2x^2} + \frac{3}{4x^4} - \frac{5}{8x^6} \cdots \right)$$

and (*b*) of the observation that interaction following equation 1 is confined to the narrow band about $d = ac^2/(b^2+c^2)$, with standard deviation $e = bc/\sqrt{(b^2+c^2)}$. It is then sufficient for the purpose of the second integration to observe that the average value of $1/2x$ is $(b^2+c^2)/2ab$, which is usually significantly less than unity so that only the first term of the expansion is needed. On this revised basis the recalculated probability,

$$P = [(1+n^2)/\pi]^{\frac{1}{2}} \, \tfrac{1}{2}(1+\operatorname{erf} nm) \, e^{-m^2} \tag{2a}$$

differs from equation (2) only by the multiplying factor $(b^2+c^2)/2ab$. The values of m, n, d and e remain as before, and the trend lines of figure 2 are lowered and somewhat steepened.

The main effect of the correction on the worked example is to cause the observed failure rates for oil storage tanks and ships to be consistent with defect populations which are 4 times larger, i.e. $\frac{4}{3}$ defect per structure for tanks, and 0.53 defect per structure for ships. The relevance of the comparison is unimparied, bearing in mind that defect rates for modern welded structures are shown by non-destructive methods to be somewhat lower than these figures. The contention is strengthened, that quite small improvements in the r.m.s. scatter in the range of interest are likely to lead to much reduced failue probabilities.

Van der Veen is right to draw attention (this discussion), as a result of his discussion with Boyd, that figure 1 substantially broadens the scatter band likely to have emanated from the products of any one steelworks because it represents a world-wide sample. Nevertheless, it is necessary to make use of such a broad sample, in order to be consistent with the world-wide population of casualites, which would have been associated with the collective products of these steelworks, and possibly more. Another demonstration then arising from the analysis is the powerful effect that would have resulted from unification of the notch ductility rules of the world grouping of Classification Societies, by means of which the proportional scatter was first reduced during that period. This in itself is a tribute to the work of Van der Veen and Boyd, and others that they would wish to be remembered. It is fitting that the debt of society to this distinguished group should be openly acknowledged.

P. B. Hirsch, F.R.S. (*University of Oxford*)

I should like to comment on the use of failure probability analysis of cracked structures. There are at least two reasons for using an analysis of this type: (1) It is desirable to have some way of quantitatively estimating the failure probability of a structure or vessel, rather than relying entirely on experience and safety factors. One simple way in which fracture mechanics is used by the designer is that for a given value of toughness and stress a critical crack size is worked out, and if this is larger than any cracks which might possibly exist or could be detected, one is reasonably happy about the reliability of the structure. But there is a spread in values of yield stress and toughness in a given material, this may lead to weaker spots being present in the structure with unacceptably high probability.

In principle, given that the spread of toughness values and of stresses are known, and given that the distribution of defects is also known, the failure probability could be estimated using the analysis of the type developed by Becker & Pederson, and also used by Wells (2.1). The variation of the failure probability with time may be evaluated given data on fatigue crack growth rates under operating conditions. If n.d.t. is used to determine defect distributions, the probability of missing defects must also be taken into account. The estimates are of course only as good as the input data, and at present data on defect distributions and on the variability of materials parameters are lacking. It seems desirable to obtain such data and to characterize materials not only by mean values of parameters such as toughness, but also with distribution curves.

The second reason for this type of approach is that it serves as a sensitivity analysis. It shows up the importance of particular parameters, e.g. how important the problem of fatigue crack growth is and the bearing it might have on inspection programmes, what the effect of the spread of values of toughness is, etc. In principle this kind of analysis could lead to a 'sharpening-up' of design parameters and safety factors. I do not underestimate the difficulty of obtaining the necessary data, and the complexity of the problem, e.g. there are alternative failure modes, but I do feel that some effort should be made in developing this approach.

J. F. Lancaster (*Kellogg*)

Boyd's paper on ship steel was a very memorable one because it represented one of the first good correlations between large scale failure experience and a small scale test, in that case the Charpy-V-notch test. 47 J (35 ft lbf) was the figure that Boyd concluded was the Charpy value required for the arrest of a brittle crack in the ship plate. This figure was built into Lloyd's specifications and subsequently the rate of failures went down quite dramatically. I think this though, should not make us as metallurgists, too ready to predict failure from small scale tests. May I give you an example? In our particular line of business we design plant operating at 900 °C and under internal pressures of 3.5 MPa (500 lbf/in^2). The basis for design is the stress for rupture in 10^5 hours and although it is a perfectly true statement people take it that because this is the basis for design the life of the tube will be 10^5 h say 5–10 years; after that it is finished and you have to throw it away. In fact when you look closely at what the designer does, it is not quite that way. He takes the average stress for rupture in 10^5 h at a given temperature, but then he applies a temperature safety factor to his. This in effect multiplies the life of his design by a factor of 10. He also uses other approximations including a mean diameter formula which biaxial tests have shown to be very conservative. Thus, in practice, he may actually be designing the tube for a life of between 100 and 1000 years while nominally according it a 10 year life. This is a case where small scale testing is being projected to determine the life characteristics of equipment in an unduly conservative way. I feel that much more work needs to be done in objectively surveying the actual rates, and modes of failure in service in order that we can determine what is the reliability of equipment and in order that theoretical formulae may be correlated with real behaviour. I was very glad to see that Wells (2.1) had made a useful and interesting beginning in this endeavour.

A. R. Flint (*Flint & Neill*)

Wells (2.1) has referred to the need to allocate the best material to the most demanding structures and Baker (Keynote) asked whether it was advantageous to reduce scatter of yield strengths. It is evident from application of reliability theory that the 'best' British steels from the designers' viewpoint are those having maximum variability for these provide the greatest reliability (least risk of failure) for a structure. This arises as a result of the practice at the mills of ensuring that the rejection rate of non-compliance with the guaranteed minimum yield strength is similar for different mills which results in the use of higher mean strengths in those which produce the more variable product. It may be simply shown that in such circumstances the nominal reliability increases with the coefficient of variation. There is thus no incentive to reduce scatter with the present definition of 'guaranteed' values related to the grades of steel. However the rationalization of safety factors for structual design which is in prospect *will* have the effect of penalizing the steels with low mean strengths and hence those mills providing the best control will be encouraged to raise the levels. It is hoped that this may be done with economic advantage.

R. R. Barr (*British Steel Corporation*)

The limitations to the application of crack opening displacement for general structural steels should be recognized if the concept is not to be misused. In the linear elastic range the conditions for unstable fracture initiation are fairly rigorously defined by K_{Ic}, but there is a need to clarify the useful range over which the c.o.d. concept can be used to characterize initiation and to calculate tolerable defect sizes. At temperatures up to the transition range, with limited plasticity accompanied by cleavage fracture, the c.o.d. can be used as a measure of tolerance for defects. Beyond this region (i.e. on the upper shelf) with a fully fibrous fracture, it is not possible to measure a c.o.d. value which has any relevance to the conditions at the crack tip. The general yielding fracture mechanics theory relies upon the c.o.d. measurement fully characterizing localized conditions of strain acting in the vicinity of the crack tip. The c.o.d. at onset of fibrous crack extension, δ_i, does not relate to conditions for fracture in a large structure

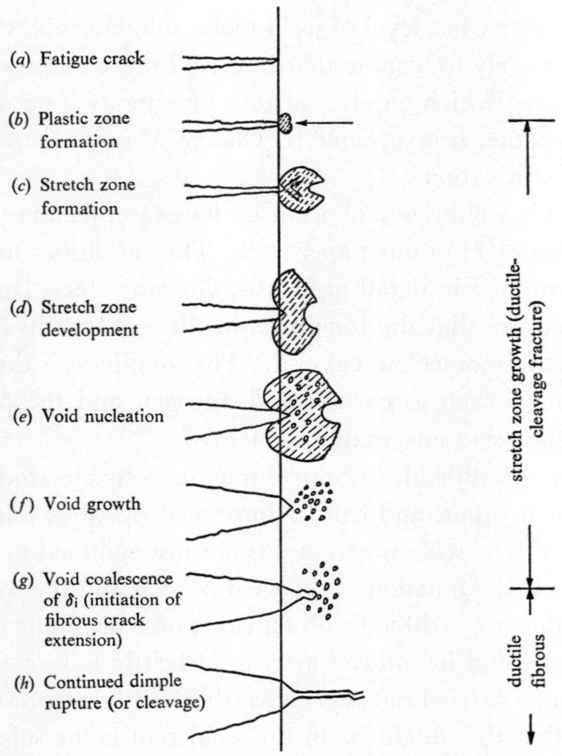

FIGURE 1. Successive stages in crack extension of a fatigue crack in a low strength steel by 3 point bending, (for clarity, the plastic zone is omitted in (*f*)–(*h*)).

and, if used, gives a very significant under-prediction of acceptable defect size. In the circumstances where a δm value, the c.o.d. at maximum load, is measured this is a function of test piece size and machine stiffness, and fracture is controlled by tensile instability. Whenever a maximum load type of failure is observed in a full thickness c.o.d. test conducted at the appropriate loading rate and temperature, the most realistic failure parameter is one of plastic collapse based on the attainment of the flow stress (or yield stress) in the remaining uncracked section of the structure.

These comments do not refer to the high strength materials where unstable fracture can occur by microvoid coalescence, but to the structural steels where fracture is accompanied by significant amounts of plastic deformation.

The sequential events which occur at the crack tip have been examined in detail at Swinden Laboratories (see figure 1). Stretch zone formation is the consequence of localized yielding and shear ahead of the crack tip. Finally, there is void nucleation, void coalescence and crack extension.

Thus c.o.d. should only be used over a limited, but usually important, temperature range to measure the significance of defects.

I wish to refer now to the statement that sufficient notch toughness is required in parent plate to provide a reserve of toughness to account for the reduction in toughness in the h.a.z. due to the thermal effects of welding. This means that at normal design conditions the fracture initiation toughness of the parent plate is usually well in excess of what is required. Attention has therefore to be focused on weld metal and h.a.z. toughness. Interpretation of the c.o.d. measured in fracture toughness tests can become a problem when examining these regions and these problems are discussed in our paper.

It is important to consider what level of weld metal toughness is really necessary in a particular structure and not merely to assume that it should meet the same minimum levels as for parent plate. This is an area which requires further investigation because the selection of weld metal toughness levels whether it is specified by Charpy V-notch impact strength or c.o.d. can affect the economics of fabrication.

Also low toughness in the weld root can occur in, for example, a vertical-up weld using a full weave as illustrated in figure 11 of our paper (2.2). The conditions for occurrence of low root toughness have been examined in detail at B.S.C., General Steels Division Laboratories. The main conclusions reached are that the toughness in the weld metal is controlled by dynamic strain ageing due to the thermomechanical cycle. The toughness is therefore dependent on the level of interstitial elements such as carbon and nitrogen and the parent plate composition does not affect the toughness to any marked extent.

In these circumstances it is difficult to control toughness by electrode composition but it can be controlled by weld heat input, and can be improved by stress relief. The existence of low root toughness is believed to be widespread and is not just confined to quenched and tempered steels or to British Steels. It is a function of the weld process and the welding consumables. The interpretation of its significance is difficult. Practical experience indicates that it is not a serious problem otherwise there would have been a series of brittle failures associated with low root toughness. Wide plate tests carried out at our Motherwell laboratories support the view that one should not assume that the toughness in the weld root is the sole criterion describing the integrity of the weldment. Detailed information on these aspects will be published shortly.

T. Kanazawa (*University of Tokyo*)

I would like to introduce the testing method widely used in Japan to evaluate the fracture toughness of materials under the condition of large scale yielding.

Using various size specimens for bend and tensile tests, a comprehensive cooperative research work has been conducted by the T.M. committee of the Japan Welding Enginering Society. It has concluded that the fracture stress may be predicted from the critical c.o.d. value for an appropriate size bending specimen based on c.o.d. concept.†

Taking account of this fact, we have no standard testing method and specimen size at present, but a c.o.d. bend specimen with only a mechanical notch of 0.2 mm width is generally accepted in Japan, to avoid the difficulties for producing a notch by fatigue for the practical and industrial test.

† T. Kanazawa, S. Machida & T. Miyata 1974. Present status on the evaluation criteria of fracture toughness for structural steels and their melted joint in Japan. Conference on the Prospects of Advanced Fracture Mechanics, Delft.

Of course, we know there are some differences in the critical c.o.d. values between the specimens with and without a fatigue notch. But at present we feel that such differences should be included in the safety margin, deduced from the statistical data based on the experimental results, for bend specimens with and without fatigue notches.

When specially designed machines are available to produce notches by fatigue faster, easier and cheaper, we shall not hesitate to adopt c.o.d. specimens with fatigue notches.

B. F. Dyson (*National Physical Laboratory*)

I would like to take issue with Branch over his conclusion (2.4) that the physical metallurgist has given insufficient attention to creep ductility. This surely is incorrect since the reduction in creep ductility by intergranular cavitation has been the reason for most of the physical metallurgical interest in this field over the last 20 years.

I want to take up two areas of research that Branch raised in his paper. The first concerns the influence of extraneous factors, which Baker also brought up in his keynote lecture, on the service performance of materials. One such extraneous factor is plastic deformation. This can occur during fabrication or during plant 'start-up' or 'shut-down' when it is then termed thermal fatigue. In Nimonic 80A, plastic deformation has a deleterious effect on the subsequent creep ductility due to the creation of intergranular cavities during plastic deformation (1). These cavities are created at the intersection of slip bands with grain boundaries (2) and cause much earlier intergranular cracking during subsequent creep. The design of plant on the basis of uniaxial rupture data is plainly inadequate if certain critical parts of the structure are subject to static or cyclic plastic strains and the material responds in a similar manner to Nimonic 80A. The second point is in answer to Branch's request for experimental information on the component of the stress tensor which correctly describes the deformation resistance at very low creep rates. In Nimonic 80A and at stresses which give lives up to 10000 h the von Mises criterion adequately describes the minimum creep rate and is therefore in agreement with most of the published high strain rate data.

References

(1) Dyson, B. F. & Rodgers, M. J. 1974 *Metal Sci.* **8**, 261.
(2) Dyson, B. F., Loveday, M. S. & Rodgers, M. J. Submitted *Proc. R. Soc. Lond.*

The following five contributions were received after the meeting.

J. D. Harrison (*The Welding Institute*)

I am sure that Wells (2.1) is right and that one of the most important requirements in the future will be to develop a probabilistic approach to fracture mechanics. It seems to me that currently the most difficult problem in this field is the prediction of the probability distribution of the size of pre-existing defects. This assumes great importance when we consider that we are usually concerned with extreme value statistics involving the overlapping of the tails of two distributions, defect size and toughness. Wells has assumed that the defect size is distributed as one half of a normal distribution. Becker & Pederson, referred to by Wells, assumed an exponential distribution; but distributions of size are usually assumed to be log-normal. The data available to choose between these distributions are almost non-existent. This may seem a minor point but changes in the assumed distribution of defect size could change the apparent risk of failure by orders of magnitude. I would plead for a major effort to study the actual distributions of defect size in real structures as part of a more general attack on probabilistic fracture mechanics.

Secondly I would like to dispute a point made by Forsyth (2.3) .We find that the vast majority of fatigue failures studied by us are caused by the presence of the weld itself rather than by weld defects in the normally accepted sense. Of course any weld could be considered as a blemish on the unsullied product of the material manufacturer; but on the whole fatigue of welded structures appears to be a problem of design. In disagreement with Dr Forsyth we do find that fatigue strength reduction factors can be assigned to welded joints and this is the basis of current design in welded bridges and other types of structure.

L. P. Pook (*National Engineering Laboratory*)

It is surprising to see it suggested by Forsyth (2.3) that fracture mechanics are not applicable at short crack lengths, surely what is meant is that it is difficult to get the applied mechanics right in the situations of interest.

Fracture mechanics, especially the stress intensity factor, has contributed a great deal in that it has provided a clear view of fatigue crack growth. Unfortunately after about 25 years active work there is no agreed standard method of concluding a fatigue crack growth test under constant amplitude loading never mind variable amplitude loading. This is relatively unimportant when the mean life of a structure has to be predicted, perhaps for fracture analysis purposes. However it becomes serious when making a safety case in that it is not possible to make worthwhile estimates of scatter in structure life from the scatter observed in fatigue crack growth tests.

L. M. Wyatt (*Central Electricity Generating Board*)

I would like to augment statements by Baker (2.5) and Branch (2.4) on crack propagation in creep. They rightly describe some of the factors which make this difficult to assess, but there in one important factor they do not mention. This is that the law which governs the propagation of a crack in creep is very dependent on the method of creep deformation. In ductile materials which deform by disolocation movement, creep deformation quickly blunts the tip of the crack and it is probably safe to consider the area of the residual section when calculating the time to failure. On the other hand, when grain boundary diffusion is the dominant mode there is a very strong tendency to form cavities of the kind shown by Dyson (this discussion) ahead of the creep crack and creep propagation is proportional to a very high exponent of the stress intensity at the tip of a crack. The mode of creep deformation is very dependent on stress. In addition, however, such factors as high solution temperatures undergone in a heat affected zone or the presence of relatively large contents of tramp elements tend to promote the tendency to deform by grain boundary diffusion. By avoiding both these conditions the danger of failure by creep propagation can be very much reduced.

With the greatest respect I cannot endorse the principles of materials selection advocated by Flint (this discussion) and I would like to bring to the attention of the Conference the reason why his proposal can be potentially very dangerous. This is that the 'guaranteed minimum value' that he mentions is not the minimum value that can occur within the material or within the parcels of material supplied. It is the minimum value obtained by testing a limited number of specimens. When this is borne in mind it becomes clear that the minimum value actually existing in the material is likely to be much lower than the guaranteed minimum value if there is a wide scatter in material properties. If there is a small scatter the chances of a low value are very much reduced. Moreover the lowest value may in the former instance be well below that allowed for by the design safety factor. Therefore, the chances of failure are much greater with a wide scatter of material properties than with a small scatter.

H. G. Wolfram (*West Gate Bridge Authority*)

Wells (2.1) covers a wide spectrum of structural design philosophy in relation to materials properties. I should like to comment briefly on only one aspect of this paper.

I was struck particularly by the statement:

'While it is the function of mechanical testing to identify the limits of safe operating conditions in structures, it is no advantage in the final synthesis to overemphasize one property, particularly when this is at the expense of other properties.'

I come from a part of the world where in the last 15 years three major bridges collapsed; one during construction and two in service. One collapse was caused by brittle fracture, one through a combination of circumstances and one through navigational mishap.

I think the converse of Wells's statement is also true, namely that one must not underemphasize any one property.

I think we have reached the stage in structural design and knowledge of the behaviour of structures that, provided we make full use of the knowledge of the properties of materials and the tools of structural analysis, we can have a reasonable degree of confidence in the safety of our structures.

I am thinking particularly of the developments in the fields of brittle fracture and weldability, the refined methods of structural analysis and the developments of non-destructive testing.

But there is one area where knowledge is either insufficient or the available knowledge has not yet penetrated to the design engineer, that is in the understanding of the phenomenon known as 'lamellar tearing'.

I was once involved in the design of a fully-welded multi-storey frame, where cracks occurred in highly restrained beam-column connections. Fortunately, the metallurgist whom I consulted was able to identify the cracks, and identified them as 'lamellar tearing'. This was some 12 or 13 years ago when 'lamellar tearing' was a new term to structural engineers.

More recently, we found lamellar tearing in the saddle of a tower for a cable-stayed bridge. In both cases the problem was resolved by gouging and rewelding and, in the second case, this was followed by stress-relieving.

I have been told that the best way to avoid lamellar tearing is by clever designs which do not result in highly restrained T-joints or cruciform joints. Maybe we engineers are not ingeneous enough, but it is not always possible to avoid these problems purely by clever design.

The yield capacity of highly restrained joints can be improved by buttering up the joints with low yield strength weld metal. This also has its limitations. What is needed, is a deeper understanding of the mechanism of failure by lamellar tearing and methods of quantitative analysis of behaviour of joints where the through-thickness properties of the material are important.

If these methods do not yet exist then I would suggest that this is a property which urgently requires full investigation. If the methods do exist then there is perhaps scope for wider dissemination of the information.

P. J. E. Forsyth (*Royal Aircraft Establishment*)

In reply to comments by Harrison (this discussion) I would say that I did mention the fact that fatigue strength reduction factors could be assigned to welded joints; I questioned only the precision of such procedures.

Pook (this discussion) has objected to my rejection of fracture mechanics for the treatment of small cracks. I would still maintain that what I said is substantially correct. The stage I crack extends by the joining of slip plane damage and is formed in the first place by the action of the plastic deformation itself. It frequently lies in a surface zone where the spread of plastic deformation is considerably greater than the dimensions of the crack. Under these circumstances I cannot see how one is justified in claiming that fracture mechanics could be made to work, Even the small stage II crack may lie in a zone of plastic deformation and the same objections would apply. As a metallurgist, and a failures investigator, I eagerly await any advances in applied mechanics that would help to make service life predictions on components containing short cracks.

2.5 Weldability and its implications for materials requirements

By R. G. Baker
Division of Materials Applications, National Physical Laboratory, Teddington, Middlesex

[Plate 12]

The potential economic and technological advantages of welding over other joining techniques have made it the principal fabrication process for a wide range of engineering products. However, the achievement of good weldability makes demands on material quality which the material supplier must optimize with the required mechanical behaviour and an acceptable cost.

The physical metallurgist has contributed significantly in providing a sound basis for materials selection based on an understanding of the potential welding problems and the factors responsible for them. The principal welding problems in steels, nickel alloys, aluminium alloys and titanium alloys are briefly outlined and the influence on material specifications discussed.

1. Introduction

Welding has the general advantages of low cost, saving in weight and great flexibility of design. For the majority of large metal constructions, it is the only joining process which is economically realistic. In many applications, including bridge and line pipe construction, welding has permitted an evolution of design and of construction methods which would otherwise have been technically impossible. However, the potential advantages of welding can only be realized if unexpected problems can be avoided during construction and if the resulting structures have sufficient reliability during the whole of their service lives.

The consequent demand for metals which can be welded simply and cheaply without impairment of performance poses an open-ended question to suppliers. 'Satisfactory weldability' depends on choice of welding process, the skill and control exercised during fabrication, and on the service performance expected of the resulting structure as well as on material quality. Not infrequently, material quality suitable for welding by any process, in any type of design, for any purpose, could only be achieved at an expense, which for the vast majority of applications, would be quite unnecessary and regarded as exorbitant.

The position is evidently very complex, demanding a degree of understanding on the part of the design engineer, material supplier and constructor of the implications of potential welding problems in design and on material quality. It follows that material specifications must often allow a choice to be made from a range of welding qualities, so that performance, reliability and cost can be optimized for any specific application.

The only alternative to a mass of empirical rules for material selection is a basic understanding of potential welding problems in relatively simple terms, so that a rational approach to material selection, design of welded structures and assurance of reliability can be achieved. It has been the task of the welding metallurgist to provide this understanding and the classical physical metallurgy approach has proved invaluable. Virtually all the potential welding problems arising with materials presently used in construction have been defined, at least qualitatively, in terms of the metallurgical factors affecting them. This has permitted a rational approach to material selection, the development of welding procedures of defined reliability, and problem solving. A great deal remains to be done in quantification of the factors affecting weldability, in the evolution of specifications, in the integrated development of welding technology and materials, and in relating service reliability quantitatively to the quality of materials, welded details and components. However, the advantages to the construction industry already achieved by the physical metallurgy approach in welding are incalculable.

Some of the more important potential welding problems will be examined in outline to illustrate the metallurgical factors responsible and the implications to the material supplier. In some applications, such as bridges, pipelines and offshore structures, welding is already well accepted and indeed indispensible. In others, such as aircraft, welding, though increasingly used, is not the dominant joining process. The potential advantages of using it are enormous, but the degree of reliability which presently attends welding design and technology in the materials and structures involved is still insufficient in relation to the required reliability of the structure. For this reason, the emphasis has been placed more on the potential welding problems in steels than on the other engineering materials involved in the subject matter of this Conference.

2. The general effects of welding

Most of the welding processes in common use involve the local application of an intense heat source to produce local melting. Often filler metal is added to the melted pool to produce the required weld shape and dimensions after solidification. The effects of welding are accordingly most apparent in the part of the joint which was melted; the 'weld metal', and the 'heat affected zone' (h.a.z.) adjacent to it. The weld metal is essentially a small chill casting which has been cooled very rapidly. Its composition is derived at least partly and with thin materials sometimes wholly, from that of the parent material, whilst its metallurgical structure depends critically on the rate of cooling associated with the process and procedure. This is generally high because of the mass of cold metal surrounding the joint, but it can vary quite widely with process as shown in figure 1, which compares three common types of welding process. In the manual metal arc (m.m.a.) process, an electric arc is struck between a consumable metal wire, coated with bonded mineral particles, and the prepared joint. The wire melts and blends with melted parent material to form the weld pool, while the mineral coating melts to form a protective slag coating over it. This process is used for making the circumferential welds in line pipe, many of the positional welds in bridges and offshore structures and for many other critical and non-critical purposes. It has the advantages of flexibility, low capital cost and the availability of a wide range of consumable types. It is suitable for steels and nickel based alloys but not for the more reactive metals such as aluminium and its alloys or titanium.

The submerged arc process is one of the most commonly used automated processes for constructional steel. It is used extensively for longitudinal welds in line pipe and in the 'tubulars' for offshore structure. It is also used for shop welds in bridge work; for example in web to flange joints in girders. In this process, consumable wire is supplied through a nozzle, the arc being made between it and the joint, beneath a pile of granular flux which is progressively supplied from a hopper ahead of the arc. The advantages are a very rapid deposition rate and good weld shape and properties. There are many other welding processes (see table 1 and other publications (Houldcroft 1967; Welding Handbook, vol. 2; Baker 1973)). Those most relevant to our discussions are the two mentioned, together with processes in which shielding of the weld pool is achieved by an inert gas such as argon. The arc can either be made between an inert tungsten electrode and the joint, feeding consumable wire from the side (t.i.g. process) or between consumable filler wire and the joint (m.i.g. process). The two latter processes are suitable for Al and Ti and their alloys and can be used with steel and Ni-base alloys with advantage in some cases.

Metal adjacent to the weld is both heated and cooled extremely rapidly so that there is the risk of a change in properties as a result of phase changes, grain growth or ageing in the heat affected zones. Local strain takes place in both heat affected zone and weld metal during the heating and cooling cycle, and this results in high tensile stresses remaining in the weld after it has cooled. These are of yield point magnitude at the centre line of the weld in a direction parallel to it. If there are pre-existing defects adjacent to the weld the local concentration of strain may be very considerable and the combination of the strain temperature cycles can radically affect the properties of the material locally. One of the implications of this is discussed in another Conference paper (Barr & Burdekin (1975)).

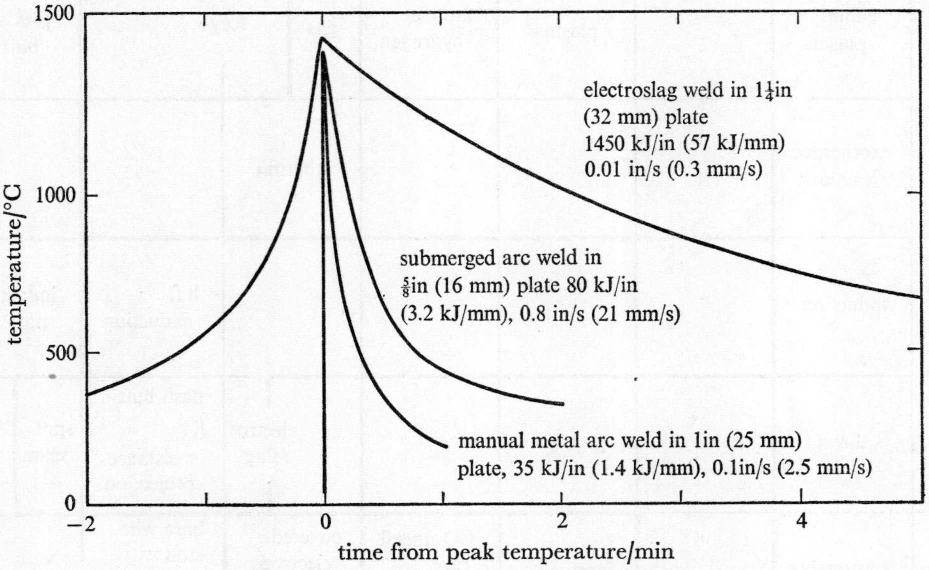

FIGURE 1. Comparison of the thermal cycles experienced in the heat affected zone when welding steel by three different processes, manual metal arc, submerged arc and electroslag.

Welding processes which involve melting always introduce a risk of gaseous contamination of the weld pool. Generally the processes used are efficient at excluding nitrogen. Controlled amounts of oxygen are usually beneficial in the weld pool in steel because of the refining action, subsequent harmful effects being avoided by deoxidation. With aluminium and titanium, oxygen must be excluded from the pool altogether because of the high reactivity. Thus gas shielded techniques are used, slag producing processes leading to excessive oxygen contamination. Hydrogen is very difficult to avoid because hydrogen containing compounds present in fluxes, rust, grease, etc. decompose in the welding arc to atomic hydrogen, which is absorbed very quickly into many molten metals including steel and titanium. Atomic hydrogen diffuses rapidly and may subsequently reduce the ductility of either the heat affected zone or the weld metal or both to a point which can even lead to cracking under the action of residual stress as discussed below.

Table 1. Grouping of welding processes according to heat source and shielding method (after Houldcroft 1967)

source of heat		shielding method: vacuum	inert gas	gas	flux	no shielding	mechanical exclusion
no heat or heat by conduction		cold pressure	thermo compression bonding				cold pressure hot pressure
mechanical		explosive				explosive	friction ultrasonic
thermo chemical	flames plasma		plasma	atomic hydrogen	gas	forge	pressure butt
	exothermic reactions				thermit		
electric resistance	induction					h.f. induction	induction butt
	direct				electro -slag	flash butt h.f. resistance projection	spot seam resist -ance butt
electric arc	consumable electrode		inert gas metal arc	CO_2 metal arc gas/flux metal arc	covered electrode submerged arc	bare wire stud spark–discharge percussion	
	non-consumable		inert gas tungsten arc			carbon arc	
radiation	electro magnetic					laser	
	particle	electron beam					

3. The potential welding problems

3.1. *Introduction*

The properties of the h.a.z. and weld metal usually differ from those of the parent metal because of the differences of composition and thermal history. This is not necessarily a disadvantage. In butt welds, a higher weld strength is usually tolerable and often beneficial, as long as it is combined with adequate ductility. In fillet welds, either a higher or lower strength is often acceptable without prejudice to the behaviour, assuming an adequate weld thickness. Subject to these qualifications, welding problems can arise either through the presence of defects of technological origin, such as lack of fusion, or of metallurgical origin, such as weld cracking, the impairment of strength, ductility or toughness in the welded region, or because some other important service characteristic, such as corrosion behaviour, is adversely affected to a point where useful life is shortened. The elements of some of the more important problems are now briefly described.

3.2. *Solidification cracking*

When weld metal begins to solidify, columnar grains with cellular or cellular dendrite substructures (Calvo *et al.* 1960) grow from the fusion boundary towards the central region of the weld pool. Many of the alloying elements and impurities are rejected ahead of the growing crystals and become concentrated in the last liquid to solidify. Their presence depresses the freezing temperature of this liquid, sometimes by as much as several hundreds of degrees

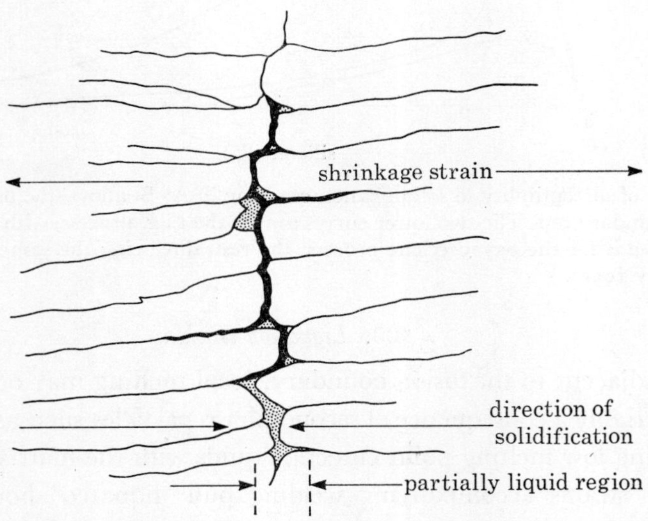

Figure 2. Schematic diagram illustrating the mechanism of solidification cracking.

Celsius, below that of the first liquid to solidify. While weld solidification is taking place, the surrounding material is progressively cooling and this gives rise to contraction strains across the weld. These strains matter little until solidification is almost complete and the growing crystals begin to meet in the centre of the weld pool. The further persistence of low melting point films may, however, then lead to such low strength and ductility that the contraction strains produce cracking, the amount of residual liquid being insufficient to fill the voids thus produced. This is illustrated schematically in figure 2, while figure 3, plate 12, shows an example of solidification cracking in practice. This form of cracking is a very general problem in welding metallurgy.

Evidently the alloy and impurity content of the weld metal will have an important influence. In general, alloys with wide freezing ranges might be expected to have higher intrinsic susceptibility, and impurity elements, with a high partition coefficient between liquid and solid metal, to be particularly deleterious, such as phosphorus and sulphur in steel.

The degree of strain applied to the joint, however, will depend heavily on the welding procedure, and thus the necessary degree of purity to avoid cracking will depend on this. Figure 4 shows the influence of different welding processes and procedures on the degree of solidification cracking in Al–Si alloy (Pumphrey 1955).

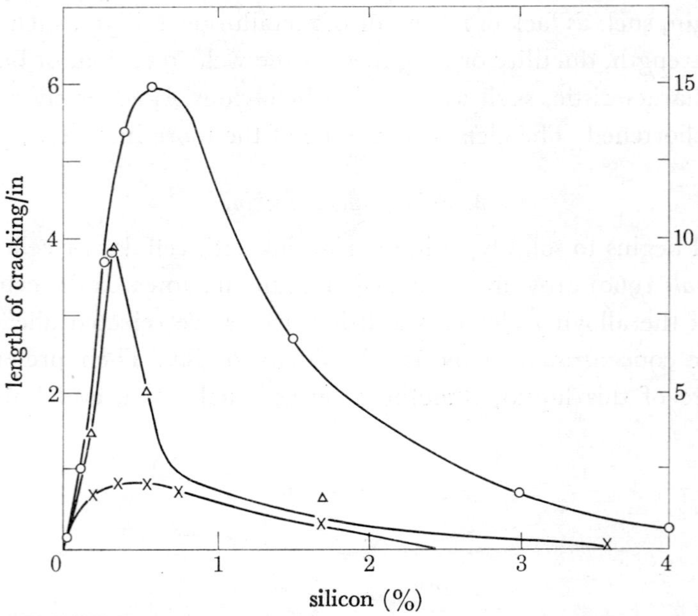

FIGURE 4. Variation of susceptibility to solidification cracking in Al–Si alloys, the ordinate showing the extent of cracking in a standard test. The two lower curves are for the t.i.g. process, with different degrees of restraint. The upper curve is for the oxyacetylene process, the restraint being the same as for the lower t.i.g. curve (after Pumphrey 1955).

3.3. *Liquation cracking*

Immediately adjacent to the fusion boundary, local melting may occur at the grain boundaries. This is invariably a consequence of second phase particles such as non-metallic inclusions or carbides forming low melting point eutectic liquids with the matrix. Micro cracking results when the plastic strains accompanying welding pull 'liquated' boundaries apart to leave

---

DESCRIPTION OF PLATE 12

FIGURE 3. Solidification cracking in the first run of a double-run submerged arc butt weld in C–Mn steel. (Courtesy, The Welding Institute; magn. × 4.)

FIGURE 6. Liquation cracking accompanying 'burning' in the heat affected zone of welded rail steel to BS 11. There is clear evidence of sulphuric eutectics at the grain boundary. (Courtesy, The Welding Institute; magn. × 400.)

FIGURE 7. Hydrogen induced toe cracking, with a little underhead cracking also, adjacent to a fillet weld in a C–Mn steel. (After Wilkinson; magn. × 7.)

FIGURE 11. An example of lamellar tearing adjacent to a T butt weld in BS 4360 grade 50 steel. (After Farrar & Dolby 1972.)

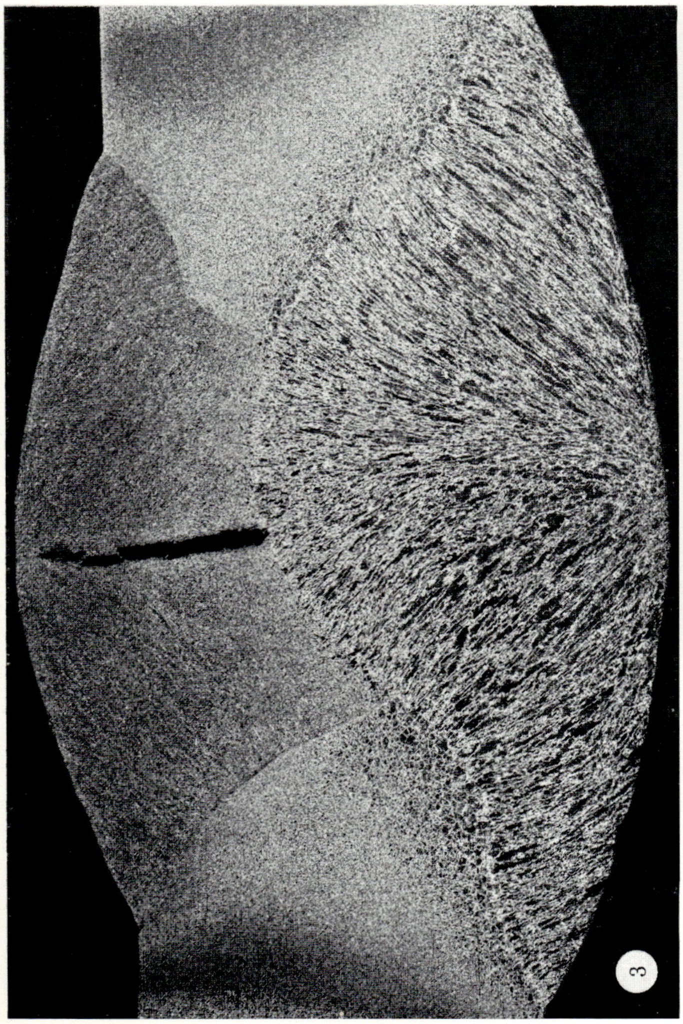

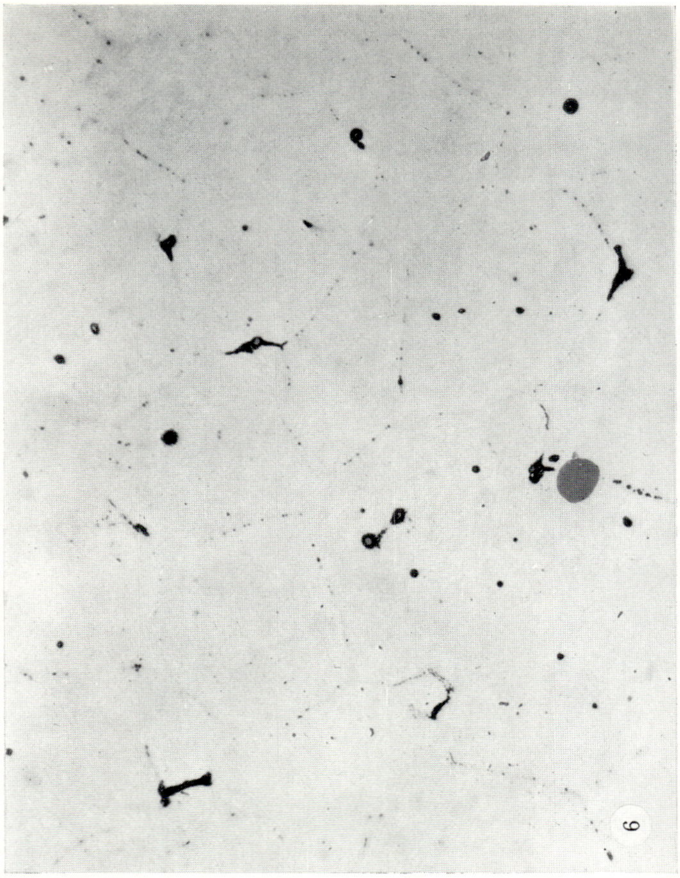

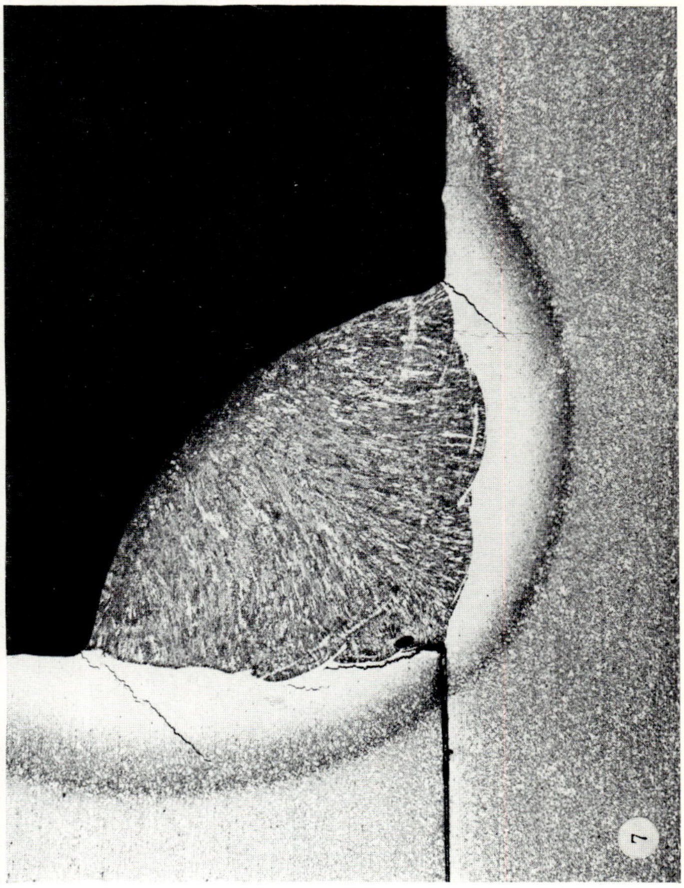

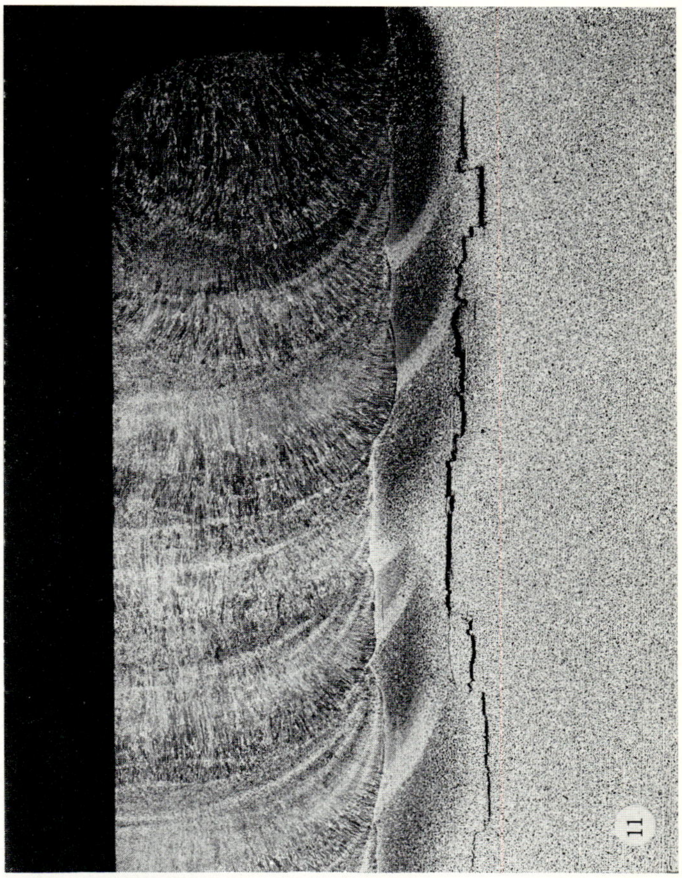

FIGURES 3, 6, 7 and 11. For description see opposite.

unhealed cracks when the weld cools. This is illustrated schematically in figure 5. The problem may occur in ferritic steels due to the formation of sulphide eutectics by the process of 'burning' (Boniszewski & Watkinson 1964) but liquation may occur in other types of materials including nickel based and aluminium alloys. Figure 6, plate 12, illustrates liquation cracking and sulphide eutectic formation at grain boundaries in the h.a.z. of a 0.4 %C rail steel. The importance of the problem depends very much on the context. In ultra high strength steels, such as those used in undercarriage components or in rocket motor casings, the critical defect size may be very small, and the size of the micro cracks appreciable in relation to this. Similarly, if burning is extensive enough in low strength ferritic steels to surround the micro cracks by a region of very low ductility, again the problem may be significant. If, however, the cracks are small compared with the critical defect size, and are not surrounded by a brittle matrix, then they can usually be ignored quite safely. The result of this conditional qualification means that the criteria used for material selection will depend on the process and application. For example, in using welded ultra high strength steels for critical applications, liquation cracking must be avoided altogether but in ferritic steels for general applications it is only sufficient to avoid appreciable burning during welding in most cases. This can lead to quite different requirements for limitation of P and S contents.

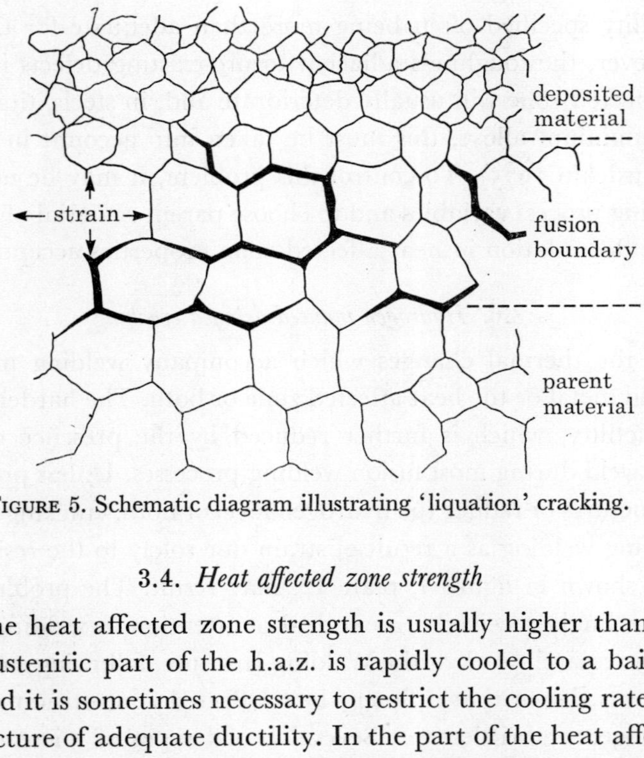

FIGURE 5. Schematic diagram illustrating 'liquation' cracking.

3.4. *Heat affected zone strength*

In ferritic steels, the heat affected zone strength is usually higher than that of unaffected parent metal. The austenitic part of the h.a.z. is rapidly cooled to a bainitic or martensitic microstructure. Indeed it is sometimes necessary to restrict the cooling rate by artificial means to obtain a microstructure of adequate ductility. In the part of the heat affected zone reaching peak temperatures below the eutectoid, there is a possibility of reduction of strength through ageing. However, in steels the time spent is so short relative to the rate of diffusion that the effect is usually negligible.

In materials where there is no phase change in the heat affected zone, however, there is a real risk that softening will occur. In aluminium and nickel alloys, grain growth will occur in the heat affected zone, while in hardened alloys precipitates may dissolve. Both these effects may cause softening. In precipitation hardened alloys, overageing may occur in the outer part of the h.a.z. This can be an important cause of softening in aluminium alloys since ageing temperatures are comparatively low (Pumphrey 1955; Welding Handbook, vol. 4). Figure 1 shows that the rate of cooling is very slow when the temperature is in the ageing range for aluminium alloys. This is a serious limiting factor in the welding of high strength aluminium alloys (Select Conference 1970; Baker 1970). Binary alloys relying for their strengthening on solid solution hardening can be welded satisfactorily, but alloys relying for strengthening on precipitation may suffer a loss of strength in the heat affected zone. This restricts the alloy compositions which can be successfully welded and sometimes demands special heat treatment following welding.

3.5. *Weld toughness and ductility*

The degree of ductility actually required in the heat affected zone depends largely on the intended application of the welded component. If a forming process is involved after welding or if the weld detail has to withstand substantial plastic deformation during use, the requirement for substantial ductility is clear. However, welds are frequently situated at regions of stress concentration in the structure and may themselves introduce shape discontinuities. Thus, even in 'elastically' designed structures, plastic deformation may occur locally in the vicinity of welds under design load conditions. Normally, ductility presents no problem in welding and a severe bend test is frequently specified as part of the quality assurance procedure during construction; the ductility specified often being more than adequate for the design and service requirements. However, the toughness adjacent to pre-existing defects in the sub-critical and transformed heat affected zone will usually deteriorate and, in steels, titanium alloys and even in high strength aluminium alloys, this must be taken into account in design against brittle fracture (Barr & Burdekin 1975). To control this problem, it may be necessary to place constraints on the welding process variables and to choose parent material of sufficient toughness to make the inevitable degradation in heat affected zone properties acceptable.

3.6. *Hydrogen induced delayed cracking*

In ferritic steels, the thermal changes which accompany welding may lead to hardened structures in the weld metal or the heat affected zone or both. The hardened structures have an intrinsically low ductility, which is further reduced by the presence of hydrogen which is introduced into the weld during most fusion welding processes. Unless precautions are taken to modify the microstructures or reduce the hydrogen level or both, cracking may occur at ambient temperatures following welding as a result of strain due solely to the residual welding stresses. Severe cracking, as shown in figure 7, plate 12, may result. The problem is widespread and serious and has accordingly received a great deal of attention in welding research (Coe 1973; Baker *et al.* 1971). In research work at the Welding Institute, following earlier work at Birmingham University, it was shown that different heat affected zone structures have different susceptibilities to hydrogen embrittlement. Several broad types of microstructure were defined with different degrees of susceptibility to hydrogen cracking. The least susceptible type of hardened structure was the plain martensite structure found in low carbon steels such as 9% nickel. Structures containing twinned martensite, however, suffered a very drastic reduction in strength in the presence of hydrogen, while certain bainitic structures showed intermediate behaviour. The work is briefly reviewed elsewhere (Baker 1968) but figure 8 summarizes the results of constant load rupture tests on notched specimens of different types of steel heat treated to simulate h.a.z. microstructures. Most of the common structural steels fall into the category of the QT35 and 9% nickel steels shown in the diagram, but the high strength steels of the medium carbon type show a high degree of susceptibility, epitomized by the 0.34% carbon steel illustrated in the diagram (the notch tensile strength was reduced from 2240 N/mm^2 to less than 150 N/mm^2 by about 5 parts/10^6 of hydrogen).

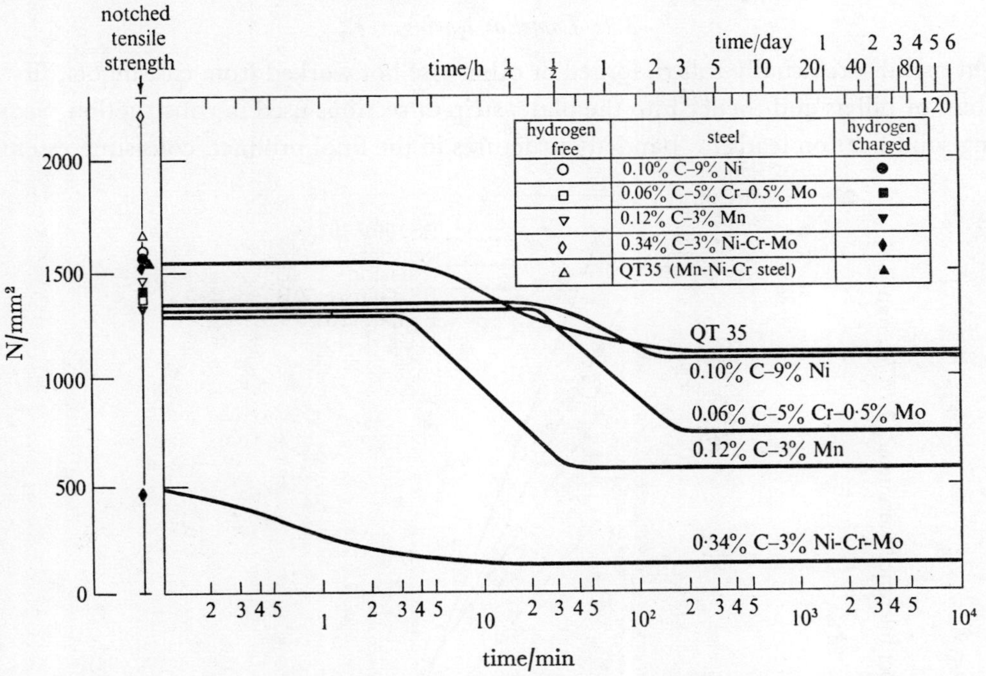

FIGURE 8. Constant load rupture strength of different steels in the quenched condition to simulate h.a.z. microstructures, and charged with hydrogen. (Courtesy, The Welding Institute.)

The relation of the cracking susceptibility to microstructure permits a rational approach to the control of the problem through the use of continuous cooling transformation data. Other work showed that the embrittlement mechanism was temperature dependent, involving absorption of hydrogen atoms on dislocations. The progressive increase in entropy with temperature leads to an associated fall in the free energy of adsorption and thus to a decrease in embrittlement (Graville *et al.* 1967). Above 250 °C the embrittlement is negligible.

With high strength steels the solution is accordingly simple but costly. The joint must either be held at a temperature exceeding 250 °C during welding and then raised to a higher temperature to reduce the hardness by tempering and, incidentally, disperse some of the remaining hydrogen by diffusion, or the joint must be held at a temperature during welding sufficiently high to ensure that the h.a.z. will transform to a bainite of lower hardness and susceptibility (Bowne 1946).

In structural steels such as those used in offshore structures, line pipes and bridges, the problem is less severe but more complex. Figure 9 shows the variation of defect tolerance in hydrogen charged specimens with hardness. There is a progressive rise in defect tolerance with decreasing hardness but the steels shown fall into a reasonably sized envelope. General experience suggests that control of h.a.z. hardness to a maximum level of 350 H_v is usually effective in preventing cracking. Recommendations are available which enable the fabricator to establish a procedure, taking into account the numerous welding procedural variables which can affect cracking. However, the critical heat affected zone hardness for cracking will depend strongly on joint geometry, welding procedure and also residual stress level (which in turn will be a function of parent material strength). Thus great precision is superfluous in written recommendations and procedural trials are usually to be recommended to ensure that the procedure is effective for the application (Coe 1973; Cane *et al.* 1973).

3.7. *Lamellar tearing*

Wrought metals are usually rolled, forged or otherwise hot worked from cast ingots; first into slabs, blooms or billets and thence into the plate, strip or sections used in construction. Segregation during solidification leads to 'banded' structures in the final product, consisting essentially

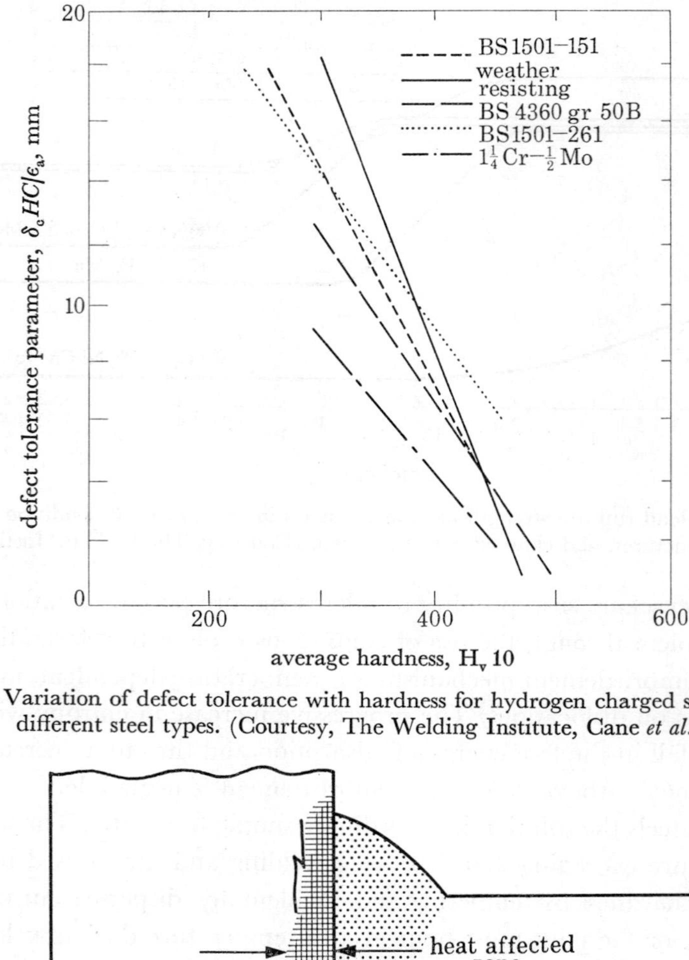

FIGURE 9. Variation of defect tolerance with hardness for hydrogen charged specimens of several different steel types. (Courtesy, The Welding Institute, Cane *et al.* 1973.)

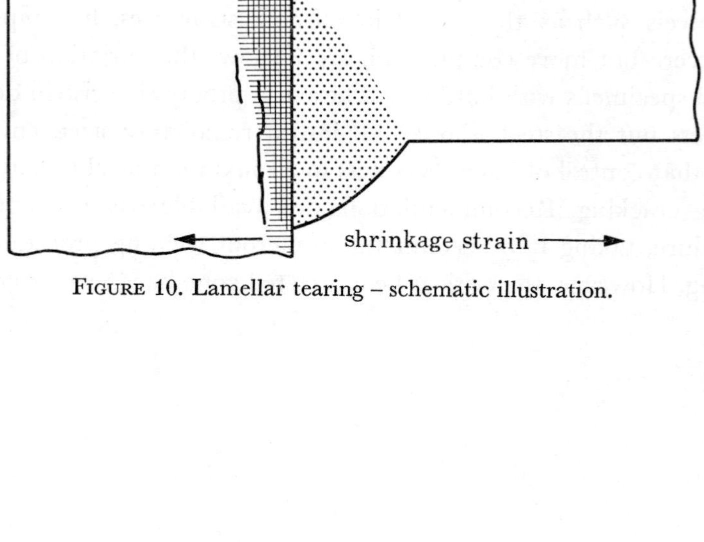

FIGURE 10. Lamellar tearing – schematic illustration.

of layers of material of different composition and properties. Non-metallic inclusions will also tend to be worked into planar shapes or clusters. These factors confer directional properties on wrought metals. The ductility is particularly affected and may be very low, measured through the thickness of the material. Often the directional characteristics of the metal are unimportant and may sometimes even be advantageous, but in T-joints and in highly restrained stiffened details, found, for example, in offshore and certain bridge structures, a high degree of strain may occur in the through thickness direction as shown in figure 10. This can lead to the characteristic 'lamellar tearing' in which cracks initiate at lamellar inclusions and are linked by tearing as shown in figure 11, plate 12. Research has shown that the major microstructural factor involved in lamellar tearing of steels is the presence of lamellar inclusions, singly or in groups on the same or neighbouring planes (Farrar & Dolby 1972).

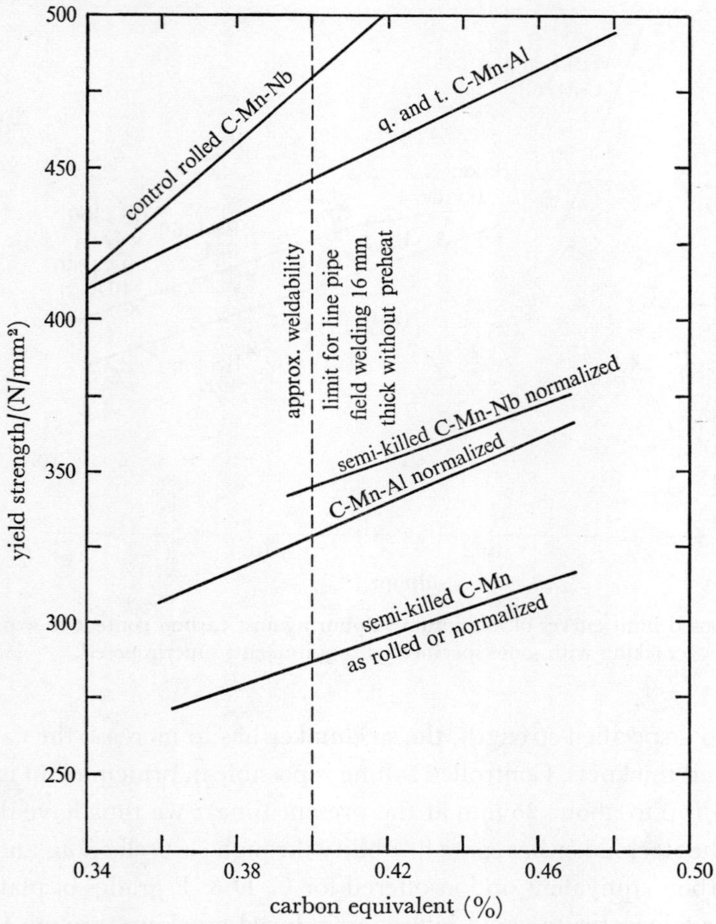

FIGURE 12. Curves plotting minimum yield strength against composition to show how different steel types compare with respect to weldability and strength. Curves have been taken from Roberts (1967), with the exception of that for control rolled steel, but must be regarded as approximate only!

4. Weldability and specifications

4.1. *Introduction*

The principal problems having been outlined together with their mechanisms, the implication on material specifications will now be examined very briefly.

4.2. *Steels – structural*

Steels for structural purposes are specified in BS 4360 (BSI 1972) which deals exclusively with normalized and as-rolled steels and classification is primarily on the basis of yield strength. This places severe constraints on the steelmaker. The principle problem in welding these steels is hydrogen induced cracking, and it is obviously a considerable advantage to the fabricator if he need take no special precautions, such as preheating. The risk of hydrogen induced cracking is largely related to hardenability and for convenience a 'carbon equivalent formula' based on hardenability is defined as follows (BSI 1974):

$$\text{carbon equivalent} = \%\text{C} + \frac{\%\text{Mn}}{6} + \frac{\%\text{Cr} + \%\text{Mo} + \%\text{V}}{5} + \frac{\%\text{Ni} + \%\text{Cu}}{15}.$$

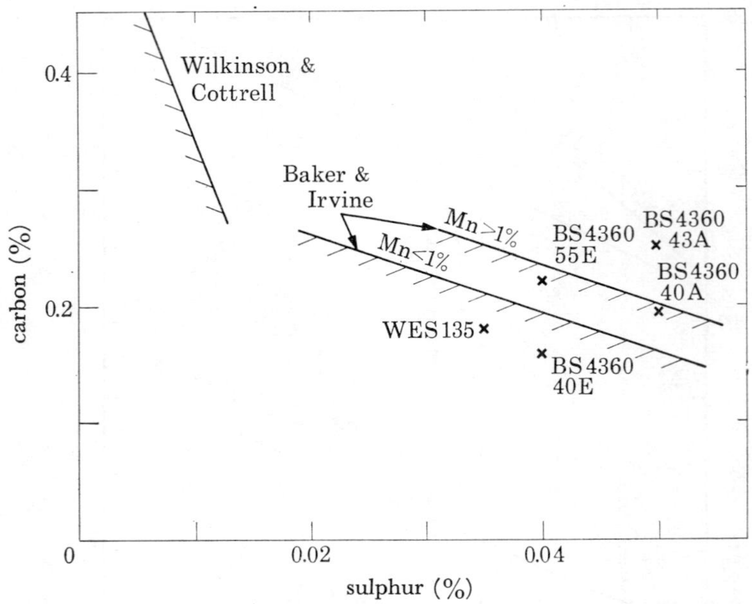

FIGURE 13. Proposed limit curves of maximum sulphur against carbon content to avoid liquation cracking with some specificated requirements superimposed.

In order to supply to a specified strength, the steelmaker has to increase the carbon equivalent with both strength and thickness. Controlled rolling is possible in principle but is only applicable for thinner material (up to about 25 mm at the present time); we thus have the situation outlined in figure 12. The steelmaker has some flexibility through cast selection, and this is reflected in the controlled carbon equivalent option offered for C, D & E grades of plates. For the rest, it is necessary for the fabricator to avoid hydrogen induced cracking through the judicious use of preheat and other means as recommended in BS 5153 (BSI 1974). Paradoxically, with higher strength steels (of modest thickness) of the controlled rolled and quenched and tempered types, the position is very much better, and this has led to the formulation of the Japanese WES 135 specification (Kihara *et al.* 1971) of which we shall hear more from Kanazawa and his colleagues, and the introduction of the so-called 'micro alloyed steels' into Europe for line pipe and storage tanks.

Few of the BS 4360 steels would show serious liquation cracking, even when welded under high heat input conditions. The specification is broadly in line with several recent recommendations (by, for example, Kihara *et al.* 1971; Baker & Irvine 1971) for reduced sulphur levels as shown in figure 13. In fact, the majority of steels, supplied to the BS 4360 specification, have sulphur levels well below the maximum permitted, so the actual position is better than the permitted maxima might suggest.

The BS 4360 specification gives no assurance against lamellar tearing. For many structural applications, very little through thickness ductility is necessary, but for certain types of design, such as box stiffening or moment connexions, BS 4360 may be inadequate without supplementary requirements. In offshore structures, the tubular connexions and associated stiffening at the 'nodes' make particularly heavy demands on resistance to lamellar tearing. This demands a restriction of the inclusion content. The number of sulphide inclusions can be reduced by restricting the sulphur level. However, this alone is insufficient and must be accompanied by control of oxide inclusions. Generally this has been tackled in specifications by an assurance of through thickness ductility by tests on product coupled with restrictions of maximum sulphur content. There has been some controversy concerning the level of through thickness ductility necessary to avoid lamellar tearing, even under the most severe conditions. The matter is complicated by the fact that there is no viable model relating severity to through thickness ductility, and because hydrogen induced cracking and lamellar tearing may be mutually exacerbating. However, there have been few cases of lamellar tearing reported where the through thickness ductility has been found to exceed 10% and none, to the author's knowledge, in material showing a mean through thickness ductility exceeding 20% (where adequate precautions have been taken to avoid hydrogen induced cracking). A possible supplement to the structural specification is provided in table 2.

TABLE 2. TENTATIVE RECOMMENDATION FOR A SUPPLEMENTARY SPECIFICATION TO AVOID LAMELLAR TEARING

grade	S%	t.t. ductility†	U.S. grade	remarks
I	0.035	10% (5)	C	critical web to flange joints
II	0.025	15% (10)	C	moment connexions – box stiffeners
III	0.010	25% (20)	A	critical high restraint connexions

† Ductility is mean of three tests, value in brackets is the minimum allowable for any one test.

4.3. *Steels – for line pipe*

Large diameter line pipe is produced by forming plate into a cylindrical shape which is then welded longitudinally by submerged arc welding, or by spirally welding coiled plate or strip by submerged arc welding. The pipes are joined together in the field by a special form of manual or semiautomatic welding, and the nature of the procedure demands a very low heat input with an associated rapid cooling rate (Barkow 1962). This places more stringent requirements on the hardenability than would normally be expected, demanding a carbon equivalent level of 0.4% or less. This implies a carbon level in the vicinity of 0.1%, with a manganese level of about 1.4%. Up to proof stresses of about 350 N/mm^2 (50000 lbf/in^2), this can be obtained by normalizing but above this strength controlled rolling is necessary to achieve the required properties. The pipeline steels are normally ordered to API specifications. These are suitable for small diameters but for diameters greater than about 45 cm and thicknesses greater than about 6 mm, much more stringent requirements may be necessary to achieve adequate field weldability.

TABLE 3. SUMMARY OF POTENTIAL WELDING PROBLEMS AND THEIR IMPLICATIONS

processes	structural steels	medium carbon u.h.s. steels	aluminium and binary alloys	aluminium heat treatable alloys	nickel and solid solution alloys	heat treatable nickel alloys	titanium and alpha alloys
PART 1. THE PROCESSES†							
oxy-acetylene	B	D	B	E	B	E	E
m.m.a.	A	C	C	D	A	E	E
m.i.g.	A (CO_2 or A+O shielded)	A	A	C	A	C	A
t.i.g.	A (for critical runs)	A	A	C	A	C	A
submerged arc	A	E	E	E	E	E	E
resistance	D	E	A	C	A	D	A
electron beam	D (because of cost)	A	A	C	A	C	A
PART 2. THE POTENTIAL PROBLEMS‡							
solidification cracking	B	C	B	C	A	B	A
liquation cracking	A (except with low Mn and high heat input)	D (for critical components)	C	C	C	C	A
h.a.z. hydrogen cracking	C (preheat may be necessary)	D (preheat and postheat required)	—	—	—	—	usually A D in repair welds chemical plant
weld softening	A	C	A (usually not significant)	D	A	D	A
weld embrittlement	B	C	A	C	A	D (may have to be welded in the annealed state to avoid cracking)	A
lamellar tearing	C-only with certain highly restrained structures	A	A	A	A	A	A
post welding h.t. cracking	A	A	A	A	A	D (in high strength alloys)	A

PART 3. THE IMPLICATIONS

implications to supplier	Need for guaranteed C equivalent and sulphur levels in specification at least on options. Through thickness ductility required where lamellar tearing is a problem. Controlled rolling or q. and t. an advantage and necessary for line pipe.	S and P must be restricted severely for critical components. Need to optimize alloy content to achieve best combination of weldability and strength.	Restriction on alloy content and on residual and trace elements.	Restriction to Al–Zn–Mg system and very lean ternary alloys only.	Control of trace elements desirable.	Special treatment necessary for all the heat treatable alloys which can be welded at all.	Weldability mainly restricted to the alpha type alloys at present.

† Suitability of process: A, generally suitable; B, suitable for non critical applications; C, suitable for some, but not all alloys; D, suitable only with special circumstances; E, not recommended.

‡ Incidence of problem: A, not normally a problem; B, potentially a problem – easily controlled by means other than through material; C, serious problem, control has implications on material specifications; D, controlled only with difficulty.

4.4. *Ultra high strength steels*

The major implications on the specifications for high strength medium carbon steels concern susceptibility to liquation cracking, as a result of burning in the heat affected zone. As the carbon level goes up, more stringent restriction of phosphorus and sulphur is necessary to avoid significant burning during the rapid welding cycle as shown in figure 13. Wilkinson & Cottrell (1958) have suggested that for medium carbon steels $\%S + \%P \ngtr 0.007/\%C$. This is plotted in figure 13 for the case $\%P = \%S$.

4.5. *Aluminium alloys*

Many of the higher strength aluminium alloys currently used are not fully weldable because of a high susceptibility to solidification and liquation cracking, coupled with loss of strength and ductility in the weld region. The lower strength alloys, depending for strength on solid solution, present fewer problems, although the processes which can be used satisfactorily are more limited than with steels. The Al–Zn–Mg alloys appear the most promising of the high strength precipitation hardening alloys. They have already been used for military bridgework and show promise for other applications also. Potential problems include solidification cracking, irrecoverable joint softening, poor fracture toughness and stress corrosion cracking. Optimization of composition to avoid these suggested two broad alloy types (Select Conference 1970); a 230 N/mm^2 alloy of nominal composition Al–4.5Zn–1.2Mg and a 310 N/mm^2 alloy of nominal composition Al–$3\frac{3}{4}/4\frac{1}{2}$Zn–2/3Mg.

4.6. *Nickel based alloys*

Nickel based alloys of the solid solution type can be welded by gas shielded and m.m.a. processes without special problems. Care is necessary to exclude sulphur containing materials from the joint since this can lead to liquation cracking. The precipitation hardening alloys need special precautions to avoid softening and loss of corrosive resistance at the joint. The very strong alloys must be welded in the softened condition to avoid cracking during welding due simply to the ductility being insufficient to accommodate the welding strain (Welding Handbook volume 4; Evans 1962).

4.7. *Titanium and its alloys*

Special procedures to avoid weld contamination are necessary when using m.i.g. or t.i.g. welding processes. These include improved gas shielding, cleanness of joints and high purity gas. Alpha alloys are readily welded subject to these qualifications. The alpha-beta type may be greatly affected – many being severely embrittled and special consideration is required in selecting these for welding. Beta alloys are easily weldable, but the strength of joints is low (Welding Handbook volume 4; Faulkner & Voldrick 1959).

5. Summary and conclusions

Potential problems to be expected in welding structural materials have been briefly reviewed. The review has been restricted to arc processes for reasons of space and the emphasis has been primarily on steels reflecting the greater dependence on welding in these aspects of engineering in which steels are employed. The subject matter of the paper is conveniently summarized in table 3.

This paper falls far short of being a comprehensive review. Inevitably the author has oversimplified and done less than justice to the many welding metallurgists who have contributed to the understanding and solution of the problems discussed. He would like to apologize for any significant omissions and emphasize that they were unintentional. The writer takes full responsibility for the views expressed, but would like to acknowledge gratitude and appreciation to all those who contributed to their formulation; particularly his previous colleagues at the Welding Institute and the Research Organisation of British Steel Corporation General Steels Division.

References (Baker)

Baker, R. G. 1968 Weld cracking – a modern insight. *Br. Weld. J.* **15**, 283–295.

Baker, R. G. 1970 Al–Mg–Zn light strong weldable alloys for industry. *Met. Constr.* **2**, 65–68.

Baker, R. G. 1973 *The welding of pressure vessel steels.* London: Climax Molybdenum Co. Ltd.

Baker, R. G., Dolby, R. E. & Watkinson, F. 1971 The assessment of cracking problems. *Proceedings of Conference on Weldability of Structural and Pressure Vessel Steels,* 16–19 November 1970, pp. 90–103. London: The Welding Institute.

Baker, R. G. & Irvine, K. J. 1971 An empirical specification for welding. *Proceedings of Conference on Weldability of Structural and Pressure Vessel Steels,* 16–19 November 1970, pp. 280–283. London: The Welding Institute.

Barkow, A. G. 1962 Gas and oil pipeline practices. *Weld. Res. Com. Bull.* No. 76, April. New York: Weld. Res. Com.

Barr, R. R. & Burdekin, F. M. 1975 Design against brittle fracture. *Phil. Trans. R. Soc. Lond.* A (this volume).

Boniszewski, T. & Watkinson, F. 1964 Examination of hot tearing in the heat affected zone of ferritic steels. *Br. Weld. J.* **11**, pp. 610–619.

Bowne, L. F. 1946 *Weld J.* **23**, 234s–241s.

Calvo, F. A. C., Bentley, K. P. & Baker, R. G. 1960 Growth substructures in weld metal. *Acta Metall.* **8**, 898–899.

Cane, M. W. F., Dolby, R. E. & Baker, R. G. 1973 A slow bend test for HAZ hydrogen cracking susceptibility and its correlation with welding experience. *Weld. Res. Int.* **3**, 1–16.

Coe, F. R. 1973 *Welding steels without hydrogen cracking.* London: The Welding Institute.

Evans, R. M. 1962 Joining of Nickel base alloys, BMIC Report 181. Defense Metals Inf. Center, B.M.I. Columbus.

Farrar, J. C. M. & Dolby, R. E. 1972 *Lamellar tearing in welded steel fabrication.* London: The Welding Institute.

Faulkner, C. C. & Voldrick, C. B. 1959 Interpretive report on welding titanium and titanium alloys. *W.R.C. Bull.* December, No. 56. New York: Welding Research.

Graville, B. A., Baker, R. G. & Watkinson, F. 1967 Effect of temperature and strain rate of hydrogen embrittlement of steel. *Br. Weld. J.* **14**, 337–342.

Houldcroft, P. T. 1967 *Welding processes,* B.W.R.A. Welding Series. Cambridge University Press.

Kihara, H., Suzuki, H. & Horikawa, K. 1971 Japanese specifications for weldable high strength steels with special reference to WES-135. *Proceedings of Conference on Weldability of Structure and Pressure Vessel Steels,* 16–19 November 1970, pp. 116–134. London: The Welding Institute.

Proceedings of Select Conference on Al–Zn–Mg Alloys held 23–25 September 1969. Abington, Cambridge: The Welding Institute.

Pumphrey, W. I. 1955 Research in the welding of aluminium and its alloys. *Res. Rep.* No. 27, Aluminium Dev. Assoc., July.

Roberts, J. E. 1967 The ducol series of structural steels. *Proc. ISI/BISRA Joint Conference on Stray Tough Structural Steels, I.S.I. Publ.* **104**, 103–109.

Specification for the Metal-Arc Welding of Carbon and Carbon-Manganese Steels, BS 5135. London: British Standards Institution.

Weldable Structural Steels, BS 4360 1972 London: British Standards Institution.

Welding Handbook, volume 4, *Metals and their weldability.* New York: A.W.S.; London: Cleaver-Hume Press Ltd.

Welding Handbook, volume 2, *Welding processes.* New York: A.W.S.; London: Cleaver-Hume Press Ltd.

Wilkinson, F. J. & Cottrell, C. L. M. 1958 *Weld Metal Fab.* **26**, 171–184.

2.6 Corrosion, design and materials: general and pitting

By G. Butler and H. C. K. Ison

Division of Materials Applications, National Physical Laboratory, Teddington, Middlesex

[Plates 13–16]

A brief description is given of the corrosion process and type of attack with particular reference to metal behaviour in aqueous solutions. Some factors which have to be taken into account in the design of plant are described and examples given of the choice of materials for particular services. The methods of corrosion prevention are briefly reviewed and some predictions made on future prospects for improving the corrosion behaviour of metals and alloys.

1. Introduction

In the eyes of the metallurgist and engineer, corrosion has been a Cinderella in not being given the same status as mechanical properties. Fortunately, alloys designed for particular demanding structural and engineering applications frequently have a composition which gives them a high corrosion resistance. Indeed two of the best known corrosion-resistant alloys were found by accident, the stainless properties of chromium steel by direct observation (Monypenny 1951) and Monel a 'natural alloy' smelted directly from a mixed ore (Gaines 1912).

The attitude of metallurgists to alloy design is illustrated by the fact that Rosenhain (1914) wrote that corrosion prevention resolves itself into a question of finding a suitable protective coating, and 50 years later in a book on alloy design Wakeman (1966) considered that corrosion behaviour points to a need to specify the actual chemical reactions which need to be inhibited, corrosion is not very structure sensitive and prediction of attack involves a great deal of chemistry, electrochemistry, and chemical kinetics. The greater demands of the chemical industry have led to the development of alloys resistant to chemical environments, e.g. the Hastelloys. In the gas turbine field the Nimonic series were developed primarily for resistance to high temperature creep although a secondary factor was their resistance to oxidation.

The choice of a material for resistance to corrosion is complex because the reaction between a metal and its environment is very far from simple. Apart from a knowledge of the intrinsic resistance of the metal or alloy, it is necessary to predict the electrochemical and chemical kinetics of the corrosion process and the effect upon them of parameters such as temperature, heat transfer, pressure, fluid velocity, fluid mixing and so on. Consideration must be given not only to the choice of a resistant metal but also to lay-out, design and whether the metal should be protected to eliminate corrosion or reduce it to an acceptable level. A factor of considerable importance in these decisions is the planned life and the relation between the initial capital outlay and the probable future costs of maintenance.

2. The corrosion process

With the exception of the noble metals – and to some extent copper – which are found as such in nature, most metals have to be *won* from ores and hence are unstable and, given the opportunity, will return to their natural state. Thus it is not surprising that corrosion products have a similar chemical composition and physical structure to those of naturally occurring ores: in the case of iron the products of corrosion may be lepidocrocite, goethite, haematite, magnetite and so on.

It is now generally accepted that corrosion is an electrochemical phenomenon involving a corrosion cell in which current flows from a cathode, at which a reductive reaction is taking place, to an anode, at which dissolution of the metal is occurring. The two reactions are complementary and both must take place for corrosion to occur. The cathodic reaction may involve reduction of oxygen to form hydroxyl ions, hydrogen evolution, utilization of hydrogen in the metabolism of certain bacteria or the reduction of metal ions. The anodic reaction involves the dissolution of the metal, which either goes into solution as hydrated metal ions or forms a solid metallic compound on the surface. The latter may consist of a loose friable layer on the metal surface permitting corrosion to proceed unchecked or, under favourable conditions, will form a coherent thin film on the metal reducing the rate of dissolution to a very low value.

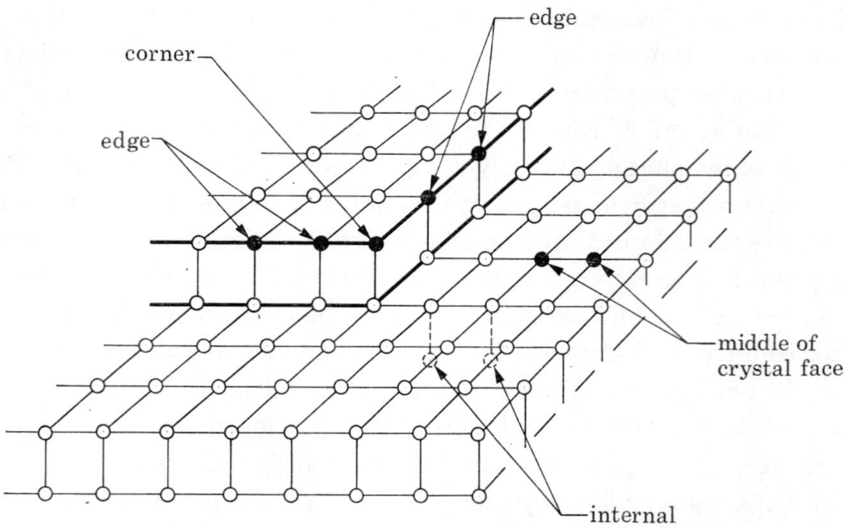

Figure 1. Simplified representation of a metal surface showing sites of differing susceptibilities to attack.

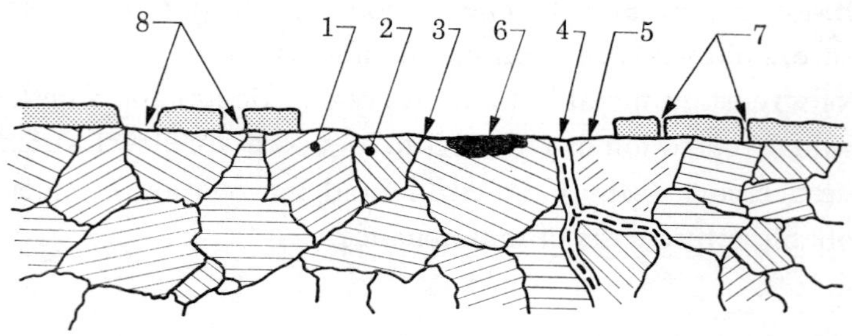

Figure 2. Features of a metal surface which can give rise to corrosion cells. 1, 2, grain orientation; 3, grain boundary; 4, grain boundary precipitation; 5, depletion of alloy in one component; 6, inclusions; 7, porosity in oxide film; 8, discontinuities in oxide film.

Corrosion cells can be formed in a variety of ways. Even the surface of a single crystal will not be perfect and atoms at corners or edges will have a greater tendency to go into solution than atoms in the middle of a plane (figure 1). In practice we are concerned with polycrystalline materials in which cells can form because of heterogeneities introduced by grains of different orientation, by grain boundaries with or without precipitates, by differences in composition in the case of alloys, by porosity or breaks in more-noble surface films, by differences in stress concentration and so on (figure 2). Cells can also be set up by variations on the metal surface of the liquid composition, e.g. salt concentration, pH value and oxygen concentration, and also of variations in physical conditions such as temperature, water velocity, and stray currents.

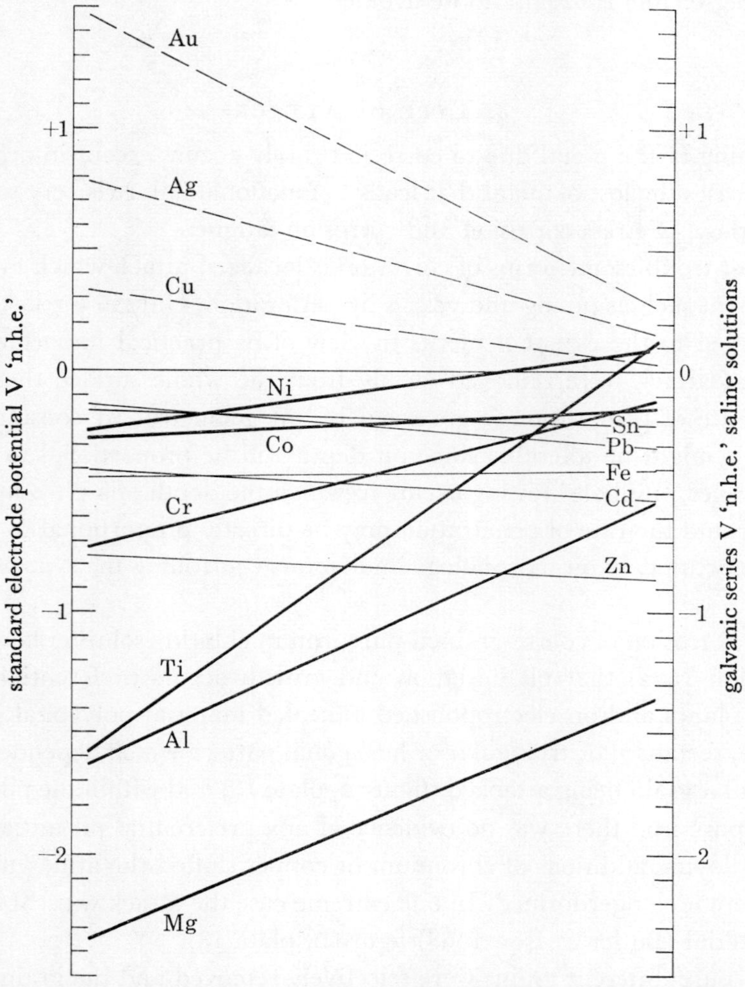

FIGURE 3. Comparison between the standard electrode potentials of metals and their galvanic potentials in saline solutions.

Corrosion cells can also be formed between dissimilar metals. The corrosion potentials of metals and alloys measured in, say, sea water or salt solutions, can be arranged in a galvanic series (figure 3), in which a metal high in the series (noble or electropositive) will generally have a greater resistance to corrosion than one low in the series (base or electronegative) (Butler & Ison 1966; Butler, Francis and McKie 1969). This galvanic series should not be confused with the standard electrochemical potential series, in which the potentials refer to that of the metal in a solution of its own ions at unit activity (Kaye & Laby 1973). Although the latter is often quoted in papers and text books on corrosion, it has little relevance to practical corrosion conditions as may be seen from the change in position of metals in figure 3. In a bimetal couple current will flow from the metal or alloy high in the galvanic series to the one lower in the series, usually decreasing corrosion of the former and increasing that of the latter. This is a useful guide to the compatibility of metals, the further apart the metals are in the series the greater the chance of a corrosion cell being formed. This should not be taken as an absolute guide, however, since the ease with which a reaction occurs, and the relative areas of the two metals will play a part. Thus oxygen reduction is difficult on stainless steel and in many applications it is compatible with aluminium alloys, whereas oxygen reduction on copper is easier and the aluminium–copper couple is usually to be avoided.

3. Type of attack

Uniform thinning of the metal due to corrosion rarely occurs except in acid solutions. In a large number of cases the loss of metal that leads to functional failure is very small, the extreme examples being those of stress corrosion and corrosion fatigue.

One of the most troublesome forms of corrosion is localized attack which eventually leads to failure of equipment such as piping and vessels by perforation. Only in the last decade or so has pitting been studied to the extent it merits in view of its practical importance. If the metal dissolved by local attack were removed evenly from the whole surface the effect would be negligible. The rate of penetration is governed by pit geometry. At constant corrosion rate hemispherical pits might be tolerable since pit depth will be proportional to the cube root of time. Often, however, 'pinhole' pitting occurs in which the depth of a pit is appreciably larger than its diameter and the rate of penetration may be directly proportional to time and lead to rapid failure. Unfortunately pit morphology and factors controlling the type of pitting are little understood.

Studies on the corrosion of coarse-grained pure iron in chloride solution have shown (Butler, Stretton & Beynon 1972) that pit initiation and growth occurs preferentially along defined crystallographic planes and on electropolished annealed material polygonal pits were formed showing a square, rectangular, triangular or hexagonal pattern which depended on the orientation of the particular grain being attacked (figure 4, plate 13) and within the pits faceting occurs. In this work on pure iron there was no evidence of any preferential pit initiation in the grain boundaries but alloying additions of chromium or copper shifted the attack to these areas and polygonal pits were no longer formed. In one extreme case the attack was catastrophic even on an annealed material (Butler & Ison 1968) figure 5, plate 13).

After long exposure different grains were selectively removed and the grain boundary could remain unattacked (figure 6, plate 14) or be selectively etched depending on whether the corroding solution was acid or alkaline. This would indicate some chemical or physical heterogeneity in the grain boundary region. The results are of practical interest in indicating that the tendency to pitting may be reduced either by giving the sheet material from which equipment is to be fabricated a certain preferred texture or by reducing the grain size as far as possible to ensure that there is a low probability of encountering grains of an orientation favourable to pit penetration.

Localized attack may also occur in crevices due to the limited access of electrolyte to the metal within the crevice. This results in the formation of a concentration cell due to differences

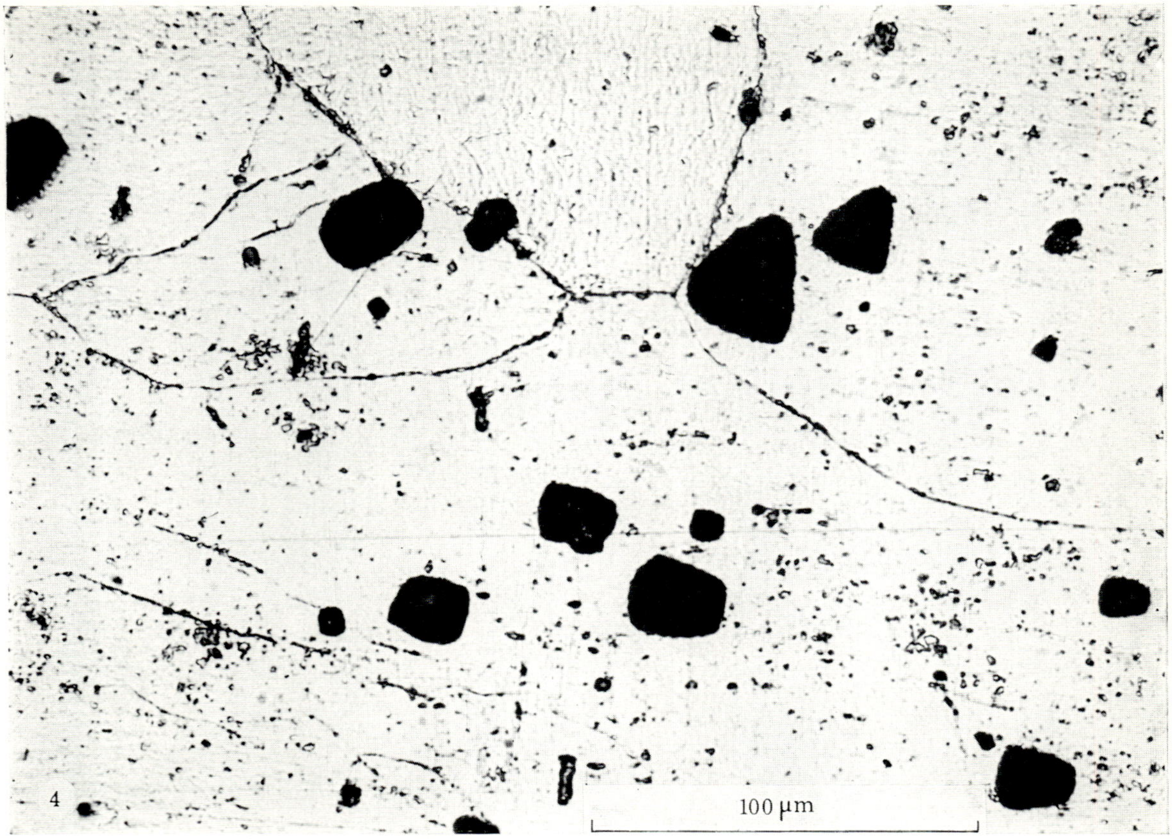

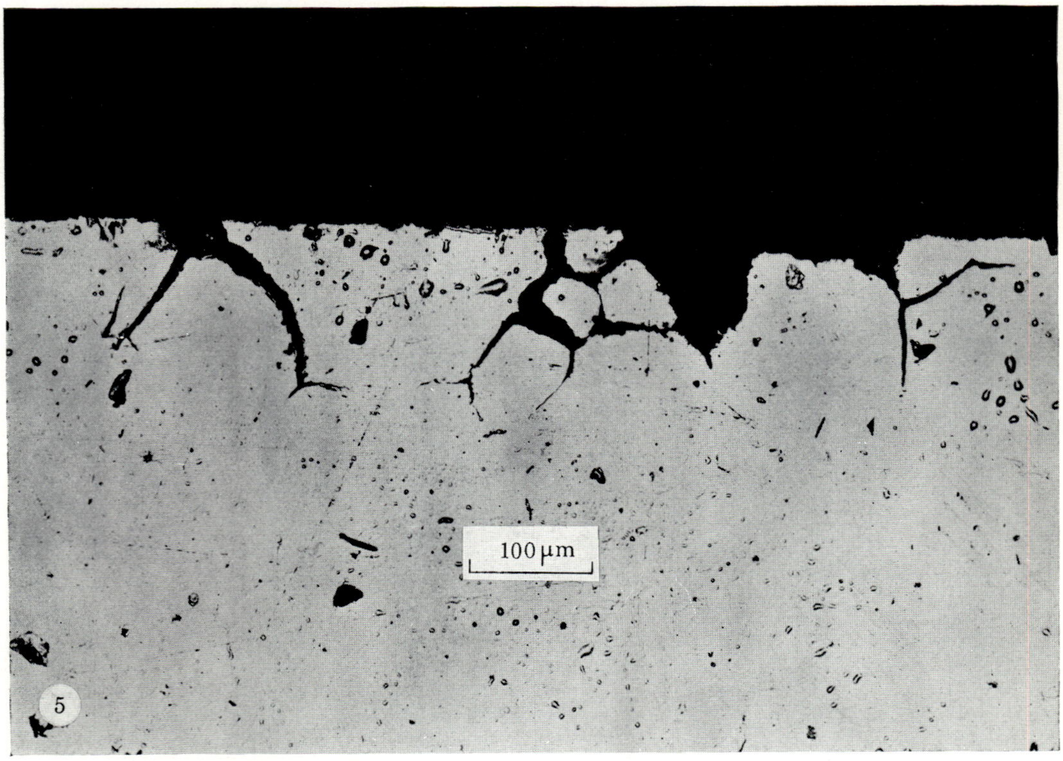

FIGURE 4. Polygonal shape of pits formed on pure iron in solutions containing 100 parts/10^6 Cl′ ion. On the grain in the lower half the pits are rectangular, on the grain in the upper left area they are also rectangular but with a different orientation, and on the grain in the top right hand corner the pits are triangular.

FIGURE 5. Catastrophic grain boundary corrosion in an iron chromium alloy exposed to 0.5 M NaCl solution.

(*Facing p.* 228)

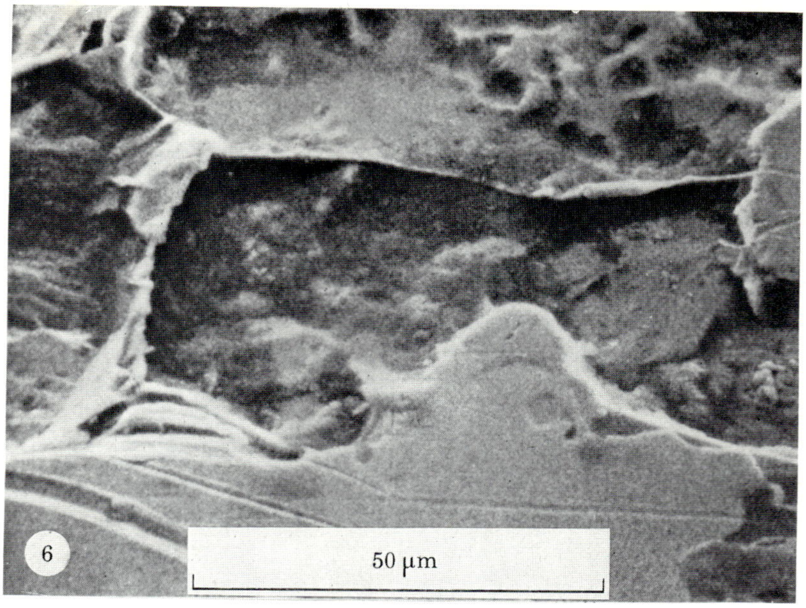

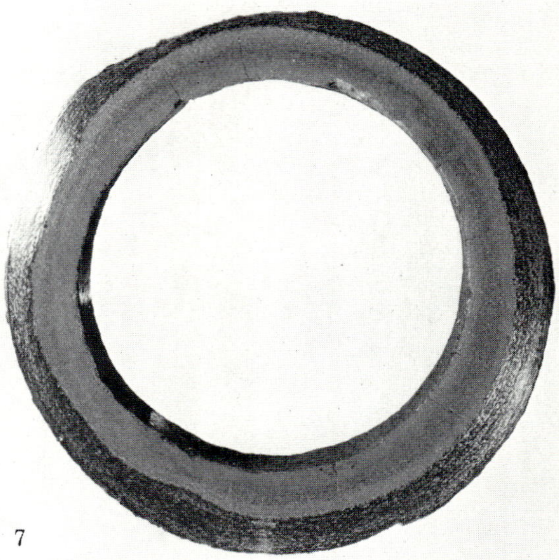

7

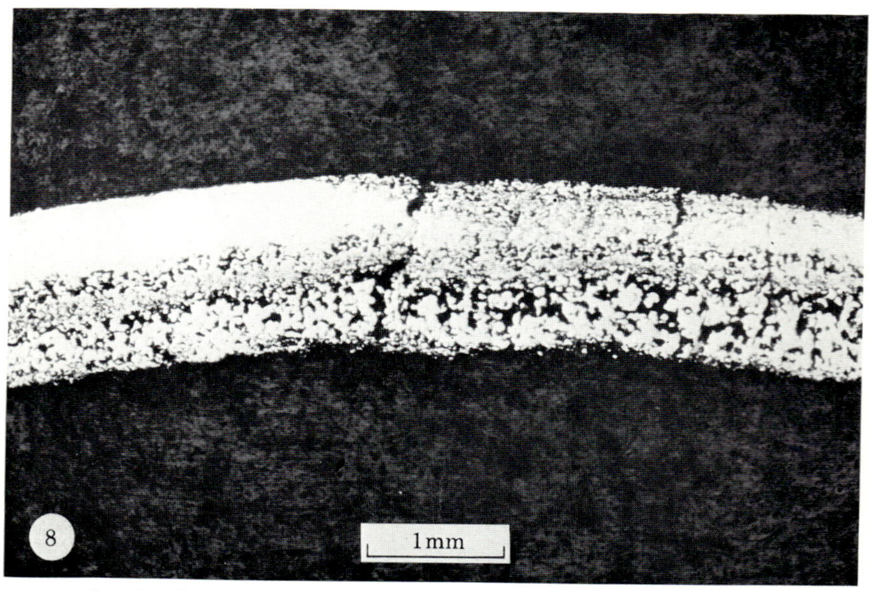

FIGURE 6. Stereoscan showing form of corrosion attack on a 5 % chromium–iron alloy in a solution containing 1000 parts/10^6 Cl^- at pH 9. In the lower part negligible attack has taken place but in the attacked upper part protruding metal ridges marking the grain boundaries may be discerned.

FIGURE 7. Graphitization of a 10 cm diameter cast-iron water main.

FIGURE 8. 'Plug' and 'layer' type dezincification found on a brass towel rail after 11 years exposure to a soft water.

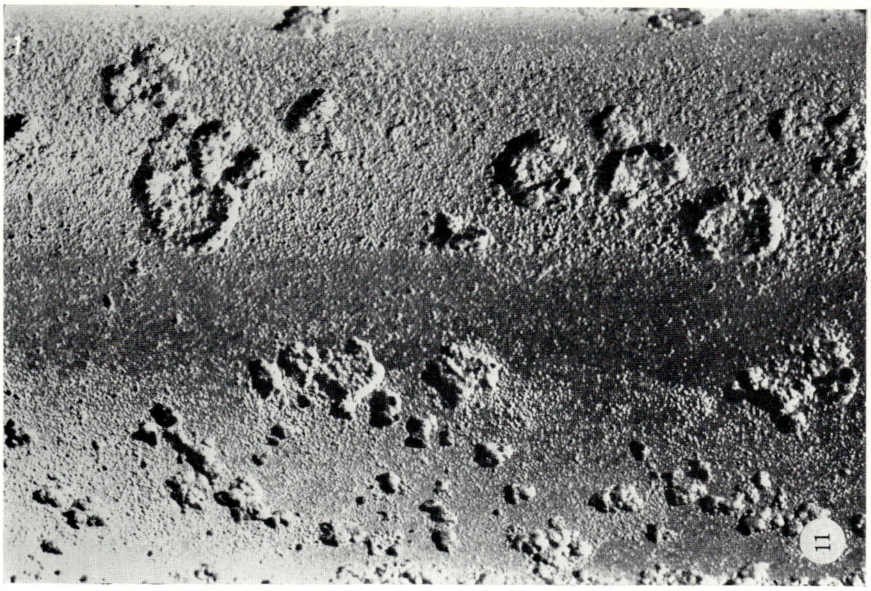

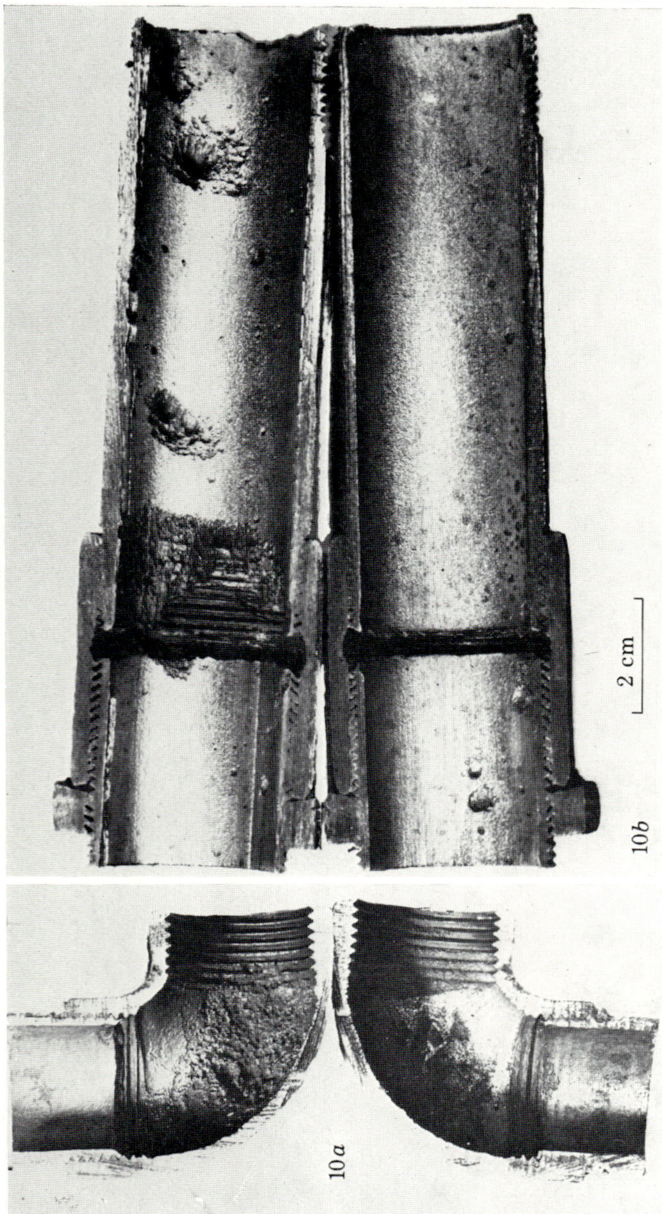

FIGURE 10. Enhanced attack produced by increased turbulence at (*a*) a sharp bend and (*b*) a junction in a pipe carrying mains water.

FIGURE 11. 'Horse-shoe' pit formation on boiler tube. (Magn. ×1.2.)

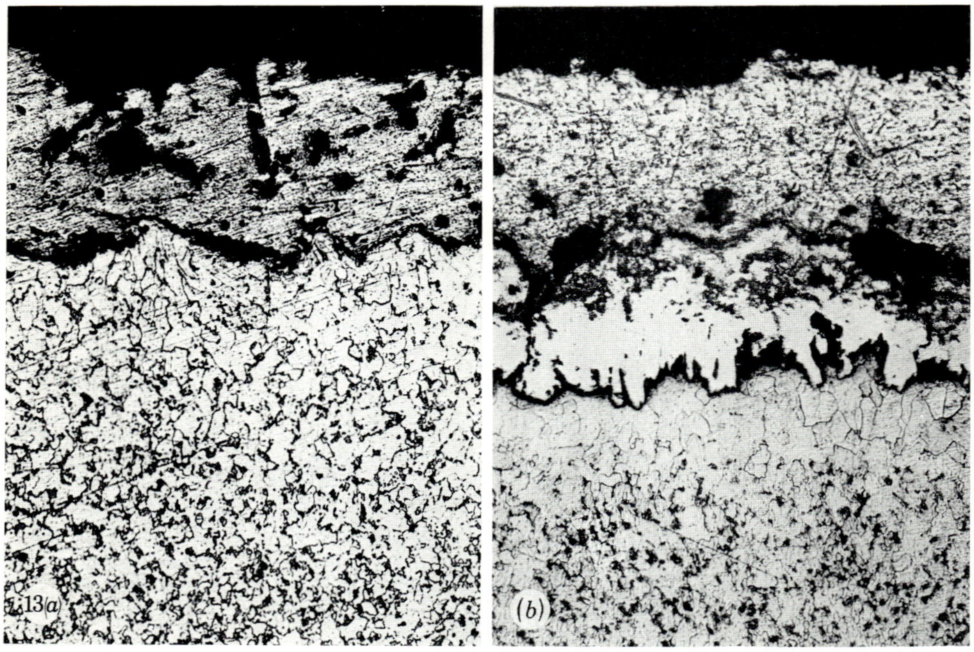

FIGURE 13. Microsections of aluminium coated tubes (*a*) without heat treatment and (*b*) heat treated. (Magn. × 133.)

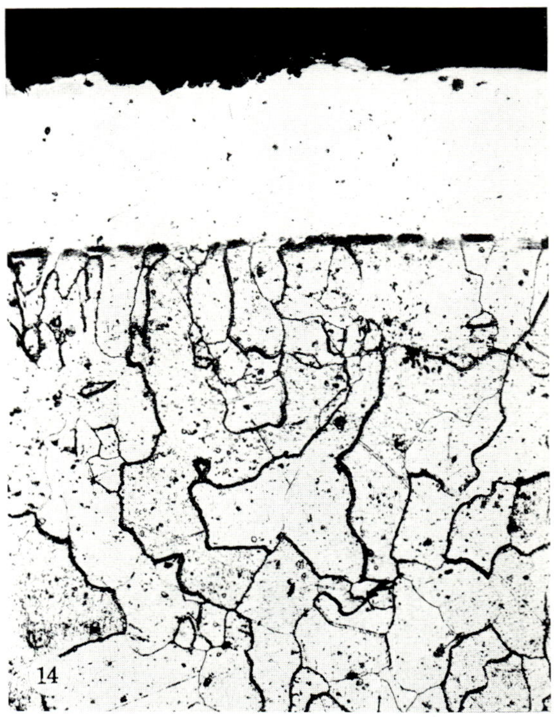

FIGURE 14. Microsection of chromized tube. (Magn. × 133.)

in salt, hydrogen ion or, more commonly, oxygen concentration. If the crevice is not blind and leakage takes place, e.g. between a boiler tube and the end plate, concentration of salt can occur within the crevice further enhancing attack as, for example, in caustic embrittlement. Metals such as stainless steel and titanium which need a supply of air or oxygen because their resistance relies on a thin oxide film are particularly prone to this type of attack. Crevices between dissimilar metals, e.g. between a stainless steel and a copper alloy, can lead to rapid catastrophic attack.

Selective attack sometimes referred to as de-alloying, can occur with some alloys particularly in soft or saline waters. Graphitization of cast iron† is the oldest example of this form of corrosion (figure 7, plate 14); the iron corrodes giving rise to iron oxides in a matrix of graphite and although the contours of the original alloy are preserved the final material has a very low mechanical strength and eventually fractures. Brasses can undergo dezincification by selective leaching of zinc (figure 8, plate 14), the less noble component, with the result that the alloy is converted into a porous mass of copper of poor mechanical strength. Both forms of attack are accelerated by the creation of a galvanic cell between the noble component remaining behind, graphite or copper respectively, and the less-noble metal. Dezincification commonly occurs in two forms; a 'plug' type, in which the rate of attack is small but the rate of penetration may be as high as 5 mm per year, commonly found in sea-water, or the 'layer' type, in which uniform attack occurs with a penetration rate of up to 0.08 mm per year. Dezincification of α brass can be prevented by small alloying additions of arsenic (Bengough & May 1924), but with the α–β brasses this is not effective and the best that can be achieved is the reduction of the β phase to small discrete areas (Nielsen & Rislund 1973).

In flowing conditions mechano-chemical attack can occur due to the combined influence of chemical attack and erosion of the corrosion product by gas bubbles or suspended solids. This gives rise to pits elongated in the direction of flow with sharp edges, undercut on the downstream side, and free from corrosion product. Cavitation, characterized by hemispherical pitting and severe and rapid wastage of metal, can occur under turbulent flow. It is caused by the formation and collapse of vapour-filled cavities by sudden changes in pressure or by vibration, the impact of the liquid being capable of keeping the affected surface free from corrosion product or protective films. Cavitation attack is most commonly found on ships' propellers, on diesel-engine liners and on components of centrifugal pumps.

4. Design

In general the physical factors to be considered will be the transfer of momentum, heat, and material to and from a fluid: the fluid composition may vary throughout the plant or equipment, the function of which will determine the extent of the changes. Even in the absence of impressed movement or stirring, some relative movement between fluid and solid will occur as a result of density differences produced by evaporation, temperature variations, and even by a corrosion process itself. In the case of impressed movement, the flow pattern may be laminar or turbulent according to the forces applied. In all cases, however, there will be a boundary layer at the fluid/metal interface within which maximum transfer will occur but whose thickness will depend on whether we are concerned with transfer of momentum, heat or material (Levich

† This term should not be confused with that used by metallurgists which relates to the formation of graphite from Fe_3C.

1962). All three can have an influence on corrosion but will interact one with another. The interplay between these variables is complex and insufficiently understood even in systems with the simplest geometry.

The corrosion technologist will be largely concerned with the diffusion boundary layer within which transport of reactants or products will take place by diffusion through the fluid, whereas outside this layer transport will occur more rapidly by the motion of discrete volumes of the fluid. The thickness of the diffusion boundary layer will decrease with increase in flow from a value of about 0.5 mm under conditions of natural convection to a very small value when the flow is highly turbulent. With steady flow this can increase the rate of corrosion by several orders of magnitude (Butler & Stroud 1965).

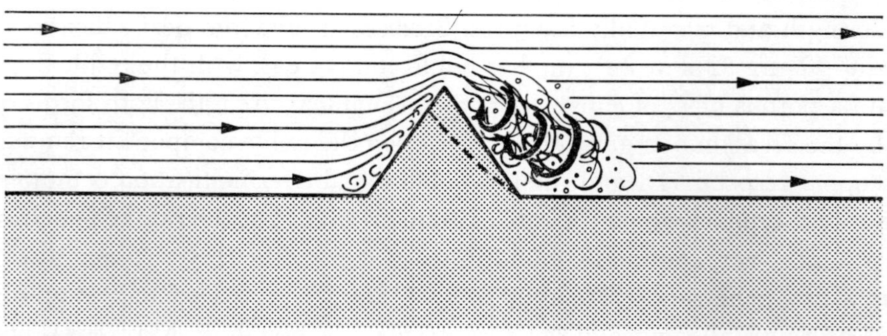

FIGURE 9. The effect of an asperity in producing increased turbulence. The broken line indicates the probable area of increased corrosion.

Turbulence can be induced or increased by asperities (figure 9) sudden changes in cross section and sharp bends (figure 10, plate 15), weirs or irregular deposits on the metal surface. The effects of these can be twofold: the difference in the rate of supply of reactants will give rise to a concentration cell thereby increasing attack in one region and decreasing it in another, or the mechanical effect will give rise to impingement or cavitation attack. Hence, wherever possible, sharp changes in cross section or deposition of particulate matter should be avoided.

In general, equipment should be designed without sharp edges and recesses and should have smooth contours that will ensure easy drainage and maintenance. This also provides for metal or paint coating of good quality and uniform thickness if these should be necessary. Crevices should be avoided as far as possible, particularly with metals and alloys which rely on a thin preservating film for their protection, and welded joints are usually preferable to riveted constructions.

The attack on a metal under conditions of heat transfer is often extremely severe. Transfer of heat can take place by conduction, convection, nucleate boiling or film boiling. Conditions leading to film boiling should be avoided since heat transfer then takes place through a poorly conducting vapour blanket and this results in very high surface temperatures which in the extreme can cause melting of the metal. In nucleate boiling the transfer takes place through the formation of vapour bubbles at discrete points on the surface. Boiling from such nuclei can persist over long periods of time and lead to an appreciable concentration of dissolved solids thereby causing increased aggressivity and local attack. Such a mechanism explains the formation of pits which in plan (figure 11, plate 15) have a 'horse shoe' or circular shape (Butler & Ison 1961). When a thin layer is deposited on a metal, 'wick boiling' (Macbeth 1971) can take place from the bottom of pores in the layer and gives rise to a very large concentration factor –

a million may not be unrealistic. Deposits of corrosion product or scale are obviously undesirable, particularly at high heat fluxes although some protective action may be achieved at low heat fluxes.

5. Choice of materials

A large variety of metals and alloys is available ranging in corrosion resistance and cost from iron and steel to the noble metals. Iron and steel are widely used owing to their good mechanical properties, ready availability, and low cost. Furthermore, their mechanical properties can be markedly improved by small alloying additions of $\leqslant 1$–$2\,\%$. Unfortunately, these do little to improve the corrosion resistance of the iron and steel which is lower than that of almost any other metal. One notable exception is copper which has a marked effect on the behaviour of steel on exposure to the atmosphere (Buck 1919), to acids (Williams & Komp 1965; Cleary & Green 1967), and to oxygenated saline solutions at 200 °C (figure 12) (Butler & Ison 1957) even at a concentration as low as 0.03 %, although maximum benefit is obtained with about 0.25 %. The evidence for its effect on the corrosion of steel in natural waters is, however, conflicting. We also find copper additions of several per cent in many more highly alloyed materials and here also it will improve corrosion resistance, particularly towards acid attack.

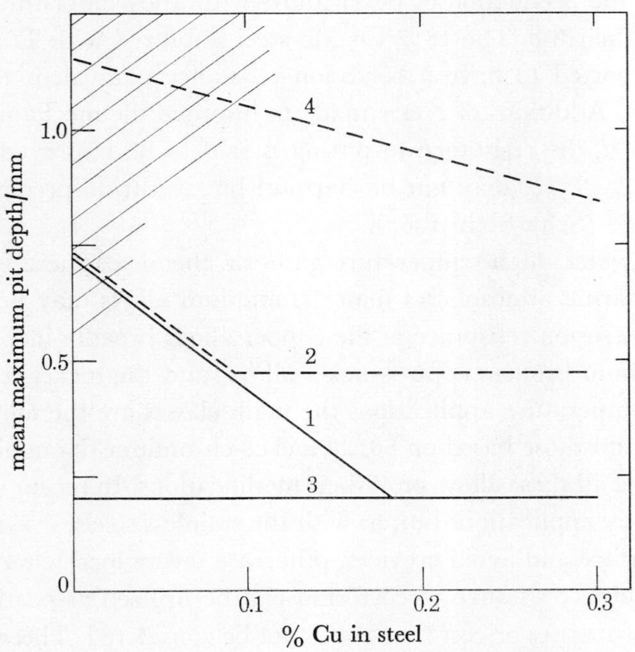

FIGURE 12. The influence of copper content of the steel in reducing pit penetration on boiler tubes in saline solutions. Seamless tubes: curve 1, 4 cm^3/l oxygen; 2, 15 cm^3/l oxygen; 3, 4 or 15 cm^3/l oxygen. Welded tubes: curve 4, 4 cm^3/l oxygen. Shaded area, occasional tube perforations in this region.

Copper itself has a special position in the galvanic series as a 'near-noble' metal and resists corrosion in a variety of environments, particularly in natural waters. Problems can arise with piping carrying a carbon film, formed from the lubricant used in the drawing process, but these have now been largely eliminated by abrasive cleaning following manufacture.

Alloying additions of elements such as chromium, aluminium, and to a certain extent, nickel, which owe their corrosion resistance to the formation of a passive oxide film, can transfer this resistance to other metals when added in sufficient quantity. Thus the corrosion resistance of steel alloyed with 12% or more chromium has given rise to the development of probably the most important family of alloys, the stainless steels. The term 'stainless' can be misleading since many of these alloys are prone to pitting attack in chloride solutions, to crevice attack, and to stress corrosion cracking. Nevertheless, these alloys find wide use in chemical plant and are being increasingly used in domestic water systems. Heating stainless steel, e.g. in welding, can give rise to precipitation of chromium carbide in the grain boundaries and depletion of chromium in the adjacent thin layers. These layers then become anodic to the rest of the surface and intense local attack results in intergranular penetration and eventually to grains falling apart. This led to the addition of stabilizing elements such as titanium or niobium which form carbides preferentially and leave the chromium in solid solution. The addition of nickel not only maintains the austenitic structure and improves the mechanical properties but it also contributes towards the corrosion resistance. Small additions of molybdenum are particularly beneficial in increasing the resistance of stainless steels to the possibilities of pitting and crevice attack.

The ferritic stainless steels, although almost free from susceptibility to stress corrosion, have not historically found favour in agressive environments because their general corrosion resistance is lower than that of the austenitics. In recent years, however, the advance in steel-making technology has enabled the production of purer alloys with a low interstitial content and this has renewed interest in their use. The 18:2 Cr–Mo steel stabilized with Ti and with C and N as low as possible is reported to have a corrosion resistance equivalent to that of the best austenitics (Myers 1972). Additions of N are made to improve the mechanical properties and at concentrations $\leqslant 0.06\%$ the resistance to pitting is said to be improved (Moskowitz *et al.* 1967): in solution up to 0.3%N may not be harmful but if nitride precipitation occurs the resistance may be lowered (Scharfstein 1967).

For resistance to sea water, high-temperature gases or chemicals there is a wide choice of non-ferrous alloys. In marine atmospheres many aluminium alloys may be used, and in immersed conditions the corrosion resistance of the copper alloys broadly increases in the order: brasses, bronzes, aluminium bronzes copper-nickel alloys, and the nickel-copper alloys of the Monel type. For high-temperature applications the main classes are the nickel–molybdenum–iron alloys (Hastelloys) and those based on 80:20 nickel–chromium (Nimonics). In the chemical industry use is made of all these alloys or of their modifications. In recent years titanium and its alloys have found many applications but, as with the stainless steels, it is necessary to maintain aeration at their surface and avoid crevices, otherwise severe local attack may take place.

Many metals with a high resistance to corrosion can be utilized as coatings where on the grounds of mechanical properties or cost they could not be considered. These can be applied in many ways: molten metal, pack powder, spraying, plating, deposition from the gas phase, plasma arc, welding, explosive bonding and so on. Hot-dipped or heat-treated coatings, e.g. of zinc or aluminium, lead to the formation of a transition zone (figure 13, plate 16) in which alloying with the basis metal occurs which can improve adhesion but may produce some brittleness.

In chromizing, application of chromium from the vapour phase gives an alloy layer on the surface (figure 14, plate 16), with a corrosion resistance equal to that of a stainless steel. However, in its production a disadvantage is the high temperature that is necessary and, furthermore, if the steel is not stabilized and the carbon carefully controlled at a very low level the chromized component can be embrittled.

Coatings should be free from pores or breaks particularly with cathodic coatings such as nickel or chromium, where pitting of the underlying steel can be enhanced. With anodic coatings, such as zinc, a measure of sacrificial protection is given to the steel.

Under more severe conditions bimetals can be used enabling advantage to be taken of the corrosion resistance of the two bonded metals to different environments on either side of a metal wall, e.g. the fire side and water side of a boiler. An alloy chosen for its mechanical properties may be clad with the pure parent metal which may have a higher corrosion resistance as is the case, for example, with aluminium and its alloys. In other cases cladding with a noble metal or alloy can be used where, on the grounds of cost, the use of the solid material would be entirely uneconomic.

6. Corrosion prevention

So far in this paper emphasis has been placed on constructional materials rather than the environment. In design, however, consideration should also be given to the widely used methods of corrosion prevention: painting, corrosion inhibition, and cathodic or anodic protection.

Paints are used to provide a protective barrier between the metal and its environment. It is not always appreciated that paint films do not necessarily act by preventing the access of oxygen and water to the metal surface – indeed diffusion of these reactants may proceed fast enough for corrosion to continue unchecked – but owe their action to the formation of a corrosion inhibitor within the paint film (Mayne & van Rooyen 1954). Needless to say, the design requirements include the provision of adequate access to all parts of the structure for paint applications.

It is often necessary to modify the environment by water treatment or the addition of inhibitors. In this the additives act by reducing or preventing the anodic or cathodic action or by laying down a protective film or layer. The amount of inhibitor required is generally greater in multi-metal systems or where access is limited, as in crevices.

An important method of preventing corrosion is by applying a current to change the potential of the metal to a value sufficiently negative to prevent metal ions from dissolving (cathodic protection), or to a value sufficiently positive for a passive film to develop (anodic protection). Cathodic protection may be achieved either by connecting the metal to be protected to a less noble metal which dissolves sacrificially, e.g. steel to zinc, and anodic protection by connexion to a more noble metal, e.g. titanium to platinum. The same function can be obtained by supplying direct current of the appropriate sign from an external source. It is essential that the environment be sufficiently conducting and that the design is such that no part of the structure is shielded and inaccessible to current flow.

7. Future prospects

In this brief description of the causes and forms of corrosion and some of the methods whereby it can be prevented, an account has also been given of the development of alloys, fortuitously or otherwise, that have improved corrosion resistance. For the future there is the possibility of using computer methods for the design of alloys as in the Phacomp method (Woodyatt, Sims & Beattie 1966) used for gas-turbine alloys. It is most likely that resistant properties will be sought by the incorporation in an alloy of elements that will produce more stable, inert surface films and more effort could be applied to the production of thin resistant coatings without the disadvantages of the silicides, borides and glasses. Another approach is the development of alloys on which a small amount of uniform corrosion occurs with the formation of an adherent product that prevents or reduces further attack, e.g. the Corten type steels under atmospheric conditions.

Technological advances in the preparation of purer metals and alloys have already led to the production of stainless steels with improved corrosion resistance and similar developments may be expected with other metals and alloys. The production of metals or metal coatings with specified textures seems to be worth further study in relation to corrosion behaviour.

Much future research will still be directed towards the improvement of our knowledge of the environmental factors involved and of their control. For example there is insufficient knowledge of the influence of metal composition on its pitting propensity towards a given water and hardly any data on the influence of temperature on the corrosion process. Search will continue for new inhibitor formulations as well as improved methods of corrosion monitoring.

As the knowledge of corrosion phenomena advances, the designer can take advantage in reducing the allowance for metal wastage, for example, by reducing wall thickness. This leads to a demand for stricter corrosion control, and so the story goes on.

References (Butler & Ison)

Bengough, G. D. & May, R. 1924 *J. Inst. Metals* **32**, 81.

Buck, D. M. 1919 The influence of very low percentages of copper in retarding the corrosion of steel. *Proc. Am. Soc. Test. Mater.* **19**, 224.

Butler, G., Francis, P. E. & McKie, A. S. 1969 The corrosion potentials of some common metals in oxygenated solutions at 20 and 150 °C. *Corros. Sci.* **9**, 715.

Butler, G. & Ison, H. C. K. 1957 Model boiler tests on the influence of the copper content of the steel on the corrosion of tubes in artificial sea water. *Trans. Inst. mar. Engrs* **69**, 121.

Butler, G. & Ison, H. C. K. 1961 *First International Conference on Metallic Corrosion* (409). London: Butterworths.

Butler, G. & Ison, H. C. K. 1966 *Corrosion and its prevention in waters* (6). London: Leonard Hill.

Butler, G., Ison, H. C. K. & Stretton, P. 1968 Catastrophic grain boundary corrosion in a chromium–iron alloy. *Corros. Sci.* **8**, 281.

Butler, G. & Stroud, E. G. 1965 The influence of movement and temperature on the corrosion of mild steel. III. Chloride solutions. *Br. Corros. J.* **1**, 110.

Butler, G., Stretton, P. & Beynon, J. G. 1972 Initiation and growth of pits on high purity iron and its alloys with chromium and copper in neutral chloride solutions. *Br. Corros. J.* **7**, 168.

Cleary, H. J. & Greene, N. D. 1967 Corrosion properties of iron and steel. *Corros. Sci.* **7**, 821.

Gaines, R. H. 1912 Monel metal. *J. ind. Engng Chem.* **4**, 354.

Kaye, G. W. C. & Laby, T. H. 1973 *Tables of physical and chemical constants.* 14th edn. (216). London: Longman.

Levich, V. G. 1962 *Physicochemical hydrodynamics.* U.S.A.: Prentice-Hall.

Macbeth, R. V. 1971 Boiling on surfaces overlaid with a porous deposit: heat transfer rates obtainable by capillary action. *Publ. No. R.*711. Atomic Energy Establishment, Winfrith.

Mayne, J. E. O. & van Rooyen, D. 1954 The mechanism of the corrosion – inhibitive action of paints with special reference to basic pigments. *J. appl. Chem., Lond.* **4**, 384.

Monypenny, J. H. G. 1951 *Stainless iron and steel.* London: Chapman & Hall.

Moskowitz, A., Saltzman, G. A., Pinnow, K. E. & Redmerski, L. S. 1967 Influence of residual and minor elements on the pitting and atmospheric corrosion resistance of austenitic stainless steels. *A.S.T.M., Spec. Tech. Publ. No.* 418 (3). Philadelphia: U.S.A.

Myers, J. R. 1972 Stainless steel tube for potable water service: review of literature. *Tech. Cttee T-7B National Association of Corrosion Engineers, U.S.A.*

Nielsen, K. & Rislund, E. 1973 Comparative study of dezincification tests. *Br. Corros. J.* **8**, 106.

Rosenhain, W. 1914 *Physical metallurgy* (329). London: Constable.

Scharfstein, L. R. 1967 Effects of residual elements on the general corrosion resistance of austenitic stainless steels. *A.S.T.M., Spec. Tech. Publ. No.* 418 (90). Philadelphia, U.S.A.

Wakeman, D. W. 1966 *Modern theory in the design of alloys* (91). London: Iliffe.

Williams, E. & Komp, M. E. 1965 Effect of copper content of carbon steel on corrosion in sulphuric acid. *Corrosion* **21**, 9.

Woodyatt, L., Sims, C. T. & Beattie, H. 1966 Prediction of sigma-phase occurrence from compositions in austenitic super alloys. *Trans. Am. Inst. Min. Metall. Engrs* **236**, 519.

2.7 The contributions of physical metallurgy and of fracture mechanics to containing the problem of stress-corrosion cracking

By B. F. Brown
U.S. Naval Research Laboratory, and Senior Research Scientist, The American University, Washington, D.C.

[Plates 17 and 18]

The features in the province of the physical metallurgist – grains, precipitate particles, inclusions and lattice imperfections – have sizes approximately in the range of 1 μm–0.1 nm. The nature and properties of these microfeatures determine the engineering properties of alloys, including their resistance to stress-corrosion cracking. No basis has developed for confidence that knowledge of these microfeatures can be used to exclude the possibility of stress-corrosion cracking. Neither have we learned how to sum up the effects of these microfeatures to predict stress-corrosion cracking behaviour, nor is there any basis for predicting how soon we may be able to do this summing up analytically. In the meanwhile we need experimental means to do this summing up, and fracture mechanics appears to be an eminently suitable way to treat the stress factor. Other fundamentally different approaches may be found to be both valid and needed for special areas of the problem. Thus the role of fracture mechanics is to provide engineering data, and the role of physical metallurgy is to permit orderly alloy development through understanding of the data from fracture mechanics (and possibly other) tests. The two approaches are complementary, not competitive, and neither can fulfill the role of the other.

Non-redundant load path designs may require specially competent attention to both stress-corrosion cracking and brittle fracture.

Introduction

The invitation of the organizers of this conference to discuss the contributions of physical metallurgy and of fracture mechanics to stress-corrosion knowledge is especially welcome in view of what the author believes to be a widespread and prolonged misunderstanding of the roles of these two disciplines in the containment of the problem.

The practitioner of physical metallurgy contributes to the development and selection of alloys for engineering structures through his knowledge of the microstructure of alloy families and the contribution of various microstructural features to engineering performance. Microstructure is used here somewhat more broadly than usual and includes crystal structure, morphology of constituent grains, precipitates in the most general sense, including preprecipitation states of aggregation, texture, and the various types of crystal imperfections, including what Dr Rosenhain would possibly have termed mosaic structure. These features, having sizes spanning the range from about 1 μm–0.1 nm, determine the engineering properties of alloys.

Dr Rosenhain had the benefit of the art of the microscopy of metals due to Sorby's pioneering work a century ago, and while Rosenhain was still active X-ray diffraction had become a standard tool of the physical metallurgist. Since his day the number of experimental tools developed to aid the physical metallurgist has increased following the exponential pattern common to so much of science. During the 1950s, and following the lead of a contemporary of

Dr Rosenhain, A. A. Griffith, the methodology of (macroscopic) fracture mechanics was developed to the stage of engineering applicability, including eventually stress-corrosion cracking. The aim of this paper is to examine, in the light of all these developments, the contributions and roles of physical metallurgy and of fracture mechanics in controlling the problem of stress-corrosion cracking in present day engineering.

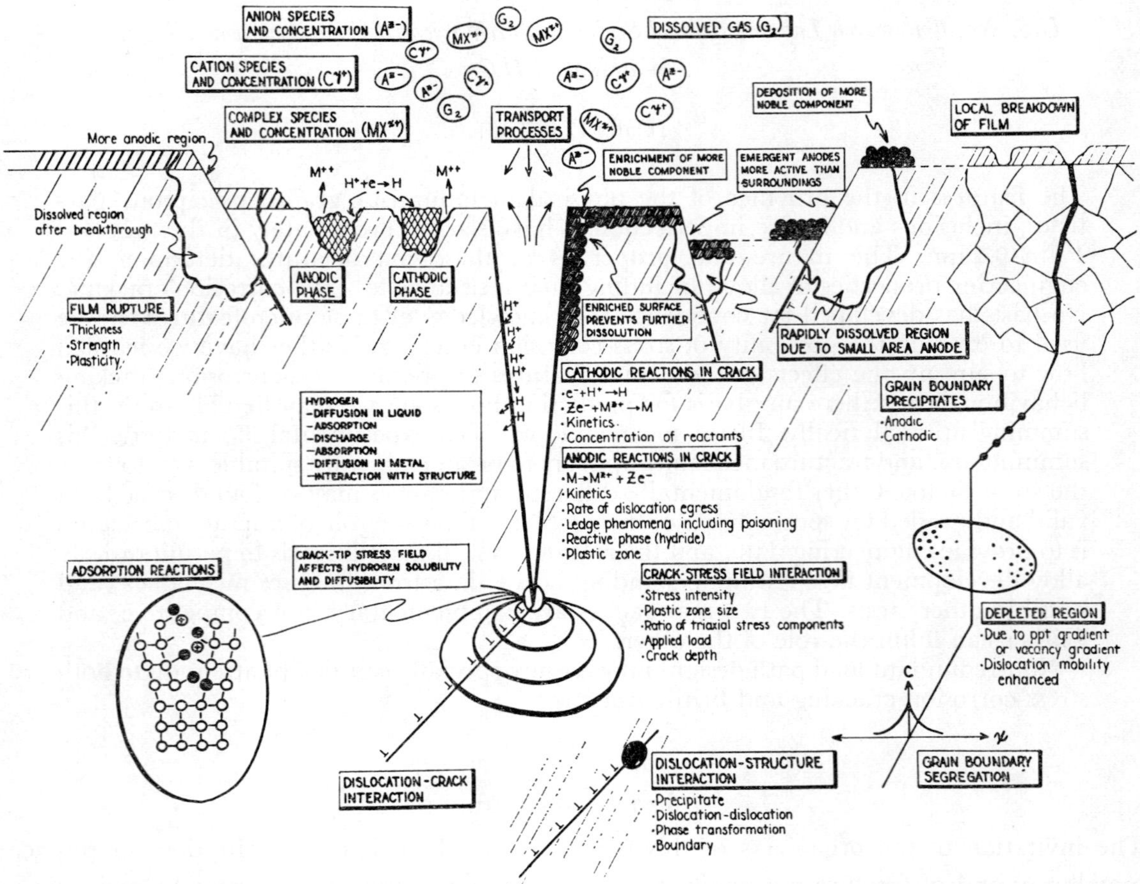

FIGURE 1. Schematic summary of microfeatures believed possibly involved in stress-corrosion cracking. Figure kindly supplied by Staehle (from Staehle 1969).

Most of the features and processes known or thought to be involved in stress-corrosion cracking in various systems are shown schematically in figure 1. There was a period when there appeared to be an unwritten creed in some circles that if the physical metallurgist or metal physicist became sufficiently clever in studying the 1 μm–0.1 nm features involved in cracking, measures could be devised to banish the problem, somewhat as the development of the vaccine banished the scourge of smallpox from much of the earth. Nothing has appeared on the horizon to support that outlook.

Since we have no reason for confidence that we will be free from the threat of stress-corrosion cracking but must somehow learn to cope with the effects of the microfeatures of figure 1, we must learn how to sum up those effects. We have absolutely no way to estimate how long it will take us to do this analytically. It could easily be the end of the century.

It is here that we can look for guidance from the progress which has been made in the closely related field of purely mechanical brittle fracture, the problem which caused the breaking apart of welded freighters in World War II, the bursting of the fuselages of the Comet III aeroplane, and numerous other catastrophes. It was known that the features familiar to the physical metallurgist are involved in brittle fracture – grain size, crystallography, precipitate particles, inclusions, and dislocations. The difficulty was that despite large and lengthy research programmes no one found a way to sum up analytically the contributions of these 1 μm–0.1 nm features to predict brittle fracture of engineering structures.

Fortunately, a second approach was made employing what might be called macro-mechanical metallurgy, to use R. F. Mehl's phrase, and including fracture mechanics in that term. The current technological bases for avoiding brittle fracture are the fruit of this macro-mechanical metallurgy approach.

These fracture control measures came about by the demonstration by Pellini and Puzak of a ductile-to-brittle transition temperature which, properly determined, affords positive protection against brittle fracture of ship plate and similar steel structures. We shall see below that an analogous approach may become useful in controlling stress-corrosion cracking in certain fields. For alloys which do not undergo a sharp transition, the fracture mechanics of Irwin and Kies affords a means of controlling the brittle fracture problem. Both the transition temperature method (standardized as ASTM E208) and the fracture mechanics method (standardized as ASTM E399) for controlling brittle fracture rely only on macroscopic specimens to sum up experimentally the effects of all the physical metallurgical microfeatures.

Through the use of these macroscopic methods we know quantitatively what changes in fracture toughness can be achieved by changing ferrite grain size, or by special vacuum melting, or by lowering the size of intermetallic compounds in high-strength steels. Turned the other way round, modern physical metallurgy enables us to understand the engineering properties of materials in terms of microfeatures, even though the properties must be measured macroscopically. Thus in brittle fracture the role of measurement is played by macro-mechanical metallurgy, and the role of physical metallurgy is to understand the reason for the measured properties in order to improve alloys in a non-random way. Physical metallurgy and macro-mechanical metallurgy are then not competitive approaches to the problem, but rather complementary approaches. Technological advances in the field require both.

The designer of submersibles used in the oil and gas fields in the North Sea specifies his hull steel by 'NDT + 60', meaning that the nil-ductility transition temperature must be at least 60 °F ($33\frac{1}{3}$ °C) lower than the minimum service temperature. A vast amount of research went into the development and validation of the 6-character specification inside the quotation marks above by which the designer sets his mind at ease regarding brittle fracture. The success in controlling the brittle fracture problem either by the transition temperature method or by fracture mechanics, as appropriate, has caused stress-corrosion cracking to replace brittle fracture as the most worrisome threat to structural integrity facing the designer, fabricator, and maintenance engineer in a wide range of technologies.

Fracture mechanics and stress-corrosion cracking

Returning to figure 1, in addition to the features of physical metallurgy we see ionic species, passivation, film rupture, and various other complexities of electrochemical corrosion. The inescapable conclusion is that if in the case of brittle fracture, where only physical metallurgy and mechanics are involved, we had to resort to a macroscopic specimen to sum up experimentally the contributions of all the microfeatures, there is surely even more reason to use the same approach when electrochemical corrosion has been added.

The macroscopic specimen approach and the physical metallurgy approach are not alternatives to solving the stress-corrosion problem. As with brittle fracture, they represent complementary approaches, one to produce characterization of engineering significance and the other to understand the reasons for the values measured.

But what sort of macroscopic test? After all, Roberts-Austen identified tensile stress as a necessary condition for this form of cracking using macroscopic specimens almost 90 years ago.

If one wishes to determine the stress at which a stress-corrosion crack will stop growing, or at which it will reinitiate, or if one is concerned with the effect of stress on cracking kinetics, it is necessary to quantify the stress effect in the presence of a crack. We do not have a valid way to do this except using fracture mechanics. This does not mean that to get stress-corrosion information of engineering utility one must use a fracture mechanics test. If I were asked to evaluate the effects of changing condensate composition on the cracking possibilities in Admiralty brass feedwater heater tubing, for example, I would probably prefer to cut a ring from the tubing, slot it to make a 'C-ring' specimen, and stress it as in service in the proposed environment. But the only reason for that preference would be that the material having the proper texture would be available for easy specimen manufacture. There is no reason that the same information could not be obtained using a fracture mechanics test, which would have the added advantage of being incorporable into a data bank of homogeneous information.

Three objections have been raised to the use of fracture mechanics to measure stress-corrosion characteristics: (1) The fracture mechanics specimen is provided with an initial crack-like flaw of controlled location and dimensions, and it has been the experience that designers, fabricators, and vendors may express umbrage at the implication that their product may be flawed. But the U.S. aerospace programme alone proved the wisdom of including an evaluation of the performance of materials containing a reasonable flaw. (2) Some of those experts who have been aware that the elastic fracture mechanics of Irwin and Kies is not useful to measure the brittle fracture characteristics of tough materials (because of the enormous size of specimens required to provide the necessary constraint to cause brittle fracture) have concluded that fracture mechanics is not therefore suitable for studying stress-corrosion cracking in the same alloys. But *all stress-corrosion cracks are brittle*, that is, they occur at stresses below that required for general yielding and therefore propagate in an elastic body even though there is *local* plasticity which is probably essential to the cracking process. Since the body as a whole is in the elastic range, elastic fracture mechanics is appropriately used to treat stress. (3) But the matter of local plasticity must be addressed. Indeed so, but the local plasticity is caused by and is therefore related to the surrounding macroscopic elastic field, even though we may not know the exact relationship. Thus the local plasticity is included with the other physical metallurgy features summed up by the macroscopic specimen, though one must bear in mind that room temperature creep is observed in *some* alloys stressed as precracked specimens.

Since the fracture mechanics approach is to observe the response of a cracked body to loading, experimentally one inserts a mechanical crack initially in order to obtain a crack of controlled location and geometry, commonly by machining a notch and fatiguing it. This procedure has special merit in assessing the stress-corrosion characteristics of weldments, especially for medium strength steels in seawater. In such materials stress-corrosion cracks usually initiate from corrosion pits, the location of which is determined by several electrochemical factors and may be unrelated to the relative vulnerability of the underlying metal to stress-corrosion cracking. If one works with a small sample of weldment, it is entirely possible that no pit forms at the locations which are most vulnerable to cracking, yet a large welded structure may have some pits at the sites of maximum vulnerability. This argument has given rise to the practice of siting precracks at the weld centreline, at the fusion line, in the heat-affected zone, and in the prime plate for stress-corrosion tests.

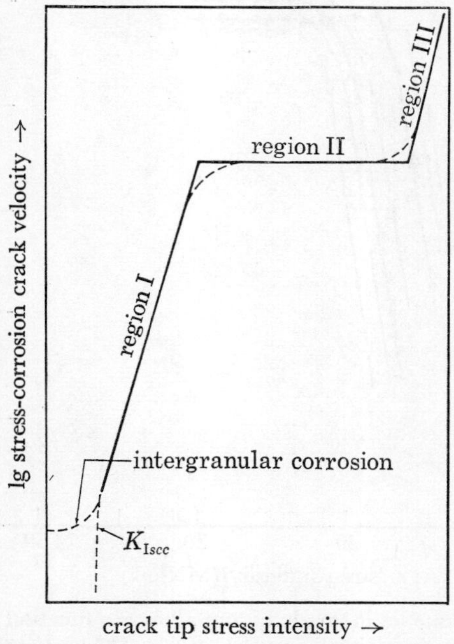

FIGURE 2. Schematic summary of the effect of stress-intensity K on stress-corrosion cracking kinetics. In some alloys region III is missing, and region II sometimes appears only as a point of inflexion. (From Speidel 1971.)

Having a cracked specimen, the elasticians have given us a theoretical method, and the continuum mechanicians an experimental method, to determine and express the stress intensity by a single parameter K. At sufficiently high values of K a stress-corrosion crack grows, and over the K range designated region I in figure 2, the cracking rate is approximately an exponential function of K. At higher K values there may be a plateau labelled region II in which some transport process is rate controlling and the cracking rate is independent of K. For some metals there may be a region III at the highest K levels in which cracking kinetics are again K-dependent. At sufficiently low K levels there may be a genuine threshold labelled $K_{\text{Is.s.c.}}$. Such a genuine threshold has been proven for titanium alloys, based on electrochemical measurements (Brown 1970). There are theoretical reasons to believe that such a genuine threshold exists for high-strength steels. But for aluminium alloys there is uncertainty about the existence of such a threshold (note figure 3).

The value of a number such as $K_{\text{Is.c.c.}}$ lies in its ability to predict the combinations of stress level, flaw size and shape which would be expected to produce stress-corrosion cracking through the Irwin equation

$$K^2 = \frac{1.21\pi\sigma^2 a}{\phi^2 - 0.2(\sigma/\sigma_y)^2},$$

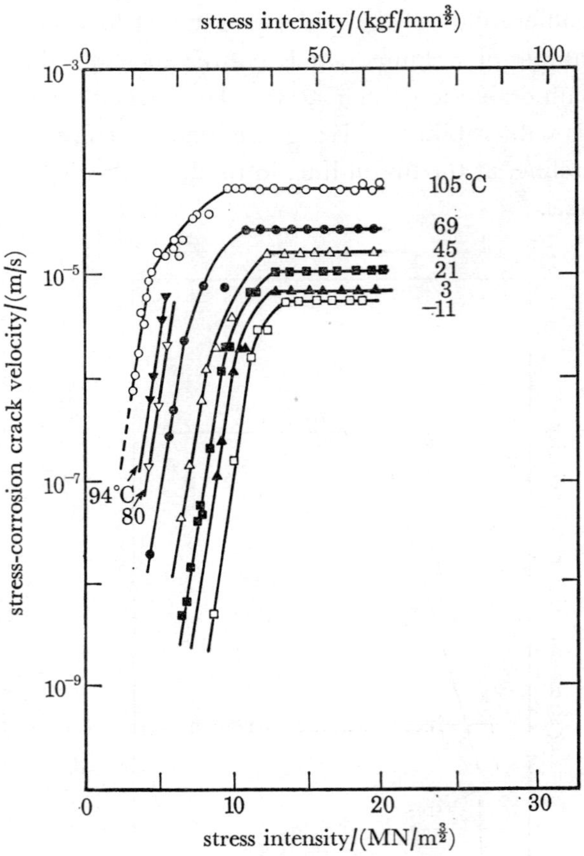

FIGURE 3. Stress-corrosion cracking kinetics in an aluminum alloy as a function of stress intensity at several temperatures. The activation energy in region I is much greater (27 kcal/mol; 113 kJ/mol) than in region II (3.8 kcal/mol; 15.9 kJ/mol), a fact which would not be suspected in the traditional undifferentiated total-time-to-failure data. (From Speidel 1971.)

Alloy 7079-T651; 2.5 cm thick plate; crack orientation TL; 3 m aqueous KI solution; potential −450 mV *vs* E_{H_2/H^+}; pH 6.

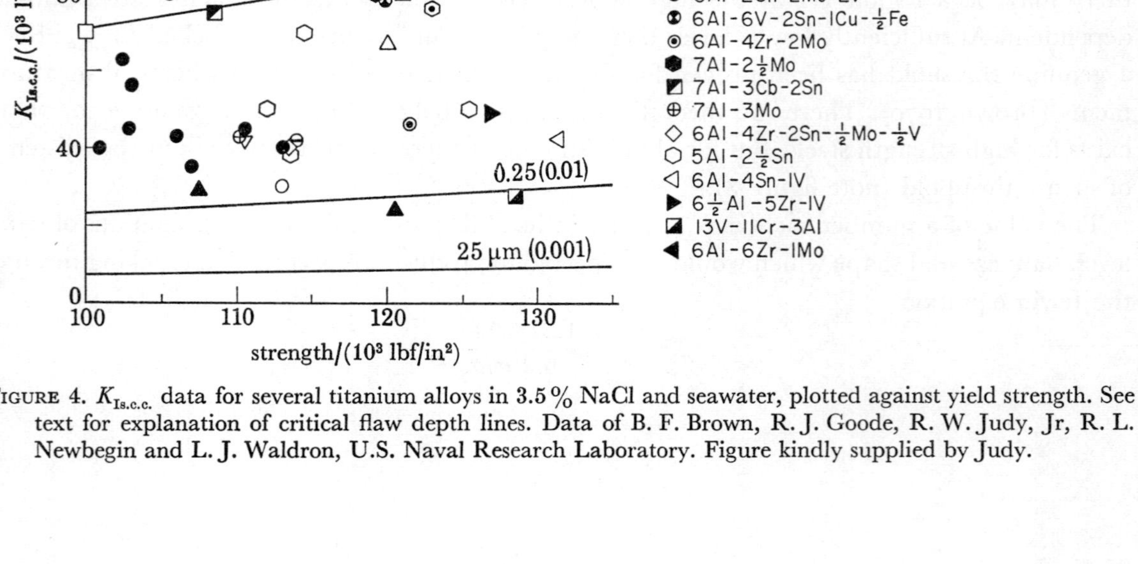

FIGURE 4. $K_{Is.c.c.}$ data for several titanium alloys in 3.5% NaCl and seawater, plotted against yield strength. See text for explanation of critical flaw depth lines. Data of B. F. Brown, R. J. Goode, R. W. Judy, Jr, R. L. Newbegin and L. J. Waldron, U.S. Naval Research Laboratory. Figure kindly supplied by Judy.

where σ is the remote stress normal to the surface crack of depth a. σ_y is the yield strength, and ϕ is a shape factor. For $K = K_{Is.c.c.}$, $\sigma = \sigma_y$, and for a crack at least 10 times as long as deep, the equation becomes

$$a_{cr} = 0.2\ (K_{Is.c.c.}/\sigma_y)^2.$$

$K_{Is.c.c.}$ for a number of titanium alloys of varying yield strengths are shown in figure 4. In this figure are lines for four values of critical flaw depth, a_{cr}, with the simplifying assumptions noted immediately above. The significance of these lines may be illustrated as follows: If one has the same environment for which $K_{Is.c.c.}$ was determined, and if he cannot be certain of inspecting out cracks as deep as 0.01 in (0.25 mm), he had better not use the two alloys having $K_{Is.c.c.}$ values which lie below the 0.25 mm line. Another way to look at the data of figure 4 is that the load carrying capability of a member having a standard surface flaw is linearly proportional to $K_{Is.c.c.}$.

Most of the data of figure 4 were taken using cantilever loaded bend bars. There have been two studies (Judy & Dahlberg 1968) in which by the use of the Irwin equation the predictability of cantilever beam $K_{Is.c.c}$ data have been checked using large plates stressed in simple tension and provided with small surface flaws of known geometry. Both studies have confirmed the significance of the $K_{Is.c.c.}$ number in panels bearing reasonable resemblance to large structures. There is a second kind of evidence of the validity of $K_{Is.c.c.}$ data based on service failures. In the early days of the U.S. aerospace programme, rocket motor chambers of high-strength steel were observed to undergo stress-corrosion cracking during hydrostatic testing with tap water. Estimates of $K_{Is.c.c.}$ indicated that a critical flaw depth for that steel in tap water would be about 25 μm. Because of incomplete control over heat-treating atmospheres in large furnaces in those days, there were numerous oxide intrusions at grain boundaries approximately 25 μm deep, and although the nominal hydrotest stress was somewhat less than the yield strength, in those days out-of-roundness was such as to elevate the stress locally to the yield level, undoubtedly cracking either alongside or through some of the oxide intrusions to provide the precrack. Thus all the criteria for stress-corrosion crack growth were met. To avoid the problem, the pressurizing fluid was changed to an inert oil. The conclusion from these experiments and experience, more meagre than I would prefer, is that if carefully measured fracture mechanics stress-corrosion data indicate a likely problem, it is a rash engineer who would disregard them in a critical circumstance.

It is nonetheless too early to know how fracture mechanics data on stress corrosion will fare in engineering design to contain the problem. There are reasons to believe that it may not be the best way to control the problem in certain technologies, such as steel pipeline, where the pioneering work of Parkins has shown that determining the electrochemical potential regime in which cracking occurs, and avoiding that regime, rather than measurement of cracking resistance itself in that regime, may be the way to go about controlling the problem. If so, this approach is exactly analogous to the use of transition temperature to avoid brittle fracture of ship plate.

Physical metallurgy

The engineering solution to cracking by certain environments, such as methanol or mercury around stressed titanium alloys, has been to simply prohibit the combination. In some circumstances, such as N_2O_4 in titanium bottles for space power plants, the engineering solution has been to add an inhibitor. But such solutions are not appropriate for aircraft or marine vehicles, for which the containment of the stress-corrosion problem had to be sought by the metallurgical route. There has been much progress in this way (Blackburn, Smyrl & Feeny 1972).

The phase α_2 (Ti_3Al) was found to promote aqueous stress-corrosion cracking, which explains why an alloy such as Ti–8 %Al–1 %Mo–1 %V is so highly susceptible (see figure 4). Susceptibility in the titanium alloys containing aluminium starts at about 5 % Al, but oxygen and tin contribute to the susceptibility. The 8–1–1 alloy mentioned, which consists of $\alpha + \alpha_2 + \beta$, can be solution treated at about 1000 °C and quenched to produce α' or $\alpha + \alpha'$, with greatly increased resistance to cracking. But as the merits of various titanium alloys were better assessed, many of them such as the 8–1–1 alloy, Ti–7 %Al–2 %Nb–1 %Ta, Ti–5 %Al–5 %Zr–5 %Sn and a few others have become effectively obsolete. The alloy Ti–6 %Al–4 %V, preferably with oxygen less than 0.13 %, and even more preferably less than 0.10 %, has largely replaced the susceptible alloys in critical components for manned vehicles. New alloys of low susceptibility, based on present understanding, are expected to appear.

In the field of aluminium alloys, the texture of wrought alloys is a major tool in containing the stress-corrosion cracking problem. Not only is the textured alloy less susceptible than the equiaxed alloy, but additionally vulnerability is effectively restricted to one direction with respect to the texture, and design may be used to shield the vulnerable direction from sustained stress and corrosive environments. Manganese, chromium and zirconium in Al–Mg–Zn alloys are effective in pinning grain boundaries and thus retaining the texture after heat treatment. Iron and nickel are similarly effective in Al–Cu alloys. The research submersible Aluminaut was made of the Aluminum Association alloy 7079-T651, which is probably the most susceptible of any commercial alloy but which was nevertheless specified for its yield strength. From considerations of stress-corrosion cracking vulnerability, the vehicle was designed to be made of a series of rolled rings, with the vulnerable end-grain tucked under in each ring so as to be inside the hull rather than being washed by seawater. This vehicle had extensive service with no cracking problems. In view of Roberts-Austen's finding that tensile stress is necessary for stress-corrosion cracking to occur, one might reasonably ask why there should be any concern about the problem in the hull of a submerged hull, where the design stresses are compressive. The answer is that most stress-corrosion cracks in aluminium structures are due not to design stresses, but to stresses caused by heat treatment or assembly. In fact, one class of European submarines made of a special steel was scrapped because of the appearance of the stress-corrosion cracking problem due to assembly stresses. The principal metallurgical tool to avoid stress-corrosion cracking in Al–Mg–Zn alloys is overaging to produce the -T73 temper, even though this heat treatment entails some loss in strength.

There is a wide range of susceptibility of various heat-treated alloy steels in seawater. Strength is one factor which affects the degree of susceptibility, but even at a given strength level there is a wide range of susceptibility. The important work at the Welding Institute in Abington has shown that if twinned martensite is present, the $K_{\mathrm{Is.c.c.}}$ is low, and the crack path is along the prior austenite grain boundaries. In less susceptible steels, meaning that cracking occurs only at higher K levels, the crack path is transgranular, often called quasi-cleavage; the fewer large intermetallic particles present, the more resistant is the steel. Hence precipitation hardening steels and maraging steels can be made which are more resistant than tempered martensitic steels, meaning that cracks grow only at still higher K levels, and the fracture mode is one of microvoid growth.

The foregoing is cited as indicative of the body of knowledge, here greatly condensed, which

FIGURE 5. Specimens of the precipitation hardening steel PH 13-8 Mo after being tested. See text for details. Experiments demonstrate that electrochemical factors can override mechanical factors in selecting site of initiation of stress-corrosion cracks. Photograph courtesy of R. L. Newbegin, U.S. Naval Research Laboratory.

(*Facing p.* 242)

FIGURE 6. Consequence of stress-corrosion cracking in a structure having a non-redundant load path. Photograph courtesy of H. C. Burnett, National Bureau of Standards.

FIGURE 7. Stress-corrosion crack (lighter area, upper right hand corner) which triggered the collapse of the Silver Bridge shown in figure 6. The stress-corrosion crack extended downward about 0.075 in (1.9 mm) before changing to brittle fracture. Photograph courtesy of H. C. Burnett, National Bureau of Standards.

has been developed by comparing macroscopic stress-corrosion characteristics with the microstructure (in the most general sense) of the same alloys as elucidated by the many elegant tools available to the modern physical metallurgist. Just as in Dr Rosenhain's day, a division of labour is required, with equally important roles for the physical metallurgist and the macro-mechanical experimentalist.

Electrochemistry

It is easy for the metallurgist or the engineer to assume that control over the physical metallurgy and the mechanics of a stress-corrosion cracking test covers the essentials. But to match professional level metallurgy and mechanics with amateur level corrosion science invites confusion. Note the specimens of the precipitation hardening stainless steel PH 13-8 Mo tested in salt water and shown in figure 5, plate 17. The lowermost specimen was tested at a high K level, which with the participation of the chloride ions ruptured the protective oxide film at the bottom of the precrack and initiated stress-corrosion cracking, which was halted before complete specimen separation. The middle specimen was loaded at a lower K level, and after 1300 h a stress-corrosion crack initiated, but not at the precrack. The initiation was under the wall of the cell which surrounded the central portion of the specimen and contained the salt water. Careful examination of this and of replicate specimens revealed small crevice corrosion pits under the cell wall, which initiated stress-corrosion cracks from an almost smooth surface. Even if these small pits had been as sharp as a fatigue crack, the K level there would have been much lower than at the machined and fatigued notch. The factor which was different in the crevice corrosion pits was that in these stagnant conditions under the cell wall the stainless steel and salt water combined to form hydrochloric acid according to the reaction (Peterson, Lennox & Groover 1970):

$$3Cl^- + Cr^{3+} + 3H_2O = Cr(OH)_3 + 3H^+ + 3Cl^-;$$

$$pH = 1.60 - \tfrac{1}{3}\lg[Cr^{3+}].$$

Thus the low pH in a crevice, due to the hydrolysis of chromium corrosion product, overcame the mechanical disadvantage of the lack of a precrack. The top specimen serves as a test of the effectiveness of electrochemical conditions in crack initiation. Saturated ferric chloride was selected as a convenient way to lower the pH to the range inside an active corrosion pit in the stainless steel, and upon application of the solution to the specimen many cracks initiated immediately from the smooth surface. Hydrochloric acid was found to be equally effective.

The full implications of local solution chemistry to the technological control of stress-corrosion cracking have yet to emerge. But it is abundantly clear that both the understanding and the control of stress-corrosion cracking requires professional level attention from the metallurgist, the applied mechanics specialist, and the corrosion scientist-engineer.

Non-redundant load path designs

There is a class of design which requires a special level of sophistication in knowledge about stress-corrosion cracking and fracture toughness. This class of design is illustrated by the 'Silver Bridge' (figure 6, plate 18), the Hinckley Point A steam turbine (Kalderon 1972), and the rocket motor chamber mentioned above. A study of the collapse of the bridge (Bennett & Harold Mindlin 1973) attributed the failure to the following sequence: A stress-corrosion crack initiated in the eye of an eyebar of which the bridge was built; the stress corrosion may have been due to sulphur compounds in the air. The bridge was built long before the development of fracture mechanics or its application to stress-corrosion cracking, but an after-the-fact test of the steel indicated that in a sulphide solution the critical flaw size to initiate stress-corrosion cracking would be about 0.008 in (0.2 mm), credibly present in the bridge. The fracture toughness indicated that a flaw to set off brittle fracture would need be no larger than 0.05–0.08 in (1.2–2 mm) deep (spread due to inhomogeneous strength). The size of the fatal stress-corrosion crack was actually about 0.075 in (1.9 mm) deep (figure 7, plate 18). Note the generic similarities in the three structures cited: (1) There was stress-corrosion cracking in mild environments, (2) in areas in which the stress was elevated locally (by the eye of the eyebar, by the bore plus keyway of the turbine, and probably by slight out-of-roundness of the chamber), (3) in steels having low crack tolerance, (4) in non-redundant load path designs. The lesson is that if the designer cannot afford a failure, he is denied non-redundant load path designs unless he has complete control over both stress-corrosion cracking and fracture toughness.

Conclusions

(1) The specifications, standards, and codes needed to control the problem of stress-corrosion cracking require characterization by macroscopic tests. Fracture mechanics offers the possibility of playing a role in this characterization, though probably other methods may also be both valid and needed in special areas.

(2) Understanding the characteristics of a given alloy and the development of improved alloys by other than random methods require the contributions possible through the techniques of physical metallurgy. The macroscopic testing and the physical metallurgical studies are complementary, not competitive, approaches. Both are essential.

(3) The combination of fracture mechanics and other macroscopic tests with physical metallurgy has produced and continues to produce useful descriptive physical metallurgy knowledge about the relation of structure to stress-corrosion cracking characteristics.

(4) For engineering purposes standardized tests are required which can be called out in specifications after the validity of the standardized tests is established.

(5) In the development and validation of standard stress-corrosion cracking tests, the contributions of professional corrosion science are needed to minimize ugly surprises and confusion. In trying to achieve valid standard tests we could do worse than to observe Francis Bacon's maxim, 'Truth will sooner come out of error than from confusion'.

(6) Non-redundant load path designs require exacting attention to fracture toughness and stress-corrosion cracking unless brittle fracture can be either excluded or tolerated.

The preparation of the final manuscript was supported by funds from the U.S. Naval Sea Systems Command, PHM Ship Acquisition Project.

References (Brown)

Bennett, J. A. & Mindlin, Harold 1973 *J. Testing and Evaluation*, **1**, 152.
Blackburn, M. J., Smyrl, W. H. & Feeny, J. A. 1972 *Stress corrosion cracking in high strength steels and in aluminum and titanium alloys* (ed. B. F. Brown). Washington, D.C.: Naval Research Laboratory.
Brown, B. F. 1970 *J. Materials* **5**, 786.
Judy, R. W. Jr. & Dahlberg, E. P. 1968 Report of N.R.L. progress, May 1968, p. 30.
Kalderon, D. 1972 *Proc. Instn mech. Engrs* **186**, 341.
Peterson, M. H., Lennox, T. J. Jr. & Groover, R. E. 1970 *Materials Protection* **9**, 23.
Speidel, M. O. 1971 *The theory of stress corrosion cracking in alloys* (ed. J. C. Scully), p. 289. Brussels: N.A.T.O.
Staehle, R. W. 1969 *Fundamental aspects of stress corrosion cracking* (eds. R. W. Staehle, A. J. Eorty & D. van Rooyen), p. 7. Houston, Texas: N.A.C.E.

2.8 Weldability and toughness specifications for structural steels in Japan – with special reference to WES-135 and -136

By H. Kihara†, T. Kanazawa‡, and H. Tamura§

† *The Japan Welding Engineering Society, Chiyoda-ku, Tokyo* 101, *Japan*

‡ *University of Tokyo, Bunkyo-ku, Tokyo* 113, *Japan*

§ *Tokyo Institute of Technology, Megro-ku, Tokyo* 152, *Japan*

During the last decade, the specifications, WES-135 and -136 played important guiding roles for developing and standardizing weldable high strength steels and structural steels for low temperature applications.

In WES-135 which was established in September 1970, equivalent carbon content, C_{eq}, is adopted as an indication of the weld cracking susceptibility for high strength steels. In order to accurately estimate the cracking susceptibility, a new parameter, P_{cm}, is adopted as an indication of the carbon equivalent since March 1971.

On the other hand, the toughness requirements specified in WES-136 are based on the correlation between small-scale tests and large-scale brittle crack propagation arrest tests as expressed in terms of K_c-value. In this standard, steel plates are classified into two classes of G and A, where G is generally accepted for welded structures where hazard of brittle fracture is anticipated and A is used for arresting a propagating crack. In order to establish revised toughness requirements for G class steels, comprehensive cooperative research works have been conducted since 1971.

In this paper, the descriptions are focused on the weldability and toughness requirements of WES-135 and -136, their connexions with other standards, and the points under continuous examination.

Introduction

The present form of WES-135 'Specification for high strength steel plates for high pressure vessels and other welded structures' was established in September 1970 as an extensive revision of the 1964 draft. In this specification, equivalent carbon content, C_{eq}, is adopted as an indication of the weld cracking susceptibility for high strength steels. In order to accurately estimate the susceptibility, however, a new parameter, P_{cm}, was extensively examined by the JC and the P_c committees of the Japan Welding Engineering Society and this parameter P_{cm} is adopted as an alternative indication of the carbon equivalent since March 1973.

On the other hand, the current WES-136 'Evaluation criterion of structural steels for low temperature applications', a separate specification of the Society, was revised in March 1973 as an extensive revision of the 1964 draft. The toughness requirements specified in WES-135 are established on the same concept as WES-136. The specifications for toughness are based on the correlation between small-scale tests and large-scale propagation arrest tests as expressed in terms of K_c value. In this standard, steel plates are classified into two classes of G and A, where G is generally usable for welded structures where hazard of brittle fracture is anticipated and A is used for arresting a propagating crack. In ordinary structural steels, however, in-service failures of steel structures are more often caused by fracture initiation from pre-existing crack under large scale yielding. Consequently, G class steels in this specification should be revised to account for fracture under large scale yielding. In order to establish revised toughness requirements for G class steels, the TM committee of the Japan Welding Engineering Society has conducted comprehensive cooperative research works since 1971 and the conclusions of this committee is forthcoming soon.

International introductions of WES-135 and -136 were already presented by Kihara *et al.* at the Annual Assembly of the International Institute of Welding in 1964, on the original edition. In this paper, therefore, the descriptions will be focused on the weldability and toughness requirements of WES-135 and -136, their connexion with other standards, and the points under continuous examination.

Weldability specifications for high strength steels for welded structures

Requirements for equivalent carbon content and maximum weld hardness

It goes without saying that the most important factor to be considered in estimating the weldability of high strength steels is weld cracking susceptibility. In WES-135-1970, cracking test which is practically difficult to perform is omitted and equivalent carbon content and maximum weld hardness are specified in relation to weld cracking.

For the purpose of introducing an adequate alloying composition for the weldability in the specification of steels, equivalent carbon content would be most convenient. This is recommendable in view of the fact that high strength steels are still in the process of being developed and there is a possibility of new grades of steels containing various alloying elements in various combinations appearing here-after.

Many formulae for equivalent carbon content have been proposed as an indication of weldability of steels. As a formula of equivalent carbon content for maximum weld hardness, the following one, which has been well-proven in Japan, is adopted in WES-135;

$$C_{eq} = C + Si/24 + Mn/6 + Ni/40 + Cr/5 + Mo/4 + V/14\ (\%) \qquad (1)$$

Maximum hardness of heat-affected zone in weldment is intimately related to weld cracking and loss of weld ductility. Thus a weld cracking test is often substituted by the measurement of maximum hardness as a simple method of evaluating the weldability of steels. Generally speaking, as maximum weld hardness and martensitic structure increase, cold cracking becomes more liable to happen. For example, Mn–Si steels with $H_{max} < 350$ Vickers are supposed to be weldable without preheating.

Maximum weld hardness which indicates the critical limit for weld cracking tends to increase somewhat with increase of the strength of steel. When various grades of steels are involved, the relation between maximum weld hardness and cracking becomes blurred, making it difficult to estimate weld cracking susceptibility in terms of maximum weld hardness alone.

Nevertheless, it remains true that decreased maximum weld hardness would be desirable for prevention of weld cracking. Considering that higher preheating temperature is adopted for higher strength steels and that welding electrodes which keep the hydrogen content in weldment at lower level are commonly used, WES-135 stands on the position that the upper limit of maximum weld hardness together with strength level of steels may be specified at somewhat higher values.

The WES values of equivalent carbon content and maximum weld hardness thus specified for different steel grades are summarized in table 1 together with V-notch Charpy and the other specified values.

TABLE 1. SUMMARY OF SPECIFICATION FOR HIGH STRENGTH STEEL PLATES FOR HIGH PRESSURE VESSELS AND OTHER WELDED STRUCTURES (WES-136-1970)

grade	heat treatment†	chemical composition (%) under *C*	*P*	*S*	WES carbon equivalent (%) under	maximum hardness 10 kg Vickers under	proof stress kgf/mm² over	tensile strength kgf/mm²	V Charpy§ testing temp. °C	absorbed energy kgf m over
HW 36	AR, N or QT	0.20	0.035	0.040	(*c*)‡ 0.47	390	36	53 ~ 65	+15	4.8
									0	4.8
									−5	4.8
HW 40	AR, N or QT	0.20	0.035	0.040	(*c*) 0.49	400	40	57 ~ 69	+15	4.8
									0	4.8
									−5	4.8
HW 45	AR, N or QT	0.18	0.035	0.040	(*a*) 0.44	380	45	60 ~ 72	+10	4.8
					(*b*) 0.48	400			−5	4.8
					(*c*) 0.50	410			−10	4.8
HW 50	AR, N or QT	0.18	0.035	0.040	(*a*) 0.46	390	50	62 ~ 74	+5	4.8
					(*b*) 0.50	400			−10	4.8
					(*c*) 0.54	420			−15	4.8
HW 56	QT	0.18	0.035	0.040	(*a*) 0.53	420	56	68 ~ 82	+5	4.8
									−10	4.8
									−15	4.8
HW 63	QT	0.18	0.035	0.040	(*a*) 0.57	435	63	72 ~ 86	0	4.0
									−15	4.0
									−20	4.0
HW 70	QT	0.18	0.030	0.035	(*a*) 0.63	450	70	80 ~ 95	−5	3.6
									−15	3.6
									−20	3.6
HW 80	QT	0.18	0.030	0.035	(*a*) 0.69	470	80	88 ~ 105	−5	2.8
									−20	2.8
									−25	2.8
HW 90	QT	0.18	0.030	0.035	(*a*) 0.76	480	90	97 ~ 115	−10	2.8
									−25	2.8
									−30	2.8

† AR, as rolled; N, normalized; QT, quenched and tempered.
‡ (*a*) QT steel 50 mm under; (*b*) QT steel 50 ~ 75 mm; (*c*) others.
§ Upper: plate thickness 13 ~ 20 mm; middle: plate thickness 20 ~ 32 mm; lower: plate thickness 32 mm over.

Figure 1 illustrates the relation, based on the latest results of steel production, between equivalent carbon content and maximum heat affected zone (h.a.z.) hardness in steel plates. Maximum hardness increases linearly with equivalent carbon content C_{eq} but it tends to saturate above $C_{eq} > 0.50\%$.

Requirements for weld cracking susceptibility

As pointed out in the preceding section, regulation for weld cracking susceptibility is vitally important in the specification of weldable high strength steels. It is well known that there are hot crackings and cold crackings depending on the temperature of their occurrence. For example, end cracking in one-side welding and microcracking of high strength steels, etc., fall under the category of hot crackings. Thus WES-135, considering the prevention of hot crackings, specifies low contents of *C*, *P* and *S*.

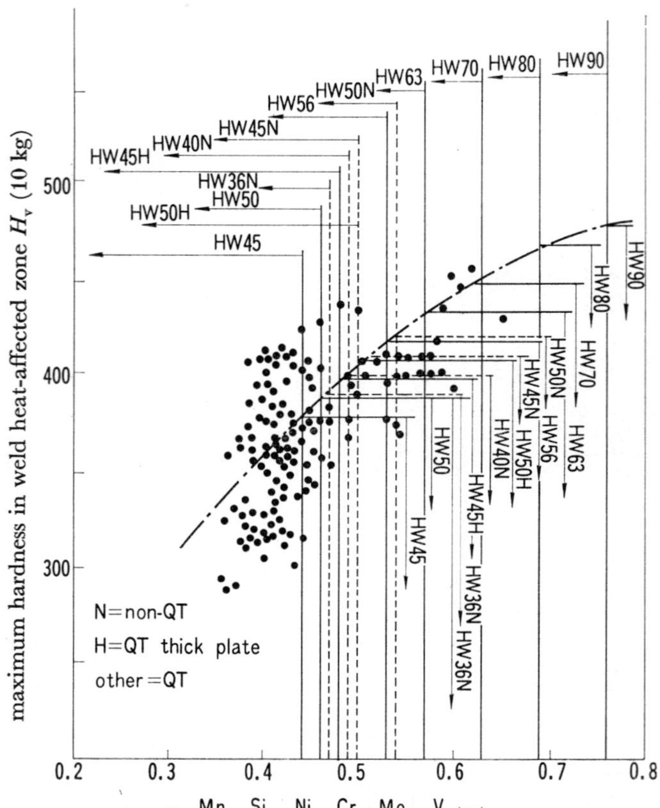

$$C_{eq} = C + \frac{Mn}{6} + \frac{Si}{24} + \frac{Ni}{40} + \frac{Cr}{5} + \frac{Mo}{4} + \frac{V}{14}\ (\%)$$

FIGURE 1. Relation between equivalent carbon content, C_{eq}, and maximum hardness (10 kg V.h.n.) in heat affected zone.

Since the weld crackings occurring in welded structures of high strength steels are predominantly root cracking and toe cracking, however, it is strongly demanded that some regulations for the prevention of cold crackings should be considered in the specification of high strength steels.

Most of the conventional formulae for equivalent carbon content, including the formula (1), are intended for estimation of h.a.z. hardenability. Meanwhile, cold crackings in welding depend not only on the h.a.z. hardness but also on the other factors such as welding conditions, hydrogen content and restraint. Therefore it is desirable to work out a formula which can give a direct measure for cold cracking susceptibility.

Lately, based on the results of oblique Y-slit cracking tests at room temperature for about 200 charges of various kinds of steel plates (from mild steel up to 100 kgf/mm² class high strength steel) with various thickness (from 19 mm up to 50 mm). With use of low hydrogen electrode, Ito & Bessyo (1968) have proposed a new weld cracking susceptibility parameter P_c including hydrogen content of weld metal and plate thickness as follows:

$$P_c = C + Si/30 + Mn/20 + Cu/20 + Ni/60 + Cr/20 + Mo/15 + V/10 + 5B + t/600 + H/60\ (\%), \qquad (2)$$

where H is the hydrogen content of weld metal (cm^3/100 g) and t, the plate thickness (mm).

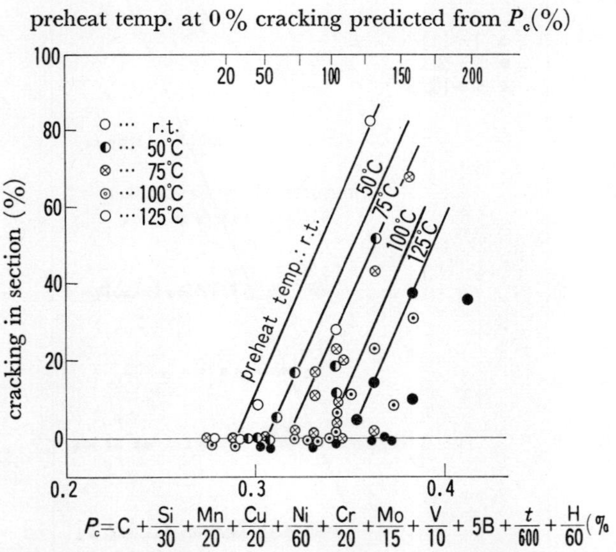

FIGURE 2. Relation between cracking percentage in section of oblique Y-slit cracking test and cracking parameter, P_c (plate thickness 16 ~ 40 mm).

A parameter considering only the term for chemical composition, P_{cm}, which is called weld cracking parameter for chemical composition, may be given as follows:

$$P_{cm} = C + Si/30 + Mn/20 + Cu/20 + Ni/60 + Cr/20 + Mo/15 + V/10 + 5B\ (\%). \qquad (3)$$

The advisability of introducing this P_c value into WES-135 has been discussed for about two years by the JC committee of the Japan Welding Engineering Society. Thereby it has been checked whether or not the P_c value which has been established with steel plates produced by one particular steel maker can be applied to the products of other makers.

The relation between cracking percentage in section of oblique Y-slit cracking test and measured values of weld cracking susceptibility, P_c, under various preheating temperatures are shown in figure 2. From this figure preheating temperature to prevent weld crackings for various kinds of structural steels may be estimated from calculated P_c values.

Summarizing the results of numerous cracking tests carried out by the JC committee, the relation between P_c values and preheating temperature is shown in figure 3. From this figure the relation between preheating temperature, T, and P_c value can be expressed by:

$$T/°C = 1440P_c - 392. \qquad (4)$$

Since it is confirmed that P_c value could be a good measure of weld cracking susceptibility for high strength steels, the P_c committee was organized to introduce P_c values into WES standard as an alternative specification for WES-135. In this specification, weld cracking parameter for steel composition, P_{cm}, given by formula (3) has been adopted instead of P_c, so that standard can be specified only by chemical composition of steels. Assuming that plate thickness is 50 mm for all grades of steels, and, based on the statistically measured values, hydrogen content is 2.0 $cm^3/100$ g for HW 45 ~ 63 and 1.4 $cm^3/100$ g for HW 70 ~ 90, the sum of the terms of plate thickness, $t/600$, and of hydrogen content, H/60, in formula (2) are given in table 2 as the value of α for each grade of steels.

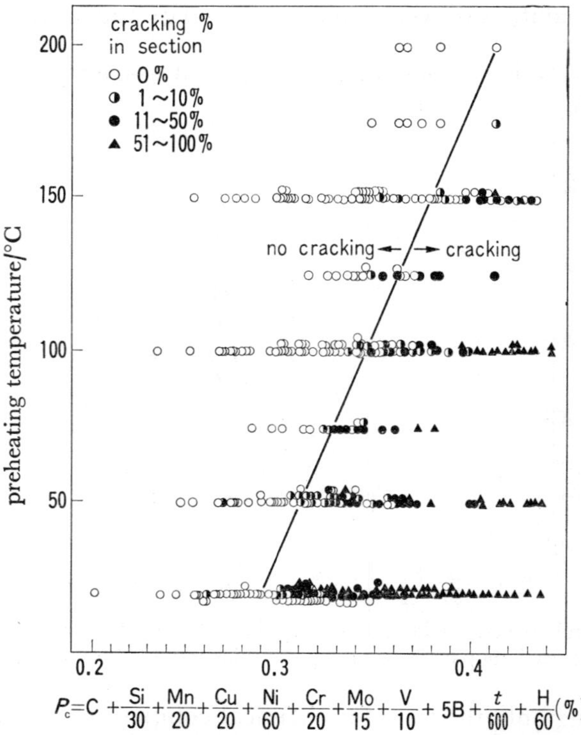

FIGURE 3. Relation between cracking parameter, P_c, and preheating temperature (the JC committee, plate thickness 16 ~ 50 mm).

Meanwhile, the lowest preheating temperature to prevent cracking in oblique Y-slit test, referring to the conventional J.L.P.A. storage tank standard, has been established for each grade of steels and the value is substituted into formula (4) to decide P_c value. By subtracting α from P_c value P_{cm} has been specified for each grade of steels as given in table 2.

TABLE 2. SPECIFICATIONS OF P_{cm} VALUE

grade	P_c(%)	constant α(%)	P_{cm}(%)
HW 45	0.39	0.11	0.28
HW 50	0.39	0.11	0.28
HW 56	0.41	0.11	0.30
HW 63	0.41	0.11	0.30
HW 70	0.43	0.10	0.33
HW 80	0.45	0.10	0.35
HW 90	0.46	0.10	0.36

The specification in table 2 can be applied as an alternative rule upon the request from the customer until sufficient data can be collected, and thereby either the value of C_{eq} or of P_{cm} specified for quenched and tempered steel has only to be satisfied.

The applicable range of P_{cm} covers structural steel plates with proof stress more than 45 kgf/mm^2 and plate thickness less than 50 mm as specified under WES-135; HW 36 and HW 40 are omitted because of their scanty data as quenched and tempered steels.

In this manner evaluation of weld cracking susceptibility evaluated according to this alternative rule from the oblique Y-slit cracking test of quenched and tempered steels has been established. Cracking in actual structures, however, is largely affected by details of welding procedure and restraint; and therefore it is needless to say that these factors should be additionally taken into account in the practical welding fabrication. Elaborate study on this point has been conducted by the Society of Steel Construction of Japan (JSSC 1972).

Toughness specifications for high strength steels for welded structures

Requirements for toughness specified in WES-136

When a brittle crack originated from a notch in a test piece is arrested at the length, C, under the constant tensile stress, σ, due to the change of temperature distribution in the crack propagation direction, the following formula, neglecting the effect of finite width of test piece, will be obtained as the results of double tension tests for various kinds of structural steels:

$$K_c = \sigma\sqrt{(\pi c)} = K_0 \exp(-k/T) \tag{5}$$

where K_0 and k are material constants and T, the temperature at crack arresting point (K).

For the sake of simplicity, it is assumed that in formula (5) K_0 alone comes under the mechanical effect due to plate thickness. This assumption may be justified in view of the fact that the results of double tension tests on different test pieces from the same material machined to various thickness turn out nearly parallel to one another, when they are plotted in terms of lg K_c and $1/T$. Taking 30 mm as the standard plate thickness, K_c value for an arbitrary plate thickness, t, may be expressed as follows:

$$K_c(t,\, T) = \sigma\sqrt{(\pi c)} = f(t)K_0(30)\exp(-k/T), \tag{6}$$

where $f(t)$ is the plate thickness factor and $K_0(30)$, K_0 for 30 mm plate thickness.

As a small size test, pressed-notch Charpy test which gives fracture surface appearance of brittle fracture arresting type may be expected to have relatively good correlation with double tension test. Therefore, using the fracture appearance transition temperature, ${}_pT_c$, in pressed-notch Charpy test, formula (6) may be re-written as follows:

$$\begin{aligned} K_c(t,\, T) &= \sigma\sqrt{(\pi c)} = f(t)K_0(30)\exp(-k/{}_pT_c)\exp(k/{}_pT_c - k/T) \\ &= f(t)K_c(30,\, {}_pT_c)\exp(k/{}_pT_c - k/T) \end{aligned} \tag{7}$$

where $K_c(30,\, {}_pT_c)$ is K_c for 30 mm plate thickness at temperature, ${}_pT_c$(K).

Experimental investigations for various kinds of steels has been conducted to obtain the correlations between k, $K_c(30,\, {}_pT_c)$ and ${}_pT_c$. The following experimental formulae are adopted as the conclusions:

$$k = \begin{cases} 3100 & \text{for } 273\text{ K} \leqslant {}_pT_c \\ 3100 + 22.5({}_pT_c - 273) & \text{for } 233\text{ K} \leqslant {}_pT_c \leqslant 273\text{ K} \\ 2680 + 12({}_pT_c - 273) & \text{for } 77\text{ K} \leqslant {}_pT_c \leqslant 233\text{ K}, \end{cases} \tag{8}$$

$$K_c(30, {}_pT_c) = \begin{cases} 260 & \text{for } 273\ \text{K} \leqslant {}_pT_c \\ 260 + 2.75({}_pT_c - 273) & \text{for } 233\ \text{K} \leqslant {}_pT_c \leqslant 273\ \text{K} \\ 150 & \text{for } 77\ \text{K} \leqslant {}_pT_c \leqslant 233\ \text{K}. \end{cases} \tag{9}$$

Based on the TM committee organized by the Japan Welding Engineering Society, the present WES-136 has been revised in March 1973 to extend the applicable range of plate thickness from 6 mm ~ 50 mm to 6 mm ~ 100 mm. Figure 4 shows formula (8) still holds up to 100 mm thick plate. In this figure, k_{cal} is the calculated k value from formula (8) with the experimentally obtained ${}_pT_c$, and k_{exp} means the k value obtained by double tension test.

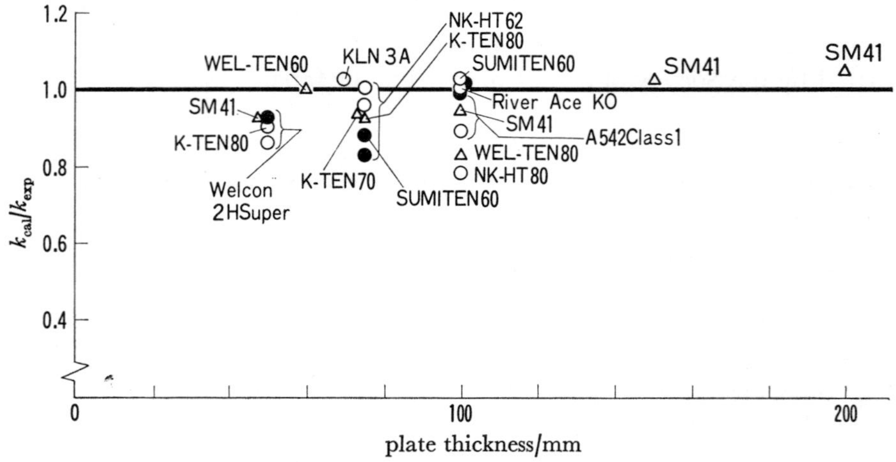

FIGURE 4. Relation between k value and plate thickness (WES-136-1973). k_{cal}, calculated by using the value of ${}_pT_c$ at $\frac{1}{4}t$; ○, top side of ingot; ●, bottom side of ingot; △, unknown whether top or bottom.

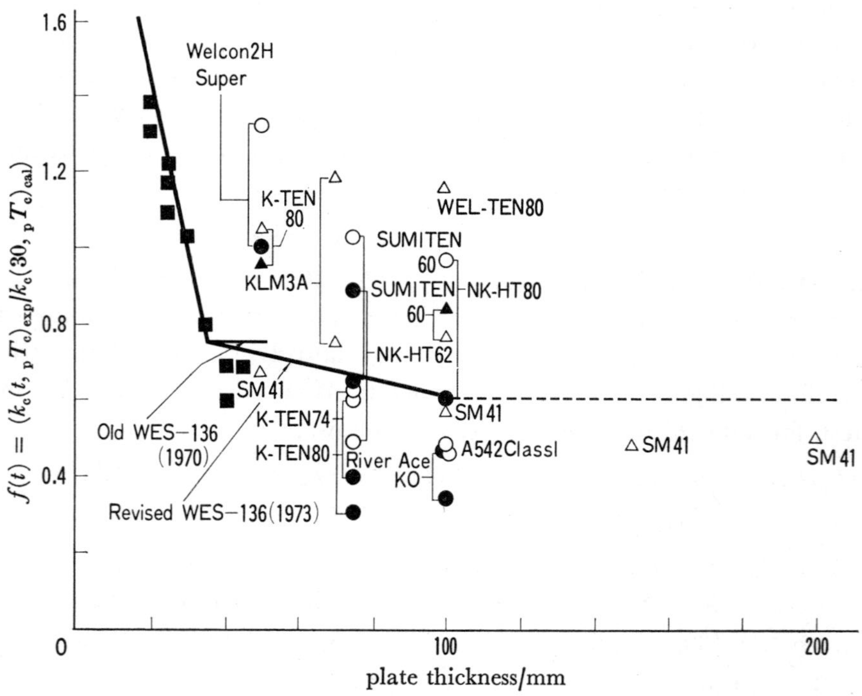

FIGURE 5. Relation between plate thickness factor, $f(t)$ and plate thickness (WES-136-1973). k_c $(30, {}_pK_c)_{cal}$ calculated by using the value of ${}_pT_c$ at $\frac{1}{4}t$. ○, from ${}_vT_{rs}$; ●, from ${}_pT_c$ (top side of ingot). △, from ${}_vT_{rs}$; ▲, from ${}_pT_c$ (unknown whether top or bottom side of ingot). ■, reproduced from WES-136(1970).

Putting T equal to ${}_{\mathrm{p}}T_{\mathrm{c}}$ in formula (7), plate thickness factor, $f(t)$, may be expressed by the following formula:

$$f(t) = K_{\mathrm{c}}(\mathrm{t}, {}_{\mathrm{p}}T_{\mathrm{c}})/K_{\mathrm{c}}(30, {}_{\mathrm{p}}T_{\mathrm{c}}). \tag{10}$$

Figure 5 shows the relation between $f(t)$ and plate thickness. In this figure, $K_{\mathrm{c}}(30, {}_{\mathrm{p}}T_{\mathrm{c}})_{\mathrm{cal}}$ means the calculated value from formula (9) using the experimentally obtained ${}_{\mathrm{p}}T_{\mathrm{c}}$ or ${}_{\mathrm{v}}T_{\mathrm{rs}}$†, and $K_{\mathrm{c}}(t, {}_{\mathrm{p}}T_{\mathrm{c}})_{\mathrm{exp}}$ is the value obtained from double tension test. From these results, plate thickness factor, $f(t)$, is proposed as follows:

$$f(t) = \begin{cases} 1-0.05(t-30) & \text{for} \quad t \leqslant 35 \text{ mm} \\ 54/65-3t/1300 & \text{for} \quad 35 \text{ mm} \leqslant t \leqslant 100 \text{ mm}. \end{cases} \tag{11}$$

In formula (7), σ and T shall be considered as the service stress and the lowest service temperature for the welded structure concerned. From formula (7) together with formulae (8), (9) and (11), we can calculate ${}_{\mathrm{p}}T_{\mathrm{c}}$ as the required toughness for the material used, assuming an appropriate value for arrested crack length. In WES-136, the following values are adopted as the most appropriate ones; assumed crack length of 10 mm (total length 20 mm) for class G steel and 100 mm (total length 200 mm) for class A steel.

Adoption of V-notch Charpy instead of pressed-notch Charpy test in WES-136 is based on the following: (1) pressed-notch Charpy test cannot yet be recognized as a universal testing method; (2) V-notch Charpy is widely adopted in the standards of various countries as well as Japan, etc.

Correlation between fracture appearance transition temperature of pressed-notch Charpy, ${}_{\mathrm{p}}T_{\mathrm{c}}$, and that of V-notch Charpy, ${}_{\mathrm{v}}T_{\mathrm{rs}}$, is established as follows:

$${}_{\mathrm{p}}T_{\mathrm{c}}(^{\circ}\mathrm{C}) = \begin{cases} {}_{\mathrm{v}}T_{\mathrm{rs}}(^{\circ}\mathrm{C})+18 & \text{for} \quad -40\,^{\circ}\mathrm{C}\,(233\text{ K}) \leqslant {}_{\mathrm{p}}T_{\mathrm{c}} \\ 1.13\,{}_{\mathrm{v}}T_{\mathrm{rs}}(^{\circ}\mathrm{C})+25.5 & \text{for} \quad -196\,^{\circ}\mathrm{C}\,(77\text{ K}) \leqslant {}_{\mathrm{p}}T_{\mathrm{c}} \leqslant -40\,^{\circ}\mathrm{C}\,(233\text{ K}) \\ {}_{\mathrm{v}}T_{\mathrm{rs}}(^{\circ}\mathrm{C}) & \text{for} \quad {}_{\mathrm{p}}T_{\mathrm{c}} = -196\,^{\circ}\mathrm{C}\,(77\text{ K}). \end{cases} \tag{12}$$

Therefore, it has been decided to use formula (12) for conversion of ${}_{\mathrm{p}}T_{\mathrm{c}}$ to ${}_{\mathrm{v}}T_{\mathrm{rs}}$. But, in order to determine ${}_{\mathrm{v}}T_{\mathrm{rs}}$, the percentage shear of fractured surface must be calculated and here lies a great possibility of a personal error slipping in. Thus it would be more appropriate to adopt absorbed energy as the criterion. From the fact that the absorbed energy at ${}_{\mathrm{v}}T_{\mathrm{rs}}$ is nearly equal to a half of the maximum absorbed energy, the energy transition temperature, ${}_{\mathrm{v}}T_{\mathrm{rs}}$, (absorbed energy at the temperature is equal to a half of the maximum absorbed energy) has been adopted for the present WES-136.

From what has been explained in the above, it follows that if ${}_{\mathrm{v}}T_{\mathrm{rs}}$ is determined from ${}_{\mathrm{p}}T_{\mathrm{c}}$ corresponding to the required K_{c} value, V-notch Charpy test is performed at this temperature (${}_{\mathrm{v}}T_{\mathrm{rs}}$) and the obtained absorbed energy exceeds 50% of maximum absorbed energy value, the material may be considered to satisfy the required K_{c} value.

It goes without saying that, in real structure, the assumed crack length mentioned above, C, should be related to the width of the region of tensile welding residual stress, the extent of the region where high stress concentration by structural discontinuity exists, or the width of weld bead, etc. Especially for class G steels, the value of C may be greatly influenced by these factors.

† ${}_{\mathrm{v}}T_{\mathrm{rs}}$ is the fracture appearance transition temperature of V-notch Charpy test and relation between ${}_{\mathrm{p}}T_{\mathrm{c}}$ and ${}_{\mathrm{v}}T_{\mathrm{rs}}$ is given by formula (12).

Studies on toughness requirement for class G steels in WES-136

As stated in the introduction, the TM and the BE committees have conducted comprehensive cooperative research work since 1971 in order to establish revised toughness requirements for class G steels based on c.o.d. concept.

One of the difficulties met by these committees is how to correlate c.o.d. data with V-notch Charpy test. The most appropriate procedure is now being searched for by trial and error methods. So far several tentative proposals have been discussed, one of which will be briefly described below (Kanazawa, Machida & Miyata 1974).

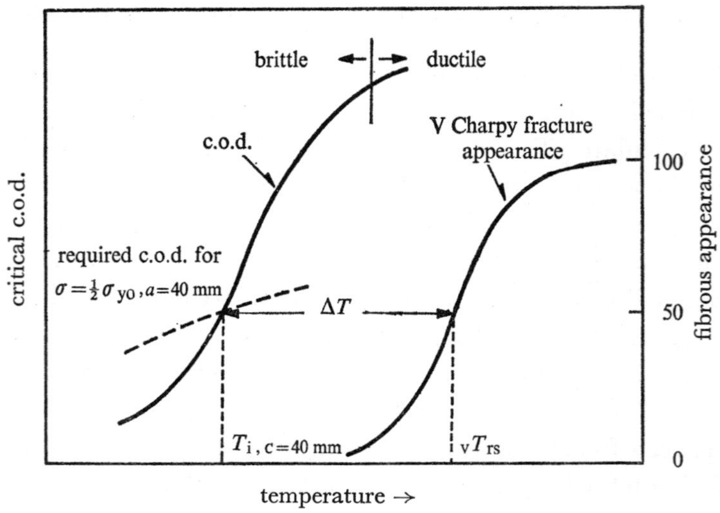

FIGURE 6. Schematic representation of critical c.o.d. and V-notch Charpy transition.

Usually, transition phenomenon of the critical c.o.d. does not occur in the same temperature range as that of V-notch Charpy test. Though the definition of transition temperature, T_i, is rather arbitrary in the case of c.o.d. test, we can define it as a temperature corresponding to a certain level of c.o.d. value. Tentatively this level of c.o.d. was taken as the value calculated through the Dugdale model with design stress of a half of yield stress at room temperature and half crack length of 40 mm.

Taking the fracture appearance transition temperature of V-notch Charpy test, $_vT_{rs}$, the shift of transition temperature ΔT, as shown schematically in figure 6, was investigated in detail. Considering the fact that the critical c.o.d. should be affected by plate thickness and the effect of strain rate as well as notch acuity depends on the strength level of steels, ΔT may take the form given by

$$\Delta T = T_{i,\,c=40\,\mathrm{mm}} - {}_vT_{rs} = f\,(\text{plate thickness, strength}). \tag{13}$$

After trial and error procedure with due allowances for available data, it was found that formula (13) was not good enough but f should be a function of $_vT_{rs}$ besides plate thickness and strength level of steels. In place of formula (13), the finally obtained empirical relation is expressed as follows:

$$\Delta T/\mathrm{K} = (0.0047\sigma_{y0} - 0.46)\,{}_vT_{rs} + 11.4\sqrt{t} - 102 \tag{14}$$

and thus

$$T_{i,\,c=40\,\mathrm{mm}} = (0.0047\sigma_{y0} + 0.54)\,{}_vT_{rs} + 11.4\sqrt{t} - 102, \tag{15}$$

where σ_{y0} is the yield strength at room temperature and t, the plate thickness (mm).

The temperatures, $T_{\mathrm{i,\,c=40\,mm}}$, defined from c.o.d. tests compare well with those predicted from formula (15) as shown in figure 7, which includes mild steel up to 80 kgf/mm² class high strength steels (parent material as well as welded fusion line).

When the welding residual stress and/or structural stress concentration does not exist, $T_{\mathrm{i,\,c=40\,mm}}$ is considered to be a temperature where the material is tough enough to tolerate a flaw of 80 mm in length at stress level of $\frac{1}{2}\sigma_{\mathrm{y0}}$. When the design stress is taken as $\frac{2}{3}\sigma_{\mathrm{y0}}$ through severer defect control, the above requirement corresponds to the maximum allowable flaw size of 40 mm in length.

Making conservative allowances for the scatter of data in figure 7, examples of the relation between the minimum service temperature and the required V-notch Charpy transition temperature are shown in figure 8 with varied plate thickness and strength level of the materials.

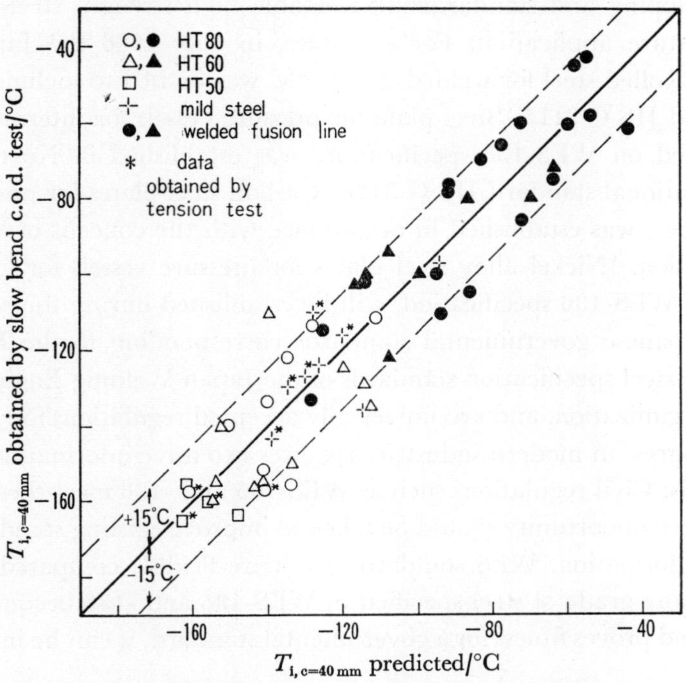

FIGURE 7. Prediction of $T_{\mathrm{i,\,c=40\,mm}}$ from Charpy impact test.

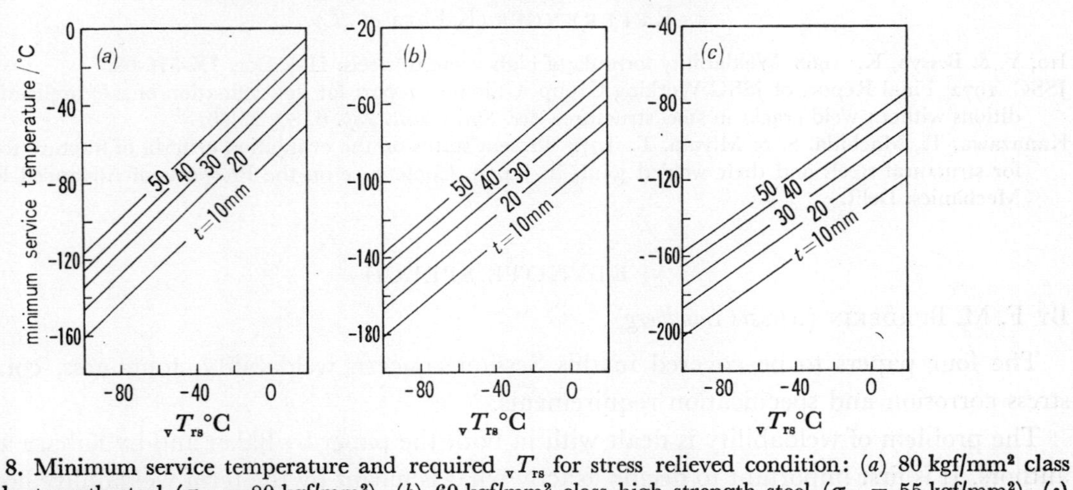

FIGURE 8. Minimum service temperature and required $_{\mathrm{v}}T_{\mathrm{rs}}$ for stress relieved condition: (*a*) 80 kgf/mm² class high strength steel (σ_{y0} = 80 kgf/mm²), (*b*) 60 kgf/mm² class high strength steel (σ_{y0} = 55 kgf/mm²), (*c*) mild steel (σ_{y0} = 30 kgf/mm²).

The current WES-136 has unsatisfactory aspect that it takes no account for brittle fracture initiation characteristics. To date, the c.o.d. approach or other related concept seems most promising but it has still some difficulties in providing a useful and simple engineering design basis.

But it is necessary to find better expressions correlating c.o.d. test with V-notch Charpy test and to establish appropriate procedure to take into account the parameters involved in the fabrication of welded structure.

Concluding remarks

During the last decade, the specifications, WES-135 and WES-136 have played important guiding roles for developing and standardizing weldable high strength steels and structural steels for low temperature application. For examples, in July 1966 the Japanese national standard JIS G 3106 'Rolled steel for welded structures' was revised to include WES-135 and a new national standard JIS G 3115 'Steel plate for pressure vessels for intermediate temperature service', also based on WES-135 specification, was established in November 1968. In January 1972 a new national standard JIS G 3126 'Carbon steel plates for pressure vessels for low temperature service', was established in accordance with the concept of WES-136 and a new national specification, 'Nickel alloy steel plates for pressure vessels for low temperature service', also based on WES-136 specification, will be established during this year.

The JIS are the Japanese governmental standards corresponding to the BS of the U.K. WES-135 and -136 are steel specification standards of the Japan Welding Engineering Society, a non-governmental organization, and are universally accepted regulations for Japanese industrial circles. Rapid progress in modern industries requires extensive information in addition to governmental standards. Civil regulations such as WES-135 and -136 meet these needs. It goes without saying that every opportunity should be taken to improve existing standards by making use of all up-to-date information. WES standards are more flexible compared to the governmental regulations. If any grade of steel specified in WES-135 and -136 becomes 'popular' in practical application and proves fitness for a governmental standard, it can be incorporated into JIS standards.

References (Kihara *et al.*)

Ito, Y. & Bessyo, K. 1968 Weldability formula of high strength steels. IIW Doc. IX-576-68.

JSSC 1972 Final Report of JSSC Working Group, Guidance report for determination of safe preheating conditions without weld cracks in steel structures. *Soc. Steel Constr. Jap.* **8**, 80, 22/50.

Kanazawa, T., Machida, S. & Miyata, T. 1974 Present status on the evaluation criteria of fracture toughness for structural steels and their welded joints in Japan. Conference on the Prospects of Advanced Fracture Mechanics, Delft.

Keynote speech

By F. M. Burdekin (*Messrs Sandberg*)

The four papers to be covered in this Session concern weldability, toughness, corrosion, stress corrosion and specification requirements.

The problem of weldability is dealt with in both the paper by Baker and by Kihara and coauthors. It is first important to decide exactly what is meant by the term weldability however, I would define weldability as the capability of different materials to be joined by a range of different welding processes with the object of giving satisfactory joints which meet the following objectives:

1. Free from harmful defects.
2. With adequate mechanical properties to carry loads which may be applied, that is with adequate strength and toughness.
3. With adequate resistance to long term environmental effects, e.g. creep, corrosion/stress corrosion, and fatigue.

Fusion welding gives rise to the possibility of a number of problems which may occur in any materials being welded, including the possible occurrence of defects, the presence of material in the weld metal when added fillers are used, changed material in the heat affected zone due to the thermal cycle, and the presence of residual stresses.

The paper by Baker is devoted entirely to weldability and covers a range of different materials, welding processes, and a detailed outline of a number of the problems which may be encountered. The paper by Kihara is concerned with certain aspects of weldability as dealt with in two steel specifications but does not cover the wide range of materials, processes and problems outlined by Baker. Table 3 of Baker's paper gives an excellent summary of the potential welding problems which may occur with different welding processes in different materials and their likely severity. He outlines in his paper the basic mechanism of a number of cracking problems such as solidification cracking, liquation cracking, hydrogen induced delayed cracking, and lamellar tearing. He also gives consideration to the mechanical properties of weld metal and heat affected zones outlining factors which may affect these. It is interesting to note that in table 2 of Baker's paper tentative recommendations are made for specification requirements to avoid lamellar tearing, based on limiting sulphur content, on minimum through thickness ductility in tensile tests, and on ultrasonic testing. As far as I am aware there are no national specifications which provide the option of specifying material with increased resistance to lamellar tearing although of course a number of the major purchasers have been able to negotiate material of tested quality to reduce lamellar tearing problems. It would be interesting to have the comments of the steel suppliers on the effects of implementing an additional specification as recommended in Baker's table 2. I suspect that the answer will be that the implementation of such a recommendation is only a matter of cost and that provided the customer is prepared to pay sufficiently the mills would be able to supply. However an indication of the possible cost implications would be helpful. Also on the problem of lamellar tearing I see no reason in principle why this problem should be a phenomenon confined to steel although Baker's indication in table 3 is that this problem is not normally a problem in other materials. Perhaps it is because the other materials are supplied at a high level of purity, or perhaps it is because the welded joints are not commonly made on the surface of such materials, but it would be useful to have guidance from the non-ferrous metals industry of their experience in this respect.

Both Baker and the Japanese authors deal at some length with the problem of hydrogen induced cold cracking and it is interesting to note the differences in detail between the approaches taken in the different countries. The British approach for carbon manganese steels is to make use of the concept of carbon equivalent, based on the well established formula $C + Mn/6$ etc. Account is taken of the effects of welding process, heat input, pre-heat and thickness on the cooling rate in the heat affected zone, and of hydrogen level in the resultant weld. This approach is now conveniently summarized in the recently published British Standard

BS 5135. By use of this Standard it is possible to derive welding procedures for different materials which would form the basis of a procedure test to confirm satisfactory welding conditions. Where previous experience has shown such conditions to be satisfactoy the procedure test may perhaps be omitted. The Japanese approach as in the previously established versions of specification WES 135 is also based upon a carbon equivalent principle for different steels, and upon controlling the maximum hardness in the heat affected zone to a limiting value. The detailed carbon equivalent formula has minor variations from the formula used in the British approach but this is of no major significance as the resulting total carbon equivalent criteria are adjusted accordingly. In table 1 of the Japanese paper the various steels contained within the WES 135 specification are summarized indicating a range of strengths from 530 to 1150 N/mm^2. The carbon equivalent throughout the range of steels by the Japanese formula range from 0.44 to 0.76 and these steels are apparently available in thicknesses up to 50 mm in all cases and higher in some cases. The problems of welding steels with an ultimate strength of 1000 N/mm^2, with a carbon equivalent of 0.76 raise considerable trepidation unless one is concerned with relatively small components under strictly controlled shop supervision.

In recent times the Japanese authors have been considering a modified approach to the problem of hydrogen induced cold cracking on the following lines. A somewhat different carbon equivalent formula is proposed based upon a parameter for chemical composition called P_{cm}. This attributes different factors to the different elements of the normal carbon equivalent formula and although again the criteria for judging the resultant total is different it would be useful to have some comment on the relative factors adopted for the different elements in respect of their different effects on hardenability. The revised Japanese approach then proposes a new weld cracking susceptibility parameter, P_c, which adds to the compositional parameter P_{cm} terms for the thickness and hydrogen content. This cracking susceptibility parameter is plotted against the percentage cracking observed in a restrained slit cracking test for different pre-heat levels in figures 2 and 3 of the paper. There is no mention of welding processes or of heat input in these two figures however, and some further explanation from the authors is warranted as to how such correlations can be obtained without consideration of welding process and heat input. Presumably the graphs were derived for particular cases but it would be important to clarify their range of applicability. It would also be helpful to have some joint discussion of the relative conclusions implied from the British and Japanese approaches in respect of the level of pre-heat recommended for particular steels and welding conditions.

The Japanese approach to specifications for structural steels is worthy of note. The two specifications described in the Japanese paper are specificatons developed by the Welding Engineering Society but which have now been adopted as national standards under the JIS system. The flexibility offered by the development of standards by a non-government organization such as the Welding Engineering Society is a feature which should merit discussion in other countries.

The toughness specifications in the WES 135 specification are expressed in terms of Charpy V notch energy absorption as summarized in table 1 of the paper. It should be noted that as the strength of the steel increases through the range available, the Charpy energy requirement decreases and even though there are adjustments in the testing temperature the specification offers lower Charpy energy for the higher strength steels. This is in direct conflict to the expectations from fracture mechanics analyses where with higher strength steels used at higher design stresses, a higher level of toughness would be expected for given defect size conditions.

The toughness approach in the WES 136 specification divides steels into two classes, G considered to be generally usable for welded structures where brittle fractures have to be considered, and A for the arrest of a propagating crack. Both grades are said to be based upon crack arrest conditions, in which an extended series of correlations between tests is employed. The general grade is based upon calculations for the toughness necessary to arrest a crack of total length 20 mm, and the A grade upon a toughness to arrest a crack of total length 200 mm, although it is not clear at what stress level these calculations apply. The calculations are based upon an exponential relation of toughness at arrest with temperature, correlated to the fracture appearance transition in a pressed notch Charpy test, further correlated to fracture appearance in a V notch Charpy test, and then finally correlated to an energy absorption transition in the Charpy V notch test. The scatter which occurs in any type of fracture test is notorious, and it must be assumed that the various correlations between tests adopted in the WES 136 specification are based upon mean correlations. The possible effects of this scatter upon the eventual final results of the correlations are a feature which it would appear fruitful to examine statistically and it would be helpful to know whether the authors have carried out such overall statistical analyses.

The Japanese authors indicate in their paper that they are considering altering the specification for the general grade to be based upon a fracture initiation criterion using crack opening displacement concepts, and some introductory ideas on this are presented in the paper.

Turning now to corrosion the paper by Butler & Ison presents an excellent review of the general principles and mechanisms of the corrosion process. This review is based upon the generally accepted electro-chemical theory in which a reductive reaction occurs at the cathode, and metal dissolves at the anode. The various interacting features which may give rise to corrosion cells are pictured in figure 2 of this paper which may be compared with figure 1 of the paper on stress corrosion by Brown where additional complexities are also introduced.

The type of corrosive attack by uniform thinning is rare in metals except under acid attack. Problems more commonly arise by localized attack which may take the form of pitting, of crevices, of selective attack, or in the extreme case of stress corrosion. The features controlling the occurrence of pitting are described with reference to the shape of pits and to preferred growth in respect of crystals and micro-structure. The operation of concentration cells at crevices is described. Selective attack such as graphitization in cast iron, or de-zincification in brass can cause serious loss of mechanical strength without apparent changes in volume being evident. Stress corrosion presents the insidious development of cracks which I shall discuss in detail shortly.

Although design features may be important in some cases in controlling corrosion, and in avoiding cavitation and erosion problems, the choice of materials for particular environments is a most important feature of corrosion control. This immediately presents a compromise in terms of requirements for mechanical properties, corrosion behaviour and cost. The inherent behaviour of particular types of materials in different environments is well established and due note must always be taken of experience in this way. For example the susceptibility to pitting of stainless steels in sea water and the behaviour of dissimilar metals in galvanic contacts are well known. It would be helpful to have clarification of the effect of copper additions on the corrosion behaviour of structural steels, in particular as to what benefit is obtained when such steels are fully immersed as opposed to operating under alternate wet and dry conditions.

The various methods of prevention of corrosion apart from the choice of basic material for the environment amount to either control of the environment chemistry, control of galvanic conditions by cathodic or anodic protection, or surface protection by paint, metal coatings or natural oxides. Since corrosion occurs at a surface condition there would appear to be a strong case for trying to develop materials which have an inherent surface protection for the required environments without necessarily incurring the full cost of such material throughout the thickness. Thus the development of materials with a stable inert surface film, either as a chemical compound of the basic material or bonded to it would appear to be a line for future research and prospects.

The approach to stress corrosion described by Brown is one which he has pioneered. The term $K_{Is.c.c.}$ is virtually synonymous with Brown's name for the early research work in this field which he carried out. An analogy is drawn between the application of fracture mechanics to solving the problems of brittle fracture, and the complexities of the problems of stress corrosion. Brown believes that because of the intricate complexities it is essential to test macroscopic specimens and to seek a macroscopic test to sum up the many contributions of the various features which may be involved.

The $K_{Is.c.c.}$ threshold stress intensity factor for stress corrosion cracking is obviously a valuable method of rating the worst possible performance of different materials in particular environments. It would be interesting to have further examples of the application of the $K_{Is.c.c.}$ numbers by fracture mechanics analyses to practical situations. Figure 4 of Brown's paper indicates critical flaw sizes for a number of titanium alloys in sea water at different strength levels. The lesson which I would draw from the very small defect sizes at the bottom end of the figure would be that I should not contemplate using such materials in such an environment. It would seem from the experience in the aircraft undercarriage field however that this is not necessarily always the case and that it may be possible to take other action to overcome the low threshold susceptibility to stress corrosion cracking. Just as the approaches to prevention of corrosion may be by surface protection, and by cathodic or anodic protection, as well as by initial material selection presumably it is possible to take preventive action against stress corrosion by similar means. Any comments which may be raised in discussion concerning the pitfalls of the alternative approaches of surface protection or galvanic protection would be a useful clarification. With stress corrosion which requires a combination of various features to casue cracking, one of the contributory factors is tensile stress. Since the phenomenon of stress corrosion cracking develops from the surface one would assume that the superposition of compressive residual stresses would be a helpful feature and it would be helpful to learn whether peening of the surface or other means of introducing compressive residual stresses are regarded as satisfactory precautionary measures.

The difficulty of obtaining stress corrosion information within a reasonable time is one which should receive attention in discussion. For example it has been suggested by some research workers that the minimum $K_{Is.c.c.}$ values for high strength steels can be derived relatively quickly by tests in a hydrogen sulphide environment. The relevance of the $K_{Is.c.c.}$ approach to situations where stress corrosion occurs by the hydrogen induced mechanism is another feature worthy of discussion. The significance of growth rates of stress corrosion cracks for practical applications is again an area where discussion is welcomed. While the crack growth rate may well be a useful factor in comparing susceptibility of materials it would seem to be of little comfort to know that the crack will grow to a critical size in 3 years rather than 3 weeks if one is considering a bridge structure or long term structure of indefinite service life. A more general contribution from the floor is invited to discuss the relative merits of the various alternative means of avoiding stress corrosion.

Finally I would like to draw attention to the need to consider the interaction between all the possible various modes of failure which may occur and to try to ensure that a sense of perspective is maintained in considering failure modes. Where one has a number of interacting factors all of which must occur for failure to take place, it would appear that the statistical probability assessment is a valuable line of approach. The application of such an approach to welding problems, brittle fracture, stress corrosion and corrosion appears to be fields worthy of future research.

In conclusion I should like to draw attention to the major contribution which physical metallurgy has made to each of the four subjects which I have discussed in examining and assessing failures and in learning from them to develop new materials and principles to avoid further failure.

Discussion (Chairman C. EDELEANU (*Imperial Chemical Industries*)
T. KANAZAWA (*University of Tokyo*)

First of all, I should like to emphasize that WES 135 is just a material specification for quality control. So, as Burdekin (Keynote) pointed out, it does not contain any statements about the influences of real welding procedure.

It goes without saying that, for application purposes, we should take account of many important factors affecting the behaviour of materials in a real structure. For example, in order to avoid weld cracking in real structures, we must have the critical cooling time determined from weldment cracking parameter, P_w, which is expressed as follows:

$$P_w = P_{cm} + \frac{H}{60} + \frac{K}{40\,000},$$

where H is the concentration of diffusible hydrogen content under the welding condition used ($cm^3/100$ g) and

K is the intensity of restraint of real welded joint in structure (kg-mm/mm)

We compare the P_w value thus obtained with the cooling time in the actual steel structure obtained from the welding procedure depending on heat input, preheating condition and plate thickness etc. If the critical cooling time to avoid weld cracking obtained from P_w is less than the cooling time obtained for real structures we do not expect weld cracking (K. Satoh, S. Natsu, *et al.* 11W Doc. IX–834–73).

Next I would like to explain briefly the basis of the Charpy requirement specified in WES 135. It may seem curious that the required absorbed energy level decreases as the strength level of steels increases, but as you can see from table 1 in our paper, the testing temperature decreases as the strength level of steels increases. So the absorbed energy should be increased with the strength of steels at the same temperature. And also it should be remembered that the specification of WES 135 is only applicable for structures exposed above -10 °C.

Lastly, I would like to introduce an alternative proposal to establish the revised toughness requirements for class G steels in WES 136, based on the c.o.d. concept.

In this proposal, we define the transition temperature in a c.o.d. bend test as the temperature, T_i, at which stable shear fracture (thumb nail fracture) starts to appear at the tip of the notch. The critical value of c.o.d. δ_i, corresponding to T_i temperature, seems to depend on the strength level of steels.

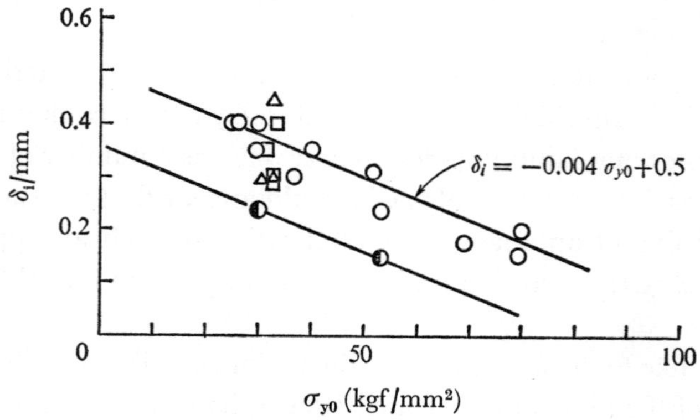

FIGURE 1. Relation between δ_i and σ_{y0}, ○, base metal; △, weld metal; □, fusion line; ◐, fatigue notch.

Figure 1 shows the relation between δ_i and the yield strength of steels at room temperature, σ_{y0}. The value of δ_i decreases linearly as the yield strength increases and expressed for a 0.2 mm width mechanical notch in a bend specimen as follows:

$$\delta_i = -0.004\sigma_{y0} + 0.5. \tag{1}$$

Using the relation expressed by (1), the T.M. committee investigates how to correlate the transition temperature T_i, with the V-notch Charpy fracture transition temperature, ${}_vT_{rs}$. Considering that T_i should be affected by the strength level of steels, notch acuity as well a plate thickness, the following formula is proposed for mechanically notched specimens after a trial and error procedure:

$$T_i = 0.782\,{}_vT_{rs} + 0.114\sigma_{y0} + 4.72\sqrt{t} - 16.68\ (\mathrm{K}), \tag{2}$$

where ${}_vT_{rs}$: fracture transition temperature in V-notch Charpy test (K)

σ_{y0}: yielding strength of steel at room temperature (kgf/mm²)

t: plate thickness (mm)

The T_i experimentally obtained compares well with the predicted temperature from equation (2).

Making conservative allowances of 20 °C for the scatter of the data shown in figure 2 and also 20 °C for the temperature rise for a natural crack (fatigue notch) compared with a mechanical notch of 0.2 mm width including some other safety margin, namely taking 40 °C as the total allowance, we can calculate examples of the relation between the minimum service temperature and the required V notch Charpy transition temperature as shown in figure 3 with varied plate thickness and strength level of materials.

In this proposal the effect of the strength level of steels is not so effective compared to the proposal shown by figure 8 in our paper owing to the small coefficient in the term of strength level in equation (2).

On the other hand, the examples shown by figure 8 in our paper do not include the allowances for a natural crack (fatigue notch) and other safety margins. We must take account of this fact to establish the revised specifications for WES 136.

We are now discussing the final proposal for the revised specifications including the proposal stated in our paper (method based on a certain mechanical level of c.o.d. value), the proposal just introduced above (method based on transition temperature of c.o.d. bend test T_i) and other probable methods based on the allowable defect size depending on the strength level of steels etc.

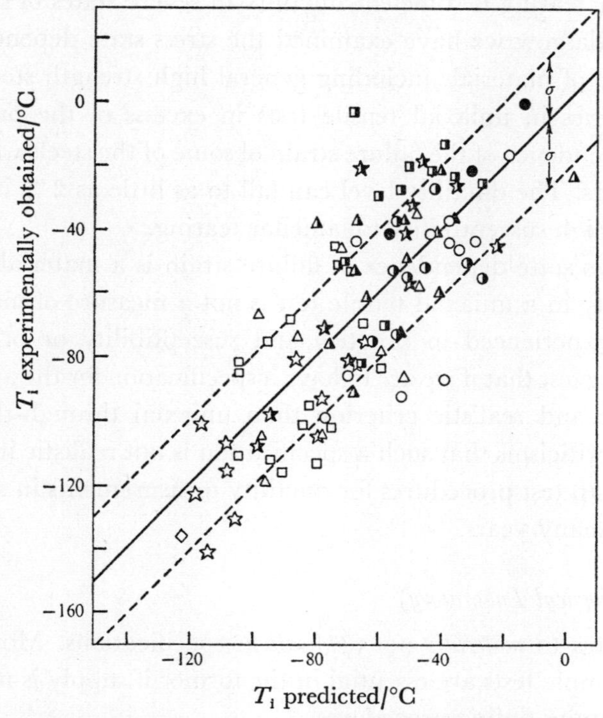

FIGURE 2. Comparison between T_i experimentally obtained and T_i predicted by equation (2). Open symbols, base metal; half filled symbols, fusion line. ○, M.S.; △, HT50; □, HT60; ◇, HT70; ☆, HT80.

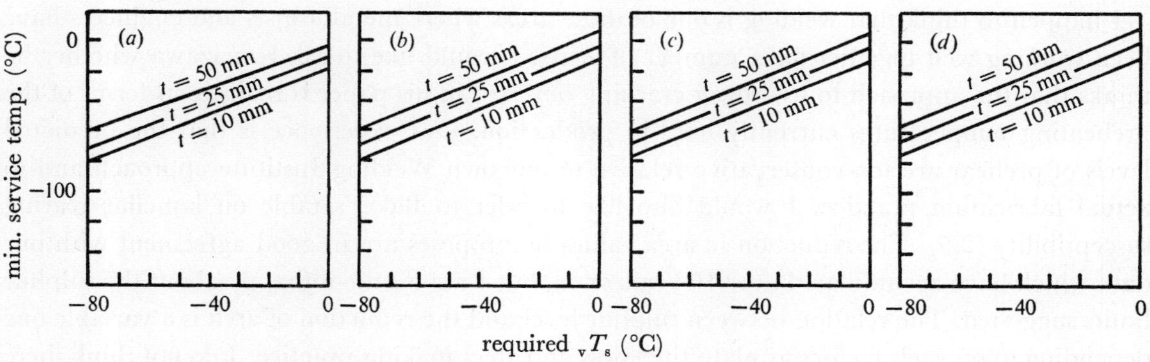

FIGURE 3. Minimum service temperature and required $_vT_{rs}$: (a) mild steel (σ_{y0} = 30 kgf/mm²), (b) HT50 (σ_{y0} = 40 kgf/mm²), (c) HT60 (σ_{y0} = 55 kgf/mm²), (d) HT80 (σ_{y0} = 70 kgf/mm²).

M. J. COWLING (*Glasgow University*)

I would like to make a few comments on the subject of lamellar tearing in relation to Baker (2.5) and with particular reference to his table 2.

I am surprised that this tentative supplementary specification for the avoidance of lamellar tearing is still being pursued, with its criterion of through-thickness ductility in a uniaxial tensile test. A substantial through-thickness ductility value is not a sufficient or unique measure of the lack of susceptibility of a material to fail by lamellar tearing in a welded joint. Structural failures due to lamellar tearing do nor normally occur at sites where the stress state is uniaxial tension. Failure usually occurs in regions of considerable constraint, for example at T-joints, where the stress state is considerably higher than uniaxial tension. Hence the requirement for the avoidance of lamellar tearing is sufficient ductility in severe states of stess.

At the University of Glasgow we have examined the stress state dependence of the ductile failure strain in a variety of materials including general high strength steels, all of which had through-thickness ductilities in uniaxial tensile tests in excesss of the proposed 25 % 'safe' limit. The stress state dependence of the failure strain of some of the steels was found to be much more severe than in others. The ductility level can fall to as little as 2 % in moderately severe stress states indicating a high susceptibility to lamellar tearing.

In conclusion, the stress state dependence of failure strain is a material characteristic. The through-thickness ductility in a uniaxial tensile test is not a measure of material behaviour in the severe stress states experienced in practice and susceptibility or otherwise to lamellar tearing. I would like to suggest that if we are to have a specification for the avoidance of lamellar tearing a more scientific and realistic criterion than uniaxial through-thickness ductility is required. In response to criticisms that such a specification is not realistic in the quality control situation, simple and cheap test procedures for ductility measurements in severe states of stress have been available for many years.

R. G. Baker (*National Physical Laboratory*)

My table 2 (2.5) relates to *materials* not *applications* specifications. More complex tests are necessary in the latter; simple tests are essential in the former if supply is not to be interrupted by costly and time consuming testing procedures.

R. Dolby (*Welding Institute*)

I happen to think that welding is one of those areas where metallurgists and engineers have been working well together for a number of years. I would like to ask Kanazawa whether he thinks that the approach to hydrogen cracking outlined in his paper is realistic in terms of the preheating temperatures currently used in production. Our experience is that the predicted levels of preheat are too conservative relative to our own Welding Institute approach and to actual fabrication practice. I would also like to refer to Baker's table on lamellar tearing susceptibility (2.5). The reduction in area values he proposes are in good agreement with our data which have accumulated over 10 years or so, but I am a little unhappy about the sulphur limits suggested. The relation between sulphur level and the reduction of area is a variable one depending upon such factors as plate thickness and steel making practice. I do not think there is a unique relation for aluminium treated steels between sulphur level and through-thickness ductility, and this is evident when comparing data given in some of the papers presented at this Conference, namely those by Roberts (3.1) and Van der Veen (3.4).

M. J. Cowling (*University of Glasgow*)

Baker (this discussion) defended the uniaxial tensile tests of the material requirement test. I submit that it takes just a few minutes longer and a few pence more to make a Bridgman notch specimen than a plain uniaxial tensile specimen, so enabling the stress state to be tested.

J. E. Roberts (*British Steel Corporation*)

The uniaxial tensile test with a reduction of area criterion is an adequate quality control procedure to assess the success of steelmaking practices aimed at clean steel to avoid lamellar tearing in practice. The notched specimen suggested by Cowling (this discussion) will be less satisfactory for this purpose because the plane of failure is pre-ordained and may not coincide with the most susceptible region of the plate thickness. A parallel specimen is more searching as a quality control test. It obviously does not provide all the data for material choice, and I agree with Dr Baker that more complicated tests are necessary to increase the understanding of lamellar tearing phenomena – including not only notches, but also the effects of hydrogen level, and applied restraint etc., as have indeed been mentioned in Cotton's paper.

The proof is that many thousands of tonnes of steel have been supplied in this and other countries, checked for quality by the simple through-thickness tensile test, followed by successful and trouble-free fabrication.

B. F. Brown (*American University*)

In the field of stress corrosion cracking many hard scientific facts are known, but it is characteristically difficult to apply them to the real world of engineering problems, where the stress may not be known, where the environment may be imperfectly known, and where (thanks to work such as was carried out by Edeleanu 25 years ago) the metal itself changes the environment in very significant ways, not all of them anticipated. The problem of stress corrosion cracking owes some of its difficulty to the multiplicity of responses of a given alloy to various chemical environments. Should we categorize our alloys according to their response to worst-case environments? I think not, but I would be interested in other opinions.

Another problem is the enormous time span required for valid testing according to present understanding, and as a rule we cannot count on the results of a rapid test. This fact exacerbates the problem of assembling a 'data bank' of stress corrosion cracking characteristics.

The pattern of controlling the stress corrosion cracking problem in engineering is to select different control measures for different engineering situations. For example, nitrogen tetroxide of high purity has caused serious cracking problems in titanium alloys. Here the problem was solved by adding 2 % water. (We do not know why water has this inhibiting effect.) With methanol and titanium we always see cracking (if even slight traces of chloride are present), and here the solution has been to prohibit any contact between methanol and titanium. The problem of cracking of titanium alloys in the atmospheric environment was solved metallurgically by keeping the aluminium and oxygen low, adding some molybdenum if possible. By such a metallurgical approach, and keeping the tensile stresses low, we have successfully used large numbers of titanium alloy flotation spheres in the sea, where the inhibitor approach (i.e. inhibition of the sea water) is not feasible.

For fundamentally new structues, I should like to point out the value of corrosion considerations begin taken early in the conceptual design stage. A good example of this approach is the research submersible *Aluminaut*. Because of the need for high yield strength material, the aluminum alloy 7079–T6 was selected, even though this alloy was known to be highly susceptible to stress corrosion cracking in the marine environment. But this susceptibility is highly anisotropic, and in the conceptual design phase fabrication procedures were developed whereby only the most resistant aspect of the texture of the alloy was contacted by sea water.

Additionally a sophisticated paint system was applied, and this was supplemented by cathodic protection. Finally, maintenance procedures called for thorough removal of chloride after each campaign and before lay-up. The result of pyramiding these partial measures was successful performance of the vehicle, and I should think this multiple-measure approach will be seen a great deal in difficult engineering siuations.

Now coming to stress corrosion cracking as studied by fracture mechanics, this approach would seem to be appropriate even in ductile alloys, because stress corrosion cracks even in such alloys are brittle. Therefore it is appropriate to use brittle (elastic) fracture mechanics to quantify the stress contribution. This procedure yields a K number, from which a critical crack size or a defect tolerance parameter can be estimated. The earliest practical example of this approach was the 2nd stage of the old Polaris missile, which was made from a steel from which the critical crack size for stress corrosion cracking in tap water was a very small fraction of a millimetre. At one period, 100 % of the production of these steel casings failed during pressure testing with tap water, which caused stress corrosion cracking during the specified hold time. Since the critical crack was too small to be either eliminated or detected, the solution was to substitute a more tolerable fluid – oil – as the pressurizing liquid. For steels which have an extremely large indicated defect tolerance parameter (for example, if $K_{Is.c.c.}$ is equal numerically to or exceeds the yield strength), we think we know the scientific meaning of such a quantity, but we do not really know the engineering implications. I would suspect that for a long time to come we must seek to be educated in this aspect of the problem by measuring $K_{Is.c.c.}$ values on both failure material and material which did not fail in service.

Another thing which will lead us to more rapid progress in such technological problem areas will be more cross-discipline conferences such as the present one.

G. P. Rothwell (*National Corrosion Service*)

Butler & Ison (2.6) and Forsyth (2.3) have emphasized the division of corrosion phenomena into three types; uniform attack, localized corrosion such as pitting, crevice corrosion and exfoliation, and conjoint mechanochemical action, including corrosion fatigue and the several aspects of stress-corrosion cracking. And in real load-bearing plant, of course, the transition from localized attack to a stress-corrosion or corrosion-fatigue situation is often merely a matter of time. From the papers and the earlier discussion, it is evident, and indeed it comes as no surprise, that the major problems in the practical use of a great many materials lie in these latter fields which arguably form the most important area of the interface between chemical metallurgy and engineering at the present time.

The reason they assume such importance is that we can do so little in a general sense to avoid them: the phenomena are very specific to particular material/environment systems, and palliatives where they exist, are usually very specific also. Obviously, if one owns the problem before one owns the hardware, a wise choice of material coupled with attention to design may solve the problem, provided of course that a suitable material exists, provided that one can actually design with it and provided that one can *afford* to design with it. If the extent of the environment is limited and the situation permits, modification of the environment, including the use of inhibitors, may enable the problems to be avoided. Again, in particular situations, careful use of coatings, or impressed current methods of corrosion control may provide useful palliatives, but in both of these cases any other than the most careful application may make the corrosion effects many times worse, and so an additional reliability factor must inevitably be considered.

What does emerge, however, is that groups of engineers are more and more having to design with materials under conditions of risk from these types of attack. Chemical engineers appear to have been living with the problem for some time, but the situation is becoming more general, as more and more stringent demands are made on designers and engineers. We should ask ourselves two questions. If materials must be used in risk situations, how much risk can we tolerate? And how can we monitor the behaviour of materials under these conditions? Current n.d.t. techniques can tell us little about defects in the critical size range for many modern materials, and even in cases where they might be useful on size grounds alone, there is the added complication of the random nature of the phenomena. Again, conventional corrosion-monitoring techniques are of very limited value in such situations.

Our approach should, I suggest, be fourfold. First, we would all agree that work must continue on the development of resistant alloys, and in understanding the processes concerned and their control. Secondly, I think that the metallurgist must pass on the knowledge he gains to the engineer, swiftly, and in a form which the engineer can use; and I think we may be failing in this at present in a wider field than just localized and stress-assisted corrosion. Thirdly, we should move towards a rational consensus of how far we can live with materials in these risk situations, not so much for the benefit of those who are already doing so with some success, but as a guide to designers who find increasing pressure to begin to use materials under such conditions. Finally, we must make great efforts to advance monitoring techniques of all kinds, for to live with a risk when we can make no meaningful measurements is to compound that risk to an unacceptable degree.

O. A. Kerensky (*Freeman Fox and Partners*)

I should like to enquire about the use of self-sealing steels under water.

H. C. Cotton (*British Petroleum*)

I have attended discussions where the question of whether or not corrosion fatigue studies could be made adequately in artificial brine or whether it is necessary to collect fresh natural sea water was argued. Perhaps we should be collecting the effluent of the various molluscs and sea anemonies which surround points of high s.c.f. in order to find out whether calcium stopping of cracks is a significant effect under such conditions.

Protection of platforms below the splash zone is usually by means of sacrificial zinc anodes. In the splash zone Monel cladding is often favoured to protect risers, the platform itself being provided with a substantial corrosion allowance as well as being painted.

F. Watkinson (*Lloyds Register of Shipping*)

Two papers refer to the problem of hydrogen cracking. Baker (2.5) referred to a British Standard in which there is a method for deciding recommended welding procedures. Kanazawa (2.8) has already explained that there is a method which is related to the P_{cm} formula in his paper and can be used for deciding welding procedures which will be safe from hydrogen cracking. The two methods are different in detail and appear to be substantially different in the recommendations which they make for apparently otherwise similar situations. The method proposed in the Japanese papers seems to result in much higher preheat temperature being required than from the method specified in the British Standard. Perhaps this is because the British Standard deals only with carbon-managnese steels while the Japanese method aims to deal with a very wide range of low alloy as well as carbon-manganese type steels and so has to take a more conservative line. The British Standard, for example, takes full advantage of the limited hardenability of carbon-manganese steels while this can not be the case with steels having a wide range of hardenability. Burdekin (Keynote) I think has fallen into the same trap as many other people in being confused by the carbon equivalent numbers which appear in Professor Kanazawa's paper, and I would repeat that the importance of different carbon equivalent formulae is not the number itself, but what those numbers relate to, which is the welding procedure recommendation, perhaps the level of preheat required.

J. H. van der Veen (*Hoogovens, Netherlands*)

In the context of the query for experience with sea-water resistant steels, I would mention that after the Dutch Government's decision to close the sea-arms in the south of our country, Cr–Al alloyed sea-water resisting steels (destined for the locks) have been tested in the Stevin laboratory at Delft to determine the corrosion fatigue resistance of components made of such steels. The endurance limit for corrosion fatigue turned out to be zero and it was decided to use a (normal) fine grain notch tough high-strength steel with a yield point of at least 360 N/mm^2 and let the corrosion hazard be taken care of by, as far as I remember, zinc-epoxy protective coatings and thorough subsequent inspection.

W. J. Murphy (*U.S. Steel Corporation*)

Bare Cor-Ten steel derives its significantly better atmospheric corrosion resistance compared to other steels through the formation of an adherent and protective oxide layer on its surface which is dependent upon the alternate wetting and drying conditions that occur in the atmosphere. Thus Cor-Ten does not offer any substantial advantages over presently used steels under conditions of immersion in sea water. Furthermore such weathering steels offer no advantages with respect to the corrosion-fatigue problem that has been encountered in the welded tubular structural members of offshore platforms.

I should like to point out that U.S. Steel produces a steel somewhat similar to Cor-Ten steel, called Mariner steel for sea-water applications such as piling. The main advantage of this steel over carbon steel is its substantially improved corrosion resistance in the splash or tidal zone of the piling where alternate wetting and drying occur.

O. A. Kerensky (*Freeman Fox and Partners*)

Is there any advantage in painting Cor-Ten steels?

W. J. Murphy (*U.S. Steel Corporation*)

The surface of Cor-Ten steel offers an excellent base for paint where a painted appearance is desired. When the paint film is broken, e.g. by a mechanical scratch, the oxide layer that forms is highly adherent and does not tend to fall off and thus carry with it adjacent paint; thus the continuity of the paint is preserved.

There are also applications for galvanized Cor-Ten steel. For example, for the curtain walls of buildings where the surface exposed to the atmosphere can be used in the bare condition, while the interior surface can be galvanized to provide protection against the relatively severe corrosion conditions due to condensation.

G. Butler (*National Physical Laboratory*)

The remarks of Murphy (this discussion) concerning the necessity of alternate wet and dry conditions for the corrosion resistance of Cor-Ten alloys would not apply to all 'weathering-steels' – certainly not the simpler copper-bearing steels. The influence of alloying additions of copper in improving the corrosion resistance of mild steel in the atmosphere – particularly where significant pollution by sulphur gases SO_2 and H_2S occurs – is well established. However, such additions have also a beneficial effect when mild steel is completely immersed in mineral or organic acids. We have also demonstrated the very marked effect of alloying additions of 0.2 or 0.3 % copper on the corrosion of mild-steel boiler tubes in oxygenated saline solution (cf. ref. in paper by Butler & Ison). Such additions have also proved to be of advantage in materials for coal wagons, where, of course, alternate wet and dry conditions do apply.

The common factor in all these applications is the presence of some acidity: even in unpolluted regions water precipitated as rain could have a pH as low as 5.4 due to adsorption of carbon dioxide from the air. Even in natural waters with a slightly alkaline pH value a beneficial effect of copper might be expected. It has been shown by many investigators including ourselves that the pH value in pits can be less than 2 and some acid attack with hydrogen evolution would be expected. However, the evidence for the effect of copper in steel on its behaviour in natural waters is very conflicting. In some cases reports of a negative effect can be attributed to the bad design of experiments and erroneous evalution of the data obtained. Under atmospheric conditions more than 50 % of the beneficial action has already been achieved with a copper content of 0.03 to 0.04 % in the steel and hence the standard steels against which experimental steels are to be compared should have a very low copper content, preferably 0.01 % or less. This has not been the case in many investigations. Secondly, copper may not necessarily be effective in reducing the corrosion rate but in reducing the rate of pit penetration, e.g. it may act by deflecting attack sideways and making attack more general, as in the Teddington tests in boilers. Hence, mass loss measurements are not adequate in themselves and should be supplemented by measurements of pit depth. Thirdly, the effect obtained may be quite small, say a 10 % reduction in the rate of pitting, and within the general spread of results. In this case, it may be necessary to expose a large number of samples and to treat the experimental data statistically. It is obvious that such a small effect could lead to a considerable saving in material, bearing in mind the large amount of mild steel used. In some recent work, we have shown that an alloying addition of 0.27 % copper to high-purity iron could lead to a doubling of the time to perforation of thin sheet in pitting solutions.

Incidentally, the beneficial effect of copper in steel on its corrosion behaviour was first patented in this country in 1822 but little further work was done until the second and third decades of this century. I am sure that Dr Rosenhain would not have been very impressed by this slow rate of technology transfer.

R. G. Baker (*National Physical Laboratory*)

With regard to hydrogen cracking there is close agreement between U.K. and Japanese workers on the relative importance of the different factors. The differences in approach are for purely pragmatical reasons. Hydrogen cracking can involve more than a metallurgical mechanism and, because of this, a matrix approach has been adopted, hydrogen potential of consumables being taken into account as a separate variable. This seemed to us, in the U.K., to offer less chance of ambiguity.

I would also like to comment that the Japanese Y-groove test is very severe. Figure 9 of my paper (2.5) does suggest that this would cause cracking at a lower hardness. Thus, this might account for the greater severity of the Japanese requirements in part at least.

B. F. Brown (*American University*)

I should like to respond to Burdekin (Keynote). Coatings, often incorporating inhibitors such as soluble chromates, are widely used especially on military aircraft to mitigate stress corrosion cracking, even though these coatings are recognized to be an imperfect solution to the problem. Peening the surface is called for in U.S. military specifications for aircraft, even though the benefits are uncertain. In high strength steel the one parameter which overrides all others in determining stress corrosion cracking behaviour in the atmosphere is strength, so that a decarburized (or otherwise low carbon) surface is recognized to be a metallurgical composite having superior stress corrosion resistance. In situations where the adverse effect of decarburization on fatigue resistance is not of concern, I think we will see practical application of this knowledge. The Japanese have already exploited this knowledge in the development of high strength bolts with reduced strength surface layers.

G. Butler (*National Physical Laboratory*).

It would be desirable if data were available for the engineer to make a corrosion allowance in designing a construction. Unfortunately, this can only be done when the environment and probable mode of attack can be defined with some precision. This applies even to attack by chemicals when the corrosion products are soluble, as well as to attack where a coherent and adherent film or layer of corrosion product is formed evenly over the surface. A great deal is known about atmospheric corrosion, particularly of mild steel, and the effect of atmospheric pollution in industrial areas or in maritime regions. Care must be taken, however, in judging the type of atmosphere prevailing. Even in a rural atmosphere which, in general, is unpolluted a micro-climate where the pollution is appreciable can be created on the leeward side of a chimney expelling sulphur gases and particulate matter such as soot.

In the vast majority of cases, however, the corrosion problem involves localized attack in pits and crevices and, for example, in stress-assisted attack the metal lost by dissolution may be very small indeed. In such cases it is unlikely that the corrosion allowance can be quantified and the probability of attack may be reduced only by design, protective measures and good house-keeping and maintenance. The latter is only possible if the operating personnel have a corrosion awareness. Thus, the corrosion of many medium and high pressure boilers can be controlled by correct water treatment monitored either by the operating firm, when testing facilities are available, or by a water treatment firm with the necessary expertise. There are, however, many cases in which personnel operating medium and low pressure boilers in smaller establishments do not have the awareness or facilities to operate anything other than very simple water treatment. It is in such regions that the search must be made for materials with improved corrosion resistance or even more simple water treatment which is fool-proof and allows a greater margin of error. An equivalent case could be made out for many types of equipment other than boilers.

E. D. Hondros (*National Physical Laboratory*)

I think it would be very instructive if we were to hear of specific instances of close metallurgy engineer interaction and I would like to mention some work which I believe demonstrates this.

As part of our long-standing interest in surfaces and interfaces, we have in recent years paid close attention to the chemistry of grain boundaries involved in various forms of brittle fracture. In this respect, we believe in the value of the Rosenhain scientific methodology of projecting information obtained on the micro-scale to the macro-mechanical scale. Sophisticated surface electron spectroscopy techniques have been developed which permit the fracture, in tension or by impact, of specimens *in situ*, in an ultra-high vacuum chamber, followed by a quantitative analysis of all the species present in the exposed surfaces to a depth of one to three atom distances. An example of the examination procedure comes from a $2\frac{1}{4}$ Cr 1 Mo pressure vessel steel sample which, in the as received state, contained a number of fine stress corrosion cracks. This was cracked to completion in the apparatus and a scanning Auger electron image of the field of interest with a resolution of 5 μm is able to show a high sulphur region in the reaction front close to the tip of the crack, indicating the effect of the aggressive environment. The virgin intergranular surfaces exposed in fracture reveal a high phosphorus level, demonstrating that in this steel, in the temper embrittled state, the phosphorus-coated grain boundaries are preferential paths for the extension of the stress corrosion crack, perhaps by a localized chemical reaction at the crack tip, or because the phosphorus has weakened the cohesion of the boundaries.

In short, this means that the bulk phosphorus level is too high for this steel and for the heat treatment it has undergone. The stress corrosion life of the steel should be greatly increased if the steel is not in the temper brittle state, i.e. we should tell the engineer that the composition specification of this steel in residuals should be reviewed and in particular, the bulk phosphorus tolerance level should be amended to a lower value.

F. M. Burdekin (*Messrs Sandberg*)

We have reviewed the possibilities of several different types of failures in a wide range of structures. However, these modes of failures are not going to occur in all structures, but some of them are going to occur only in certain structures and certain materials, and many structures will go on being designed and made without even hearing of some of the problems we have discussed. Nevertheless I also get involved in investigations of a large number of failures in engineering components. Very often these failures have occurred because the designers or the fabricators did not consider the metallurgical aspects that have eventually contributed to the failure, and this again is a plea for cooperation between metallurgists and engineers at an early enough stage.

The following two contributions were received after the meeting:

W. O. Alexander (*University of Aston*)

It is not perhaps well known that when Dr Rosenhain was in consulting practice the old Broughton Copper Company was one of his clients.

A particular problem with which he was concerned was to ascertain the cause of, and eliminate, the cracking which developed during the manufacture of alpha aluminium bronze periscope tubes. This was a phenomenon which had baffled metallurgists since World War I. Unfortunately there is no record of how far Rosenhain had been able to solve this problem, the essence of which was severe intergranular cracking occurring up to the second stage of cold drawing. All tubes surviving these first stages of production performed satisfactorily in service.

It was some years later that the trouble was found to be due exclusively to traces of hydrogen picked up during the melting operation leading to the formation of a micro voiding known as a primitive gas network which proved to be remarkably stable during subsequent hot punching and initial cold drawing - the reasons appeared to be:

(*a*) the size and distribution of the gas bubbles,

(*b*) that the polygonal cell form on which these bubbles were distributed survived hot punching and first cold drawing operations without significant morphological change. Recrystallization after hot working and one interstage anneal revealed that the α crystal structure had recrystallized across the gas voids.

Fracture has occurred without any significant ductility during the first heavy drawing load, i.e. the second drawing. The only equivalent micro voiding which the writer has since seen is the type which can occur as the result of radiation damage. Apart from hydrogen embrittlement of steels, which has been extensively studied, I am confident that inadequate attention has generally been given to the role of the total residual gas contents of many metals or alloys where they are subjected to high stresses whether at room temperatures or elevated temperatures and/or also subjected to environmental corrosion attack.

I say this because very few investigators are in a position to carry out complete or accurate gas analyses of the metals they test. Secondly I am not happy that the observations that some of the gases may be present in solid solution or out of solution as minute spheroidal voids have been properly observed and assessed. The observation of this phenomenon requires very delicate polished relief attack otherwise the primitive gas network is never revealed in the micro structure.

Thirdly it is a well known phenomenon in the aluminium industry that hydrogen which can form blisters during production of strip and sheet can be forced back into solid solution by subsequent rolling.

In concluding I would also add that study of this phenomenon in periscope tubes also taught the writer that many metal working processes are themselves a form of non-destructive testing. The stresses involved in working are far greater than are ever likely to occur in certain types of service. I understand that the ill fated *Thetis* submarine was lifted from the sea bed by its periscope tubes.

L. M. WYATT (*Central Electricity Generating Board*)

The point was made by Butler & Ison (2.6) and also by the keynote speaker that 'uniform thinning of the metal due to corrosion rarely occurs'. I would like to point out that this statement is only true when consideration is limited to attack by aqueous liquids. If one widens the consideration to corrosion by gases such as, for example, CO_2 or flue gases, one gets a very much closer approach to uniform thinning and the corrosion proceeds very much more in accordance with mathematical laws which can be reasonably reliably derived. Under these circumstances, the procedure of using a corrosion allowance is essential and the allowance can fairly reliably be calculated throughout the life of the component.

While I am on the subject of corrosion, I would like to refer to weathering steels. Besides the features mentioned during the discussion, these steels sustain a very considerable disadvantage on environmental grounds. Most environmentalists looking on a structure made of weathering steel throw up their hands in horror and say 'This looks like a pile of rusty iron' (which indeed, it is) and refuse to allow its use at any price.

3. MATERIALS DEVELOPMENT PRESENT AND FUTURE

3.1 Development of normalized structural steels

BY J. E. ROBERTS
British Steel Corporation, Ladgate Lane Laboratory, P.O. Box 74, Middlesbrough, Cleveland TS8 9EG

[Plate 19]

The principles governing the standard range of normalized steels are well established, namely improved uniformity of properties compared with steels which finish rolling at random temperatures, and the use of low carbon fine grain steels where needed to meet requirements for ease of welding, strength, and notch toughness. This paper therefore, looks at the prospects of extending the range of notch toughness guarantees at low temperatures (below −50 °C); the achievement of resistance to lamellar tearing where stresses are applied perpendicular to plate surfaces; and the response of steels to stress relieving procedures. In practice, steels exist to meet these situations, but the ability to predict their behaviour and to understand all the physical metallurgy involved, is incomplete and warrants further study as an aid towards avoiding unforeseen problems in production or service.

The basic requirements for normalized structural steels are the same as for steels produced by other routes; they must provide adequate strength, be resistant to other forms of failure, such as brittle fracture, be capable of fabrication into the finished structure without undue problems for the fabricator and must then perform satisfactorily in service, usually for many years. Last, but by no means least, they must be economic – this ought to be based on total cost over service life, but frequently initial steel cost is a prime consideration. Normalized steels are most likely to be used where either hot forming operations are necessary (because of thickness or the need to make complicated shapes) or where notch ductility is of particular concern rather than high yield strength. There is a region of plate thickness (between approx. 15 and 40 mm), where, for many applications, normalizing or controlled rolling are acceptable technical alternatives and the choice is likely to depend on the availability of rolling mill time compared to heat treatment furnace capacity. The controlled rolled steels are the subject of a separate paper to this conference.

Normalizing can, in principle, be applied to structural sections as well as plate, but in practice there are serious problems in avoiding or correcting distortion of heat-treated sections and while these can be overcome, the extent to which normalizing is actually applied is very low. When these problems are overcome, the metallurgical requirements are the same as for plates, so this paper will refer to plate steel developments.

To achieve the characteristics briefly outlined above, it is first necessary to understand the controlling factors, and define alternative ways of influencing these factors in the right direction to obtain an economically attractive product. The role of physical metallurgy is to provide this information.

Two papers were published in 1967 by Irvine, and by Duckworth which included the physical metallurgy principles which were, or could be, applied to structural steels, and since that date much effort has been expended on trying to refine metallurgical structure–property relations. In general, this work has not altered the principles that can be used, but has tried to improve the accuracy of the relations established.

The majority of commercial steels are, therefore, based on well-established principles:

(1) The need for a comparatively low carbon content in the interests of weldability and notch ductility.

(2) Grain refinement to improve notch ductility and enhance yield strength.

(3) Precipitation hardening effects using V where the fine vanadium carbide particles formed during cooling from the austenitizing temperature lead to a significant increase in strength with only a small loss in toughness.

Steels utilizing these principles are included in various national specifications, such as BS 4360, and cover the range of strengths likely to be used in structural applications. Developments not fully covered are:

(1) Extension of notch ductility guarantees to temperatures below −50 °C.

(2) Attention to the corrosion, or weathering, behaviour of unpainted steels to reduce initial and maintenance costs.

(3) Provision of enhanced strength in the 'short transverse' or 'through thickness' direction of plate steels to avoid lamellar tearing in welded joints.

(4) The property changes occurring during stress relief or welded structures, and the possibility of minimizing these changes by choice of initial steel composition.

Improvements in notch ductility

Relations established between composition, structure and properties are reasonably accurate for predicting yield stress, and ultimate tensile strength, thus assisting the choice of economic alloy combinations, but accurate prediction of impact transition temperature is less satisfactory. There has been a tendency for the physical metallurgist to devote a lot of attention to a 50% fracture appearance definition of transition temperature (50% cleavage and 50% shear on a Charpy V notch impact test specimen) which, with the possible exception of very thick plate of limited usage, is a criterion of doubtful value for structural applications, where fracture initiation is the normal controlling factor, and an energy criterion is more appropriate.

Attempts to explain the energy characteristic by theoretical approaches have not been too successful even when use is made of flow stress (for, say, 5% strain) and the effects of grain boundary cementile particles on fracture initiation and extension across a ferrite grain – for example, the modification by Almond, Timbes & Embury of the equations derived by Petch (figure 1) still does not give results matching experimental values over a reasonable range of carbide thickness. Recourse must then be made to empirical relations such as shown in figure 2. The production metallurgist is thus back to seeking economic means of producing fine-grained low carbon steels!

Petch
$$R\sigma_y > \frac{8\mu\gamma_p}{\pi(1-\nu)kl^{\frac{1}{2}}}$$

$$\epsilon T_c = \sigma_0^* + c - \left(\frac{4R\mu\gamma_p}{k} - k\right) l^{-\frac{1}{2}}$$

Almond *et al.*
$$R\sigma_y > \left(\frac{k^2 l}{4t^2} + \frac{8\mu\gamma_p}{\pi(1-\nu)t}\right)^{\frac{1}{2}} - \frac{kl^{\frac{1}{2}}}{2t}$$

FIGURE 1

Use has been made of aluminium, nitrogen, vanadium and niobium as the most common routes to grain refinement, together with attention to the quantity and form of inclusions which influence notch ductility. The latter requires a reduction in total inclusion content, and the avoidance of stringer inclusions. Although the problems with sulphides are more intense with the low rolling temperatures of controlled rolled steels, even in normalized steels the effect of aluminium in promoting type II sulphides leads to the steels having planes of weakness which,

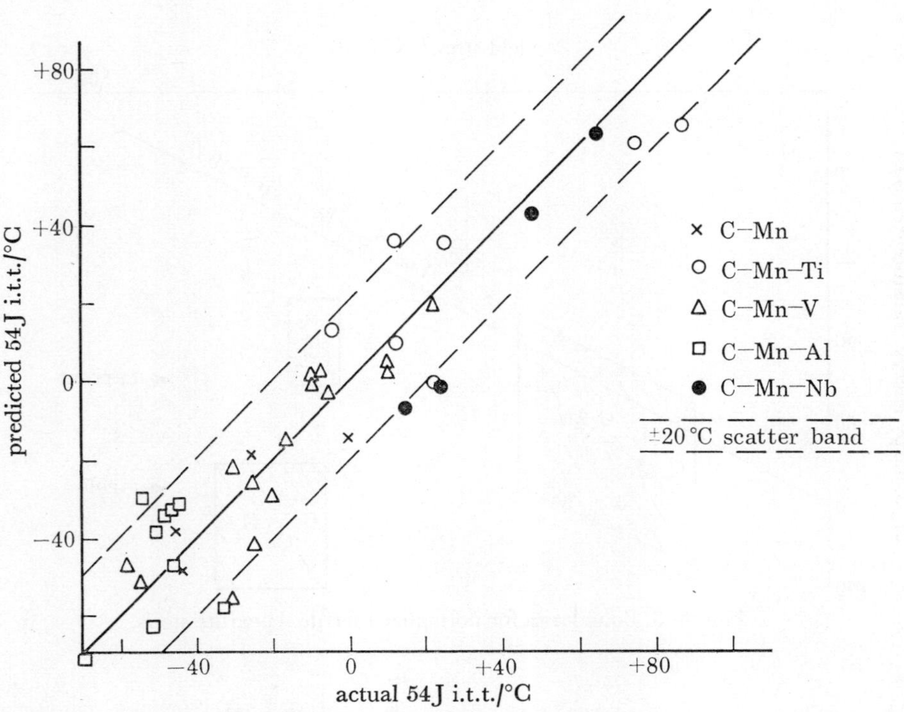

FIGURE 2. Comparison of actual 54J i.t.t.s with those calculated from derived regression equation, 54J i.t.t. $= 85-16.4\, l^{-\frac{1}{2}}+2.28\times 10^{-3}(t\sigma_y^2)$.

in notch bend tests (slow or impact) reduce restraint at the root of the notch, and show fibrous fractures but which absorb comparatively low energy during failure. Conversely, low sulphur steels (less than 0.01 %) will have few inclusions, absorp more energy in failing, but tend to show cleavage fractures when failure does occur at low temperatures. The significance of this situation depends on the anticipated service requirements to some extent, but the more common user demand is for the high energy – clean steel option. Implementation of this requirement falls back on improved steelmaking procedures, and the need for low sulphur iron and effective sulphur removal during refining. Fortunately, the techniques are well understood, and any inhibitions arise from economic factors rather than technical. Application of these principles results in current steels being produced which have energy transition temperatures superior to the boundary for normalized ferrite pearlite steels suggested by Irvine in 1967, as shown in figure 3.

While the increasing use of aluminium as an aid to notch ductility is widespread, it is also evident that unless the aluminium level can be closely controlled, either the control of nitrogen will be inadequate, or excess aluminium will lead to a degree of grain coarsening, and hence loss of yield strength for a given composition. This effect is illustrated in figure 4 which shows that in a series of laboratory casts, aluminium exceeding approx. 0.04 % leads to a deterioration in properties. Further confirmation is given by a comparison of commercial results (figure 5) for Al free C–Mn–Nb steels, and plates with 0.03 and 0.06 % Al. Thus a desirable range of Al content of 0.01–0.03 % is indicated, but it is difficult to remain within these limits in bulk steel productions, and higher levels are bound to occur, and must be allowed for in writing practical specifications. The situation appears to be that when aluminium is present in a quantity significantly greater than that needed to combine with available nitrogen, the excess results

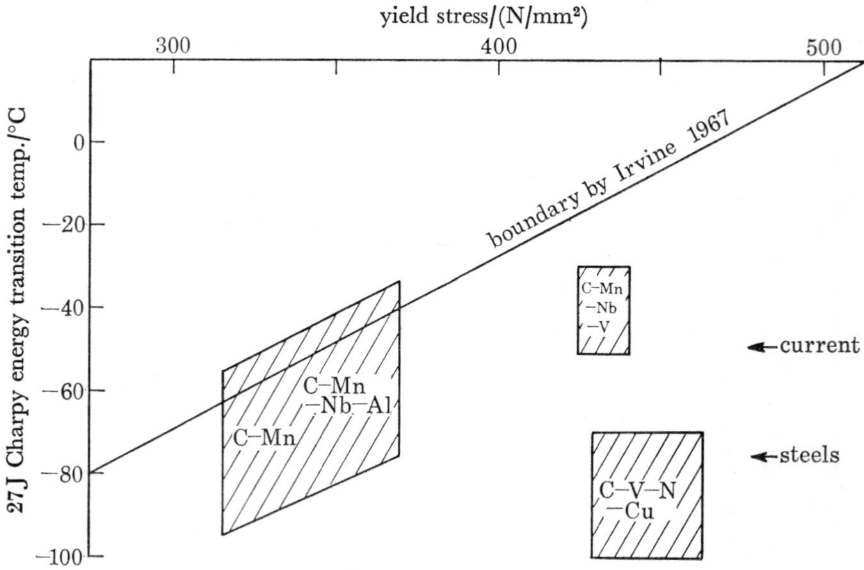

FIGURE 3. Boundaries for normalized ferrite – pearlite steels.

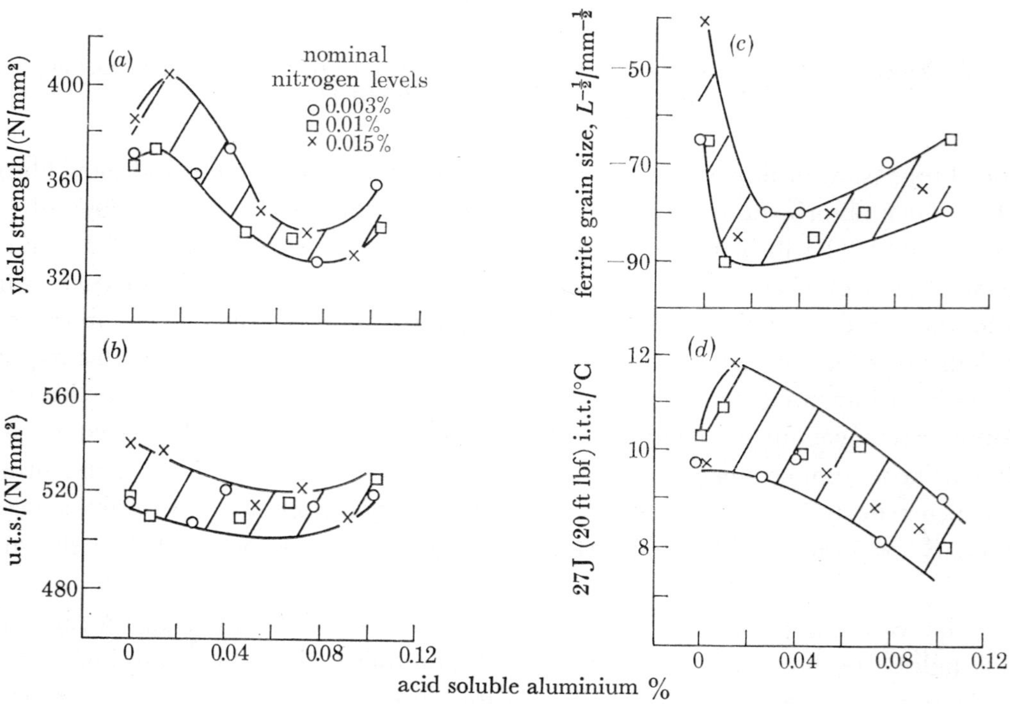

FIGURE 4. Effect of nitrogen and soluble aluminium.

in a coarser grain size (especially where grain refinement by nitrides, such as in V–N steels, is involved) and hence deterioration in properties. Precisely how this effect is produced is not clear.

Alternative grain refining procedures are incorporated in the results shown in figure 2, but although fine grain sizes are achieved, precipitation hardening effects result in high yield strengths, and also the Nb and Ti steels contain thicker grain boundary carbides (one of the factors, *t*, in the empirical relation shown) which act as crack initiators, and hence these steels have relatively poor properties in the absence of aluminium. Thus, where notch ductility is important, aluminium must be used even though some reduction in yield stress for a given carbon equivalent has to be accepted. The magnitude of the change is around 30 N/mm^2 drop in yield stress in a steel containing 0.03 % Nb, when the soluble aluminium content is increased.

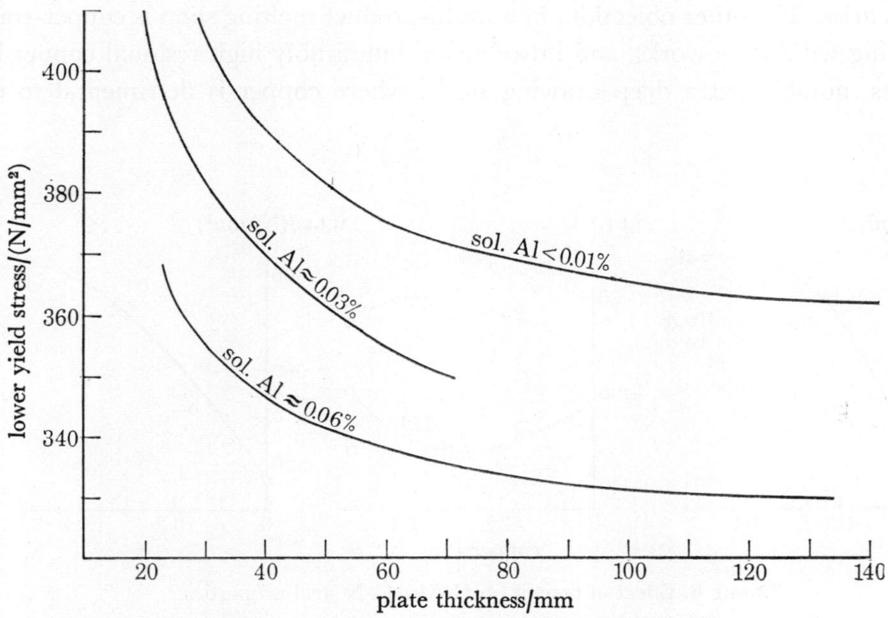

FIGURE 5. Effect of aluminium on yield stress of C–Mn–Nb plate steels.

A further approach to improving notch ductility is to reduce the carbon content, and hence pearlite fraction in the steel; but to maintain strength, some other change must be made. Control of rolling and cooling conditions is one possibility but for normalized steels, other alternatives have to be considered.

Work on as-rolled steels had shown that the presence of 'unprotected' boron at 0.003 % caused the 27*J* Charpy V-notch energy criterion to be at a temperature some 30–40 °C lower than in the absence of the boron. While so far unexplained, this effect is reproducible, and the possibility of a similar effect in normalized steels was therefore, checked – and unfortunately no benefit was obtained! The trials were on C–Mn–V steels, and a possible explanation is that in the as-rolled steels, the boron acts as a scavenger for free nitrogen, once the steels are normalized, the vanadium itself is fully effective, and thus no further benefit accrues from the presence of boron. A similar situation could exist with niobium steels, but has not been checked as far as is known.

Copper is another element which has promise of useful effects, contributing to strength by both solid solution hardening and (if enough copper is added, say $>0.8\%$) augmenting precipitation hardening, and, by lowering transformation temperature, a slight grain refining effect. It is, in fact, an essential ingredient in the 'weathering' steels, including the well established low carbon Cor-Ten A steel and of Cor-Ten B used for thicker plate. An example of the effects of adding copper to a C–Mn–V–N steel is shown in figure 6, where the presence of 0.35% Cu has increased the lower yield stress by some 90 N/mm², with a rather variable effect on impact energy transition temperature, depending on whether 27 or 54J is taken as the criterion. Overall, it is reasonable to suggest that notch ductility is maintained at a satisfactory level with copper from zero to 0.5%. In spite of this apparent benefit, copper-bearing steels are not popular in the melting shops, because of a fear of hot-shortness and, therefore, cracking during hot rolling. With low scrap steelmaking processes, this is rarely a real danger, but can be serious in high scrap processes, where tin is likely to be a high residual element, and trouble certainly does arise. The other objection, in a multi-product melting shop is copper-containing scrap circulating within the works, and introducing undesirably high residual copper levels in other products, notably extra deep drawing steels, where copper is detrimental to drawing qualities.

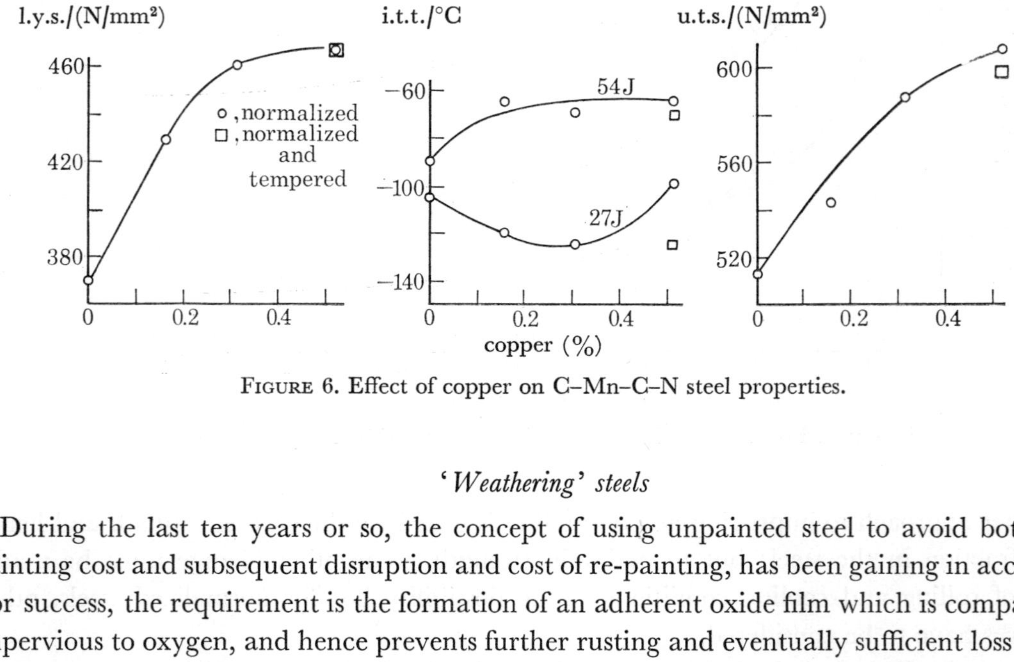

FIGURE 6. Effect of copper on C–Mn–C–N steel properties.

'Weathering' steels

During the last ten years or so, the concept of using unpainted steel to avoid both initial painting cost and subsequent disruption and cost of re-painting, has been gaining in acceptance. For success, the requirement is the formation of an adherent oxide film which is comparatively impervious to oxygen, and hence prevents further rusting and eventually sufficient loss of metal to cause failure. Copper is an essential element, as outlined above and gives maximum benefits in steel of low carbon content, and containing chromium.

High phosphorus is also beneficial, but is also detrimental to notch ductility and so in practice is not normally used in plate above about 15 mm in thickness. The effects on strength and notch ductility of copper have already been discussed and chromium is, on the whole, beneficial and similar to manganese in its effect.

What is not clear, is the precise nature of the protective oxide film, or the conditions necessary to generate a stable condition. Certainly, in damp, polluted atmospheres, stability is only achieved after several years of exposure, but whether this is simply a time effect, or whether it is a case of waiting for a particularly fortunate weather condition to occur for a significant time, is not known.

This is a boundary between the physical metallurgist, and the corrosion chemist, which could repay further study in the interests of defining satisfactory steel compositions and environmental conditions for successful application. The practical significance would be in increased confidence in applying an economically attractive principle.

Lamellar tearing resistance

The growth in heavy welded construction has lead to situations in which stress is applied in a direction normal to the rolled plate surface. This is the weakest direction of normal plate steels, and while a simple approach is to reduce the inclusion content of the steels, greater understanding is needed to define what is necessary and what is practicable. Based on cooperative work at the Welding Institute (1972), the most useful small-scale test result is the reduction of area of a tensile specimen taken in the 'through-thickness' or 'short transverse' direction and

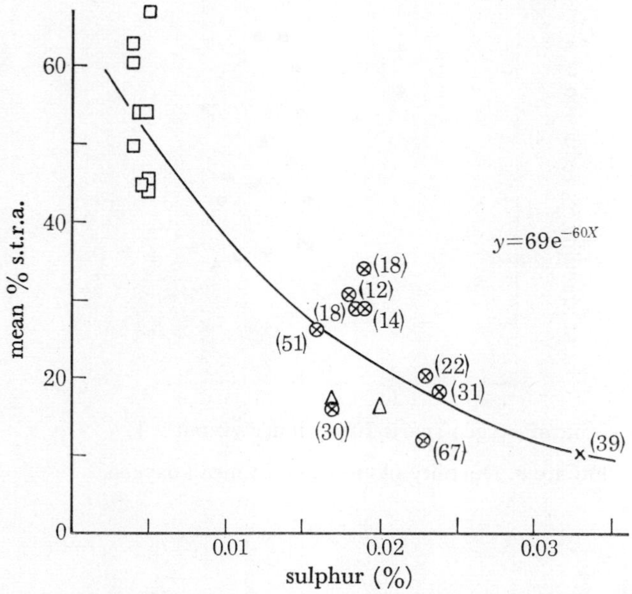

FIGURE 7. Relation between mean % s.t.r.a. and sulphur content in Si-killed Al-deoxidized (>0.012 %) continuously cast steels having type 2 sulphide morphology (figures in brackets are parts O_2 per million).

it was shown that an area reduction of 20 % or over would initially eliminate the risk of lamellar tearing, provided it was accompanied by careful control of welding consumables (notably relative to hydrogen potential). The steelmakers problem was to define first what inclusions were critical.

Usually, the lamellar tearing risk arises in restrained structures using thick plate, which also leads to a demand for good notch ductility. To achieve the latter requires, as outlined earlier, the use of aluminium. While this is beneficial for oxide control, leading to the avoidance of soft silicates and their replacement by smaller quantities of hard alumina, the aluminium also affects ingot solidification and leads to sulphur rich liquid concentrating at grain boundaries in the ingot and hence appearing in the finished product as thin films of sulphide commonly referred to as 'type II sulphides', which can create a plane of weakness. Study of inclusions and their behaviour during rolling from ingot to plate suggests a number of possible alterations:

(1) Reduce the total number of inclusions and so reduce the risk of a plane of weakness at the internal regions in a structure. For this to be effective requires for normal ingots, a low sulphur level, typically in the range 0.005 to 0.010%, as shown in figure 7 (Cavaghan & Chapman 1973; Chapman, Clark & Kirkwood 1974; Cavaghan 1973). While oxides also have a qualitative effect, a quantitative relation is difficult to establish, though a general trend such as that in figure 8 can be discerned. The probable reason is the non-uniform distribution of oxides in an ingot, and a tendency for oxide clusters to form in preferred regions of the ingot during solidification.

(2) Change the initial inclusion distribution and subsequent deformation. Faster solidification will reduce the clustering referred to above, and can be encouraged by casting thinner

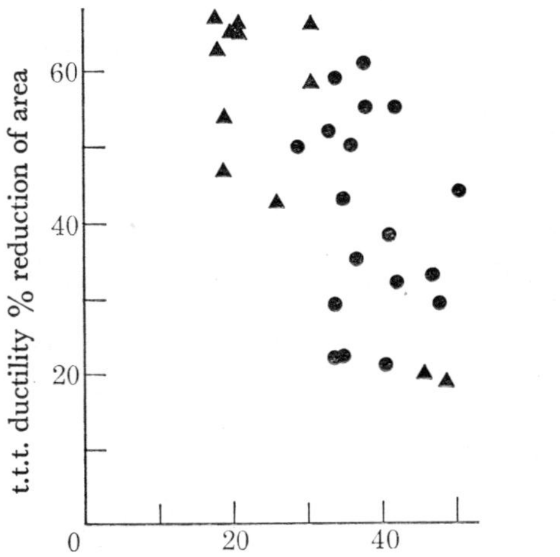

FIGURE 8. Ductility plotted against mean oxygen.

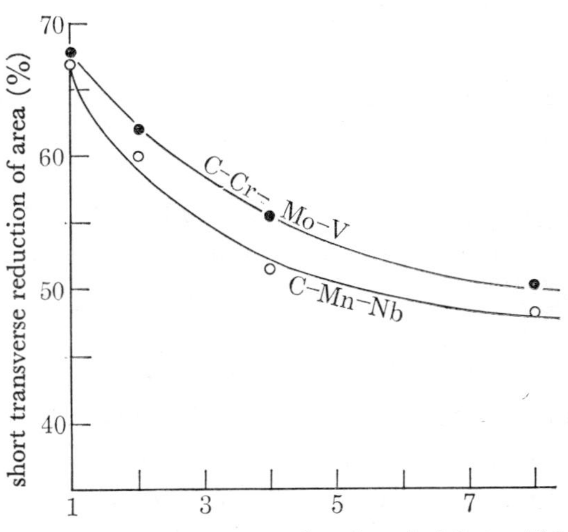

FIGURE 9. Effect of deformation on short transverse properties of unidirectionally rolled plates.

silicate stringers discrete alumina particles

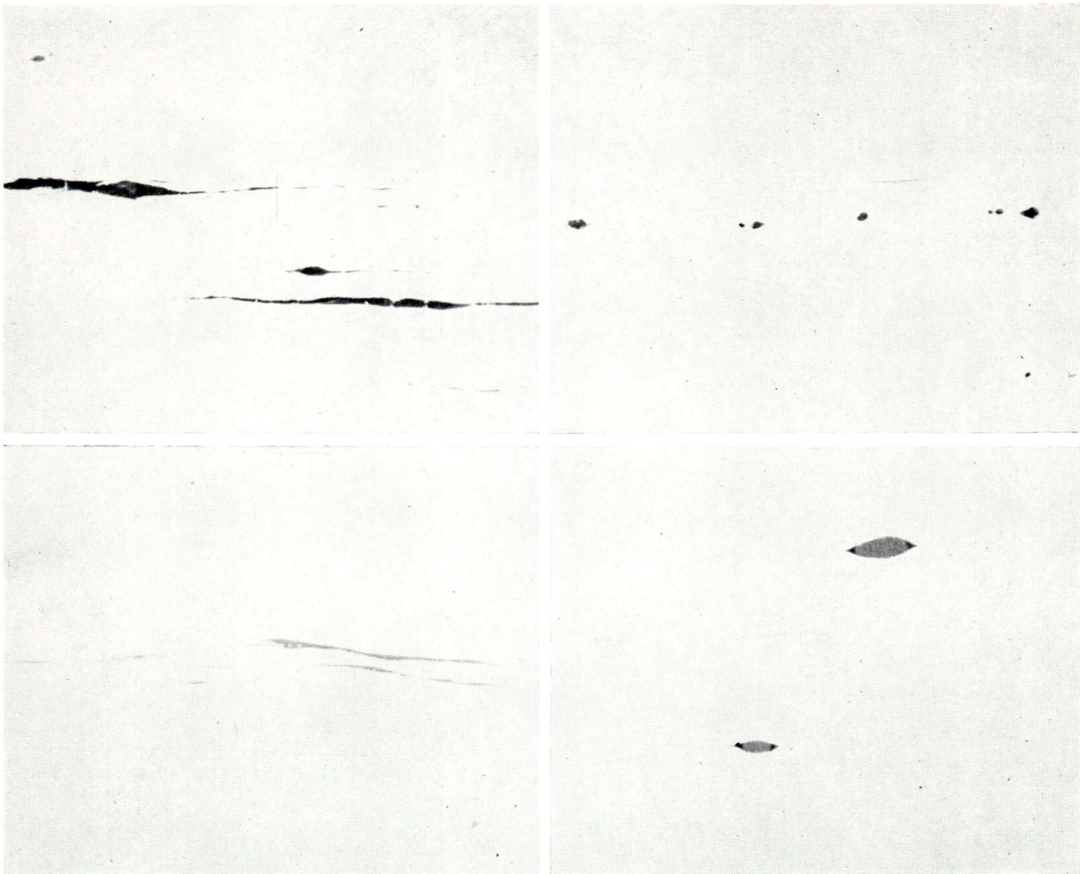

type II sulphide stringer globular sulphides resulting from rare earth metal treatment

FIGURE 10. Inclusions encountered in normalized steels.

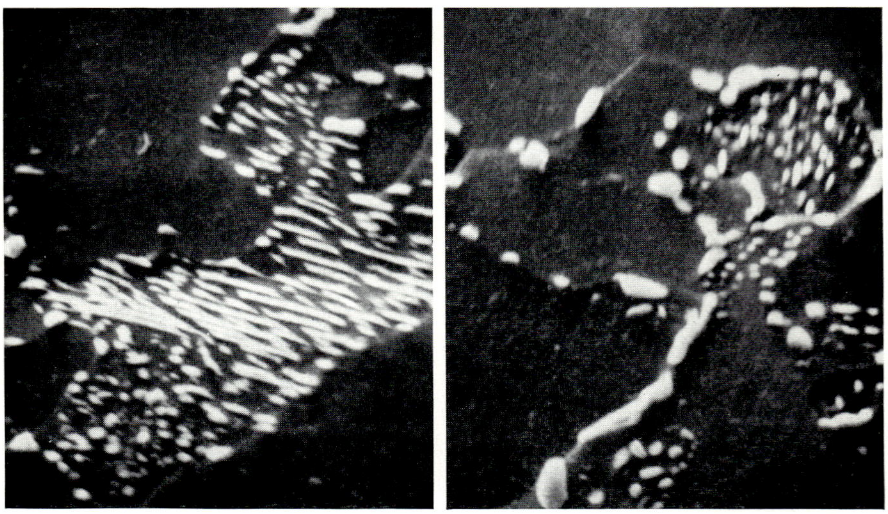

5700 Magn. ×4000

Degree of spheroidization at $H_p = 18.5$ (573 °C 120 h); some pearlite lamellae still visible.

5702 Magn. ×4000

Degree of spheroidization at $H_p = 19.3$ (650 °C 5 h); no lamellae visible.

FIGURE 11

ingots, or slabs even. There is a secondary benefit in this case, that the deformation from initial solidification to final product is reduced, and hence the extent to which initial inclusions elongate is also reduced. An indication of the effect for cast slabs is shown in figure 9 and for ingots the overall ratios must be at the high ratio end of the graph.

(3) Change the inclusion deformation behaviour. Oxides can be dealt with by using aluminium to create Al_2O_3 as already discussed. Sulphides, however, require a specific treatment, and common practice is to use rare earth metals (Henderson & Chapman 1973), which combine with sulphur in preference to manganese and form hard rare-earth sulphides and oxysulphides, which do not deform during rolling. There is, unfortunately, a tendency to form rare-earth sulphide clusters which, while giving plates of generally very good 'through thickness' ductility, can also cause local areas of poor properties. Possible means of controlling the situation for heavy plate may exist, but the practice is not common. In thinner (up to, say, 20 mm thick) controlled rolled plates, though here the objective has been to improve transverse notch ductility rather than 'short transverse' properties, it has been widely and successfully used.

In all of this work, the difficulty of defining precise relationships is considerable and probably will remain impracticable until a major change in steel casting methods is implemented. The increasing use of continuous casting may provide the necessary change, especially when allied with low sulphur levels. Treatment of continuously cast steel with rare earth additions causes many problems, and at present no satisfactory method has been proved. With a low sulphur and rapid cooling of the cast slab, r.e.m. treatment is probably an unnecessary complication. Figure 10, plate 19, shows typical inclusions illustrating the above points.

While structures showing ferrite-pearlite banding have been suspected of producing poor short transverse properties, in many cases the bands are related to inclusion distribution, probably both arising from initial boundary segregation during freezing. It is difficult to separate the two factors, but one attempt to do so by a short-time high temperature treatment which did not affect inclusion shapes significantly indicated that in the presence of significant inclusions, banding was exerting little or no additional influence on properties, but in very clean steel, its removal was beneficial. It is, however, less likely to be found in clean steels!

Influence of stress relief treatment

In many cases, normalized steels are used for fabrications where stress relief heat treatment is applied after welding. The basic objective is to reduce residual welding stresses, and to temper the hardened heat affected zones adjacent to the welds, but typically the whole structure is treated to approximately 600 °C for several hours. A consequence of the treatment is that the tensile strength of the steel is reduced, and Charpy transition temperatures are raised. Although the use of tempering had been long accepted by the metallurgist as a way of reducing the strength of a 'hard' steel, the fact that stress relieving must have similar effects had been largely ignored by design engineers until comparatively recently, but now demands are made for maintenance of specified properties after all fabrication heat-treatments have been completed. Attention has thus been directed to explaining why properties change during tempering, and whether it is practicable to devise steels resistant to change.

Two major changes in microstructure have been shown to occur: spheroidization of pearlite takes place, dependent on the combination of time and temperature, and a fairly effective parameter has been developed by Hollomon to calculate the equivalence (or otherwise) of different thermal cycles. This is defined as:

$$H_p = 10^{0.3}T(20 + \lg t);$$

T is the temperature (K) and t, the time at temp. (h).

There is growing evidence that the effect on properties reaches a maximum (at about $H_p = 19$) and thereafter does not change further. This would be compatible with completion of pearlite spheroidization, and is illustrated in figure 11, plate 19. Conversely, the effects are slight for short times at low temperatures (e.g. $H_p = 17$).

Notch impact properties have been shown to be related to grain boundary cementile particles, as described earlier in this paper, and the changes arising from stress relieving treatments lead to the formation of coarse carbides at grain boundaries, and hence a deterioration in notch ductility. A modest degree of benefit can be conferred in aluminium-free steels by the removal of free nitrogen during the stress relieving treatment, by the formation of manganese-silicon nitrides, but although this results in less change in the aluminium-free steels, the nett result is still that the aluminium treated steel will show superior properties even after stress relief, i.e. the differential in properties is reduced but not eliminated.

Some of the experimental results suggest that long time treatments may cause at least a partial recovery of notch ductility, but why is obscure.

The indication of this work is that property changes during stress relief can only be avoided if pearlite can be eliminated. This may be possible with low carbon steels (say 0.03 %C) but would introduce difficulties in manufacture, and need expensive alloy additions to maintain strength, which will inhibit widespread usage.

It must be remembered that stress relief brings about many beneficial effects which contribute more to structural safety than is lost by comparatively small changes in test piece properties, and the incentive to alter practices which have worked well for many years must be low.

Conclusion

This has not been a comprehensive review paper, but has attempted to show areas where physical metallurgy has either pointed the way to practical exploitation, or has at least explained what is happening during established commercial processes. In many instances, one is left with the impression that the relation between physical metallurgy, and steel production and engineering applications is rather like the relation between science and religion – in each case the former can explain what is happening, but rarely why, and the latter is a matter of faith and experience. Both are essential for a balanced outlook.

The author has drawn freely on the work of his colleagues, to whom he is grateful for constructive discussion and loyal effort, and also wishes to thank the directors of the General Steels Division of the British Steel Corporation for permission to present this paper.

References (Roberts)

Almond, E. A. *et al.* 1969 Proc. 2nd International Conference on Fracture. Brighton, p. 253.

Cavaghan, N. J. 1973 The need for a production of low sulphur steels by the BOS route. 80th Steelmaking Conference, Scarborough, October.

Cavaghan, N. J. & Chapman, J. A. 1973 The need for low sulphur steels. International Steelmaking Day, Paris, October, Paper 1.

Chapman, J. A., Clark, A. & Kirkwood, P. 1974 Steels for North Sea structures. Welding Institute Conference on Welding in Offshore Structures, Newcastle, February.

Duckworth, W. E. 1967 Metallurgy of structural steels; present and future possibilities. ISI/BISRA Conference on Strong Tough Structural Steels, pp. 61–73.

Henderson, W. J. M. & Chapman, J. A. 1973 The use of rare earth metals in the production of high quality plate steels. 31st Electric Furnace Conference, Cincinnati, December.

Irvine, K. J. 1967 The development of high strength structural steels. ISI/BISRA Conference on Strong Tough Structural Steels.

The Welding Institute 1972 *Lamellar Tearing in Welded Steel Fabrication.*

3.2 Controlled rolling

By W. B. Morrison and J. A. Chapman
British Steel Corporation, General Steels Division, P.O. Box 25, Middlesbrough, Teesside TS2 1HS

[Plates 20 and 21]

Controlled rolling is a means whereby the properties of steel can be improved to a level equivalent to those of more highly alloyed or heat-treated steels. The processing conditions are controlled to refine the austenite structure and thereby give fine ferrite grains. Refinement of the structure is aided by the addition of micro-alloying elements such as Nb, V or Ti. The improved strength and toughness of controlled rolled steels are shown to be a result of the fine grain size. A further increase in strength beyond that due to grain refinement can be obtained by finishing rolling at low temperatures in the $\gamma+\alpha$ or α regions. Commercial controlled rolling in various mill layouts, together with some properties obtained are described and it is pointed out that optimization of productivity is a major consideration when controlled-rolling procedures are being planned.

Introduction

Hot rolling is the most commonly used method of shaping steel. Due to the fact that steel is softer the higher the temperature, it is generally rolled at temperatures high in the austenitic region, the upper limit being governed by the practical limitations associated with high temperature reheating. In these circumstances, the mechanical properties of the steel are generally improved by a relatively low temperature treatment such as normalizing which refines the structure. About 15 years ago, Vanderbeck (1958) reported that some European steel producers were controlling the rolling process to deliberately finish rolling the steel at temperatures lower than usual in order to refine the structure and improve properties, thus removing the need for a further heat treatment stage. Doubtless low finish rolling temperatures had been obtained in many plate mills previously at least partly due to circumstances such as low rolling speeds, etc., but it gradually became widely accepted that close control of the rolling schedule was advantageous.

The principal cause of the improvement of the mechanical properties is the general refinement of the structure resulting from the refinement of the austenite during hot working. In order to optimize the rolling process it is necessary to understand the relation between the structure and the properties and how the structure is affected by hot deformation. Consequently, this review will discuss briefly the various ways rolling can be controlled to optimize properties, the mechanisms involved, the properties which can be obtained, and finally provide examples of successful commercial controlled-rolling practices together with the practical limitations and difficulties associated with controlled rolling.

Grain refinement

The major purpose of controlled rolling is to refine the structure of the steel and thus enhance both the strength and the toughness. Because of the relation between the austenite and ferrite grain sizes, refinement of the ferritic structure is largely effected by grain refinement of the austenite. For example, Priestner (1975) has shown that the number of ferrite nuclei per unit austenite grain boundary area is a constant for fine fully recrystallized grains. In plain C–Mn steels containing no stable dispersion of particles, recrystallization and grain growth can occur rapidly during processing at temperatures in excess of about 900 °C. For example, after a deformation of 25 % reduction in area (equivalent to a rolling pass) recrystallization is generally complete in less than 1 s (Morrison 1972). It is obviously important to deform a C–Mn steel at as low a temperature as possible in the austenite region if a small grain size is to be obtained, otherwise grain growth will negate the refinement caused by the deformation. By using a small section size to obtain fast heating rates to the deformation temperature and also fast cooling rates after deformation, it was shown that extremely fine ferrite grain sizes approaching 1 μm could be obtained in plain C–Mn steels (Morrison 1966). Quenching into water as soon as possible after rolling (*ca.* 1 s) indicated that the steel was fully recrystallized prior to transformation. Similar processing of steels containing a dispersion of stable particles (TiC, V_3C_4 or NbC) did not lead to further grain refinement. In fact, it would appear that a ferrite grain size of approximately 1 μm is a practical lower limit in low alloy steels.

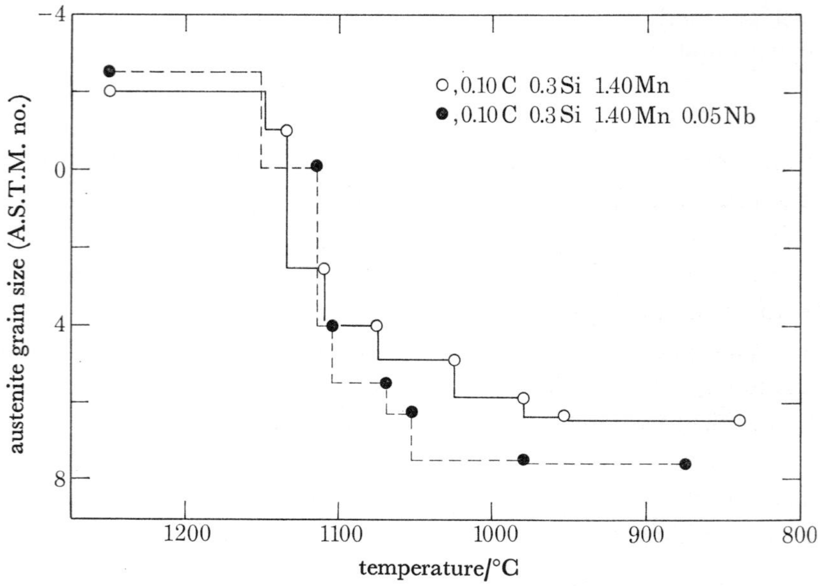

FIGURE 1. Austenite grain refinement during continuous rolling. Total reduction 90 %. After Sekine & Maruyama (1973).

The austenite grain size may be refined entirely by rolling after holding the steel at a normal soaking temperature (1200–1300 °C) or it may be refined partially by reheating to a lower than normal temperature and further refined by rolling. Figure 1 illustrates the refinement of the austenite grains in C–Mn and C–Mn–Nb steels during continuous rolling and it is seen that the austenite grain size has decreased from ASTM 2 to ASTM 7 during rolling over a temperature range of 1250–850 °C. The influence of soaking temperature on austenite grain size is shown in figure 2 for C–Mn and C–Mn–Nb steels. Thus by lowering the soaking temperature to about 950 °C for the C–Mn steel and 1050 °C for the C–Mn–Nb steel an austenite grain size of ASTM 6–8 can be obtained which is similar to that obtained by rolling (90 % reduction in area) from a soaking temperature of 1250 °C. It would therefore appear to be advantageous for maximum grain refinement to lower the soaking temperature to 1000–1100 °C in the case of a steel containing grain refining additions assuming that the time/temperature limitations would not impose too severe a restriction on the rolling schedule. There is some evidence to suggest that refinement of austenite grain size beyond about ASTM 10 may be very difficult (Priestner, Earley & Rendall 1968). Weiss *et al.* (1973) attempted to refine the austenite grain size of a Nb–Ti steel with an initial grain size of 12 μm by rolling at 850 °C. Sufficient time was allowed for recrystallization after each deformation. It is interesting to note that it required a true strain of approximately 1.2 to obtain a grain size similar to the initial value, and that at smaller true strains equivalent to single rolling passes the austenite grain size increased to values of 30–40 μm.

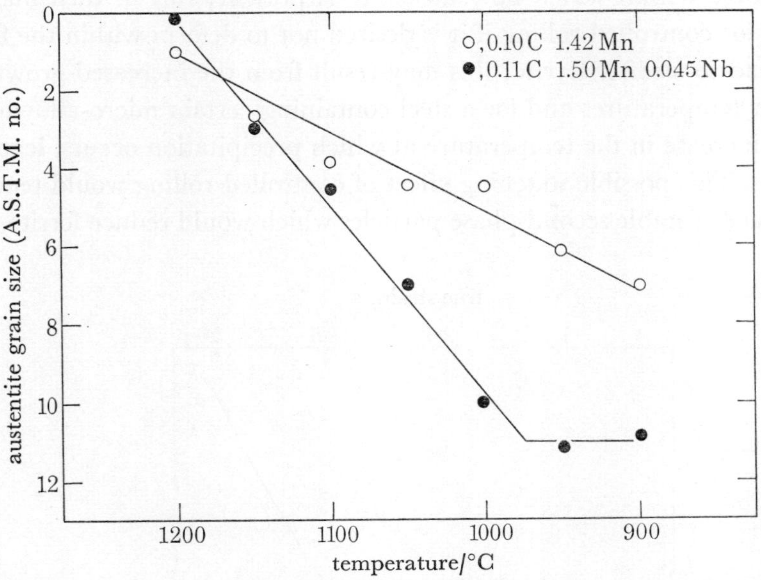

FIGURE 2. Influence of soaking temperature on austenite grain size. After Priestner *et al.* (1968).

It has been shown to be possible to refine the ferrite grain size of C–Mn steels to the same extent as in steels containing grain refining elements. However, an important effect of certain grain refining elements and in particular, Nb, is to slow down the recrystallization and grain growth of the austenite during rolling so that rolling conditions need not be so critical. The retardation of recrystallization occurs at temperatures below about 1050 °C. During deformation below this temperature, the austenite grains can elongate and remain in the elongated, deformed state until transformation to ferrite occurs. The total deformation in the temperature range where recrystallization is sluggish is typically 70–80 % and as shown in figure 3 this increases the grain boundary area only by a factor of about 2. Thus, the grain refinement resulting from the controlled rolling of a Nb-treated steel cannot entirely be a result of the increased grain-boundary area of the austenite grains. Figure 4, plate 20, shows a photomicrograph of a C–Mn steel containing elongated austenite grains which has partially transformed prior to quenching. Although nucleation of ferrite grains clearly takes place at the austenite grain boundaries, there is evidence that ferrite can also nucleate within the austenite grains (Fukuda *et al.* 1973; Priestner 1974). Thus, there appears to be evidence for increased ferrite nucleation rate in large deformed austenite grains which can be at least partly explained by nucleation within the grains but it is not yet clear whether similar observations apply to deformed fine-grained austenite. There is also some evidence for an increased ferrite nucleation rate in steels containing second-phase particles (Priestner 1974; English 1962), and it may be that optimum grain refinement would take place in a steel containing both fine deformed austenite grains and a dispersion of particles of a suitable size and type.

The increased nucleation rate of ferrite within deformed austenite has the effect of shifting the nose of the continuous cooling transformation curve to the left (Smith & Siebert 1970; Priestner & Biring 1973). Thus, controlled-rolled steels will transform at relatively higher temperatures if they retain the deformed state prior to transformation. Even if recrystallization occurs the resulting fine austenite grains should also cause an increase in transformation temperature. This particular influence of controlled rolling has two possible effects (*a*), it reduces the temperature range within which only austenite is present; this in turn may reduce the temperature range for controlled rolling if it is desired not to deform within the ferrite region, and (*b*), it could lead to a softer ferrite; this may result from the increased growth rate of the ferrite at the higher temperatures and for a steel containing certain micro-alloying elements it may arise from an increase in the temperature at which precipitation occurs, leading to larger precipitate particles. This possible softening effect of controlled rolling would tend to be minimized by the presence of stable second-phase particles which would reduce ferrite grain growth

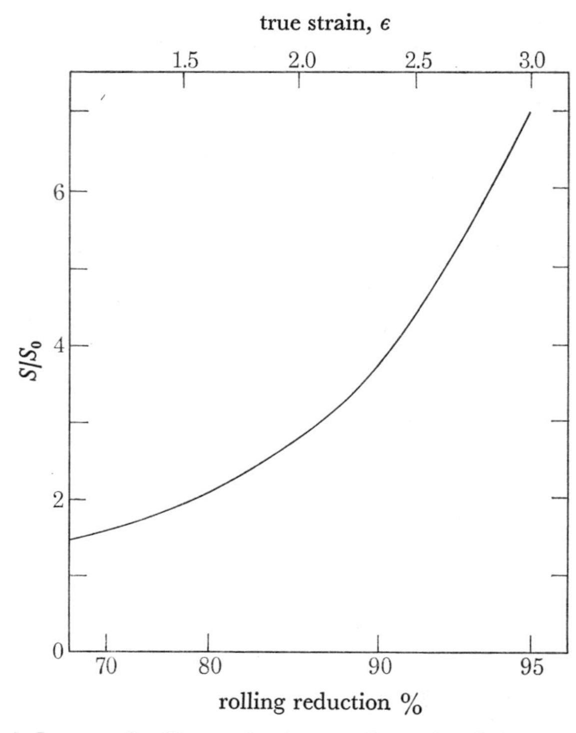

FIGURE 3. The influence of rolling reduction on the ratio of the grain boundary area (S) to the initial grain boundary area (S_0).

DESCRIPTION OF PLATE 20

FIGURE 4. Nucleation of ferrite in a deformed austenite, 0.1 C 1.52 Mn steel, finish rolled at 760 °C and water quenched. After D. Dulieu (unpublished work). (Magn. × 500.)

FIGURE 6. Duplex grain size in a commercial controlled-rolled C–Mn–Nb steel. (Magn. × 1000.)

FIGURE 7. Ferrite grains with substructure in a laboratory controlled-rolled C–Mn steel finished in the ferrite region. (Magn. × 1000.)

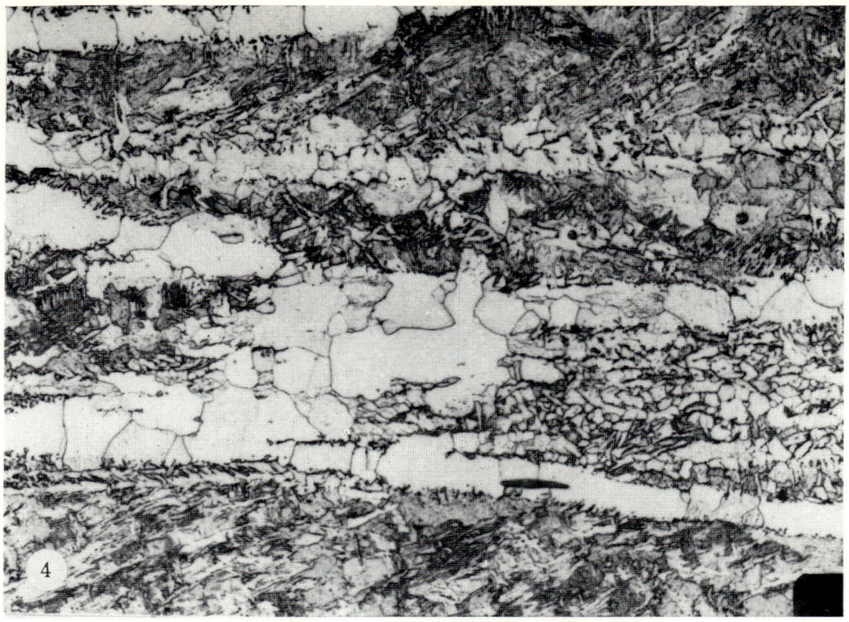

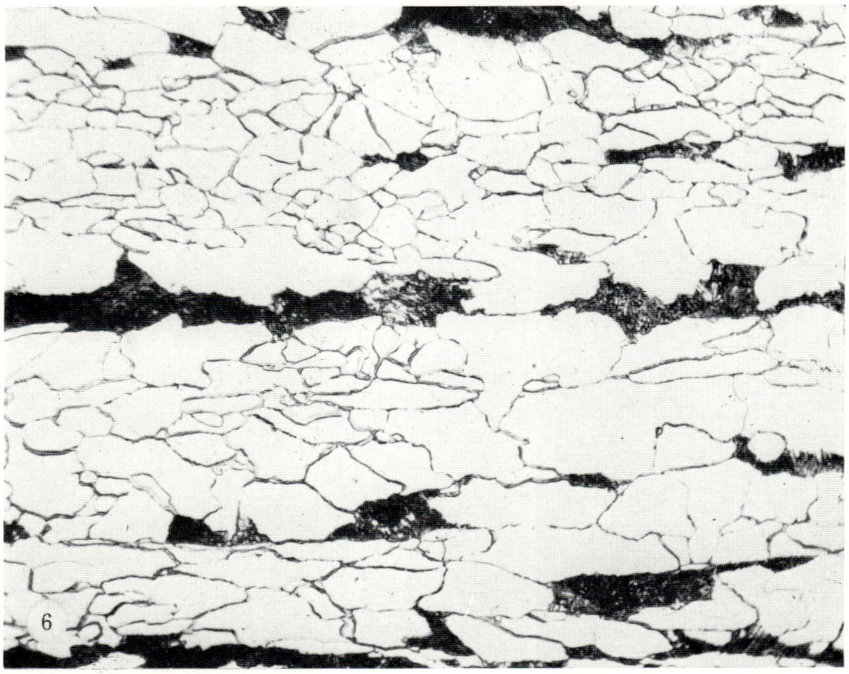

FIGURES 4, 6 and 7. For description see opposite.

(*Facing p.* 292)

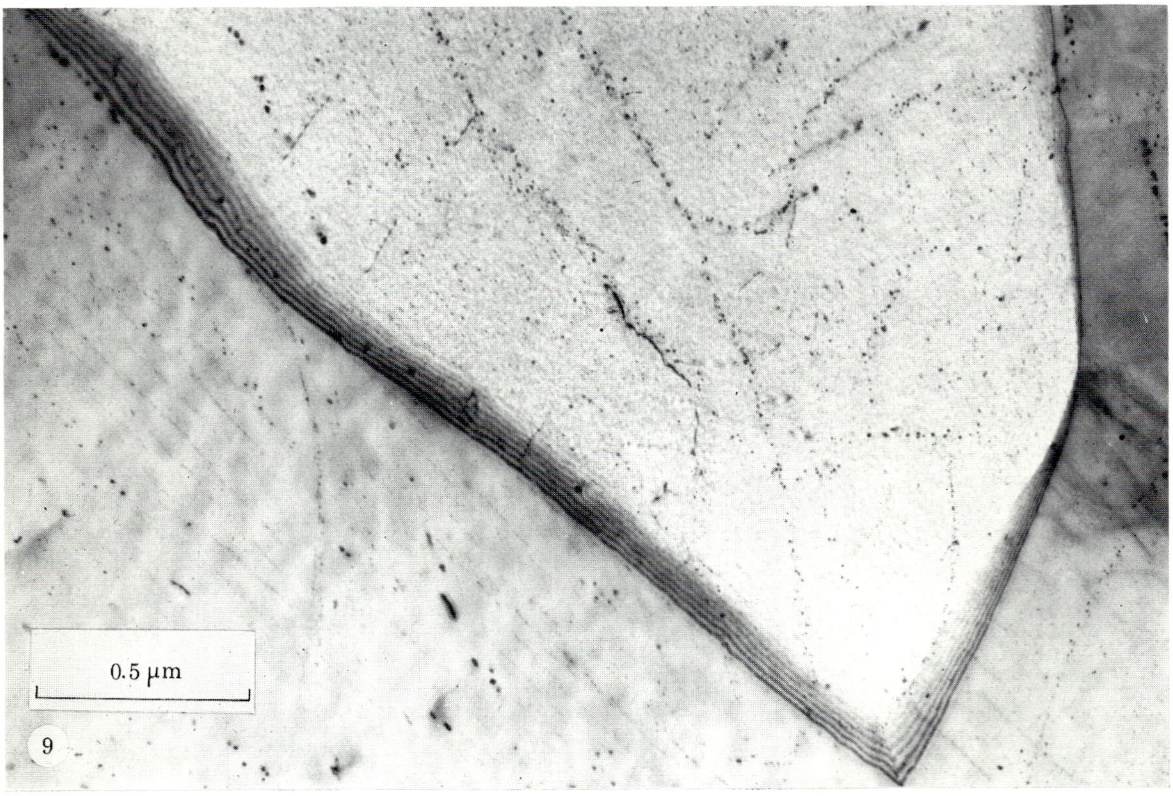

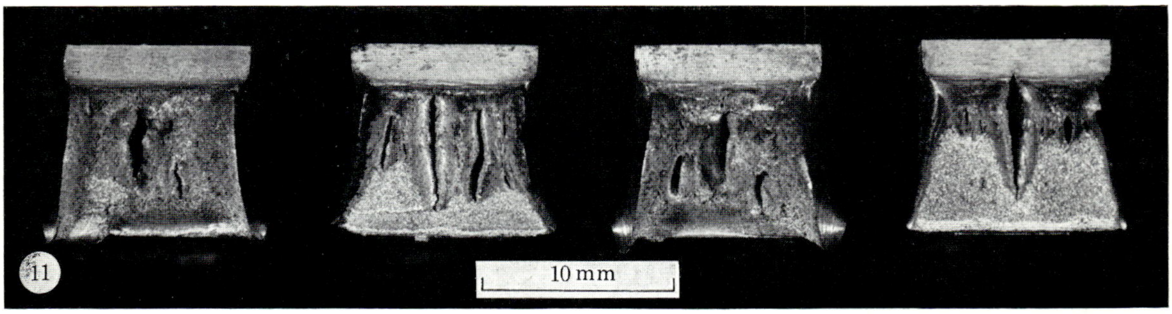

Figures 8, 9 and 11. For description see opposite.

and by the presence of certain alloying elements such as Mn, Mo, etc. which would reduce transformation temperatures.

It would be expected to follow as a result of the austenite grain refinement that the lower the finish rolling temperature the finer the ferrite grain size and figure 5 shows such a relation. Finish-rolling temperatures were varied by varying the holding time prior to the controlled-rolling passes which totalled 64 % reduction. In general, the lower the finish rolling temperature the lower the temperature range over which a significant part of the deformation occurs and in this instance the majority of the rolling was done while the steel was austenitic. Nevertheless, it is not uncommon for certain commercial controlled-rolling schedules to be completed while the steel is undergoing transformation so that the ferrite becomes deformed. The dispersion of fine particles resulting from an addition of micro-alloying elements which aids the refinement of austenite during rolling and which may, under certain circumstances, increase ferrite nucleation rate, also helps to restrict the growth of the ferrite grains after transformation. Grain sizes of about 5 μm may be obtained in thin plates of commercially controlled-rolled micro-alloyed steels (Little *et al.* 1973).

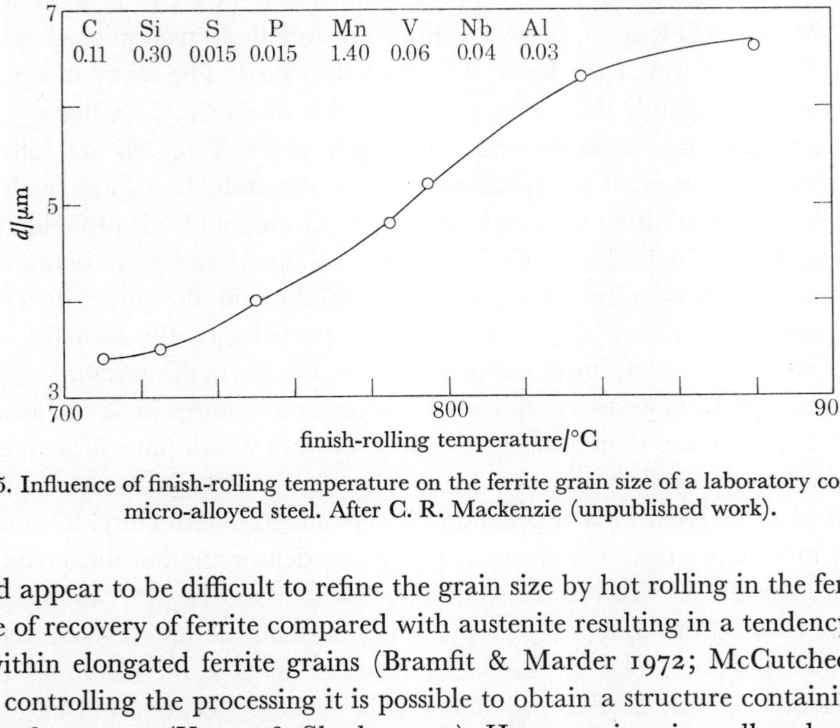

FIGURE 5. Influence of finish-rolling temperature on the ferrite grain size of a laboratory controlled-rolled micro-alloyed steel. After C. R. Mackenzie (unpublished work).

It would appear to be difficult to refine the grain size by hot rolling in the ferrite region due to the ease of recovery of ferrite compared with austenite resulting in a tendency for sub-grains to form within elongated ferrite grains (Bramfit & Marder 1972; McCutcheon & Jamieson 1973). By controlling the processing it is possible to obtain a structure containing well defined sub-grains of ca. 1 μm (Young & Sherby 1973). However, in micro-alloyed controlled-rolled steels it has been found that the ferrite matrix may be only partially recovered with few obvious sub-grains (McCutcheon & Jamieson 1973).

DESCRIPTION OF PLATE 21

FIGURE 8. Dislocation array in a commercial controlled-rolled X65 steel finished at 650 °C. After R. C. Cochrane (unpublished work).

FIGURE 9. Thin foil electron micrograph of a controlled-rolled X65 steel in which the NbC particles have formed in the austenite and in the ferrite. After R. C. Cochrane.

FIGURE 11. Examples of fissures on the fracture surfaces of Charpy V-notch test pieces of controlled-rolled steels.

Microstructure

The outstanding microstructural feature of controlled-rolled steels as discussed in the previous section is the fine grain size. Although a uniform ferrite grain size is attainable in C–Mn steels, there is a tendency to obtain a mixed or duplex grain size in controlled-rolled Nb-treated steel. This duplex structure is possibly related to the fact that Nb slows down the rate of recrystallization during controlled rolling which may lead to a mixed austenite grain size or simply to a deformed austenite, both of which may cause variable ferrite nucleation and growth rates within the material. An example of a duplex structure in a commercially controlled-rolled X65 steel is shown in figure 6, plate 20.

A common microstructural feature of wrought steels is ferrite/pearlite banding. Rolling reduction, although not the cause of banding, tends to accentuate it, particularly if it is carried out in the two-phase region. Other microstructural directionality may arise during controlled-rolling. In steels rolled into the $\gamma+\alpha$ or α regions the ferrite grains become elongated and contain a substructure as illustrated in figure 7, plate 20. In steels containing fine particles such as NbC, there is a tendency for the dislocations to be pinned and for recovery to be less marked than in plain C–Mn steels. Figure 8, plate 21, shows a thin foil electron micrograph of a controlled-rolled X65 steel in which the ferrite has been deformed. The array of dislocations is apparently fairly stable probably due to the pinning action of the NbC particles.

When steels containing micro-alloying additions such as Ti, V or Nb are rolled there is evidence to suggest that substantial precipitation occurs in the austenite region. Such precipitation is known to have an inhibiting influence on recrystallization and it is probable that this is due to the pinning effect of particles which form on the sub-grain and grain boundaries in the austenite. Any micro-alloying element remaining in solution in the austenite can, during transformation, precipitate as carbide or carbonitride particles in the form of inter-phase layers (Gray & Yeo 1968), and at lower temperatures, as a uniform dispersion. Figure 9, plate 21, shows an example of NbC particles which have formed on boundaries in the austenite and which are much larger in size than those in the form of rows which have formed during the austenite to ferrite transformation.

Controlled-rolled steels often have a pronounced crystallographic texture. This could arise in two ways: (*a*) inheritance from the austenite phase; (*b*) deformation of the ferrite.

In a steel transformed from a recrystallized austenite, the texture is rather weak and has two main components $\{100\}\langle 011\rangle$ and $\{112\}\langle 110\rangle$. If the austenite is un-recrystallized prior to transformation then the $\{112\}\langle 110\rangle$ component increases in strength. Assuming the Kurdjumow–Sachs relation, it may be deduced that this latter component would be expected to increase in intensity as the austenite became more deformed (Jones & Walker 1974). When significant amounts of ferrite are present during rolling of C–Mn and C–Mn–Nb steels, i.e. with low finish-rolling temperatures, it has been shown that the intensity of the $\{100\}\langle 011\rangle$ reflexion increases (Coleman *et al.* 1973). In a controlled-rolled C–Mn–Nb steel in which the finish-rolling temperature was relatively low, the intensity of the $\{100\}\langle 011\rangle$ component has been reported to be about $2\frac{1}{2}$ times random (Coleman *et al.* 1973). Recent work by Morrison & B. Mintz, has shown that laboratory rolling a C–Mn steel in the $\gamma+\alpha$ region (80% reduction) can give mid-thickness intensities of the conventional cold rolling texture $\{100\}\langle 011\rangle$ of about 5 times random. However, it is of interest to note that in this material texture gradients existed between surface and centre so that the average through thickness intensity of the

{100} ⟨011⟩ orientation was approximately $2\frac{1}{2}$ times random. Texture gradients have also been found by other investigators in hot band material (Takechi *et al.* 1968; Richards 1969). No texture gradients, apart from a surface effect, have been found in commercially controlled-rolled X65 steels.

Mechanical properties

Grain refinement is one of the few methods available for strengthening as well as improving impact properties. Consequently, controlled rolled steels can have a very attractive combination of mechanical properties.

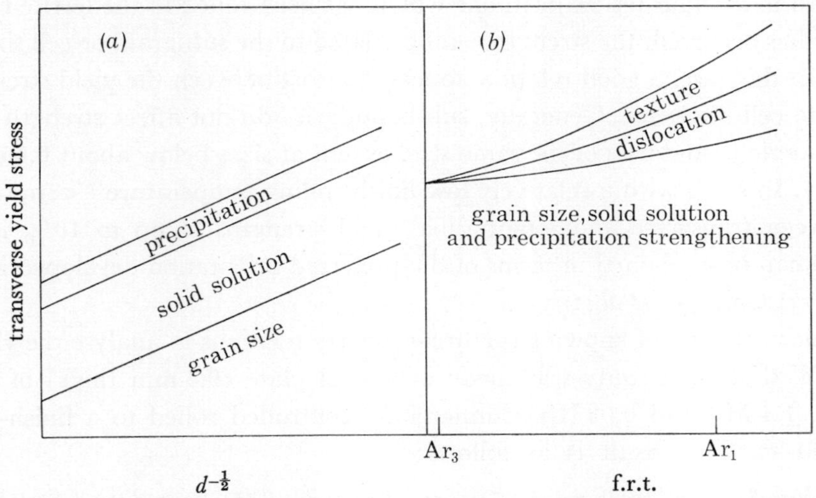

Figure 10. Contributions to the strength of a controlled-rolled C–Mn–Nb steel: (*a*) finish-rolling temperatures above Ar_3; (*b*) finish-rolling temperatures below Ar_3.

The various contributions to the strength of a typical controlled-rolled steel of the C–Mn–Nb type are presented schematically in figure 10*a* for finish-rolling temperatures in the austenite range. When finish-rolling temperatures are below the Ar_3 further increases in yield strength can be expected from dislocation strengthening, together with an increase in the transverse yield strength from texture strengthening, although there may possibly be a smaller contribution from precipitation hardening (figure 10*b*). The yield strength σ_y of controlled-rolled steels can be described as follows:

$$\sigma_y = \sigma_i + \sigma_{s.s.} + \sigma_{ppt} + \sigma_{disl} + \sigma_{text} + k_y d^{-\frac{1}{2}},$$

where σ_y is the conventional room temperature transverse yield strength, σ_i is the lattice friction stress and $\sigma_{s.s.}$, σ_{ppt}, σ_{text} and σ_{disl} are the strength contributions from solid solution, precipitation, texture and dislocation hardening respectively. The value of the term k_y which represents the grain size strengthening contribution can be taken as 18–19 $N/mm^{\frac{3}{2}}$ for as-rolled steels (Morrison 1966). The term σ_i which relates to pure ferrite has a value of approximately 32 N/mm^2 (Morrison & Leslie 1973). Values for the term $\sigma_{s.s.}$ which may be due to both interstitial and substitutional elements can be derived from data from various sources (Lacy & Gensamer 1944; Pickering & Gladman 1963; Leslie 1972). It is difficult to calculate a value for σ_{ppt} and in a steel with a normal dislocation density, this is usually assumed to be the difference between the measured and calculated σ_y values. Gray (1972) has published data indicating how the yield strength varies with size and volume fraction of NbC precipitates.

For a steel finish rolled in the ferrite region, it is very difficult to distinguish between the contributions due to precipitation and dislocation hardening. However, it has been shown (Little *et al.* 1973) to be possible to calculate reasonably accurate values for σ_{disl} by measuring dislocation densities using transmission electron microscopy and using the relation found by Keh (1965) between strength and the square root of dislocation density,

$$\sigma_{disl} = \sigma_i + 0.38 G b \sqrt{N}$$

where G is the shear modulus, b, the Burger vector, and N the average dislocation density. In a steel transformed to acicular ferrite or in one which has been rolled in the ferrite region and in which recovery has occurred, the strength can be related to the sub-grain or cell size (Young & Sherby 1973). In this case, a good relation appears to exist between the yield strength and the reciprocal of the cell diameter. Generally, sub-boundaries do not affect strength to the same degree as high angle boundaries of the same size, except at sizes below about 0.45 μm (Young & Sherby 1973). In steels with a relatively low finish rolling temperature ($< Ar_3$) a significant difference between transverse and longitudinal yield strengths of up to 10% is commonly observed which can be explained in terms of the preferred orientation developed in such steels (Little *et al.* 1973; Coleman *et al.* 1973).

As an example of the use of known structure–property relations to analyse the yield strength of controlled-rolled steel, a study was made of a steel plate (9.5 mm thick) of composition 0.13 C, 0.36 Si, 1.4 Mn and 0.03 Nb commercially controlled rolled to a finish-rolling temperature of 650 °C. The result is as follows:

$$\sigma_y(550\ \text{N/mm}^2) = \sigma_i(32) + \sigma_{s.s.}(135) + \sigma_{ppt}(35) + \sigma_{text}(45) + \sigma_{disl}(43) + k_y d^{-\frac{1}{2}}(260).$$

The relative size of the contribution from grain size is probably fairly typical of controlled-rolled steels but the contribution from the other sources can vary significantly depending on composition and rolling schedule.

The fine grain size of controlled-rolled steels also plays a dominant role in determining fracture properties. The fracture-appearance criteria (Battelle d.w.t.t. and Charpy V-notch) are improved and the energy absorbed in fracture is increased by decreasing grain size and there is also evidence to suggest that a fine, well-defined substructure can have a similar effect (McCutcheon & Jamieson 1973; Irani & Latham 1968). Various attempts have been made to analyse the Charpy V-notch impact properties in terms of structure and composition although this has not proved to be as successful as in the case of yield strength. Pickering & Gladman (1963) have shown that a decrease in grain size (or increase in $d^{-\frac{1}{2}}$) causes a decrease in impact transition temperature of 11.6 °C/mm$^{-\frac{1}{2}}$. Although this value was obtained from tests on normalized steels a similar value has recently been obtained in controlled-rolled steels. Another structural element which has an important influence on impact properties is the percentage pearlite and this has been shown to cause an increase in transition temperature of 2.2 °C per percent (Pickering & Gladman 1963). There appears to be some experimental justification for assuming that increases in strength through precipitation (Gray 1972) and dislocation (Irvine 1970) hardening leads to an increase in 50% f.a.t.t. of approximately 0.5 °C/(N/mm^2). Although elements in solid solution generally increase strength and also cause a deterioration in impact properties, the detailed influence of certain elements, e.g. N (Mintz 1973), Mn (Jolley 1968), Ni (Leslie 1972), is sometimes rather complex. The fracture surfaces of notched test pieces of controlled-rolled steel often reveal fissures and examples are shown in figure 11 plate 21. The fissures are normally associated with microstructures which are directional in nature and show banding, elongated grains, extended inclusions, etc. (Morrison 1975). Steels with such microstructures appear to contain 'planes of weakness' parallel to the rolling plane which fracture ahead of the notch to form fissures and relax the triaxial-stress condition, thus reducing the possibility of brittle fracture. The appearance of fissures is often associated with low finish-rolling temperatures and it is found that even although substantial ferrite deformation occurs the associated deterioration in impact properties due to dislocation hardening is counteracted to a certain extent by the occurrence of fissuring.

A controlled-rolled steel usually has ductile fracture properties which vary with test direction (Morrison 1975). This directionality of properties can largely be related to the shape of inclusions present in the steel. Due to the fact that much of the rolling in a controlled-rolled steel is done in a temperature regime within which the strength of the inclusion compared with the steel matrix is relatively low, the inclusions tend to be more deformed than in a steel conventionally hot rolled (Charles 1974). To improve the tensile ductility and notch-impact toughness it is necessary to either reduce the inclusion content, e.g. by desulphurization, or make special additions to the steel of elements which modify inclusion shape (Luychx *et al.* 1970).

Commercial controlled-rolling procedures

Translation of the above fundamental considerations into production practice involves solving constraints, the metallurgical difficulties of which are more or less directly proportional to the final properties required as a function of the thickness. Thus, relatively modest property requirements at a relatively thin gauge can be achieved with a minimum of metallurgical controls on the mill as, for example, in 15 mm ship plate on a single-stand mill where, after soaking at normal temperatures, the slabs can be roughed at normal speeds to a holding thickness of say three times final gauge, held to a set temperature, and rolled out to plate. The roller basically need only be told the holding thickness, end hold temperature and maximum rolling load the mill can carry. Metallurgically, the product is not demanding of the mill and the controls set are simply to ensure adequate reduction in the appropriate final rolling temperature range. Much more serious constraints for these products exist in the wider production context, noting that such products can also be readily produced normalized. To optimize total plant production efficiency therefore, involves productivity considerations like tonnage rolled per hour through the mill, slab tonnage delivered to the mill and shearing, finishing, etc. capacity; great emphasis must be placed on optimizing schedules which fit with these considerations.

Particularly in times of heavy steel demand, application of rolling schedules which maximize mill complex throughput without detriment to achieving specified properties is a major requirement and one of the central difficulties is of course the delay on a single stand mill which must be incorporated, when normal reheat temperatures are used, part-way through the schedule for deformation/temperature reasons. Several procedures exist for achieving the delay:

(*a*) *In-line, single-stand mills*

In its simplest form this presents the most straightforward practice since on a single-stand mill slabs are part rolled, held on the table till the temperature falls to the required level and then finish rolled to gauge. The length of the delay represents completely 'dead' time since nothing else can be rolled during the delay but the technique is useful and widely used for thin plates where delays are relatively short. More complex arrangements whereby one or more slabs are partially rolled and held in line on the ingoing or outgoing tables while another is being rolled have been put into practice for minimizing the delay with in-line holding on a single-stand mill and substantial improvements in productivity can be demonstrated.

(b) In-line, 2-stand mills

The position here is naturally substantially different from a single-stand mill since, with adequate space between the stands, slabs can be roughed at maximum rate and queued in front of the finishing stand such that the finishing stand works 'continuously' and the time delay on each slab is adequate to allow the required temperature drop to take place. Controlled rolling therefore is done in batches. If the interstand distance does not allow queuing, in-line delays with normal reheat temperatures will incur the possibility of 'dead time' at the finishing stand similar to single-stand mills.

(c) Off-line delays

This practice is widely followed on single-stand mills in the U.K. and abroad where final plate thickness is such that in-line delays would incur unacceptable 'dead' time. Various designs for off-line delay stations are employed to either lift the part-rolled slab above the roller table or to move the slab horizontally off-line. With off-line delays, depending on the number of holding stations and final plate thickness, controlled-rolled plate orders can either be handled in batches or mixed with normally-rolled slabs such that the holding time is a multiple of the total rolling time for normally-rolled slabs and the reheaters are programmed to deliver, say, 1 slab for controlled-rolling for every 3 slabs for normal rolling. With good matching of times, thicknesses, etc., the loss of net productivity to the mill can be minimized.

A further approach towards optimizing productivity by minimizing holding delays is to deliberately reduce the reheat temperature such that the thickness/temperature combination at the end of roughing are close to that required for entry into the finishing sequence. The approach is highly desirable from a metallurgical point of view, not only because of the fine starting grain size but also as a result of the lower temperature at the start of the 'holding period' both of which have an important effect on final properties. Other effective, but rather less satisfactory approaches, for achieving the same thickness/temperature combination at the end of roughing rely on using normal reheat temperatures but withdrawing the slab early from the furnace and allowing it to cool with or without additional water cooling prior to roughing; this cooling can be enhanced by further water cooling during the roughing sequences such that again the part-rolled slab arrives at the temperature/thickness combination for entry into the finishing sequence with minimum delay. Minimizing the delay presents serious practical problems with either practice when relatively thick plates are being controlled rolled because achieving adequate finishing deformation necessitates thicknesses around 100 mm at the delay stage at which level the rate of cooling for the body of the slab is slow. In addition care has to be taken not to incur excessive cooling at the slab ends which can cause rolling difficulties and/or difficulties in entering. The importance of adequate finishing deformation in obtaining satisfactory impact properties is shown in figure 12. Both procedures, and particularly the low reheat temperature scheme, can incur some disadvantage due to increased stiffness during roughing and the possibility therefore of lower reduction per pass. The low reheat temperature approach is practiced for achieving the more stringent line pipe properties and is also finding increased acceptance in structural plate, especially where difficult property combinations are required at higher thicknesses (Cirallero & Parrini 1971; Maekawa *et al.* 1972). Figure 13 shows some Italsider results which indicate that as the soaking temperature decreases from 1200 to 980 °C the 90% f.a.t.t. ($\frac{2}{3}$ Charpy) decreases from 0 to −50 °C while over the same temperature range there is a drop of 30 N/mm² in the yield stress.

Table 1 illustrates the results obtained on 38–40 mm plates which were controlled rolled in an early development exercise on a U.K. mill to achieve BS 4360 Grade 50D properties in

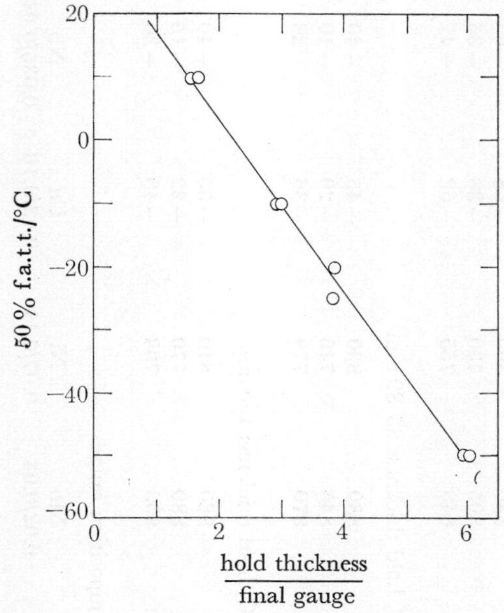

FIGURE 12. Influence of finishing deformation on the temperature for 50% f.a.t.t. (transverse), 9.5 mm X65 quality.

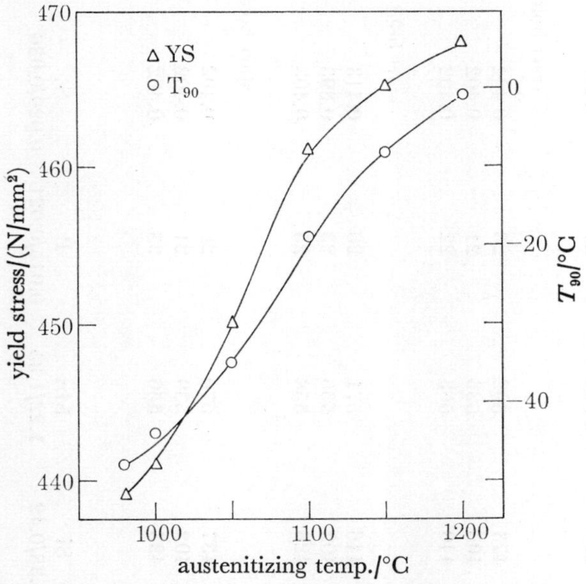

FIGURE 13. Influence of austenitising temperature ($1\frac{1}{2}$ h soak), on the yield stress and temperature for 90% f.a.t.t. of controlled-rolled C–Mn–Nb steel (Cirallero & Parrini 1971).

Table 1. Mechanical properties of C–Mn–Nb steels controlled rolled to 38–40 mm thick plates

	y.s. N/mm²	u.t.s. N/mm²	elongation % on 20 cm	c.e.v.	start hold temp./°C	end hold temp./°C	1st gauge temp./°C	temp./°C (for 27 J)	temp./°C (for 34 J)	temp./°C (for 41 J)	no. of results
				start hold < 1000 °C, hold thickness 85–89 mm							
max.	466	578	25	0.413	985	880	790	−70	−65	−65	
min.	422	556	23	0.402	940	830	730	−45	−40	−35	
av.	446	566	24	0.406	964	853	750	−57	−54	−52	6
				start hold < 1000 °C, hold thickness 100 mm							
max.	473	592	23	0.402	995	880	800	−68	−60	−60	
min.	405	536	21	0.402	940	805	730	−38	−35	−40	
av.	440	563	22	0.402	958	848	755	−52	−47	−52	3
				start hold > 1000 °C, hold thickness 85–89 mm							
max.	446	571	26	0.413	1030	880	830	−45	−40	−35	
min.	409	536	23	0.393	1005	845	745	−20	−10	−10	
av.	426	558	25	0.403	1021	870	779	−36	−28	−24	6
				start hold > 1000 °C, hold thickness 100 mm							
max.	437	570	26	0.402	1025	960	810	−55	−40	−30	
min.	402	539	21	0.402	1005	880	770	−22	−15	−10	
av.	423	556	23	0.402	1016	903	798	−40	−30	−25	7

overall composition range

C	Si	Mn	P	S	Cr	Mo	Ni	Cu	Nb	Sn	c.e.v.
0.15/0.155	0.38/0.42	1.22/1.35	0.015/0.021	0.020/0.036	0.04/0.13	0.02/0.04	0.07/0.11	0.13/0.16	0.035/0.043	0.013/0.016	0.393/0.413

Mechanical properties of C–Mn–Nb steels controlled rolled to 38–40 mm thick plate at 0.36/0.38 c.e.v.

	y.s. N/mm²	u.t.s. N/mm²	elongation % on 20 cm	c.e.v.	start hold temp./°C	end hold temp./°C	1st gauge temp./°C	V notch Charpy J at 0 °C	hold thickness	no. of results
max.	453	570	27	0.379	1025	880	815	92	89	
min.	349	511	22	0.367	985	785	775	60	89	
av.	408	541	25	0.372	1004	833	797	77	89	7

killed Nb steel with a maximum c.e.v. of 0.41. As part of the exercise 'start hold' temperature was varied in the range 940–1030 °C by a combination of withdrawing slabs early from the furnaces, water cooling during roughing and holding where required on a cradle for up to 6 min. Prime requirements were at this stage also that the schedule be easy to remember and operate, and that mill loads did not exceed 4500 tonnes. Table 2 gives further practical details and figure 14 shows the load-pass profile during rolling. It was found that the required mechanical properties could be met at 0.36 min. c.e.v.; this represents an important advance over normalized steels made for the same quality which characteristically require analysis levels close to the maxima. Adoption of the controlled rolling option thereby allows greater flexibility at the steel plant, release of normalizing capacity and general leaner analysis. Additionally, examination of these and other data suggests that semi-killed controlled rolled qualities may be used for BS 4360 Grade 50 level for say 30 mm plates to 0.41 c.e.v. max. or for Grade 50C at 0.43 c.e.v. max.; use of semi-killed steels in ingot-based production presents important

Table 2. Details of average temperatures and reductions during the final passes for C–Mn–Nb steel roughed to 100 mm and rolled to 40 mm thick

pass no.	mill setting mm	% reduction	temp/°C
1	95	5.0	870
2	85	10.5	860
3	75	11.8	850
4	65	13.3	840
5	60	7.7	830
6	55	8.3	820
7	50	9.1	810
8	45	10.0	800
9	40	11.1	790
10	38	5.0	780

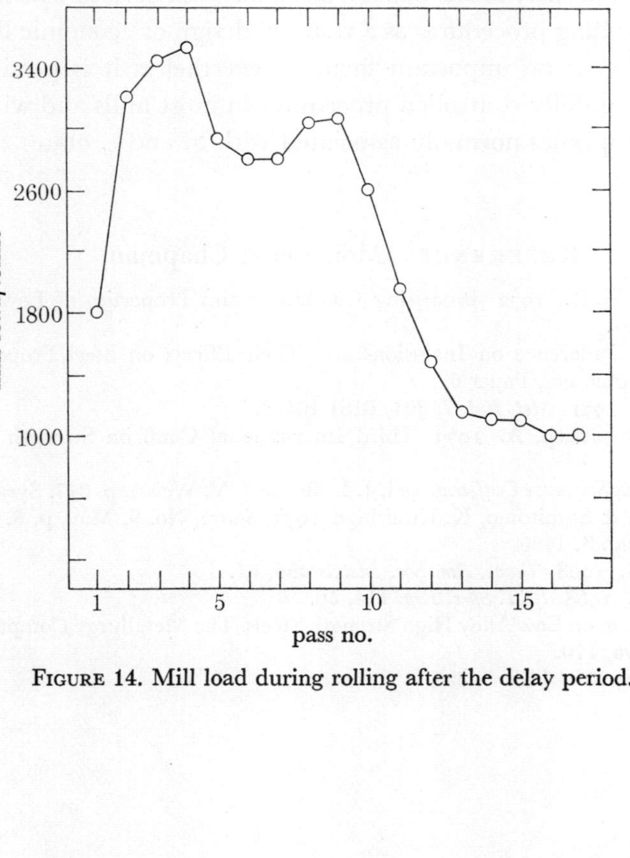

Figure 14. Mill load during rolling after the delay period.

practical gains in yield, pitside time, etc., with consequent cost savings. The results (table 1) indicate also a route for providing thick grade 55 steels at c.e.v. alloy levels (0.43 max.) substantially lower than can be met with normalized steels and at costs substantially below q. and t. steels. Impact properties given in table 1 are good and adequate for most structural requirements including BS 4360 Grade D and Lloyds DH level; further schedule and analysis development can be used both to improve the strength/c.e.v./alloy guarantee at BS 4360 Grade 50 and particularly grade 55 level, and to improve impact properties further towards the grade E levels even at 38/40 mm plate thickness.

Control of finishing temperature and finishing deformation naturally plays an important role in controlled rolling and has a significant effect on structure and properties as noted earlier. Under normal circumstances, finishing temperatures are kept as high as possible to minimize mill loadings and maximize output. Maximum and minimum set finishing temperatures in controlled rolling are usually a compromise between the metallurgical requirements and the mill capability but though some authorities claim that very low finishing temperatures should be avoided for metallurgical *and* mill reasons, it is important to recognize that rolling into the 2-phase region or ferrite region can confer substantial benefits to mechanical properties; further, by careful schedule control, mill breakdown need not be risked and use can be made of the fact that the phase forming during transformation is softer than the pre-existing phase. The property levels obtainable with relatively low-temperature rolling have already been described (Little *et al.* 1973) and it will be noted that extremely good combinations of strength and toughness were obtained. Such property combinations are adequate for many requirements normally demanding q. and t. plates.

It has been shown that there is now a reasonable understanding of the mechanism of grain refinement in controlled rolled steels and of the role of composition in aiding grain refinement and in determining the mechanical properties. The relation between strength and structure is also fairly well known but there are large gaps in our knowledge of the microstructural features which control impact properties. It has been pointed out that certain constraints are placed on commercial controlled rolling procedures as a result of design or economic limitations, and that productivity optimization is an important factor. Nevertheless it is possible to improve the properties of plates by carefully controlled procedures in most mills and with certain optimum procedures to obtain properties normally associated with q. and t. plates.

References (Morrison & Chapman)

Bramfitt, B. L. & Marder, A. R. 1972 Processing Conditions and Properties of Low-Carbon Steel. AIME Symp. Cleveland, p. 191.

Charles, J. A. 1974 BSC Conference on Inclusions and Their Effects on Steel Properties. Bodington Hall, University of Leeds, September, Paper 6.

Cirallero, M. & Parrini, C. 1971 *Met. Ital.* **7**, 301, BISI 10428.

Coleman, T., Dulieu, D. & Gouch, A. 1973 Third International Conf. on Strength of Metals and Alloys. Cambridge, p. 70.

English, A. T. 1962 *Proc. 9th Sogamore Conference* (ed. J. L. Burke & V. Weiss), p. 317. Syracuse University Press.

Fukuda, M., Hashimoto, T. & Sumitomo, K. Kunishige 1973 *Search*, No. 9, May, p. 8.

Gray, J. M. 1972 *Met. Trans.* **3**, 1495.

Gray, J. M. & Yeo, R. B. G. 1968 *Trans. Am. Soc. Metals* **255**, 61.

Irani, J. J. & Latham, D. J. 1968 *I.S.I. Spec. Rep.* **114**, 56.

Irvine, K. J. 1970 Symposium on Low Alloy High Strength Steels. The Metallurgy Companies, Nuremburg, p. 1.

Jolley, W. 1968 J.I.S.I., **206**, 170.

Jones, A. & Walker, B. 1974 *Metal Sci.* **8**, 397.

Keh, A. S. 1965 *Phil. Mag.* **12**, 9.
Lacy, C. E. & Gensamer, M. 1944 *Trans. Am. Soc. Metals* **32**, 88.
Leslie, W. C. 1972 *Met. Trans.* **3**, 5.
Little, J. H., Chapman, J. A., Morrison, W. B. & Mintz, B. 1973 Third International Conf. on Strength of Metals and Alloys, Cambridge, p. 80.
Luychx, L., Bell, J. R., McLean, A. & Korchynsky, M. 1970 *Met. Trans.* **1**, 3341.
McCutcheon, D. B. & Jamieson, R. M. 1973 *Can. Met. Q.* **12**, 105.
Maekawa, S., Miyano, K. & Shimazaki, M. 1972 *Tetsu-to-Hagane*, **58**, 1861.
Mintz, B. 1973 J.I.S.I., **211**, 433.
Morrison, W. B. 1966 *Trans. Am. Soc. Metals* **59**, 824.
Morrison, W. B. 1972 *J.I.S.I.* **210**, 618.
Morrison, W. B. 1975 *Metals Tech.* To be published.
Morrison, W. B. & Leslie, W. C. 1973 *Met. Trans.* **4**, 379.
Pickering, F. B. & Gladman, T. 1963 *I.S.I., Spec. Rep.* **81**, 10.
Priestner, R. 1974 Paper presented at Conference on Controlled Rolling Applications to Plates. Paris, October (1974). *Rev. Met.* (To be published.)
Priestner, R. 1975 *Metals Technology*. Submitted.
Priestner, R. & Biring, M. S. 1973 *Metal Sci. J.* **7**, 60.
Priestner, R., Earley, C. C. & Rendall, J. H. 1968 *J.I.S.I.* **206**, 1252.
Richards, P. N. 1969 *J.I.S.I.* **207**, 1333.
Sekine, H. & Maruyama, T. 1973 Third International Conf. on Strength of Metals and Alloys. Cambridge, p. 85.
Smith, Y. E. & Siebert, C. A. 1970 Application of Modern Metallographic Techniques. A.S.T.M., S.T.P. 480.
Takechi, H., Kato, H. & Nagashima, S. 1968 *Trans. Am. Inst. Mech. Engng* **242**, 56.
Vanderbeck, R. W. 1958 *Weld. J.* **37**, 114S.
Weiss, H., Gittins, A., Brown, G. G. & Tegart, W. J. McG. 1973 *J.I.S.I.* **211**, 703.
Young, C. M. & Sherby, O. D. 1973 *J.I.S.I.* **211**, 640.

3.3 Development of line pipe steels

By D. B. Clay and D. B. McCutcheon
Stelco, Steel Company of Canada, Ltd, Stelco Tower, Hamilton, Ontario L8N 3T1, Canada

The demand for large diameter gas line pipe with high yield strength and high notch toughness has led to increased research in high strength low alloy steels. Physical metallurgists have developed both a fundamental and an empirical understanding of the properties and microstructures of ferrite pearlite steels. As specifications become more rigorous, alternative microstructures (low carbon bainite or tempered bainite and martensite) are being used in line pipe. There is a definite need for metallurgists to develop a clearer understanding of the interrelationships between their properties, microstructures, and processing before these alternatives are completely accepted for use in line pipe.

The increasing worldwide demand for energy has led to an unprecedented expansion in the construction of large diameter gas pipelines in the past decade. Ever-increasing volumes of gas are being transported by gas pipelines over greater distances. To take advantage of the economies of scale, larger diameter lines, operating at higher pressure have been constructed to increase the throughput and lower the operating cost over the life of the lines (White 1970). These higher operating pressures have necessitated higher yield strength steels and/or greater wall thicknesses. To meet this demand, physical metallurgists have developed new high strength low alloy (h.s.l.a.) steels with a relative strength-section cost advantage over lesser strength grades. Their acceptance is reflected by the fact that the gross tonnages of skelp grades produced in the last decade have contained an increasing proportion of high yield strength grades (figure 1).

The intention of this paper is to summarize the key changes in the specifications for large diameter gas transmission lines during this growth period; outline the underlying metallurgical principles used to develop the new grades; comment on the economics of the possible processing; and review the future trends in pipeline specifications and how they will be met.

Specifications

The U.S.A. contains the largest total length of large diameter gas transmission lines and thus, has had more pipeline design and operating experience than any other country. Consequently, the basis for pipeline specifications throughout the world are those of the American Petroleum Institute (A.P.I.) – specifically API 5LX which covers eight main grades from X42 to X70, based on the pipe yield strength. Most of this specification deals with procedures for manufacturing and testing of the pipe. The sections relevant to this paper are those covering strength and toughness.

The critical change in the specification during the past decade has been the more demanding toughness requirements at higher strength levels. The pipe yield strength and wall thickness only ensure that the line can operate at a particular pressure. The yield strength does not guarantee the integrity of the line with respect to arresting a propagating fracture. The steady increase in yield strength requirements have been simply related to design increases in operating pressure. The more stringent toughness specifications have arisen from metallurgical analysis of operating failures in gas pipelines in the U.S. Studies of pipeline failures in the 1950s by the Battelle Memorial Institute concluded that the criteria for arresting a fracture was that the speed of the propagating crack must be lower than the speed of the gas decompression wave associated with the crack (AGA 1966). Battelle correlated the fracture speed to the appearance (ductile-brittle) of the fracture surfaces, and developed the drop weight tear test (B.d.w.t.t.) as a laboratory method to determine the appearance of the fracture as a function of test temperature (McClure, Duffy & Eiber 1965). This test was included in the A.P.I. specification in 1969 to ensure that fracture in a pipe steel was ductile at the operating temperature.

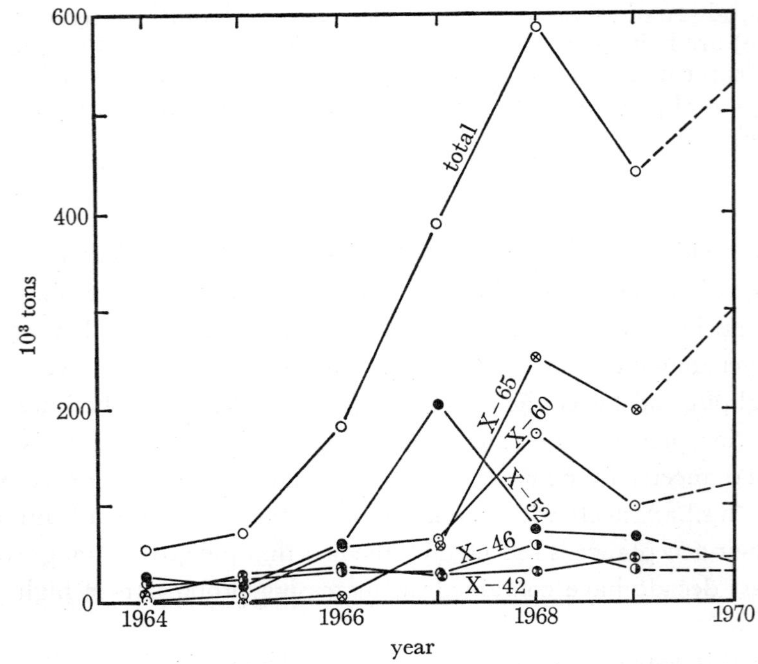

FIGURE 1. Gross tonnages of skelp grades X42 to X65.

In 1969, two shear type fractures in a U.S. gas line demonstrated that long failures could occur in a ductile manner (AGA 1969). Consequently, a new criteria was developed to ensure crack arrest – the energy absorbed during fracture should exceed the crack driving force. More recent gas pipe specifications include a minimum Charpy (C_v) energy requirement at the operating temperature, based upon empirical correlations, considering the operating pressure and the pipe geometry (AGA 1974).

There has also been increased interest in the toughness of the weld metal and heat affected zone (h.a.z.) in the pipe seam and girth welds. In the past, specifications have not been severe in this area, simply requiring that the weld regions be sound and be below a maximum hardness. To ensure weldability, various empirical carbon equivalent formulations were developed for the base metal. By and large, these are hardenability type equations which ignore the deleterious effects of microalloys and processing on the weld toughness. Increased research in the area of submerged arc welding has realized the potential for improved toughness by altering base plate, filler wire and flux compositions and process variables. Laboratory simulation of welds, h.a.z. and weld testing procedures are now being developed. Much more research is required before metallurgists fully understand the weld region in terms of its microstructure and related mechanical properties. This research is being prompted by the inclusion of weld toughness requirements in supplementary specifications.

Metallurgy of current line pipe steels

The increased demand for steels with both higher strength and toughness led to the realization that conventional chemistries and processing were not adequate. To understand the changes that have been adopted, let us review the metallurgical principles used in present line pipe steels.

The microstructures of pipe steels for grades X52 to X65 are commonly polygonal ferrite pearlite. Fortunately, the strength and transition temperature of these types of steel can be considered as being almost independent of the ductility or Charpy shelf energy.

Yield strength and transition temperature

Both the yield strength and the transition temperature can be improved by refining the ferrite grain size. This can be understood in terms of the Petch (1953) yield strength relation:

$$\sigma_y = \sigma_i + K_y d^{-\frac{1}{2}}$$

where σ_y is yield strength, σ_i the friction stress, K_y the strength coefficient, d the grain diameter, and the Cottrell (1958) approximation for fracture strength

$$\sigma_{\text{fracture}} = \tfrac{1}{2}\sigma_f + 2(\mu\gamma)d^{-\frac{1}{2}},$$

where σ_f is the frictional stress on a slip plane, μ shear modulus and γ surface energy.

The ductile-brittle transition temperature is that at which the flow and fracture stresses have the same value at the point of fracture. Since with grain refinement, the fracture stress (as a function of temperature) is raised more than the flow stress, the transition temperature is lowered (figure 2) (Hahn, Averback, Owen & Cohen 1959).

Empirical equations have been developed, following the form of the Petch equation, to predict both the strength and impact transition temperature of ferrite pearlite steels (Irvine, Pickering & Gladman 1967):

$$\sigma_y\ (\text{klbf/in}^2) = 7.8 + 4.7\,\%\ \text{Mn} + 12.0\,\%\ \text{Si} + 51\sqrt{(\%\ N_f)} + 2.5d^{-\frac{1}{2}} + \text{ppt}$$

$$\text{t.t.}\ (^\circ\text{C}) = -19 + 44\,\%\ \text{Si} + 700\sqrt{(\%\ N_f)} - 11.5d^{-\frac{1}{2}} + 2.2\,\%\ \text{P}$$

where % Mn is % manganese; % Si, % silicon; % P, % pearlite; % N_f, % free nitrogen; d, average grain diameter; ppt, precipitation strengthening increment.

Interpreting these equations in terms of the rate of increase in transition temperature, as a function of strengthening, grain refinement is the only factor that lowers the transition temperature. Solid solution strengtheners are deleterious to the transition temperature unless these elements have secondary effects, such as refining the ferrite grain size (by either lowering the transformation temperature or combining with other elements to form grain refining precipitates) or combining with free nitrogen and carbon. Similarly, precipitation and dislocation strengthening raise the transition temperature at the rate of ~7 °F per 10^3 lbf/in^2 increment in yield strength (4 °C per 15 N/mm^2) (Heslop & Petch 1958). This rate of increase can be reduced in precipitation strengthened steels by the removal of free nitrogen and carbon by the precipitates (Gray 1972). Pearlite has the indirect strengthening effect of reducing the ferrite grain size, and raises the ductile-brittle transition temperature.

As the specifications have demanded higher yield strengths and/or greater wall thicknesses, additional grain refinement has been needed to meet the transition temperature requirements. This is increasingly difficult on thicker plate because of increased inhomogeneity of the microstructure from surface to core, and the thickness dependency of the transition temperature (figure 3) (Iwasaki *et al.* 1975).

Lower transition temperatures have been achieved by using a normalizing treatment after

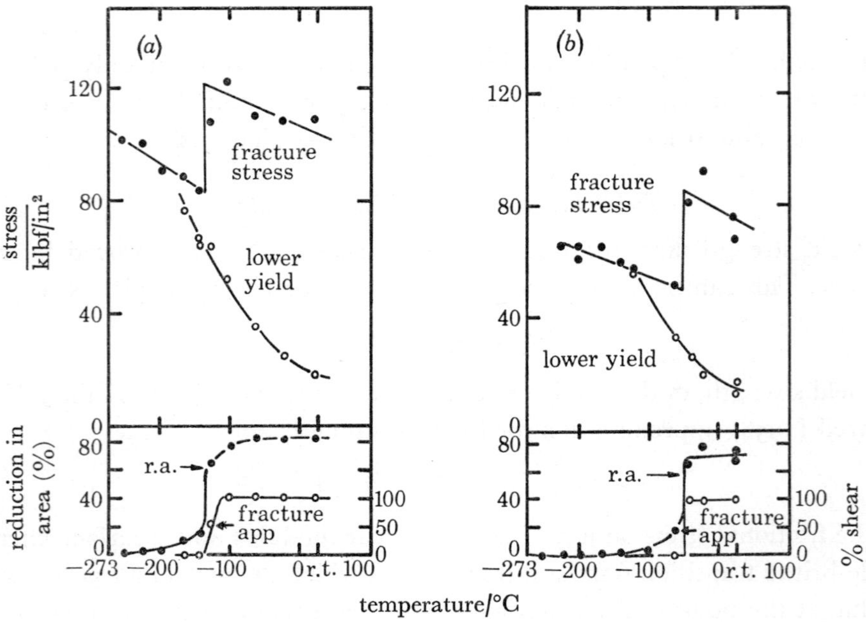

FIGURE 2. Tensile properties as a function of temperature for (*a*) coarse (0.113 mm) and (*b*) very coarse (0.409 mm) grained steels (Hahn *et al.* 1959).

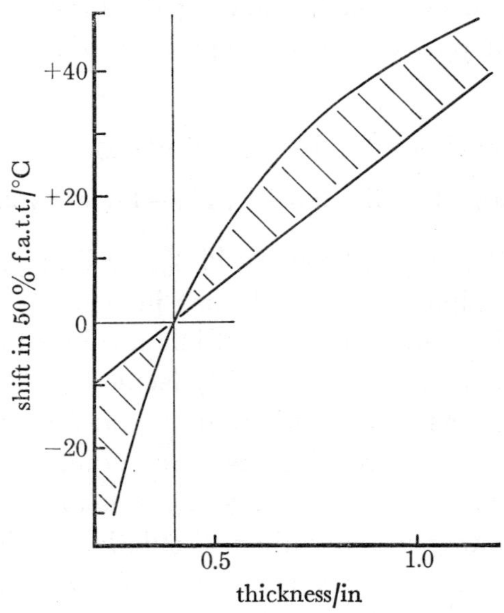

FIGURE 3. Thickness dependency of fracture appearance transition temperature in a b.d.w.t.t. after Iwasaki *et al.* (1975).

rolling, and controlled rolling and/or cooling. With normalized steels the ferrite grain size can be limited by lowering the transformation temperature by alloying and restricting the austenite grain size at the reheat temperature by the presence of second phase particles (Gladman & Pickering 1967). Strengthening over and above that due to grain refinement is achieved principally by fine second phase particles dissolved at the reheat temperature and precipitated during or after the austenite to ferrite transformation.

In controlled rolled steels, the ferrite grain size can be reduced by limiting the austenite grain size prior to transformation. Austenite grain size control is achieved by the addition of precipitation strengthening elements (principally Nb, and sometimes Ti and V). These elements retard recrystallization of the austenite and, as precipitates, limit austenite grain growth. In plain carbon steels, the austenite recrystallizes between passes even at low rolling temperatures. With steels microalloyed with niobium the deformation is cumulative below *ca.* 900 °C (1650 °F) for most rolling practices. Elongated, heavily dislocated austenite is developed which transforms to a fine ferrite grain size (Fukuda, Hashimoto & Kunishige 1973; Kozasu, Shimizu & Tsukada 1972). To optimize the properties achieved by controlled rolling, metallurgists have developed an empirical understanding of

(i) the solubility of the precipitation compounds,

(ii) their precipitation C-curves under rolling conditions,

(iii) the effect of precipitates on austenite grain growth during reheating prior to rolling,

(iv) the effect of precipitate forming elements on the recrystallization kinetics of austenite under rolling conditions,

(v) the effect of the elements in solid solution in the austenite on the transformation temperature and the austenite decomposition products, and

(vi) the amount of precipitation strengthening relative to the volume fraction and size of precipitates, and the mill processing (finishing temperature, cooling rate, coiling temperature, and transformation temperature).

Controlled rolled steels were originally developed as an alternative to normalized steels because of a lack of sufficient heat treating capacity. For thicker sections (greater than about 25 mm (1 in)), requiring a fine grain size, normalizing is often the most economic process. Over a restricted thickness and property range, normalized steels and controlled rolled steels are competitive. The cost of heat treatment is offset by the increased mill time and enriched chemistries required for controlled rolling. For stringent property requirements controlled rolling is the best alternative (Meyer, Heisterkamp & Strassburger 1970), but, severe controlled rolling is limited to those mills with adequate power and roll separating force. The adoption of thermomechanical treatment in place of heat treatment, to produce strong tough line pipe steels, is demonstrated by the construction of a number of plate and strip mills designed to control roll micro-alloyed grades at low finish temperatures.

Charpy shelf energy

During the past decade the Charpy test has gained acceptance as a measure of the energy absorbed during fracture of line pipe (AGA 1974). The fracture generally occurs along the longitudinal axis of the pipe, which for u.o.e. pipe is the rolling direction of the plate. To improve the Charpy shelf energy or ductility in this direction, physical metallurgists have analysed the fracture process in steel and, together with steelmakers, have developed effective methods to meet the more stringent transverse toughness requirements.

The Charpy shelf energy is closely related to the true strain to fracture ϵ_t in a tensile test and the u.t.s. (Reinbolt & Harris 1951). As a first approximation ϵ_t decreases hyperbolically with increasing volume fraction of inclusions (sulphides, oxides and carbides). Theoretical models have been developed to predict the ductility (ϵ_t) as a function of the morphology and volume fraction of inclusions, and the method of void nucleation (decohesion or particle cracking) (Gurland & Plateau 1963; Liu & Gurland 1968). The effect of volume fraction, shape, and type of inclusions is summarized by the data of Gladman (figure 4) (Gladman, Holmes & McIvor 1971).

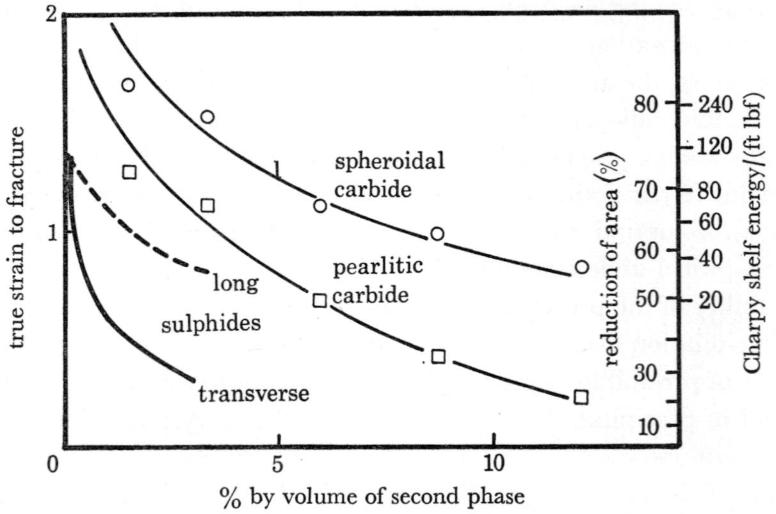

FIGURE 4. Effect of second phase particles on total ductility and Charpy shelf energy (Gladman *et al.* 1971).

To improve the transverse ductility, metallurgists have aimed at lowering the total volume fraction of inclusions and minimizing the anisotropy of the inclusions. Various techniques have been developed by chemical metallurgists to lower the inclusion content. In addition to lowering the carbon content, thereby reducing the volume fraction of pearlite, steelmakers have reduced the sulphide and oxide content by:

(i) desulphurization of blast furnace hot metal by injection of lime or calcium carbide, or plunging of magnesium impregnated coke into torpedo cars (Cameron & Dastur 1973; Fisher 1973),

(ii) limiting sulphur content by the use of selected low sulphur scrap additions,

(iii) desulphurization of the steel in the transfer ladle by rare earth silicides or calcium carbide additions (Hilty & Farrell; Luyckx, Bell, McLean & Korchynsky 1970) and

(iv) deoxidation of the steel by silicon and/or aluminium additions.

Using these techniques, the maximum sulphur content of the steel can be limited to less than 0.015%. To meet the most stringent C_v energy requirements, less than 0.005% final sulphur is required.

An economic alternative to, or complimentary process with desulphurization is the modification of the sulphide morphology by the addition of elements that combine with sulphur to form sulphides and oxy-sulphides with reduced plasticity during rolling. These elements include the rare earth metals (rare earth silicides and mischmetal) and calcium (as calcium silicate and carbide) which can be added to the ladle or mould or injected into the pouring stream. The variation in the shelf energy due to sulphide shape control by rare earth metals (r.e.m.) is shown in figure 5 (Kozasu & Tanaha 1974). The transverse shelf energy is dependent upon the amount of shape controlling element as a function of the sulphur content. Beyond an optimum addition, the shelf energy deteriorates because of the reduced cleanliness associated with the introduction of oxide inclusions and nitrides.

The transverse Charpy shelf energy can also be improved by cross rolling. Cross rolling reduces the total inclusion length in the longitudinal rolling direction (the direction parallel to the plane of fracture in a transverse Charpy test). The effect of the cross rolling ratio on the longitudinal and transverse shelf energies is shown in figure 6 (Kozasu & Tanaha 1974).

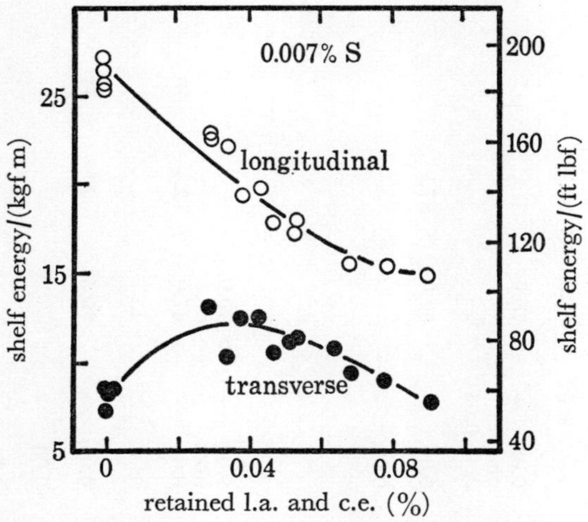

FIGURE 5. Variation of the Charpy shelf energy due to shape control with r.e.m. (Kozasu & Tanaha 1974).

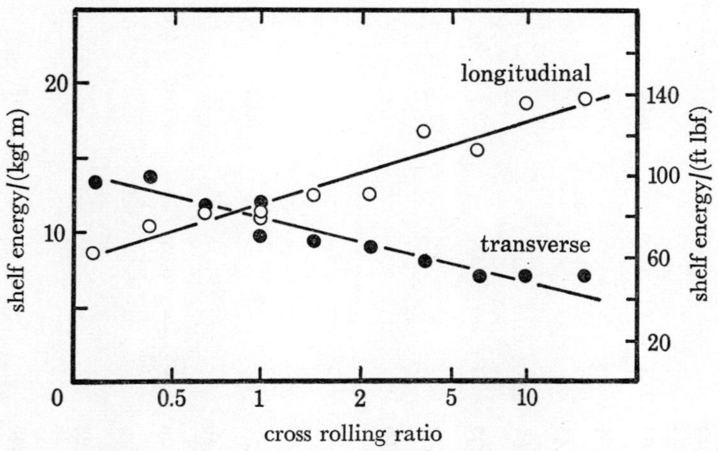

FIGURE 6. The effect of cross rolling ratio (% reduction longitudinal/% reduction transverse) on the Charpy shelf energy (Kozasu & Tanaha 1974).

CURRENT LINE PIPE STEELS

In order to meet the current specifications, present line pipe steels incorporate some or all of the metallurgical considerations described above. There is of course a great deal of diversity in the types of steel supplied within a strength range because of differences in the manufacturing equipment for each pipe producer, and because of customers supplementary specifications (e.g. weld metal toughness, C_v energy in the body of the pipe, transition temperature, etc.).

TABLE 1. COMPOSITIONS AND PROPERTIES OF STEELS USED FOR X60 AND X65 GRADE LINE PIPE

Company	Grade	composition (mass percent) C	P	S	Si	Mn	Cu	Ni	Cr	Mo	ASA	N_2	Nb†	V	Ti	Tensile properties yield klbf/in²	u.t.s. klbf/in²	Charpy data energy max ft lbf	50 % shear f.a.t.t. °F	processing	Gauge in
API Standard	X60	0.26 max	0.04 max	0.05 max	—	1.35 max	—	—	—	—	—	—	0.005 min	0.02 min	0.03 min	60.0 min	75.0 min	—	—		
	X65	0.26 max	0.04 max	0.05 max	—	1.40 max	—	—	—	—	—	—	0.005 min	0.02 min	— min	65.0 min	77.0 min	—	—		
Armco[1]	X60	0.13	0.009	0.015	0.19	1.09	0.21	—	—	—	0.004	—	0.03	0.02	—	66.2	82.9	—	—	controlled rolled plate	—
BHP[2]		0.20	0.014	0.013	0.34	1.34	—	—	—	—	—	—	0.02	—	0.02	67.5	87.7	—	< −50	controlled rolled plate	0.406
Hoesch[3]		0.12	0.030	0.025	0.30	1.45	—	—	—	—	—	—	0.05 max	0.08 max	—	> 75.0	—	—	< −40	strip strip	0.470
Mannesmann[4]		0.20	—	0.025	—	1.65	0.20	0.10	0.10	—	—	—	Cb+V+Ti < 0.06			—	—	—	—	normalized plate	0.50
Nippon Kokan[5]		0.12	0.011	0.006	0.18	1.33	0.11	—	0.18	—	—	—	0.02	0.04	—	69.0	87.8	—	—	controlled rolled plate	0.688
Sumitomo[6]		0.16	0.019	0.018	0.32	1.32	—	—	—	—	—	—	—	0.06	—	—	—	—	—	controlled rolled plate	0.531
		0.14	0.024	0.022	0 30	1.28	—	—	—	—	—	—	0.03	—	—	—	—	—	—	controlled rolled plate	0.500
BSC[4]	X65	0.12	—	0.012	—	1.35	—	—	—	—	—	—	0.03	—	—	—	—	—	—	controlled rolled plate	0.500
Hoesch[3]		0.13	0.035	0.030	0.40	1.60	—	—	—	—	—	—	0.05 max	0.08 max	—	> 75.0	—	—	< −40	strip	0.50
Italsider[7]		0.15	—	0.008	0.35	1.50	—	—	—	—	< 0.03	—	< 0.05	< 0.05	—	> 67.0	> 82.4	40	−40	controlled rolled plate	0.79
Mannesmann[4]		0.10	—	—	—	1.50	—	—	—	—	—	—	0.04	—	—	—	—	—	—	normalized plate	0.71
Nippon Kokan[8]		0.13	0.014	0.006	0.20	1.36	—	—	0.25	—	0.005	—	0.02	0.06	—	70.5	83.5	> 54	−62	controlled rolled plate	0.56
Nippon Steel[9]		0.10	0.014	0.006	0.24	1.39	—	—	—	—	—	—	0.04	0.06	—	—	—	—	—	controlled rolled plate	0.56
Nippon Steel[9]		0.09	0.013	0.008	0.25	1.25	—	—	—	—	—	—	0.04	0.04	—	—	—	—	—	strip	0.46
Stelco		0.11	—	0.010	—	1.35	—	—	—	—	—	—	0.04	—	—	> 70.0	> 82.0	> 50	< −40	controlled rolled plate	0.50
Sumitomo[10]		0.11	0.015	0.005	0.35	1.45	—	—	—	—	0.070	—	0.04	0.10	—	75.9‡	86.0	75	−80	controlled rolled plate	0.625
Sumitomo[10]		0.06	0.012	0.006	0.35	1.50	—	—	—	—	0.042	—	0.05	0.10	—	76.4‡	94.4	68	−49	controlled rolled plate	0.70
USINOR[11]		0.14	0.02	0.007	0.30	1.44	—	—	—	0.22	0.04	—	0.05	—	—	68.4	89.4	> 60	< −80	controlled rolled plate	0.75
USINOR[11]		0.14	—	0.007	0.26	1.35	0.27	—	0.22	—	0.05	—	0.05	—	—	71.9	85.0	> 60	< −70	controlled rolled plate	0.80
USINOR[11]		0.18	—	0.007	0.30	1.34	—	—	—	—	—	—	0.05	—	—	74.3	88.5	> 60	< −90	controlled rolled plate	0.59

† Nb, V, Ti additions may be used singly or in combination. ‡ Pipe results.
1, Glodowsky & Thompson 1973; 2, Thompson 1967; 3, Esche *et al.* 1973; 4, Gray 1973; 5, Taira *et al.* 1973; 6, SMI 1970; 7, Civallero & Parrini 1971; 8, Kozasu & Osuka 1973; 9, NKKK 1967; 10, SMI 1973; 11, Leclerc *et al.* 1972.

The lower strength grades (X52, X56) are generally produced by using a chemistry similar to that specified in the A.P.I. chemistry limits but without microalloying additions. If impact requirements are not severe the steel may be produced using a semi-killed practice. If the pipe is expanded after forming, yield loss due to the Bauschinger effect is minimal (Taira *et al.* 1973). The plate yield strength need not be higher than the specified pipe yield strength. If cold expansion is not used, the plate yield strength must be significantly higher than that of the nominal pipe strength. This loss of yield strength on forming is dependent on the plate yield strength and strain during forming. If low transition temperatures are required, normalizing or controlled rolling with microalloy additions are necessary to refine the ferrite grain size. Additional refinement of the ferrite grain size can be achieved in the case of strip mill products with controlled cooling. Chemistry variations (e.g. reduced carbon) may also be needed to ensure meeting weldability restrictions.

For the higher strength grades (X60 and X65) microalloying is necessary to meet the strength, transition temperature, and the weldability requirements. The types of steel chemistries used to produce these grades are summarized in table 1 for various manufacturers. Generally, the carbon contents are significantly lower than the maximum allowed in the A.P.I. specification. This ensures meeting the weldability and Charpy energy requirements and increases the potential for precipitation strengthening by raising the solubility of microalloying elements.

Each company has adopted a preference for one or a combination of the microalloying elements based on the economics of the addition, its effects on the mechanical properties and the available processing. Niobium additions are generally preferred because of lower cost and superior grain refining ability on controlled rolling. Niobium has the strongest effect of retarding recrystallization. Vanadium is generally not added on its own, but together with niobium and/or nitrogen is used to achieve higher strength levels especially in heavy wall pipe. Titanium is a difficult element to add during steelmaking because of its strong affinity for the interstitial elements oxygen, nitrogen, carbon, and sulphur (Meyer, Heisterckamp & Lauterborn 1973). It has the potential to act as both a sulphide shape controlling element and as a precipitation strengthener. The titanium addition must be greater than that required to tie up the nitrogen and combine with the sulphur to achieve appreciable strengthening. However, titanium is not as effective as niobium in refining the ferrite grain size, and thus is not suitable for low temperature applications.

Future trends in line pipe steels

Future line pipe specifications will likely continue the trend towards higher yield strength and greater wall thickness. Higher operating pressures will necessitate improved toughness at the operating temperature unless mechanical devices can be designed to prevent fracture propagation (AGA 1974).

Some companies have produced excellent strength and toughness properties with pearlite reduced and pearlite free steels by the use of microalloying additions (table 2) (Esche, Kaup & Wladika 1973; Kozasu & Osuka 1973; McCutcheon, Trumper & Embury 1974). Low transition temperatures have been achieved by austenite grain size control during reheating and heavily controlled rolling to low finishing temperatures. Finishing below the austenite-ferrite transformation temperature refines the ferrite grain size and increases the dislocation density and texture of the ferrite. This dramatically increases the yield strength without a marked

TABLE 2. COMPOSITIONS AND PROPERTIES OF STEELS FOR X70 ARCTIC LINE PIPE

		Composition (mass percent)														Tensile‡ properties		Charpy data†			
company	steel	C	P	S	Si	Mn	Cu	Ni	Cr	Mo	ASA	N	Nb	V	Ti	yield klbf/in²	u.t.s. klbf/in²	energy max ft lbf	50 % shear e.a.t.t. °F	processing	gauge in
Australian Iron & Steel[1]	acicular ferrite	0.09	0.016	0.011	0.18	1.67	—	—	0.24	0.19	—	—	0.10	—	—	70.7	98.1	64	~ −60	as-rolled plate	0.625
Hoesch[2]	NOVAR X70 PR	0.12	0.030	0.025	—	—	—	—	—	—	—	—	0.05	0.08	—	76.6	95.2	32	−40	as-rolled strip coiled at ~1100 °F	0.47
IPSCO[3]	acicular ferrite	0.04	0.011	0.014	0.04	1.61	0.21	0.12	—	0.24	—	0.006	0.006	—	—	75.1	92.4	> 114	< −46	as-rolled strip coiled at ~1150 °F	0.390
Italsider[4]	acicular ferrite	0.05	—	—	—	1.85	—	0.30	—	0.30	—	—	0.05	—	—	67.0	92.5	> 80	< −100	controlled rolled plate	0.57
																76.5	92.5	> 80	< −100	expanded pipe	
																64.0	91.1	> 80	< −100	plate	0.78
																72.5	91.1	> 80	< −100	expanded pipe	
																61.0	89.5	> 80	< −100	plate	1.19
																70.0	89.5	> 80	< −100	expanded pipe	
Nippon Kokan[5]	rearlite reduced	0.08	0.014	0.005	0.21	1.34	—	—	0.20	—	0.037	0.006	0.03	—	—	71.4	83.7	80	−49	controlled rolled plate air cooled	0.5
	acicular ferrite	0.06	0.013	0.005	0.32	1.62	0.27	0.16	0.27	—	0.024	0.006	0.035	—	—	76.5	89.0	139	−119	controlled rolled plate formed to pipe	0.625
Republic[3]	acicular ferrite	0.06	0.004	0.010	0.15	1.80	0.20	0.16	—	0.32	0.010	—	0.10	—	—	69.3	100.8	> 100	< −90	controlled rolled plate	0.5
																83.9	90.4	> 111	< −90	plate aged 1 h/1150 °F	
																78.1	101.6	> 85	< −90	U & O pipe cold expanded 1 %	
																76.8	100.8	> 90	< −90	pipe aged 1 h/1150 °F	
Stelco[6]	acicular ferrite	0.05	—	—	—	1.66	—	0.32	—	0.51	—	—	0.06	—	—	69.1	88.4	113	< −100	controlled rolled plate, air cooled	0.720
		—	—	—	—	—	—	—	—	—	—	—	—	—	—	74.4	102.8	102	< −100	controlled rolled plate, accelerated	0.720
		0.06	—	—	—	2.20	—	0.24	—	0.58	—	—	0.06	—	—	79.0	113.0	75	< −100	water cooling after rolling	0.720

† Full size Charpy tested transverse to rolling direction in plate and strip, and transverse to longitudinal axis in pipe.
‡ Tensile properties of strip determined on flattened strap.
1, Brown, 1974; 2, Esche *et al.* 1973; 3, Cryderman *et al.* 1972; 4, Civallero *et al.* 1973; 5, Kozasu & Osuka 1973; 6, McCutcheon *et al.* 1974.

deterioration in the transition temperature (Melloy & Dennison 1973; Little *et al.* 1973). The yield strength, however, is artificially high when considering formed pipe. Generally, the higher the plate strength, the greater the yield loss on pipe forming, unless the pipe is cold expanded after forming (Taira *et al.* 1973; McCutcheon *et al.* 1974). Also, the strong textures developed with heavy controlled rolling tend to reduce the through thickness ductility and contribute to splitting during fracture (Dabkowski, Konkol & Baldy 1975). This phenomenon lowers the steels ability to arrest propagating fractures as predicted by its C_v energy. Very low carbon ferrite pearlite steels have been developed, to improve ductility and weldability, at the expense of lowered ultimate strengths and very high yield to ultimate ratios.

A novel method for achieving ultra fine ferrite without heavy controlled rolling has been proposed by Webster (1965; Webster & Woodhead 1969) and Gray (1973) for high niobium (> 0.10% Nb) steels. Large Nb (C, N) particles which precipitate during hot working develop an ultra fine austenite grain size by successive recrystallizations and limit the ferrite grain size during transformation. Trials based on this approach have not been totally successful. Yield strengths were generally lower than anticipated and the ferrite grain size was not as fine as that developed by controlled rolling.

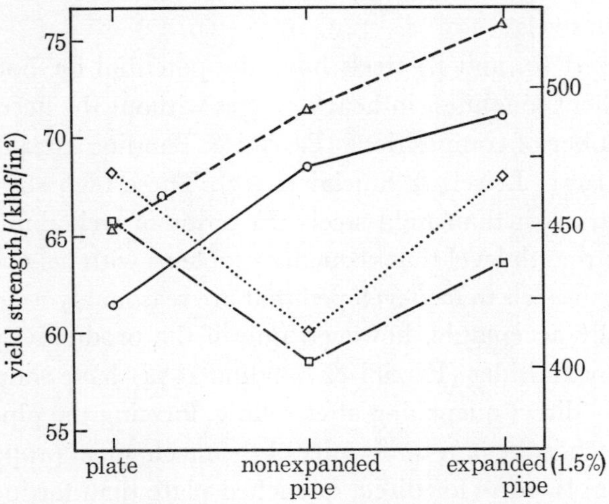

FIGURE 7. Effect of pipe making on the yield strength of acicular and polygonal ferrite steels (Civallero *et al.* 1973). △, acicular steel 18 mm; ○, acicular steel 30 mm; ◇, polygonal steel with Nb+V 18 mm; □, polygonal steel with Nb 18 mm.

It has long been recognized that the strength and toughness of ferrite pearlite steels were limited, and that ultimately, bainite and/or quench and tempered steels would be developed as economic alternatives to meet severe strength-toughness demands. Low carbon bainite, acicular ferrite steel, is a strong candidate to replace ferrite pearlite steel for heavy wall pipe at strengths greater than 70 klbf/in² (1080 N/mm²). Excellent properties can be achieved in both as-rolled plate and formed pipe (Cryderman, Coldren, Smith & Mihelich 1972; Civallero, Parrini & Pizzementi 1973; McCutcheon *et al.* 1974; Coldren 1972; Tither 1974; Mihelich & Cryderman 1972; Parrini & Bandino 1974). In the tempered condition (e.g. strip rolling) yield strengths in excess of 80 klbf/in² (1240 N/mm²) are possible together with excellent toughness. The particular advantage of this acicular microstructure is that it has a high rate of work hardening over the range of strains used in pipe making. The stress strain curve is continuous, rather than exhibiting a Luders strain region. Consequently, the yield strength losses normally encountered during pipe making are minimized (figure 7). Pipe yield strengths can be appre ciably higher than plate yield strengths if the microstructure is sufficiently acicular. This is advantageous for spiral formed pipe which does not undergo cold expansion. The spiral pipe yield strength is often less than the plate yield strength for ferrite pearlite microstructures.

There are, of course, disadvantages in producing this type of steel. In addition to controlling the volume fraction and shape of sulphide and oxide inclusions, the carbon content of acicular ferrites must be limited to achieve good ductility in the as-rolled condition. Either vacuum degassing or low carbon ferromanganese must be used to produce this carbon level at the manganese content (about 2.00 %) of these steels. Alloying costs (Mn, Mo, Ni) are substantially more than for ferrite pearlite steels. Acicular steels, like ferrite pearlite steels, must be heavily controlled rolled to achieve good toughness so there is no advantage to using them in terms of mill processing (Civallero *et al.* 1973; McCutcheon *et al.* 1974).

Presently, acicular ferrite steels are gaining commercial acceptance in line pipe. Several studies have concluded that the weldability of these low carbon grades is far superior than that predicted from carbon equivalent formulae (Cryderman *et al.* 1972; Coldren 1972; Mihelich & Cryderman 1972; Sawhill & Wada 1975). However, more quantitative analyses of the microstructure and mechanical properties are required before these steels can be used in the same manner as ferrite pearlite steels.

Quenched and tempered (q. and t.) steels have the potential for both high yield strength (>80 klbf/in^2) and excellent toughness in heavy gauges without the necessity for high levels of alloying and a large number of compositions (Parrini & Bandino 1974; Irani & Tither 1967; Tither & Kewell 1970, 1971; Lessels & Sinclair 1967). These high strength steels give lower costs per unit of yield strength than mild steels. In terms of weldability, they minimize the composition for a given strength level thus reducing problems with heat affected zone cracking, or allow high yield strength steels to be developed that are reasonably weldable. Quenching and tempering is commercially acceptable, however, only if the productivity level is not severely limited. Recent studies by Italsider (Parrini & Bandino 1974) have shown that excellent properties can be attained by direct quenching after rolling, forming the pipe and tempering. The problem with this approach is the non-uniformity of the mechanical properties along the length of the plate. This problem is worse for direct quenched plate than for indirect quenched plate because of the additional variations related to hot processing. Also, if hard spots are inadvertently developed in the pipe during processing, they could make the pipe susceptible to hydrogen stress cracking (AGA 1969). A much more thorough understanding of the relations between the parameters of hot processing, quenching rate and tempering and the mechanical properties and microstructure is necessary before this process is acceptable.

Quenching and tempering of formed pipes has also been suggested as a processing route. Although the metallurgy of this technique is similar to that of quench and tempered plate, there are unique problems associated with this proposal. Quenching from one surface will develop differential cooling rates through the pipe thickness and thus microstructural gradients. Shape could be a problem if the pipe is reheated either before or after expansion of u.o.e. pipe. At present, the mechanical-metallurgical problems and capital cost associated with the q. and t. approach give it little scope for immediate use.

References (Clay & McCutcheon)

AGA 1966 Symposium on Line Pipe Research, Dallas, 1965. American Gas Association.
AGA 1969 Fourth Symposium on Line Pipe Research, Dallas, 1969. American Gas Association.
AGA 1974 Fifth Symposium on Line Pipe Research, Houston, 1974. American Gas Association.
Brown, J. 1974 High test line pipe steel, development and production in Australia. Australian Welding Convention.
Cameron, R. & Dastur, P. 1973 Desulfurization of hot metal by use of calcium carbide injection, *A.I.S.I.* Technical Meeting, Chicago 18 (1973).
Civallero, M. & Parrini, C. 1971 Plate mill production with control rolling process of medium and high thickness plate, high strength and toughness at Taranto Works. *Proc. ICSTIS*, section 4, Suppl. Trans. *I.S.I.J.* **11**, 758–663.
Civallero, M., Parrini, C. & Pizzementi, A. 1973 Research and development into pipe steels and pipe manufacture. Symposium on Development in Gas Pipeline Technology, Melbourne, Australia.
Coldren, A. P. 1972 Mn–Mo–Cb and Mn–Mo–Cb–V–N steels for plate products up to 1 inch thick. Climax Molybdenum Company Research Report L-176-92.
Cottrell, A. H. 1958 *Trans. Am. Inst. Mech. Engr.* **212**, 192.
Cryderman, R. L., Coldren, A. P., Smith, Y. E. & Mihelich, J. 1972 L. The development of a high strength hot rolled Mn–Mo–Cb steel. 14th Mechanical Working and Steel Processing Conference, Iron & Steel Division, A.I.M.E. Chicago (January 1972).
Dabkowski, D., Konkol, P. & Baldy, M. 1975 Factors that affect splitting in plate steels. U.S. steel corporation. *Welding J.* (to be published).
vor dem Esche, W., Kaup, K. & Wladika, H. 1973 Processing and properties of low carbon steel, 1973. Metallurgical Society, *Am. Inst. Min. Met. Pet. Engrs* (New York), p. 331.
Fisher, P. 1973 Magnesium desulfurization of blast furnace iron. *Metals and Materials*, November 1973, pp. 501–503.
Fukuda, M., Hashimoto, T. & Kunishige, K. 1973 An investigation on control rolling of low-carbon killed steel. *Sumitomo Search* No. 9 (May 1973).
Gladman, T., Holmes, B. & McIvor, E. D. 1971 Effects of second phase particles on strength, toughness, and ductility. Paper presented at the BSC/ISI Conference entitled '*The Effects of Second Phase Particles on the Mechanical Properties of Steel*', Scarborough, U.K. (24–25 March 1971).
Gladman, T. & Pickering, F. B. 1967 Grain coarsening of austenite. *J.I.S.I.* **205**, 653.
Glodowski, R. J. & Thompson, J. M. 1973 Mechanical working and steel processing. XI. *A.I.M.E.*, p. 59.
Gray, J. M. 1973 Columbium as a grain refiner in hot rolled steels. Processing and properties of low carbon steel, pp. 235–247.
Gray, J. M. Metallurgy of high-strength low-alloy pipeline steels: present and future possibilities. Paper presented to the Metallurgical Task Force on the Northwest Project, Dallas (January 1972).
Gray, J. M. 1973 Technology of microalloyed steel for large diameter pipe. Paper presented at the Canadian Tri-Society Welding Metalworking Conference, Ontario, 19 September 1973.
Gurland, J. & Plateau, J. 1963 The mechanism of ductile rupture of metals containing inclusions. *Trans. Am. Soc. Metals* **56**, 442.
Hahn, G. I., Averback, B. L., Owen, W. S. & Cohen, M. 1959 Initiation of cleavage microcracks in iron and steel. *Fracture* pp. 91–116.
Heslop, J. & Petch, N. J. 1958 The ductile brittle transition in the fracture of alpha irons. II. *Phil. Mag.* **3**, 1128.
Hilty, D. C. & Farrell, J. W. Modification of inclusions by calcium. Union Carbide Corporation, Ferro-alloys Division.
Irani, J. J. & Tither, G. 1967 Quenched and tempered low-carbon steels. Strong, tough structural steels, I.S.I. Spec. Rep. No. 104.
Irvine, K. J., Pickering, F. B. & Gladman, T. 1967 Grain refined C–Mn steels. *J. Iron Steel Inst.* **205**, 161–182.
Iwasaki, N. *et al.* 1975 Characteristics of drop weight tear test on line pipe steels. 17th Mechanical Working and Steel Processing Conference, Pittsburgh, 22–23 January 1975.
Kozasu, I. & Osuka, T. 1973 Processing and properties of low carbon steel, 1973. Metallurgical Society, *Am. Inst. Min. Met. Pet. Engrs* (New York), p. 47.
Kozasu, I., Shimizu, T. & Tsukada, K. 1972 Further observations on microstructural changes of structural steels due to hot rolling deformation. *Trans. I.S.I.J.* **12**, 305.
Kozasu, I. & Tanaha, J. 1974 Effect of sulfide inclusions on the notch toughness and ductility of structural steels. Int. Symposium on Sulfide Inclusions in Steel, Port Chester, New York, 7–8 November 1974.
Leclerc, J. *et al.* High impact strength steels for pipes used at low temperatures. CIM Conference of Metallurgists, Halifax, August (1972).
Lessells, J. & Sinclair, J. 1967 Advantages and economics of quenched and tempered weldable high-strength steel. *J. Iron Steel Inst.* **205**.

Little, J. *et al.* 1973 The production of high strength plates by controlled rolling. The microstructure and design of alloys, vol. 1, pp. 80–84.
Liu, C. T. & Gurland, J. 1968 The fracture behaviour of spheroidized carbon steels. *Trans. Am. Soc. Metals* **61,** 156.
Luyckx, L., Bell, J. R., McLean, A. & Korchynsky, M. 1970 Sulfide shape control in high strength low alloy steels. *Metall. Trans.* **1,** 3341–3350.
McClure, G. M., Duffy, A. R. & Eiben, R. J. 1965 Fracture resistance of line pipe. *J. Engng Ind.* **87,** 265–278.
McCutcheon, D. B., Trumper, T. W. & Embury, J. D. 1974 Controlled rolling of acicular ferrite plate. Paper presented at *J. Int. Siderurgie,* Paris, 4 October 1974.
Melloy, G. & Dennison, J. 1973 Continuous rolling – a unique thermomechanical treatment for plain carbon and low alloy steels. *The microstructure and design of alloys,* vol. 1, pp. 60–64, Cambridge.
Meyer, L., Heisterchamp, F. & Lauterborn, D. 1973 Titanium as a strengthening and sulfide-controlling element in low carbon steels. Processing and properties of low carbon steel, pp. 297–330.
Meyer, L., Heisterkamp, F. & Strassburger, C. 1970 Alloying possibilities for increasing strength and toughness of weldable structural steels. *Symposium on Low Alloy High Strength Steels,* pp. 9–15.
Mihelich, J. L. & Cryderman, R. L. 1972 Low carbon Mn–Mo–Cb steel for gas transmission pipe. Petroleum Mechanical Engineering Conference, Petroleum Division, A.S.M.E., New Orleans (September 1972).
NKKK 1967 Nippon Kokan Kabushiki Kaisha. Test results on X65 high test line pipe. *Tech. Bull.* No. 67–1, February 1967.
Parrini, C. & Bandino, T. 1974 Controlled rolling of high strength steels: limitations, development, alternatives. Paper presented at *J. Int. de Siderurgie,* Paris, 4 October 1974.
Petch, N. J. 1953 The cleavage strength of polycrystals. *J.I.S.I.* **174,** 25.
Reinbolt, J. A. & Harris, W. J. 1951 *Trans. Am. Soc. Metals* **43,** 1175.
Sawhill, J. M., Jr & Wada, T. 1975 Properties of welds in Mn–Mo–Cb line pipe steels. Presented at the AWS Annual Meeting, Houston, Texas, May 1974. *Welding J.* (to be published).
SMI 1970 Sumitomo Metal Industries Limited. Technical data for X60/X65 line pipe. *Bull.* No. 0026, March 1970.
SMI 1973 Sumitomo Metal Industries. Tentative report on full scale burst tests at Athens by Sumitomo pipes. *Bull.* No. 034–73.
Taira, T. *et al.* 1973 Bauschinger effect in large diameter welded pipe. *Mechanical working and steel processing* vol. XI, pp. 33–69, Met. Soc. of A.I.M.E.
Thompson, J. W. 1967 Strong tough structural steels. *Iron and Steel Inst. Spec. Pub.* **104,** 173.
Tither, G. 1974 Accelerated cooling during controlled rolling of $\frac{3}{4}$-inch (19 mm) thick Mn–Mo–Cb steel plate. Climax Molybdenum Company Research Report J-4040.
Tither, G. & Kewell, J. 1970 Properties of directly quenched and tempered structural steel plate. *J. Iron Steel Inst.* **208.**
Tither, G & Kewell, J. 1971 Microstructure and mechanical properties of some direct-quenched and tempered low-carbon steel. *J. Iron Steel Inst.* **209.**
Webster, D. 1969 The effect of dispersed particles on grain boundary movement in steel. Ph.D. Thesis, University of Sheffield (U.K.), July 1965; also Webster, D. & Woodhead, J. H. Effect of 0.03 % niobium on the ferrite grain size of mild steel. *J.I.S.I.* **202,** 987.
White, J. 1970 Economics of large diameter pipelines. *Pipeline Engr.* February, pp. 30–34.

3.4 Development of steels for offshore structures

By J. H. van der Veen
Hoogovens Ymuiden B. V., IJmuiden, Netherlands

[Plate 22]

The steel for offshore jacket structures, Fe52Nb, is a silicon-aluminium killed steel, microalloyed with niobium. As the steel, in fabrication, has to be hot-formed, sometimes normalized or stress-relieved, the steelmaker cannot make use of low temperature treatments, such as controlled rolling, quenching plus tempering, etc. Therefore, the steel is delivered in the normalized condition and its strength and toughness are attained by promoting a fine ferrite grain through the action of grain growth inhibitors, such as Al-nitrides and Nb-carbonitrides. Further increase of strength and toughness, if necessary for future structures, should be pursued by the application of new, as yet unknown or not yet applied, kinds of grain growth inhibitors, in order to further refine the microstructure. These inhibitors should, if possible, be stable at almost steel melting temperatures in order to provide as an additional benefit the suppression of grain growth at the fusion line in high input welding. Maybe also a normalizing treatment at an intermediate thickness during the rolling process could be beneficial for further grain refinement of the finished plate, but most probably, not for grain-growth resistance at fusion line temperatures.

Plates used in the construction of nodes are subjected to loading in the thickness direction and, therefore, are liable to lamellar tearing. In order to avoid this, the steel should have sufficient ductility in the thickness direction, the criterion being reduction of area in a tensile test, the axis of which is perpendicular to the plate surface. Sufficient resistance, for the application at present-day nodes is obtained by reducing the sulphur level of the steel to about 0.006% and by careful deoxidation of the steel. Various ways of making this steel are put forward for discussion.

First some remarks on the development of structural steels in general. The most important property of Fe products has always been their *strength*. Structural steels, Fe-products with carbon contents less than about 0.5% are characterized by a second important, and industrially essential property: *hot and cold formability*. This makes possible, for the steel industry, to apply forging, hot and cold rolling in the manufacture of various shapes, such as plates and sections and it enables the user to apply hot and cold forming in the fabrication of structures, such as ships, boilers, pressure vessels, tubes, etc. Moreover, the formability of steel, or rather its ductility, is important for the safety of structures. It will, normally, make possible the absorption, or levelling out of local stress concentrations, by localized plastic deformation and constitute a 'plastic reserve' against 'overall' overloading of the structure.

Coming back to *strength*, it can be stated that even so-called *mild steel*, with a yield point of about 250 N/mm² is a relatively strong building material. Nevertheless, for reasons of economy, steels with higher strengths have been and are being developed. As a first step towards higher strength, the widely used, not so very complicated Fe52 can be mentioned, which has a minimum yield strength of about 350 N/mm². Other, more sophisticated high strength steels may have higher minimum yield strengths, such as 500 N/mm² and higher. The application of high strength steels can be restricted by various reasons. One of them is the fact that the *modulus of elasticity* is the same for high and lower strength steels. In structures where bending and buckling loads occur, it may happen that, depending upon the design, the material thickness is determined by considerations of sufficient stiffness, sufficient resistance against bending or buckling. In those cases the necessary yield strength of the steel is determined by the desired static strength of the structure and the already fixed plate thickness. The use of higher strength steel, in those cases, will not bring any economic advantage. Therefore, there is little application of high strength steel in ships, and much more, for example, in pressure vessels, where the steel is subjected to biaxial tension loads.

There are various *methods to increase the strength* of hypo-eutectoid steels. These can best be discussed with the help of a modified Petch formula, valid for ferrite-pearlite steels (van der Veen 1968)

$$E = f_\alpha E_{i\alpha} + f_p E_p + f_\alpha K_\alpha d^{-\frac{1}{2}},$$

where E is the yield strength of the steel; f_α, fraction of ferrite; $E_{i\alpha}$, yield strength of the ferrite; f_p, fraction of pearlite; E_p, yield strength of the pearlite; K_α, a constant; d, average diameter of the ferrite grain.

In the 'old days', before *the introduction of welding*, the simplest and cheapest method to increase strength was to increase the carbon content of the steel and thereby the percentage of pearlite (f_p). The introduction of welding put an end to this. Its development into the almost only applicable method of joining parts in steel construction, has put forward many problems and challenges in the field of welding processes, consumables and steels. In the development of weldable steels a series of phenomena were successively encountered, such as: cold cracking in the heat affected zone (h.a.z.); brittle fracture of the whole structure (Liberty ships); embrittlement of the h.a.z. by grain coarsening, due to high heat input; and lamellar tearing (rather recently).

Although welding was introduced about 40 years ago, many problems have not yet been completely or satisfactorily solved and new problems are still being encountered. Research, aiming at better processes, better steels and more know-how still needs to be continued.

The use of high strength steels in *pure fatigue loading* has always been questionable because the fatigue strength of polished bars, without welds, will normally be only about 50% of the tensile strength, whereas the static yield strength of the steel will increase from about 65% to 80–90% of the tensile strength when mild steel is replaced by a high strength steel. To make things worse, it has also been found that the *fatigue strength of welded joints*, made of high strength steels, was not substantially higher than for joints made of lower strength steels. This is attributed to a predominant, unfavourable influence of notch effects in the welded region, caused by discontinuities in the surface at toes and at the weld root and also, partly, to metallurgical discontinuities in the weld region, e.g. differences in microstructure, hardness and yield strength between weld, h.a.z. and parent plate. Therefore, the application of high strength steel in cases of pure high cycle fatigue loading will normally not be economic, unless methods can be found (better design, treatment of welded joints) by which the now prevailing adverse influences can be overcome. Only then, eventually, a physical metallurgy approach to fatigue aiming at increasing the fatigue limit of high strength steels to values of more than 50% of the tensile strength, if at all possible, will have sense.

On the other hand, when we have a *mixed load spectrum*, that is pure fatigue components plus static and low cycle components of higher magnitude, the application of high strength steels will have sense. How far this 'sense' goes is the subject of many studies that are still going on.

A lot of knowledge is still missing in this field, especially in the field of *corrosion fatigue*, a subject particularly important for offshore structures.

Up to now, the *development of weldable structural steels* has been guided mainly by two considerations. One is the desire to avoid cold cracking in the h.a.z. and the second is the necessity to attain sufficient resistance against the initiation and propagation of brittle fracture. (Newer considerations are: embrittlement of the h.a.z. by grain coarsening in high heat input welding, lamellar tearing, instable ductile fracture (gas pipe lines).)

The desire to avoid cold cracking, has led to the development of low hydrogen electrodes, and with regard to the steel, to the trend to limit the h.a.z. hardness to, for example, 350 Vickers. With regard to brittle fracture, there are practically only two means of improvement:

1. A 'fine' metallographic structure with a lowest possible C-content, that is a finest possible ferrite grain, low C pearlite with fine lamellae or other fine forms of cementite/ferrite structures, such as lower bainite or tempered martensite.

2. Solid solution of Ni (expensive).

In order to keep the h.a.z. hardness (in hand welding) below 350 Vickers and then under 'atmospheric' conditions (welding outside) and practicable conditions with regard to preheating and heat-input, the carbon-content is normally limited to 0.20% (if possible 0.18%) and the manganese content to 1.50%, which means a carbon-equivalent lower than 0.45%. In the case of Fe52 mostly used in offshore jacket structures the increase in yield stress from the mild steel value of about 250 N/mm^2 to about 350 N/mm^2 is mainly achieved, not by increasing carbon above 0.20%, but by increasing manganese to 1.50%. This increases the total pearlite contribution (f_p, E_p). The amount of pearlite (f_p) becomes larger and the strength of the pearlite increases (by finer lamellae). Furthermore, the higher manganese content will shift the γ/α transition to a somewhat lower temperature and thereby provoke a finer ferrite grain, which, together with the finer pearlite structure, is favourable in regard to resistance to brittle fracture.

Normally, this steel is killed with silicon (up to about 0.40%) and aluminium (more than 0.02%) and normalized. During normalizing the aluminium will form nitrides by combining with the nitrogen that was in solution in the steel. These nitrides will act as grain growth inhibitors and provoke a fine grain after normalizing (a little over 10 ASTM at 50 mm plate thickness). The absence of free nitrogen and the finer grain are favourable in regard to toughness and the fine grain contributes to the yield strength through the term $f_\alpha K_\alpha d^{-\frac{1}{2}}$. The silicon addition is useful and necessary. Silicon contributes to yield strength (by strengthening both ferrite and pearlite) and almost nothing to h.a.z. hardness.

Fe52 can be further improved (in the normalized condition) by the introduction of more grain growth inhibiting particles, such as Nb-carbonitrides and V-nitrides. The addition of Nb (0.02–0.05%), for example, will provoke a further grain refinement (almost $+\frac{1}{2}$ ASTM), and with controlled rolling before normalizing ($+\frac{3}{4}$ ASTM), which makes possible a decrease of the carbon-content by 0.02 or 0.03%. The Charpy-V 50% fibrous transition temperature of the Fe52Nb-steel will be up to 15 °C lower in comparison with normal Fe52. A schematic diagram of yield strength and transition temperature as a function of various factors is given and discussed in appendix 1.

Fe52Nb, as described in the preceding paragraph, has become a rather popular steel for various structures, such as bridges, storage tanks and also, recently, for the tubular jacket structures that support offshore production platforms. Apart from minor, non essential details, the *steel for offshore jacket structures* is characterized best by referring to steel FeE355 of Euronorm 113. For North Sea application the notch toughness of the steel itself is considered to be sufficiently high and with regard to yield strength, nobody among owners, designers and builders seems to be asking for higher strength. It is said that for the North Sea, Fe52, with regard to strength, is the best compromise in view of (corrosion-) fatigue considerations on the one hand, and resistance to the highest instantaneous load (the '100 year wave') on the other hand. The nominal design stress for average conditions seems to be only about 60 N/mm^2 because of the necessity to cater for 100 year storm conditions. This rather low figure partly reflects the present uncertainties about fatigue behaviour of these structures and the influence thereon of corrosion and cathodic protection. Research in this field may well lead to more certainty and improved designs and thereby to the applicability, from an economic point of view, of higher strength steels. Together with possible activities in other parts of the world (e.g. Arctic) where other sea conditions and lower temperatures prevail, this may lead to the desire to apply steels having a higher strength, at least the same weldability and a still higher resistance to brittle fracture (lower transition temperature) than the Fe52Nb steel, used up to now.

Whether *stronger and tougher weldable steels for offshore structures* can be conceived, can only be discussed after an appreciation of the essential points on which these jacket structures differ from other constructions. The jacket structure consists of tubular elements, up to almost 5 m diameter, made out of rather thick plates, e.g. 63 mm and even 130 mm thick. There are rather complicated connexions ('nodes') between the so-called 'legs' and the cross bracings, where exceptionally high stress concentrations occur (e.g. 6x). A picture of such a node is given in figure 1, plate 22. The tubular elements, depending on plate thickness and thickness to diameter ratio, are mostly made by hot forming (about 900 °C) and then longitudinally welded. Furthermore, depending on thickness to diameter ratio and/or the application of, e.g. electro-slag-welding, the elements have to be normalized, after welding. In many other cases, depending upon wall thickness, these elements or the transverse welds that connect tubular elements to each other or to the nodes, have to undergo a stress-relief treatment.

Therefore, steels that derive their strength and toughness from fine-grain metallographic structures made by some low-temperature treatment can not be proposed, as they would lose their essential properties during fabrication. This means that quenched and tempered steels and 'normal' low-temperature controlled rolled steels will generally not be suitable. (Some of the steels made by newer procedures of thermomechanical treatment, in which suppression of recrystallization during low-temperature rolling by for example Nb, and some fine precipitation are involved, could possibly exhibit an improved resistance against grain coarsening at hot forming and normalizing temperatures, but this has not yet been sufficiently established.) We will have to restrict ourselves to steels that can withstand hot-forming temperatures, for example to steels delivered in the normalized condition. These steels exist, and are used mainly for pressure vessels, when hot-forming is used for shaping the segments. Some of these steels can be found in Euronorm 113, with minimum yield strengths up to 460 N/mm^2. In certain national standards and for some proprietary steels even higher minimum yield strengths may be offered. These steels normally derive their higher strength from alloy additions, acting on strength through various mechanisms. One of these mechanisms, something like air hardening, aims at a finer cementite-ferrite transformation structure, such as lower bainite, others at solid solution hardening and grain refinement by precipitates. Generally, these steels are less weldable, in regard to h.a.z. hardenability, than the Fe52Nb, because of a higher carbon-equivalent due to

FIGURE 1. Upper leg section showing fabricated node assembly (by courtesy of Foster Wheeler John Brown Boilers Limited from *Welding and Metal Fabrication*, July 1973, p. 235).

(*Facing p.* 322)

the alloying elements. They are more destined to be welded in a closed construction shop, when more precautions are possible. To what extent these steels could be used for offshore structures, when nodes are fabricated in closed shops, with appropriate precautions, could be a subject for discussion or experiment. It must be stated, however, that with regard to notch toughness, these steels will not be better than the usual Fe52Nb. For use at lower temperatures, steels alloyed with Ni could be used. According to Euronorm 129 (not yet published) a 0.5% Ni steel for example, in the normalized condition, could be delivered with a Charpy-V notch guarantee at −60 °C, and a 1.5% Ni steel with the same guarantee at −80 °C. The yield strength of these steels, however, is not higher than that of Fe52Nb. Moreover, more precautions in welding will be necessary.

Thus, the *ideal stronger and tougher steel for future offshore structures* has not yet been invented. In my opinion the solution should be sought in the direction of a very fine low-carbon ferritic structure. We know that very strong and tough steels (especially for pipe-lines) can be made, with a carbon content of only 0.05%, by establishing a very fine grain size (12–14 ASTM) through thermomechanical treatment at rather low temperatures (down to about 750 °C). For offshore structures this method can not be used. The plates are too thick to be treated that way during rolling and, moreover, have to be hot-formed in most cases, which would destroy the achieved properties.

Other grain growth inhibitors should be discovered, apart from the already existing Al-nitrides, Nb-carbonitrides, etc., which could impart to the steel an inherent fine grain size, if possible, having a stability up to temperatures of 1200, or rather 1300 °C, so that not only finer grain normalized steels would be possible, but also the tendency to grain growth, and thereby embrittlement, of the h.a.z. in high heat input welding would be substantially diminished Research into the possibility of this solution should be pursued. There are indications that the introduction of Ti-nitride precipitates in the steel could be promising. Intermediate solutions might be possible through the insertion of one or more normalizing treatments into the total rolling procedure from slab to finished plate, at intermediate thickness. This might lead to further grain refinement, partly due to the formation of the conventional grain growth inhibitors well before the end of the rolling operation (British Patent nrs. 1 361 977/8 – Sumitomo, Osaka, Japan). An estimate on what could be achieved by further grain refinement of normalized steels is given in appendix 1.

As a special feature of the offshore jackets the complicated 'nodes' have already been mentioned. At these locations the rather thick plates may be strongly loaded in the plate-thickness direction by the connecting welds, during welding or in service. These stresses and accompanying strains, acting in the through-thickness direction, have, in many cases, caused *lamellar tearing*. This phenomenon had already been encountered in other structures, but maybe not to the same extent. Lamellar tearing is splitting up of the plate by terrace-like fracture planes, parallel to the plate surface. From a physical metallurgy point of view it is caused by inclusions. Important factors are the amount, the homogeneity of distribution and the shape, size or length, of the individual inclusions. Since, at the critical locations only Si–Al killed steel is used, only manganese sulphides and aluminium oxides are involved. Lamellar tearing under the prevailing circumstances at nodes, can be avoided by using a 'cleaner' steel, with a reduced S-content and 'not too many' Al_2O_3-inclusions. The S-content is not the only factor. At the same S-content, shorter inclusions are less harmful than long ones. This means, among other things, that high temperature rolling is better than low temperature rolling. It also means that the use of rare earths (e.g. cerium) which shortens the Mn-sulphides can be beneficial. Unfortunately, for thick plates, the use of rare earths has proven to be tricky (occasional inhomogeneity, and in continuous casting the occurrence or surface defects). Therefore, the use of rare earths is no longer recommendable. Normally, a maximum S-content of 0.006% is aimed at. This can be achieved in many ways, such as:

(1) Electric furnace steelmaking, with double slag practice.

(2) Electric steelmaking, followed by a purification treatment in the A.S.E.A.-S.K.F. vessel.

(3) Electroslag remelting (e.s.r.), after 'normal' steel making.

(4) Desulphurization of the steel in the ladle (lance + CaC or CaSi powder) after 'normal' open hearth or l.d. steel making) (Förster 1974).

(5) 'Normal' l.d. steelmaking, using desulphurized iron, selected scrap, and 'careful' deoxidation (for example by vacuum treatment or accurate oxygen determination combined with controlled aluminium addition).

As the accent of this conference should be on physical metallurgy a detailed discussion of steel making alternatives should not be attempted. The following remarks, however, could, tentatively, be made:

Desulphurization in the ladle, after normal steel making, method 4 in the above list, is claimed to have the additional advantage of extra deoxidation and transformation of Al-oxides into Ca-aluminates, which will, as another extra advantage, absorb, in a harmless way, some of the sulphur. This looks promising. On the other hand, temperature control and slag changing may cause problems and the method may be too expensive. The suggestion of 'too expensive' could also apply to the first three procedures mentioned in the above list. A further discussion on this point could be useful. According to our own experience method 5 would be the most economic and sufficiently appropriate.

In this context, it should be remarked that, according to our experience (Bellen, Spanraft & van der Veen 1971), l.d.-steel compares favourably with open hearth steel, since the H_2-content, at least in our case, is significantly lower. This, we found, will promote a more homogeneous distribution of sulphides.

Up till now, resistance against lamellar tearing cannot be checked by non-destructive testing, although research is being done to find ways and means (by ultrasonics). It has generally been accepted to test plates by some form of tensile test in the through thickness direction, the criterion being reduction of area. Although the results may vary as a function of the size and shape of the test piece, it seems that, roughly speaking, 25% reduction of area would be sufficient to prevent lamellar tearing in present-day offshore structures. Steels complying with this requirement can be made. Much higher values can be attained, but at a much higher cost. Therefore, from an economic point of view, it will be certainly worthwhile to find out which minimum reduction of area value is really necessary. Furthermore, it must be stated that reduction of area, as a function of S-content, shows a large scatter, as a result probably, of other already mentioned, secondary factors. In appendix 2 this relation is more quantitatively discussed on the basis of experimental results.

Finally, more to provoke discussion, a few words on steels with improved resistance against sea-water corrosion. The general opinion now seems to be that the use of these steels should not be considered. First of all, the improvement, especially in the splash zone, would not be large enough, and secondly there is no desire to risk weldability problems due to the extra alloying elements.

Appendix 1. Effects of various factors on yield strength and Charpy-V transition temperature of normalized Al-killed steels

The effects are schematically indicated in figure 2, based on the experience of our company (Hoogovens IJmuiden B.V.) and partly on the following formulae (Kouwenhoven 1969):

$$E = 5.4f_\alpha + (38.0 + 9.4\,\%(\mathrm{Mn}))f_\mathrm{p} + 7.2(\%\ \mathrm{Si}) + 2.6\,f_\alpha\, d^{-\frac{1}{2}}$$

$$T_{50} = 75 - 15d^{-\frac{1}{2}} + f_\mathrm{p}(274 - 44(\%\ \mathrm{Mn})),$$

where E is the yield strength in $\mathrm{kgf/mm^2}$; f_α, the fraction of ferrite; f_p, the fraction of pearlite; d, the average diameter of ferrite grain in mm; and T_{50}, the 50 % fibrous Charpy-V transition temperature in degrees Celsius.

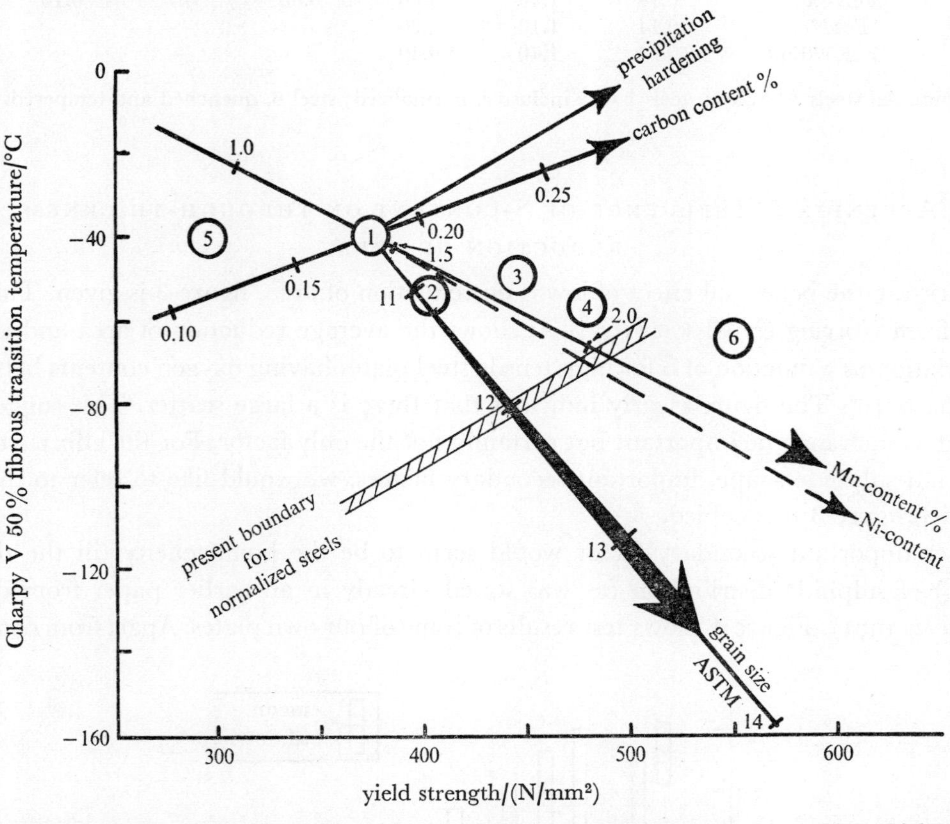

Figure 2. Diagram of transition temperature against yield strength for steels of table 1, plate thickness 25 mm.

The formulae are valid for steels with a ferrite-pearlite structure without precipitation hardening. The lines in figure 2 giving the influence of C and Mn, concern the total effects of these elements. The total effect of C consists of two contributions, an increase of f_p and a decrease of d (grain size). The total effect of Mn consists of three components, an increase of f_p, a change of the properties of the pearlite (according to the formulae) and a decrease of d (grain size).

The lines for Ni and precipitation hardening were derived from our own experience and literature (Meyer, Bühler & Heisterkamp 1971).

The positions of steels 2, 3 and 4 in relation to steel 1, are partly determined by the effect of some precipitation hardening. This effect cannot yet be predicted quantitatively from a formula or from the chemical composition. The quenched and tempered steel 6 is shown for comparison. Figure 2 suggests that a 'breakthrough' of normalized steels into the region of quenched and tempered steels, or more generally, into the region of still higher yield strengths and lower transition temperatures, must be possible if grain refinement beyond ASTM 12 can be achieved. It seems improbable, however, that along these lines yield strengths higher than 600 N/mm² can be achieved.

Table 1. Steels of figure 2.

steel number	type	C	Mn	Si	Nb	Ni	V	N
1	'Fe52'	0.18	1.40	0.30	—	—	—	—
2	'Fe52Nb'	0.16	1.40	0.25	0.02	—	—	—
3	FeE420	0.18	1.50	0.50	0.03	—	—	—
4	FeE450	0.18	1.40	0.30	0.05	0.7	0.10	0.012
5	'Fe41'	0.14	1.10	0.20	—	—	—	—
6	FeE500 q.t.	0.16	1.40	0.40	—	—	—	—

Note. All steels Al-killed; steels 1 to 5 inclusive, normalized; steel 6, quenched and tempered.

Appendix 2. Influence of S-content on through-thickness reduction of area

To illustrate the beneficial effect of low S on reduction of area, figure 3 is given. This figure is taken from Norring & Nilsson (1974). It shows the average reduction of area and the standard deviation as a function of S for high tensile steel plates having oxygen contents between 10 and 30 parts/10⁶. The figure clearly indicates that there is a large scatter. This suggests that S-content as such may be important but certainly not the only factor. For the effects of oxygen content and sulphide shape, important secondary factors, we would like to refer to the paper from which figure 3 was copied.

Another important secondary factor would seem to be the homogeneity (in the thickness direction) of sulphide distribution (as was stated already in an earlier paper from our part (Bellen *et al.* 1971). Figure 4 shows test results of some of our own plates. Apart from confirming

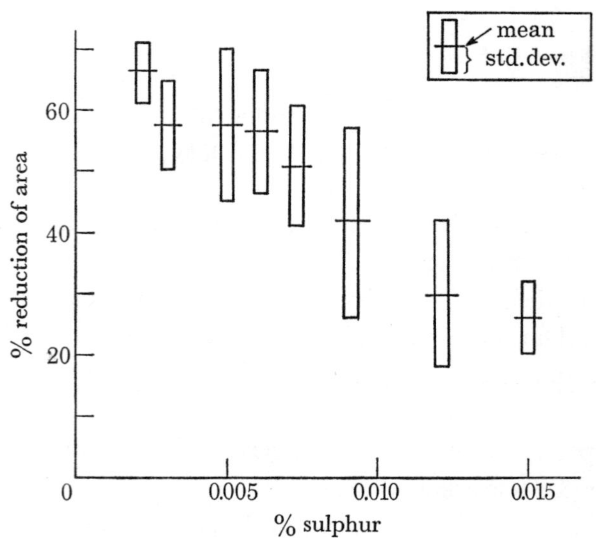

Figure 3. Reduction of area in through thickness direction as a function of S-content.

the general beneficial effect of low sulphur, this figure again shows a strikingly large scatter. Tedious inclusion counting as a function of the distance from the plate surface suggested that this scatter may be largely caused by differences in homogeneity of sulphide distribution in the through-thickness direction, and, in some cases, by a rather coarse banded structure.

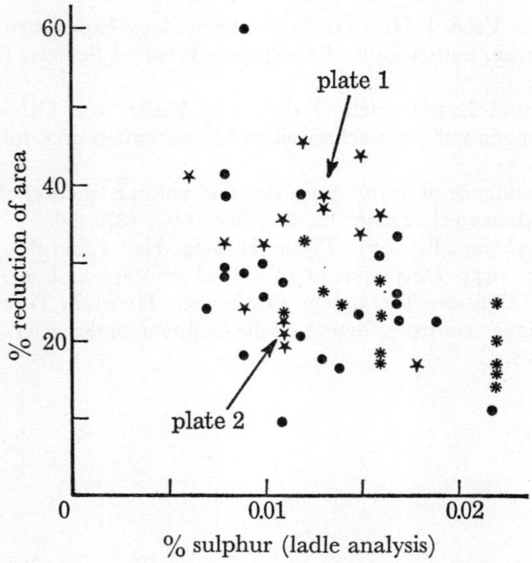

FIGURE 4. Results of Hoogovens. Averages of test pieces from same spot ★, Fe37 and Fe41; ●, Fe52; ⁕, Fe52Nb.

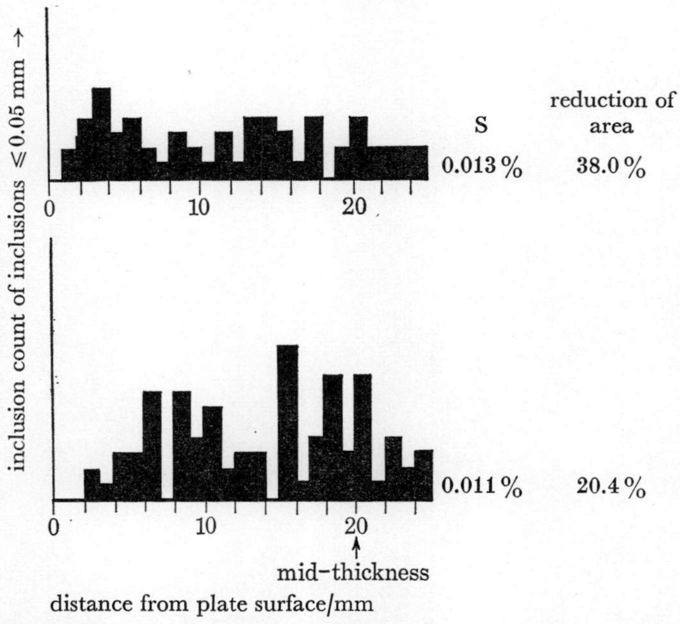

FIGURE 5. Difference in homogeneity of sulphide distribution in through-thickness direction between plate 1 (top) and plate 2 (bottom) of figure 4.

Figure 5 shows this difference in homogeneity between two plates from figure 4, marked 1 and 2 in this figure. The cause of this inhomogeneity is not yet fully understood. It would seem, however, that it is some sort of uncontrollable segregation, during instabilities in the solidification pattern, partly caused by differences in H_2-content of the steel (Bellen *et al.* 1971). This segregation will also be found in the other directions, for example in the width direction of the plate, and thus increase the scatter of test results between test pieces taken from different locations.

REFERENCES (van der Veen)

Bellen, A., Spanraft, M. J. & van der Veen J. H. 1971 Some steelmakers' experience on improving the resistance of steel plate to lamellar tearing. Symposium of the Japan Welding Society. *Cracking and fracture in welds*, 8–10 November 1971, Tokyo.

Förster, E. 1974 Desoxidation und Entschwefelung durch Einblasen von Calciumverbindungen in Stahlschmelzen und ihre Auswirkungen auf die mechanischen Eigenschaften an Grobblechen. *Stahl und Eisen* **94**, 474–485.

Kouwenhoven, H. J. 1969 The influence of ferrite grain size and volume fraction of pearlite on the lower yield stress and Lüders strain of carbon steel. *Trans. Am. Soc. Metals* **62**, 437–446.

Meyer, L. Bühler, H. E. & Heisterkamp, F. 1971 *Thyssenforschung*, Heft 1 and 2.

Norring, Kjell & Nilsson, Krister 1974 Development of a steel resistant to lamellar tearing – Z-steel, paper presented at the Sixth Annual Offshore Technology Conference, Houston, Texas, May 6–8, 1974.

van der Veen, J. H. 1968 Le laminage controlé, facteur de développement des aciers pour constructions soudées. *Rev. de Métallurgie*, janvier 1968.

3.5 A forward look at some steel developments based on physical metallurgy

By M. Cohen and W. S. Owen
Department of Materials Science and Engineering, Cambridge, Massachusetts, U.S.A.

It is projected that research on the physical metallurgy of steel during the next decade will be directed towards translating the impressive gains in our understanding of materials science into materials engineering. This will require the physical metallurgist to have an awareness of the economic and social environments in which the steel industry operates, the nature and special problems of technology transfer in the industry and the interdependencies of physical, chemical and process metallurgy. These points are illustrated by brief discussions of some areas of materials science of direct interest in the physical metallurgy of steel, some expected developments in low-alloy and in high-strength steels, and by a short review of a few of the more important problems awaiting practical solutions.

The research environment

It has been pointed out (Old 1972) that the common practice of reporting a company's research and development (R and D) expenditures as a ratio of net sales can be misleading. Instead two ratios are suggested: one, R and D expense divided by cash flow, and the other, technology-based capital investment less tax credit divided by cash flow. Typically, on this basis, a U.S. steel company might have an R and D investment ratio of 0.03 and a technology capital investment ratio greater than 0.60. Thus, about 65 % of the total cash flow is invested in technology. This is another way of saying that the steel industry is both technology and capital intensive. For a company's R and D programme to be cost-effective, it must develop technology in which it will prove profitable to invest at some not too distant time. The R and D investment ratio is so much smaller than the capital investment ratio that it is inevitable that much of the research done in a company's R and D programme will not contribute directly to a new or modified technology which will attract capital investment. In this sense, for many years research was considered to be relatively cheap and few serious attempts to assess its cost effectiveness or to direct it into profitable channels were made. In another sense, research of this kind was considered by many steel companies to be a luxury so that cuts in research budgets occurred early in periods when economies in operating costs were considered necessary.

In recent years there has been a substantial change in outlook. Basic research is no longer relatively cheap and it is realized that in many cases it is more profitable to buy, at home or abroad, the technology in which large investments are to be made. The communication of technical information within the steel industry, nationally and internationally, is remarkably unrestricted (Hiestand 1974). In the United States this situation dates back to 1922 when, in settlement of a major antitrust prosecution, the U.S. Steel Corporation agreed to make technological information available to its competition to help them expand. The selection of a technology in which a company decides to invest is influenced greatly by the local conditions: its geographical location in relation to ore and fuel sources, the nature of the products required by its market, the technology in which capital is already invested, the local environmental constraints, etc. Thus there is a need, which is being recognized increasingly by the U.S. steel industry, to direct its research towards those problems of major importance to the technology investment programme of the individual company.

Of course, some research on steel is done by industries which are large users of steel. The results of such work are used either to influence the technological development programmes of the steel companies wishing to supply the product, or to change or improve their own fabrication or heat-treatment processes. Usually, the technology capital investment ratio is not so large in the latter case as it is in the primary steel industries because the equipment is on a smaller scale. Consequently, the rate of technology transfer out of the laboratory onto the production line is more rapid. Virtually all of the steelmaking technology currently in use throughout the world is based on physical-chemical metallurgy that is at least a quarter of a century old. In contrast, heat-treatment processes in automobile plants, for example, are changing rapidly. In some cases the technology transfer interval may be as short as three to five years.

In materials research generally, the years between 1946 and 1970 were a period in which the potency of the structure-property approach of the new materials science was realized and exploited. The expansion of our knowledge of structure at all levels, our techniques for studying structures, and our understanding of the often subtle interplay between properties and structure were explosive. It also coincided with the period in which basic research, whether supported financially by industry or federal agencies, was relatively undirected. Physical metallurgists working on steel were swept up by the new science. Research in physical metallurgy done during this period was enormously productive from a scientific viewpoint. However, as the subject developed, reaching increasingly towards materials science, the couplings with chemical and process metallurgy and with engineering design were weakened. By the start of the 1970s it became clear that this trend was reversing. The economic and technological considerations in the steel industry which forced this change have been referred to earlier. There was an additional reason. The rate of development of materials science was so rapid in the postwar years that useful, although not always completely satisfying, solutions to most, but not all, the scientific problems relevant to steel technology were found by 1970. Many fascinating and difficult problems remain for the next generation of materials scientists but it cannot be claimed, at least as far as the steel industry is concerned, that solutions to those problems are quite so crucial to the development of steel technology as was the case a few decades ago. The need now and in the next decade is to develop the engineering of materials and, in so doing, to utilize the many important insights into structure-property relationships which have been acquired during the extraordinary growth of the science of materials.

At present, most research on the physical metallurgy of steel, and especially that carried out by steel-producing companies, has limited and circumscribed objectives. It is concerned with improving the yield strength or the fracture toughness of hot-worked products, the formability of cold-rolled sheet, the corrosion-fatigue resistance of stainless steel, the high-temperature creep resistance of highly alloyed steel forgings, etc. The development sector of the R and D effort is concerned with the utilization of the findings of the research and the development of a new or improved product for which, it is believed, there is a market or for which a market can be created. In this total process, two-way interactions between the materials engineers and the potential users are of crucial importance. It seems probable that this interaction will increase still further in the future. Users will have a greater influence on research planning, even in those sectors involving a large tonnage of steel in situations where there is no alternative material. This is likely to be the future situation in the construction industry, including the construction of offshore oil rigs, power plants, ships, pipelines and similar structures. Of course, other research and development carried out by steel companies is aimed at improving the cost-effectiveness of process operations within the company and, consequently, usually does not require extensive interaction with user industries. To an increasing extent, effective work on problems in this area requires the cooperative efforts of physical, chemical and process metallurgists.

In the industries producing materials, the future research directions likely to be most rewarding are often difficult to identify. The development of high-strength, low-alloy (h.s.l.a.) steels provides an example. Over the last decade or more, much imaginative and painstaking research has led to remarkable improvements in the properties of these steels (Korchynsky & Stuart 1970). Hot-rolled ferritic steels with a yield strength of 550 N/mm^2 (80000 lbf/in^2) coupled with good ductility are now available and it is probable that similar steels with strengths in excess of 700 N/mm^2 (100000 lbf/in^2) will be developed commercially in the near future. By reducing the carbon and sulphur levels and applying the results of modern research on sulphide-particle-shape control, it is possible to design these steels to have good weldability, formability and directionality of properties combined with acceptable corrosion and fatigue resistance. To take advantage of the higher available strength and to reduce the cost, it is necessary for the designer of engineering structures to achieve the required function with substantially less steel than would be needed had older conventional steels been used. This inevitably leads to difficulties with the rigidity of the structure which, in some cases, necessitates radical changes in the design. It might be argued that in these circumstances, future research on h.s.l.a. steels should be directed towards improving the rigidity of the steel itself (Kiessling 1974). In many designs, a 10% increase in the shear modulus would greatly ease the designers difficulties. However, because the elastic moduli are not structure-sensitive properties, one may question whether even this modest goal can be achieved.

A careful analysis of the total system may reveal that the h.s.l.a. steels already available cover a range of strength greater than can be used economically in large structures in which rigidity is important. If so, it would be more profitable to concentrate future research on achieving the desired combination of properties by a different and cheaper route. This might involve the utilization of a different 'structure', but is more likely to require consideration of the total processing of the steel. Microalloyed h.s.l.a. steels with the shape of the sulphide particles controlled by the addition of rare earths (cerium) or, in the case of niobium-treated steel, the addition of zirconium, must be fully killed. For many plants with a large financial investment in ingot technology, these processes are prohibitively expensive. If comparable strengths, impact shelf-energies (resistance to ductile fracture) and directional properties could be achieved in semi-killed or rimming steels without significant increases in the cost of the processing, the benefit to the construction industry could be much greater than further improvements in the properties of killed steels. Such research is likely to be concerned at least as much with the chemistry of the steelmaking process and the processing history as with the structure-property relationships of conventional physical metallurgy. The metallurgists engaged in such work would be called upon to apply a much wider range of knowledge and skill than has been the case in the recent past. Recognizing this trend, at M.I.T. we have recombined the graduate programmes in chemical, process and physical metallurgy after more than twenty years of separation.

From these considerations we conclude that even in relation to those industries which are large-scale users of steel, future directions will be dictated by the conclusions drawn from analyses of total engineering systems; analyses which will incorporate economic and environmental factors as well as considerations of hardware design and materials technology.

This conclusion is more easily recognized when the materials needs of product-oriented industries are considered: industries which use special purpose steels in combination with a variety of other metallic and non-metallic materials. The automobile industry is a good example. No longer does the automobile designer expect to find an optimum combination of properties in a single material. The function of each component in the total system is achieved by a combination, often a composite, of materials. The cost of the component is a sensitive function of the way the component is made, as well as of the cost of the basic materials. In these circumstances, steel, alone or in a composite, must compete with the whole range of engineering materials. Frequently, there is no unique solution to a particular design problem. The choice of materials is then dictated by considerations of quality control, reliability and, of course, primarily, cost. The solution adopted is very likely to vary from company to company or even from one location to another within the same company.

It is now becoming evident that three of the factors which have profound influences on the materials systems approach to functional design are the national materials policy, the content of the materials technology bank, and the rate at which stored technology can be translated into practical applications. (See, for example, the reports of the Committee on the Survey of Materials Science and Engineering, National Academy of Sciences 1974, and of the National Commission on Materials Policy 1973.) To an increasing extent, national policies are deciding the relative cost and availability of steel and the many materials with which it competes. Obvious examples include the impact that recent environmental protection regulations have had on blast furnace, open-hearth and other steelmaking processes, and the dramatic shifts in the relative costs of materials produced by the unexpected and urgent need to conserve expensive energy. The increase in the cost of polymeric materials resulting from the much increased cost of petrochemicals is likely to have an important effect on the relative attractiveness of metals, including steel, and polymers in many materials systems.

Some effects of national materials policies are less obvious. Changes in the United States policy on stockpiling of materials have resulted in significant changes in the availability of some materials. To a larger extent than is usually appreciated, national policy dictates the general problem areas to which funds made available by the federal government are applied. In materials, there has been a major redeployment of research effort to attack numerous materials problems considered to be important in relation to the development of new or improved sources of energy. One surprising consequence of this has been the re-establishment of a high priority for research on high-strength and high-alloy special steels for use in power-generation machinery.

In the past, the bank of available knowledge about achievable materials properties and viable materials process technology has been filled in an unorganized fashion. Many people who worked on the physical metallurgy which became the basis for the development of ausformed steels have been disappointed to discover that these steels have found only limited application in practice. Ausformed steels were the result of research and development motivated primarily by the desire to exploit as fully as possible the then current research trends. When considered in a total materials-system design, ausformed steels were seldom selected because of the poor stress-corrosion resistance, difficulties with processing and forming, and limited range of shapes and thicknesses in which good mechanical properties can be achieved. The knowledge and expertise resulting from the substantial research and development effort devoted to ausformed steels a few years ago has become a part of the existing materials technology bank. Whether or not at some future time ausformed steels will provide the solution to as yet unidentified materials design problems is difficult to predict.

It is becoming increasingly clear that materials research in the United States should and will be more strongly focused on specified goals than has been the case in the past. Producer industries will be influenced by the results of their materials-systems analyses in deciding what materials technologies need to be added to the technology bank. The expenditure of federal funds will be much influenced by the developing national materials policy. A possible future development connected with materials policy which might be of special importance and interest to the steel industry concerns manganese. The United States steel industry is almost completely dependent on foreign supplies of ferromanganese and for this reason there has been a great deal of interest in the possibility of extracting manganese from the nodules of metallic oxides in great abundance in some areas of the deep oceans. Substantial resources have already been expended on developing the technologies necessary to mine these deposits and to extract from them cobalt, nickel and copper (Anon 1972, 1973, 1974; Rothstein & Kaufman 1973). To extract the manganese from the oxide waste is probably relatively simple. One econometric study has concluded that if the extraction of manganese from this source were put into large-scale production, by 1980 manganese could become the next cheapest metal after iron. Of course, to warrant large-scale production, adequate markets must be available. It is clear that at present there is available insufficient understanding of the technological consequences of substituting manganese or manganese oxide for ferromanganese in the steelmaking process. Nor is enough known about the physical metallurgy of iron-manganese and manganese-base alloys to indicate whether or not useful and competitive properties can be developed in such alloys, thus enlarging the market for manganese products. If the federal government decides, as a part of the national materials policy, to encourage the production of manganese from oceanic deposits, the use of research resources to explore the chemical and physical metallurgy of alloys containing manganese follows logically. Such a decision would have an important influence on the future trends of research on steel.

From this brief review of the context in which materials research is expected to develop, it is obvious that any attempt to predict future trends by extrapolation from current interests and activities is very likely to be misleading. However, in the present circumstances, we see no alternative course of action available to us.

The application to engineering

Since the Second World War the physical metallurgy of steel, as of other metals, has been dominated by the study of structure-property relationships. The advances in our understanding of the science of many basic phenomena have been spectacular. Significant insights have been obtained into work-hardening, age-hardening, solid-solution hardening, plastic flow and fracture, to mention only a few phenomena. To apply this basic physical metallurgy to practical problems often requires a connexion to engineering which is difficult to make.

Much careful work on ferrite-pearlite annealed or normalized steels has led to the establishment of reliable quantitative relationships between the yield strength and such structural parameters as grain size, solid-solution hardening, volume fraction of pearlite, and pearlite spacing (Baird & Preston 1972). The quantitative effect of precipitation hardening by carbide particles has not been established by direct methods, but, qualitatively, the influence on strength and ductility is well understood. This and parallel work on the control of the size, shape and distribution of sulphide particles and other inclusions provided the scientific basis on which modern h.s.l.a. steels were developed. Similarly, the relative potency of the various structural strengthening mechanisms in quenched and tempered steels have been determined (Cohen 1963; Owen 1970), although the integrated effect produced by two or more of these mechanisms is not yet understood quantitatively.

The prediction of the fracture toughness of a steel from microstructural parameters is much more uncertain. The problem is compounded if the steel is to operate in a corrosive environment. For a steel, tested under non-corrosive conditions, with a yield strength greater than about 1100 N/mm^2 (160 klbf/in^2) or sufficiently thick to ensure that the specimen fractures under plane-strain conditions at a lower strength level, the fracture toughness parameter, K_{Ic}, as determined by direct application of linear fracture mechanics, is a satisfactory measure of the fundamental resistance of the steel to the propagation of a crack. This measure of fracture toughness is widely used in successful designs based upon maximum-permissible flaw concepts (Pellini 1968). However, even in these circumstances, knowledge of quantitative relationships between the fracture toughness and microstructural parameters is lacking. Attempts to obtain some general qualitative understanding of possible relationships through intermediary simple mechanical parameters such as the strain-rate dependence of the flow stress or the strain-hardening index have met with only modest success (Hahn & Rosenfield 1966). About all that is known with any confidence is that structural changes which increase the strain-hardening index are likely to improve the K_{Ic} value. This information is of limited value as a guide to the development of ductile steels because the number of different strain-hardening mechanisms, and consequently the number of microstructures in which they operate, is small.

When a steel has a yield stress of less than 1100 N/mm^2 (160 klbf/in^2) and is thinner than the critical thickness, no satisfactory measure of the fracture toughness is known, and consequently, there is no understanding of the influences of changes in the structure. This pair of coupled problems was discussed extensively at the recent Kyoto Conference. Liebowitz (1974) was of the opinion that advances in our understanding of the mechanics of fracture offer some hope that the former problem may be solved by developments of the J-integral concept or in some other way, but there appears to be no reason to believe that solutions to the structure-property problems are within sight. Some structural-hardening mechanisms which, until now, have received little attention will be mentioned later. There is, as yet, no indication of how useful they might be. Stress-corrosion failure cannot be prevented by the control of structural parameters. It is probable that, at least in the near future, the use of high-strength steels in even mildly corrosive atmospheres will be limited by the development of effective protective surface coatings (probably polymers), but the reliability of these solutions to the problem is in doubt at the present time.

The concepts of linear fracture mechanics have been extended, with as yet limited success, to explain fatigue behaviour of steels. The problem is perhaps more complex than that of static brittle behaviour but, undoubtedly, solutions to both phenomena will be found within the same theoretical framework.

High-temperature deformation (creep) is reasonably well understood. Recently, there have been substantial advances in our understanding of thermally activated deformation processes at high temperatures. Future development of high-temperature creep resistant materials is unlikely to be retarded by any lack of understanding of the basic phenomena.

Some developments possible in the future

Low- and medium-strength steels

The use of carbides precipitated in the parent austenite during controlled hot rolling to retard recrystallization of the austenite has been a key factor in the development of fine-grained h.s.l.a. ferritic steels. To date, this process has been developed empirically. It is possible that if the kinetics of nucleation and growth of the various carbides and nitrides were understood in terms of the temperature and rate of deformation and the chemistry of the austenite, ferrite grain sizes substantially smaller than those currently attainable could be obtained in commercial practice. The physical metallurgy of the problem is complex and difficult to study experimentally. However, the results obtained in a number of recent studies using either hot-torsion testing or rapidly quenched specimens coupled with systematic optical and electron microscopy indicated that experimental studies are possible and rewarding (Hansen 1975). The lower limit to the ferrite grain size seems to be a few micrometres. The Hall–Petch relation between strength and grain size holds even at this extreme of the range (Baird & Preston 1972), predicting that, in principle, steels with a yield strength approaching the theoretical strength could be obtained by this means. Grain refinement has an important advantage over other strengthening mechanisms; it enhances the ductility. Still finer ferrite spacing can be obtained by large cold deformation. In this range, however, the Hall–Petch relation is replaced by a linear relation between the strength and the reciprocal of the ferrite dimension (Langford & Cohen 1969). It is possible that similar cold-worked structures could be obtained with much smaller deformation by control of the precipitation of carbides or nitrides during the total thermomechanical treatment of suitably designed steels. Dislocation-interstitial-precipitate interactions during cold and warm working of steel in the ferritic condition is also a neglected topic. It has been known for many years that large increases in the strength of quenched steels can be obtained by dynamical strain-ageing. Unfortunately, this form of hardening is usually accomplished by a deterioration in the fracture resistance (blue-brittleness) and unstable plastic flow. In the last few years it has been recognized that dynamical strain-ageing of steels containing carbide or nitride particles is a phenomenon involving only freshly generated moving dislocations. It requires a flux of interstitial atoms from redissolving precipitate particles to the glissile dislocation (Kalish 1966; Roberts & Owen 1967). It is possible to induce dynamical strain-ageing in such a way that the steel is not excessively brittle and that unstable plastic flow occurs only at temperatures which are higher, and strain-rates which are faster, than those of practical interest. The importance of resolution of carbides and nitrides is clear, but it is not known whether or not significant differences in the dynamical strain-ageing behaviour can be obtained by using different precipitate compositions. In particular, the behaviour of the various precipitates possible in microalloyed steels have not been explored. It is possible that the exploitation of dynamical strain-ageing to develop a structure which contains an optimum dispersion of microalloy carbides or nitrides could lead to the development of steels with very attractive strength-ductility properties at temperatures and strain-rates below the lower limit of the blue-brittleness range.

Most of the h.s.l.a. steels developed to date have a ferritic substructure typical of hot-rolled steel. This dislocation substructure provides some, but probably a small, contribution to the strengthening, but it is an important source of heterogeneous nucleation sites for the precipitation processes. Recently, there has been much interest in utilizing the denser dislocation substructure developed in steels with an 'acicular-ferrite' structure obtained by suitable control of the chemistry and the cooling rates of microalloyed steels. This structure is clearly related to the bainitic structure which can be developed in low-carbon alloy steels and to the 'lathe' structure found in low-carbon and low-alloy martensites. In spite of much attention, this family of structures is not well understood. It is generally assumed that a transformation involving a lattice-shape-change (martensite transformation) is an essential step in the development, but it is unclear whether or not this shape change is the same in all cases. The relationship of the dislocation substructure to the morphology of the transformation product is also uncertain (Owen 1970). The substructure of acicular-ferrite is closely related to that produced in slowly cooled high nickel (14–18% nickel) steels of the maraging type. In low-alloy or microalloy steels accelerated cooling is required, but the chemistry can be adjusted to ensure that the cooling rates which are necessary are within the range attainable economically. The development of a high dislocation-density substructure by a transformation requiring only practical accelerated-cooling instead of cold working involving significant deformations is attractive commercially. In the future, undoubtedly there will be much study of these transformations and of the properties which can be developed in steels with this class of substructure.

Solid-solution strengthening by carbon and nitrogen and precipitation hardening by carbide or nitride particles are the hardening mechanisms operative in nearly all ferritic steels. Jack (Driver, Hanley & Jack 1972) has demonstrated the existence of another mechanism in ternary alloys of a composition in which aggregations of, for example, manganese and nitrogen atoms can occur by a transformation which has some of the features of a spinodal decomposition. In some steels these aggregates, which can form in either austenite or ferrite, are similar to g.p. zones in aluminium alloys, but in other steels they appear to be very small coherent precipitates. Transformations of this kind increase the hardness. Steels with this hardening mechanism have not yet been designed or produced in sizes large enough to allow fracture-toughness testing. Thus, it is not yet known whether or not the mechanism can be the basis for the development of a new class of structural steels.

It has long been recognized that if steels containing very little or no carbon and nitrogen could be designed and produced economically, they would have a number of advantages over the traditional carbon steels. Welding problems would be reduced and loss of ductility due to the segregation of massive carbides at grain boundaries would be avoided. Work at Berkeley (McMahon & Thomas 1973) has shown that brittleness due to interphase carbides in Fe–Cr–C steels can be eliminated by a simple heat treatment. Steels used at temperatures at which creep is significant deteriorate with time because of the instability of carbide or nitride particles. At these temperatures, carbides and nitrides coarsen rapidly primarily because of the rapid diffusion rates of the interstitial atoms in ferrite. Thus, the idea of using a precipitate containing only elements which in the iron lattice dissolve substitutionally is attractive. The only carbon-free structural steel which utilizes precipitation strengthening and is in large-scale commercial use is maraging steel in which the effective precipitates are of the Ni_3 (Al, Ti) type. This steel, used for its room-temperature and low-temperature properties, is not attractive as a creep resistant material. Decker (1969) has summarized the results of the few known attempts to use intermetallic compounds in creep-resistant steels. He concluded that if Laves phases are present in significant concentration they have little effect on creep properties but they increase room-temperature brittleness. Similar results were obtained by Mihalisin, Bieber & Grant (1968) who investigated the creep and fracture properties of a steel containing sigma phase. These undesirable properties seem to be a result of the formation of grain-boundary films of either Laves or sigma phase. Jones, Parker & Zackay (1971) showed how a continuous network of Laves phase at the ferrite grain boundary in an Fe–Ta–Cr alloy can be dispersed and caused to spheroidize by a simple heat treatment. The properties of the heat treated alloy are not unusual, but the brittleness usually associated with Laves or sigma phases is removed (Zackay, Parker & Bhandarkar 1972). Clearly, much work along these lines remains to be done. It is possible that by proper design of alloy composition and thermomechanical treatments, high-strength alloys with good creep stability can be developed from carbon-free iron alloys containing only small concentrations of elements such as tantalum. Since stainless steels with chromium as an essential alloying element are particularly susceptible to the formation of sigma phase, it might well be possible to turn this difficulty to an advantage. Developments along these lines in the future can be expected to lead to many interesting and possibly useful steels.

High-strength steels

Although detailed understanding of the relationships between structure and fracture toughness is lacking, it has been evident for some years that in order to obtain a steel with high strength combined with fracture toughness a high work-hardening rate is required. Hence, a steel with a high dislocation density and a dispersion of fine precipitate particles is required. By thermomechanical treatment of steels of suitable composition excellent combinations of strength and toughness can be obtained. The practical limit of strength for a steel with a yield stress/K_{Ic} ratio of 1.0 $in^{-\frac{1}{2}}$ appears to be about 1375 N/mm^2 (200 $klbf/in^2$). There are, of course, many drawbacks to this approach to high strength, not the least of which is the necessity to reduce the cross-section by more than 50% during the thermomechanical treatment.

A major advance in the technology of high-strength steels is likely to occur only when ways are found of utilizing different and more effective work-hardening mechanisms. Available mechanisms are not numerous. Three interrelated mechanisms deserve consideration: hardening by the formation of strain-induced martensite, Hadfield hardening, and hardening by the formation of stacking faults. The first of these has been utilized in t.r.i.p. steels. These are designed to have a microstructure after thermomechanical treatments which is predominantly martensitic but which contains a significant volume-fraction of retained austenite of a composition such that with further deformation more martensite forms by strain-induced nucleation. Large heats of t.r.i.p. steels have been made and the excellent strength and fracture-resistant properties under normal conditions of temperature and strain rate have been confirmed. Unfortunately, the fracture properties are very sensitive to strain rate and temperature. At room temperature and high strain rates, the temperature in the small plastic zone ahead of a running crack rises above the M_d temperature and the formation of strain-induced martensite is inhibited. The steel then loses toughness. To produce t.r.i.p. steels with a wider tolerance for extremes of temperature and strain rate, it is necessary to design the steel composition such that the M_s temperature is near room temperature and the M_d temperature is as high as possible. The recent ideas of Olson (1974), utilizing the strain-induced martensite hardening mechanism, provide a rational scientific basis for the development of such steels. The Olson model of nucleation assumes that the first step in martensite nucleation is faulting on planes of closest packing. The stacking-fault energy is shown to consist of both volume-energy and surface-energy contributions. When the volume-energy contribution is negative, the fault energy decreases with increasing fault thickness. This condition leads to the spontaneous formation for a martensitic embryo. The model has been elaborated to explain both thermal and strain-induced nucleation and provides clear indications of the relative importance of the various factors which affect these two types of martensite formation. From these ideas, a first attempt to design a t.r.i.p. steel which is relatively insensitive to strain-rate effects has been made. The project is not sufficiently advanced to allow definite conclusions to be drawn at this time.

The strain-hardening that occurs in Hadfield austenitic steels is unusually large and a unique phenomenon. It is much larger than in steels with the same manganese content (12 %) but lower carbon. Evidently, the low stacking-fault energy in the austenite and the unusually high concentration of carbon (1.2 %) are together responsible for the effect. Hadfield steels are often quoted as an example of the toughness and strength which can be developed by the production of strain-induced martensite during deformation but, in fact, no new phase is formed martensitically during the deformation of Hadfield steel (White & Honeycombe 1962). The mechanism deserves serious study. High-strength, high-ductility steels produced by this mechanism will be very attractive economically if manganese becomes a cheap metal.

Finally, mention should be made of the possibility of utilizing the relatively high work-hardening rate of a low stacking-fault energy austenite. Information on this effect is sparse, but there are sufficient data to suggest that the effect might be useful. There is a suspicion that metals with a low stacking-fault energy are particularly susceptible to brittle fracture and that they have a low fatigue limit. Unfortunately, most of the tests which have been done to date have used material which was insufficiently pure to exclude the possibility of extraneous effects, such as grain-boundary carbide formation, invalidating the conclusions.

Priorities

The projected developments discussed here are extensions of topics currently of interest. They do not represent the most important research topics as predicted by experts looking at the total materials needs. In fact, in the COSMAT (1974) study of research priorities, physical metallurgy of ferrous metals and alloys received a low overall research priority rating, well below that of ceramics, glasses, polymers and composites. However, some phenomena which are important in steel metallurgy were recommended for urgent attention. These included corrosion, oxidation and other surface phenomena. Undoubtedly, these will become the focus of much sophisticated research effort. The techniques essential to rapid advances are available: l.e.e.d., h.e.e.d., Auger spectroscopy, and e.s.k.a.

We have emphasized that, in the next decade, attention will most probably be given to translating the lessons learned by the study of structure–property relationships into engineering practice. Much of the research will move in this direction also. The major areas of research are likely to be: reliability and reproducibility of properties; performance of materials as assessed over their lifetimes; and problems associated with inspecting and testing materials for specific uses. Design is likely to move more and more toward the adoption of critical flaw-size concepts. This means that much research effort will have to be devoted to studies of methods of non-destructive testing, particularly to the development of reliable methods for measuring spatial flaw-size distribution. Indeed, it appears that a substantial portion of the research thrust at steel companies will be directed to this end in the next five or ten years.

REFERENCES (Cohen & Owen)

Anon 1972 *Ocean Industry*, July 1972, p. 58.
Anon 1973 *Kennecott Copper Company Annual Report*, 1973.
Anon 1974 *Chemical Engineering* 31, May 1974 p. 34.
Baird, J. D. & Preston, R. R. 1972 *Proc. Int. Symp. on Processing and Properties of Low Carbon Steels*, MD-AIME 1972.
Cohen, M. 1963 *J. Iron Steel Inst.* **201**, 833.
Committee on the Survey of Materials Science and Engineering (COSMAT) 1974 *Materials and man's needs.* Washington, D.C.: National Academy of Sciences.
Decker, R. F. 1969 Steel strengthening mechanisms. *Symposium Climax Molybdenum Company*, p. 147.
Driver, J. H., Hanley, J. R. & Jack, K. H. 1972 *Scan. J. Met.* **1**, 211.
Hahn, G. T. & Rosenfield, A. R. 1966 *Mechanics and metallurgy of brittle crack extension.* ASCE Mechanics Division, Washington DC, 1966.
Hansen, S. 1975 Structure–property relations in two high-strength, low-alloy steels. M.S. thesis M.I.T.
Hiestand, D. L. 1974 *High-level manpower and technological change in the steel industry* p. 43. N.Y.: Praeger.
Jack K. H. 1972 *Scan. J. Met.*, **1**, 195.
Jones R. H., Parker, E. R. & Zackay, V. F. 1971 *Proceedings Fifth International Materials Symposium.* Berkeley, CA. 1971, 829.
Kalish, D. 1966 The effect of strain tempering on strength and toughness. Sc.D. thesis, M.I.T.
Kiessling, R. 1974 Will steel appeal in A.D. 2000? *Metals & mater.*, February 1974.
Korchynsky, M. & Stuart H. 1970 *Symposium on low-alloy high-strength steels.* The Metallurgy Companies, Nuremberg, p. 17.
Langford, G. & Cohen, M. 1969 *Trans. Am. Soc. Metals* **62**, p. 623.
Liebowitz, H. 1974 *Proc. 1974 Symp. on the Mechanical Behavior of Materials.* Society of Materials Science, Japan.
McMahon, J. A. & Thomas, G. 1973 *Proc. Third Int. Conf. of the Strength of Metals and Alloys* **1** 20–25.
Mihalisin, J. R., Beiber, C. G. & Grant R. T. 1968 *Trans. A.I.M.E.*, **242**, 1968, 2399.
National Commission on Materials Policy 1973 *Materials needs and the environment.* Washington, D.C.: U.S. Government Printing Office.
Old, B. S. 1972 R and D investment: sharpening the focus. Management Notes, A. D. Little, Inc.
Olsen, G. 1974 A general mechanism of martensitic nucleation. Sc.D. thesis, M.I.T., 1974.
Owen, W. S. 1970 *Proc. Second Int. Conf. on the Strength of Metals and Alloys*, ASM, p. 198.
Pellini, W. S. 1968 *Advances in fracture toughness characterization procedures and in quantitative interpretations to fracture-safe design for structural steels* Naval Research Laboratory Report 6713, 3 April, p. 44.
Roberts, M. J. & Owen, W. S. 1967 The strength of martensitic iron–nickel alloys. *Trans. Am. Soc. Metals*, **60**, 687.
Rothstein, A. J. & Kaufman, R. 1973 *Proc. Offshore Tech. Conf.* Paper 1758, p. 323.
Schultz, C. E. 1971 *Proc. Offshore Tech. Conf.* Paper 1364, p. 839.
White, C. H. & Honeycombe, R. W. K. 1962 *J. Iron Steel Inst.*, June 1962, 457.
Zackay, V. F., Parker, E. R. & Bhandarkar, D. 1972 *John Dorn Memorial Symposium.* AIME, Cleveland, OH, 1972.

KEYNOTE SPEECH

BY K. J. IRVINE (*British Steel Corporation*)

At the I.S.I. Centenary Meeting six years ago I was asked to review the developments in Physical Metallurgy during this period. It is a point of some interest, that it was only during the last few years of this period that much attention had been given to structural steels. However when work started rapid progress was made in understanding and then in exploiting the physical metallurgy of these steels. The work has continued since that time and now it is possible to present a very complete picture as illustrated by the technical papers in this Session.

The relations between microstructure and physical properties, so intensively exploited since the mid-1950s, have led to significant developments in the field of high strength low alloy steels. Among the early work dealing with the quantitative effects of microstructure on properties was that of Gensamer which described the effects of interlamellar spacing on the strength of pearlite. In the early 1950s, Hall reported the quantitative relation between yield strength and grain size in mild steel, leading to the development of the widely accepted Hall–Petch relation. Also in the 1950s, considerable attention was given to aspects of precipitation hardening, and dispersion strengthening, and in this period, the basic work of Orowan, subsequently modified by Ashby, provided a working basis for the development of quantitative relations between microstructure and strength in dispersoid steels. Although there were early ad hoc investigations of the effects of nibium on strength of commercial steels, it was some time before this was rationalized.

The 1960s saw the application of much of this basic research in high strength low alloy steels. In many cases, the approach was rather pragmatic, but at least the models of strengthening were built on the lines indicated by the basic work of the 1950s. Initially, the basic properties to be studied were the yield strength and the impact transition temperature, and by 1965, the concept of embrittlement vectors had been introduced, whereby a particular strengthening mechanism was associated with a degree of embrittlement per unit increase in strength. This approach was an acknowledgement that strength alone was not sufficient and that commercial steels require a combination of properties. It was shown that pearlite had little effect on yield strength but was harmful to toughness. Consequently, attention was given to pearlite reduced steels as a means of improving toughness.

The period also saw the identification of the 'interphase' precipitation mechanism in microalloyed steels. Finally, the need for cold formability in the steels provided a stimulus for basic work on ductile fracture. Studies of the subject were considerably advanced by the work of Gurland & Plateau in 1963. It was shown that sulphide particles were important, for example with type I sulphides nucleating the ductile fracture. It was later in 1970 that the conclusions from these basic studies were applied with the introduction of inclusion modified steels.

These were some of the important advances in knowledge from which the current understanding of the important effects can be summarized as follows:

1. Basic structure/property relationships

Yield strength (room temperature)

Extension of the basic Hall–Petch relation to account for the effects of solutes, and of other strengthening contributions yields an equation of the form:

$$\sigma_y = \sigma_o + 37\ (\%\mathrm{Mn}) + 83\ (\%\mathrm{Si}) + 3000\ (\%\mathrm{N_{free}}) + 15.1 d^{-\frac{1}{2}} + \sigma_{disp} + \sigma_{disl},$$

where d is the mean linear intercept of the ferrite grains (mm²)

σ_{disp} is the dispersion strengthening contribution

and σ_{disl} is the disolocation strengthening contribution.

The dispersion strengthening contribution obtained from the Ashby–Orowan relationship is:

$$\sigma_{disp} = \frac{5.9 f^{\frac{1}{2}}}{\bar{x}} \Big/ \ln\left(\frac{\bar{x}}{5\times 10^{-4}}\right),$$

where f is the volume fraction of precipitates

and $\bar{x}$ is the mean planar intercept diameter of the precipitated particles (μm).

As dislocation density measurements are tedious and often notoriously inaccurate, the dislocation contribution to strength is often assessed on the basis of the difference between observed and calculated values.

Impact transition temperature

The impact transition temperature can be assessed from the vector concept, as shown by the following equation:

$$\text{i.t.t.} = C - V_g(k_y d^{-\frac{1}{2}}) + V_d(\sigma_{\text{disp}} + \sigma_{\text{disl}}).$$

The constant C is dependent upon solute concentrations and pearlite content. The negative vector for the grain size contribution to strength expresses the improvement in toughness accompanying the increase in strength when grain refinement is the operative strengthening mechanism. Other microstructural features contributing to strength tend to have adverse effects on toughness, though the magnitude of this effect is commonly less detrimental than the effect shown by many solutes capable of significant solid solution hardening. Although the literature contains various estimates of V_g and V_d, the values are approximately 0.8 °C and 0.5 °C per N/mm^2 respectively.

Ductility, bendability and shelf energy

These properties are similar in that they are largely controlled by the ductile fracture mechanism, which involves void nucleation, void growth, and void coalescence. Studies of second phase particles such as oxides and sulphide inclusions have shown that the mechanism involves interface decohesion at very low strains, and that ductility is largely controlled by the volume fraction of second phase particles.

The particle shape is also important with respect to the principal strain axis. Inclusions presenting a small frontal radius of curvature cause greater strain intensification, more rapid void growth, and therefore a reduced level of strain at fracture. Recent advances have highlighted the aspect of inclusion distribution and the importance of local concentrations. It is possible to get a uniform distribution of type I sulphides or at the same volume fraction, but with type II sulphides, a local concentration that has far worse properties.

Carbide cracking also plays an important role in controlling the fracture of h.s.l.a. steels, the cavities formed by carbide cracking interacting with those formed at inclusions to control the strain at which void coalescence and failure occur. These effects have been summarized by Gladman in the diagram quoted in figure 4 of the Clay & McCutcheon paper (3.3). This shows the controlling effect of sulphide particles, but indicates that when a very low volume fraction of sulphide particles has been achieved, it is also necessary to consider the amount of carbide. The general move to lower carbon steels, initially made to improve impact transition temperature, also gives benefit in terms of cold formability and shelf energy.

2. Microstructural design and control

Grain size control

The paper by Roberts (3.1) describes the control of grain size in normalized products. The effects of fine stable precipitated particles in controlling grain size is well established and understood. Roberts makes the comment that commercial steels well exceed the property possibilities outlined by myself in 1967. The 1967 diagram was drawn in a conservative way, with the experience that commercial steels were frequently 30 °C worse than the properties obtained in laboratory experiments. The significant fact is that the close control of commercial steels has enabled the full potential of these steels to be realized.

The references in the Roberts paper to the effect of aluminium indicates that this effect is not completely defined. It may well be that the effect of excess aluminium relates to the fraction of nitride that can be redissolved during soaking for hot rolling and which is subsequently precipitated in the form of fine particles during rolling. The particles which are undissolved at the soaking temperature become coarse and ineffective in controlling the fine grained austenite during normalizing.

In hot rolling, the austenite grain size is influenced by the conditions of hot rolling: high deformations and low finishing temperatures promote fine austenite grains. Additions of microalloying elements such as niobium can influence recrystallization kinetics and grain growth characteristics. These factors are dealt with in some detail in the paper by Morrison & Chapman. Compared with the reviews of several years ago it is probably in the field of controlled rolling that there has been most additional knowledge to fill in the total picture.

Alloy design

The background work referred to earlier gives a fairly comprehensive basis of alloy design to develop specific properties or property combinations. For example, a composite diagram connecting alloy composition, processing variables, and the basic strengthening models with the

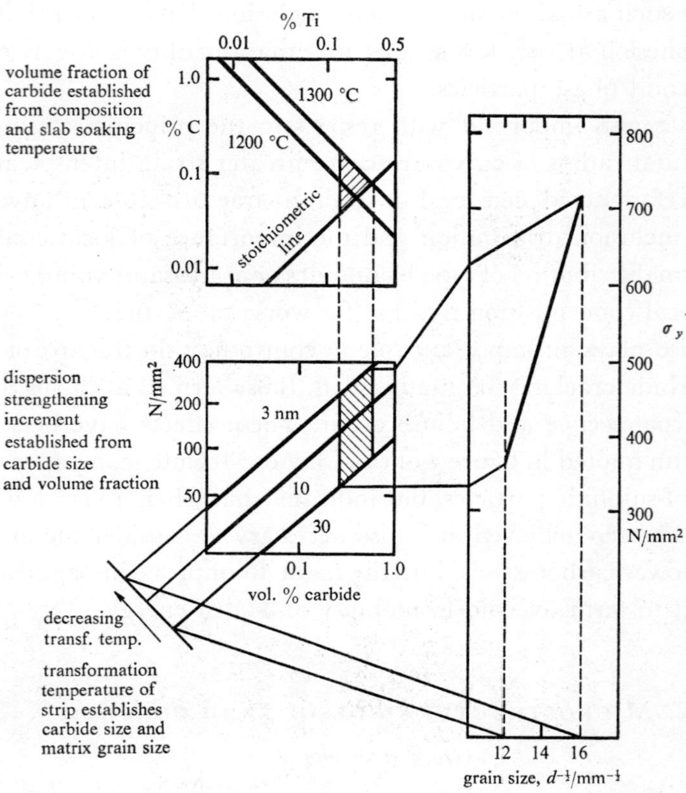

FIGURE 1. Structure/property relations used to predict the yield strength range obtainable in hot-rolled strip for low carbon, titanium-containing h.s.l.a. steel. The hatched regions of composition and particle size and the grain size range are those obtained in commercial operations (after Gladman).

mechanical properties has been prepared by Gladman (figure 1). Microalloy and carbon levels can be chosen in conjunction with appropriate dissolution temperatures to give maximum volume fractions of fine interphase precipitation. In rapid hot rolling processes such as occur in hot strip mills, there is little strain induced precipitation in the austenite during rolling, and the dissolved microalloy is fully available for interphase precipitation, the size and distribution of the interphase particles being controlled by the coiling temperature or by the transformation temperature in continuously cooled thin plate. The deformation temperature of the austenite and the austenite-ferrite transformation temperature are both involved in the control of ferrite grain size, but the grain size and dispersion strengthening components can be integrated into the mechanical property model directly.

The importance of train induced precipitation in the austenite is mentioned in the papers of Clay & McCutcheon and of Cohen & Owen (3.5). Effectively strain induced precipitation during hot rolling in the austenite temperature range reduces the volume fraction of interphase precipitate available for dispersion strengthening because of its coarse particle size.

Bainitic steels

The shortcomings of ferrite-pearlite steels are seen in thicker plate sections due to strain induced precipitation effects and the inability to apply high deformation at the lower temperatures when using thick stock. Clay & McCutcheon describe the application of low C bainitic steels.

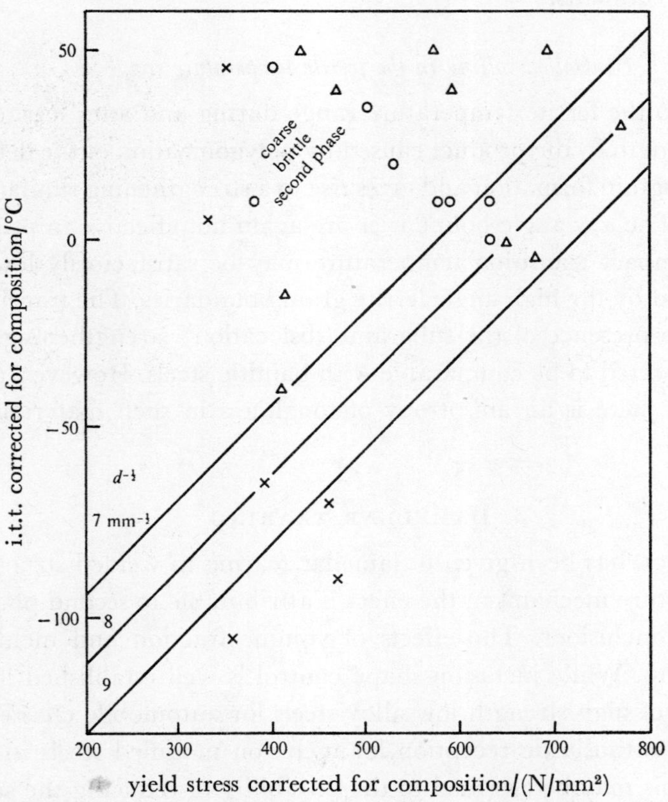

FIGURE 2. Strength–toughness relations for Bainitic steels. These are compared with predictions from strengthening mechanisms and embrittlement vectors, for ferrite-pearlite steels. ×, 0.02; ○, 0.1; △, 0.2.

The structure/property relations in bainitic steels are not fully understood, but some understanding of their properties is gained by acknowledging that the unit ferrite dimension appropriate to strength is the bainitic plate size, but the unit ferrite dimension appropriate to toughness is the bainitic colony size; the low angle boundaries are ineffective in obstructing a cleavage crack due to the low deflexion of the operative cleavage planes across the boundary. Predictions of strength and toughness from the models for ferrite-pearlite steels can be compared with the properties of bainitic steels (figure 2). At low strength levels in low carbon bainitic steels, and in higher strength steels with lower bainitic structures, the bainitic steels exhibit combinations of properties that are directly comparable with the models of the ferrite-pearlite structures. At intermediate strength levels, the impact properties are worse than predicted, the impact transition temperatures being some 100 °C higher than those predicted. The exact nature of this discrepancy is not defined in quantitative terms, but is believed to be associated with the presence of coarse brittle second phases such as carbides or high carbon martensite between the bainitic ferrite plates.

An interesting feature of the bainitic microstructure is noted by Clay & McCutcheon in connection with the Bauschinger effect. Pipe formed from bainitic plate shows an increase in yield stress whereas that formed from ferrite-pearlite structures shows a decrease in yield strength that is not completely restored by cold expansion. The effect may be associated with the existence of internal microstrains in the bainitic structure which gives an artificially low indication of strength in the first strain cycle; the proof stress is increased in bainitic steels by moderate tempering treatments.

Controlled rolling in the ferrite temperature range

Continuous rolling into the ferrite temperature range during and after ferrite formation gives rise to a high yield strength in the product caused by polygonization of the deformed sub-structure. This results in subgrain formation and gives rise to a strengthening similar to that observed in bainitic structures. The low angle boundaries are again not effective in preventing cleavage propagation yet the impact transition temperature may be satisfactorily low due to the fine ferrite grain size defined by the high angle ferrite grain boundaries. The transition temperature is increased due to the presence of the subgrain (dislocation) strengthening. The properties would therefore be expected to be comparable with bainitic steels. However, as pointed out by Clay & McCutcheon, there is an anisotropy of toughness in such materials, arising from a texture effect.

3. Lamellar tearing

Considerable attention has been given to lamellar tearing in welded structures. As with all cases of the ductile rupture mechanism, the effect is attributable to second phase particles, particularly the sulphide inclusions. The effects of volume fraction and inclusion shape have already been referred to. While inclusion shape control is well established for improving the transverse bendability of high strength low alloy steels for automobile chassis members, there has not been such an enthusiastic reception for inclusion modified steels in connection with plate steels. In part, this may be ascribed to the possibility of reducing the sulphide inclusion content by fairly well explored technologies and obtaining satisfactory properties. A further feature was the problems encountered with reoxidation of the sulphide modifiers (particularly with the rare earth treatments) which made earlier steels more susceptible to lamellar tearing rather than less so. Nevertheless, the potential benefit of successful inclusion shape control in connection with lamellar tearing is even greater than that established in connection with transverse bendability in strip or transverse shelf energy in line pipe materials.

Commercial experience in the manufacture of titanium steels is being obtained and the competitive nature of manganese as a sulphide former in titanium steels has been appreciated. The titanium content required for successful inclusion shape control increases as the manganese content is increased. At manganese contents of 1.6 %, the titanium content required to effect shape control (0.16 %) is in excess of that required for considerable dispersion strengthening in hot rolled strip. At lower manganese contents, considerably less titanium can be used, e.g. 0.5 % Mn requires approximately 0.09 % Ti. Studies of rare earth treatment have shown that a rare earth: sulphur ratio of 4:1 is optimal for shelf energy (inclusion shape control). Excess of the modifying agent can lead to a loss of toughness.

A further feature of inclusions and their effects on lamellar tearing (or through thickness ductility) is the specific distribution of inclusions not only on a micro scale but also on a macro scale as shown by Van der Veen (3.4). Inclusion modifiers can also effect significant changes in the macro-distribution of sulphides as well as in the inclusion plasticity. There is obviously scope for studies of the control of macro-distribution of inclusions that are directly relevant to the problem of lamellar tearing.

4. Future developments

A number of features have already been mentioned as being worthy of further development. These include the development of low carbon bainitic steels. The properties of thicker plate steels can be raised by use of bainitic steels. The need for low carbon contents to attain good levels of toughness ($\leqslant$ 0.03 %) in these steels requires a process route that is capable of producing such levels consistently on a commercial scale. The development of the argon-oxygen decarburization process and other similar processes in recent years has made low carbon bainitic steels a real possibility.

Another study connected with the steel manufacturing process is the control of inclusion shape and both micro- and macro-distributions to improve transverse and short transverse ductility.

The physical metallurgy background applied to the development of cheap high strength low alloy steels is capable of producing steels with yield strengths of 760 N/mm^2 (110 klbf/in^2) and having polygonal ferrite matrices. Such steels can be produced for bumper bars, chassis members, etc. but one of the factors limiting their application is the need for a rethinking of design principles to overcome the reduced resistance to elastic deformation caused by the constant modulus of elasticity and reduced section thickness. The interaction between metallurgical application and design is nowhere more clearly illustrated than it is here.

Attention is drawn to the need for detailed mechanistic studies of stress corrosion, corrosion fatigue, and fracture toughness in the paper by Cohen & Owen. Such studies are currently in hand and will develop in the near future. High strength low alloy steels are traditionally too tough for any valid assessment of fracture toughness at ambient temperatures, the concept of K_{Ic} being more relevant to the higher strength steels. In h.s.l.a. steels, c.o.d. testing is being explored and there is obviously a need for valid design criteria based on such tests. From a mechanistic standpoint, the details of failure are basically those observed in other tests, e.g. the ductile fracture mechanism, cleavage, or intergranular separation. Any attempt at relating c.o.d. or K_{Ic} must in fact recognize the operative fracture mode for meaningful metallurgical interpretation, ductile fractures involving second phase particles, cleavage being dependent on grain size, etc., and the fracture mode beng dependent to a significant extent on the triaxia stress system. Such studies will inevitably produce results with the next five years, and a detailed understanding of the role of mechanical constraint and stress system on the pertinent metallurgical features will produce the valid design concept in situations that are currently somewhat obscure.

Discussion (Chairman G. C. Smith (*Cambridge University*))

T. Gladman (*British Steel Corporation*)

Attention has been drawn to the beneficial effects of improving strength and toughness simultaneously by use of grain refinement. On an experimental basis, grain sizes of the order of 1–2 μm have been attained, but only in very thin strip heavily deformed in a single pass and quenched directly after hot rolling. The benefits to be derived from grain refinement increase progressively as the grain size itself decreases due to the functional relation between yield strength and grain size. In steels with grain sizes of 1–2 μm the contribution to yield strength may be of the order of 300–500 N/mm^2 over and above that due to solutes etc.

However, the grain sizes attainable in the laboratory are not those that are attainable in a commercial 5 cm plate. While grain sizes of say 7 μm have been attained in these commercial materials (200 N/mm^2 contribution) further refinement of the grain size is very dependent upon plant facilities to effect heavy rolling reductions, at low hot working temperatures, followed by controlled cooling.

J. H. van der Veen (*Hoogovens*)

It should be emphasized that sophisticated methods of thermomechanical treatment for grain refinement might not be applicable to thick plates, therefore investigation into the possibilities of making more stable, inherently fine grain structures, by using effective grain growth inhibitors should be continued. Of course, we do not yet know whether these inhibitors can be formed during rolling or heat treating or would have to be introduced during solidification. This approach could, in addition, provide more grain growth resistance in heat affected zones of welds made by automatic high heat input welding processes than obtained with the now available steels.

W. B. Morrison (*British Steel Corporation*)

In the case of ferrite/pearlite steels it appears that we are approaching the limit with regard to an optimum combination of strength and toughness. The maximum yield strength coupled with good fracture toughness is probably around 80 lbf/in^2 (550 N/mm^2). The most important factor influencing both strength and fracture toughness is grain size and I would agree with Gladman (this discussion) that the minimum practical grain size is approximately 4 μm in thin plates of either controlled rolled or normalized steel. Although it is possible to increase the strength of the steel by solid solution, precipitation or dislocation hardening these factors generally have a detrimental influence on fracture toughness. The best way to produce strong/tough low alloy ferrite/pearlite steels is probably to utilize the hot strip mill with controlled cooling and coiling temperatures. Further refinement beyond the grain sizes already attained is difficult. Attempts to increase the volume fraction of suitably-sized stable particles in the austenite by combined additions of Al, V, Nb and Ti together with N have not been particularly successful and it would appear that the effects of the individual elements are not simply additive when they exist in combination. Low temperature rolling of low C, low S steels in the ferrite region to produce fine subgrains has the potential of producing strong/tough steels. A similar structure, acicular ferrite, can be obtained by making alloying additions to a low C steel in order to reduce the transformation temperature to within the bainite range, combined perhaps with some controlled rolling. However the alloying costs of such a steel would make a quenched and tempered steel a more attractive proposition. Since the properties of ferrite/pearlite steels have effectively been optimized, further improvements to strength and toughness can best be obtained by producing steels with a tempered martensite, bainite or acicular ferrite type of structure. If it did prove possible to refine the ferrite grains to 1–2 μm I do not envisage any major problems with regard to mechanical properties although, because of the low work hardening rate associated with such a grain size, there may be some reduction in fracture toughness.

R. G. Baker (*National Physical Laboratory*)

Morrison (this discussion) has stressed the dependence of strength on grain size. Irvine (Keynote) pointed out that fracture toughness may well depend on colony size. Work on weld metals, which are essentially low carbon bainitics, hardened by disolocations due to rapid cooling, suggests that colonies other than bainite colonies may be important.

Attention has been drawn to the effect of residual martensite colonies retained pro-eutected ferrite and the effect of inclusions by workers at the Welding Institute and at the British Steel Corporation. To what extent are these factors important in steels generally and how far can we control them? This is perhaps the most important focus of study in the immediate future for both steels and weld metals.

S. F. Pugh (*Atomic Energy Research Establishment, Harwell*)

There is growing evidence relevant to the importance of segregation of certain impurities such as arsenic, tin, bismuth and phosphorus to prior austenite grain boundaries in having deleterious effects on the mechanical properties of bainitic and martensitic steels. Properties which are known to be adversely affected include the impact transition temperature and the resistance to stress relief cracking in the heat affected zones of welds. It seems likely that other grain boundary fracture processes may also be aggravated by the presence of segregated impurities. Such processes include stress corrosion cracking, wet fatigue and hydrogen embrittlement.

While some improvement can be achieved by refinement of the austenite grain size, serious consideration should be given to the development of techniques for producing lower levels of the deleterious elements in order to eliminate batch to batch variation and to allow the higher strength levels to be utilized efficiently.

R. Dolby (*Welding Institute*)

Several authors have referred to the difficulty of using the carbon equivalent formula for predicting hydrogen cracking in the improved low carbon micro-alloyed steels. As the original carbon equivalent formula was derived to predict the risk of hydrogen cracking for plain carbon and carbon-manganese steels we should not be surprised if it does not work in steels where carbon contents are less than 0.1 %. Although the soft heat affected zones in these steels and their associated low risk of cracking have brought about a considerable improvement in weldability, our view is that weldability in relation to hydrogen cracking of low carbon micro-alloyed steels must be assessed by self restrained welding tests and not by tests in which the applied stress is controlled, the latter approach being a current trend. There will be specialist problems in certain areas such as underwater wet welding where we have to take note of the very high cooling rates and high hydrogen levels, but on the whole hydrogen cracking is a problem which we can control.

Lamellar tearing is another problem which is understood and reasonably under control. However, current problems arise because high quality clean steel often cannot be obtained in time and plate susceptible to lamellar tearing has to be used in critical applications. There is an interesting link in Cotton (1.3) between lamellar tearing and the H_2S corrosion problem in sour oil or gas pipelines. We find that some of the dirtiest steels containing silicates do not seem to be quite so prone to H_2S corrosion as might be expected on the basis of inclusion content.

Another important problem highlighted by Irvine (Keynote) and Baker (this discussion) is toughness in welded regions, and we are now seeing that the metallurgy of weld metals, heat affected zones and parent steels is closely similar because all deal with low carbon bainitic structures in the improved low carbon steels. There has been significant progress in understanding the structure-property relations in weld metals but we need a much better understanding of the interaction between the various micro-alloying additions diluted or added in the weld metal.

Looking back, I think steelmaking techniques and steel compositions have changed radically over recent years to give us greatly improved weldability. Looking forward, this trend will continue, although at a slower rate, since many of the problems can now be controlled by proper design and proper selection of welding procedures. Where future changes in steel compositions will probably occur is in specific situations where we want to take particular advantage of, for example, high speed welding processes or high heat input welding processes.

F. M. Burdekin (*Messrs Sandberg*)

I would like the authors to comment on the following points:

(*a*) What is the effect of the different rolling and production methods for achieving higher strength steels on the residual stresses in the finished plate? This is of great importance to designers and fabricators in the bridge construction field.

(*b*) Are there any differences in inclusion shape, and susceptibility to lamellar tearing between normalized and controlled rolled steels? Do the fissures on the broken Charpy specimens shown in figure 11 of Morrison & Chapman (3.2) indicate possible lamellar tearing susceptibility?

(*c*) The possible effect of reduced sulphur level to reduce sulphide inclusions in affecting necessary preheat levels to avoid hydrogen induced cold cracking is not discussed in any detail in the papers. Further discussion on this would be appreciated.

W. B. Morrison (*British Steel Corporation*)

I would like to reply to the remark made by Burdekin (this discussion) on the relation between the appearance of fissures on the fracture surfaces of some controlled rolled steels and the occurrence of lamellar tearing. On the same theme Clay & McCutcheon (3.3.) mention that heavy controlled rolling reduces through thickness ductility and for evidence they quote a paper by Dabkowski, Konkol & Baldy, It should be pointed out that in this latter paper the through thickness ductility alluded to is the Charpy V notch ductility and not the room temperature tensile reduction in area. Work carried out at the British Steel Corporation has shown that there is no relation between the occurrence of non-inclusion-originated fissuring and room temperature uniaxial tensile reduction in area, the latter parameter being related to lamellar tearing.

In reply to Baker (this discussion) who referred to bainite colony size in low C structural steels, refinement is obtained by the same methods as for a ferrite/pearlite steel, i.e. micro-alloying and controlled rolling. If Mercer wants the properties of pipe fittings to match those of the higher strength pipe plates presently being developed then it would appear necessary to use relatively highly alloyed steels with a consequent cost penalty.

D. B. Clay (*Steel Company of Canada*)

The residual stress in pipeline steels is only of concern in the final fabricated pipe. Residual stress tests have been carried out on finished pipe but the levels of stress observed have given no cause for concern.

A. P. Miodownik (*University of Surrey*)

With regard to methods used to raise the strength of low carbon steels leading to a higher tendency to hydrogen embrittlement, it is true that a reduction in inclusion or particle content will reduce the number of available sites for innocuous hydrogen, and therefore reduce the critical hydrogen concentration which will lead to failure. However, this conclusion is only valid if corresponding changes in the number of other hydrogen trapping sites is left out of consideration. Most of the new steels rely on reducing the grain size and a general decrease in subgrain networks of dislocations; these will in general provide sufficient extra trapping sites to outweigh the loss due to a reduction in the number of particle interfaces. Given the specification of a minimum number of *total* hydrogen trapping sites, there will be no additional hydrogen cracking problems at the strength levels under consideration.

W. L. Mercer (*British Gas Corporation*)

I would like to query the method of measuring the yield strength values shown in figure 7 of Clay & McCutcheon (3.3). Were these based on standard AP1 flattened strips or were they determined by the more truly representative ring tension test? Would the difference in behaviour indicated be influenced by the test method used? The ring test gives the truer representation of strength of interest to the pipeline design engineer.

I would like to emphasize the need to maintain a proper balance in considering developments in steels for engineering use. In the case of pipelines, this means paying more attention to the development of steels for the manufacture of fittings such as valves, T-junctions, bends etc., which cannot readily exploit the remarkable developments associated with micro-alloying/controlled rolling as applied most effectively to the manufacture of pipes. The performance of these fittings is vital to the overall performance and safety of a pipeline.

D. B. Clay (*Steel Company of Canada*)

The tests used in figure 7 of our paper (3.3) were in fact carried out on a series of small tensile specimens taken through the thickness of the pipe wall – the average being shown in the graph.

The usual test is of course the flattened strip test which raises doubts as to its validity in that it does not really measure the properties of the pipe itself. In the case of u.o.e. formed pipe it probably gives an optimistic result, whereas in unexpanded pipe it will give a pessimistic result. The ideal test is the ring expansion test but this is an expensive low productivity test not suited for normal quality control purposes.

C. Edeleanu (*Imperial Chemical Industries*)

I have a feeling that we are stuck in a groove in this discussion. We have cast our minds back to the early days of metallurgy and we do not seem now to be able to think forward and talk about the future on the same scale. We seem to be concentrating on the present and the next set of experiments and moreover on some very narrow aspects of the subject we call metallurgy. Mercer (this discussion) has said for instance that pipes are not a major limitation at present but that he has problems with pipe fittings, but no one seems to have heard him or offered to help him out. Presumably he is thinking of castings, and if he is, I can say we too have problems with these items but I fear that few present day metallurgists would consider such a subject sufficiently respectable to be worth thinking about. Why is that? Castings are basically good things and potentially even better than wrought products, and it seems astonishing that all conventional metallurgical thought assumes that it is necessary to pour metal and make a casting, then beat hell out of it to make it into a wrought product before it is a legitimate material for metallurgical studies. This is even more astonishing since once this has been done you stick it together with cast metal we call weld metal.

Why can we not get off the narrow subjects which come up in meeting after meeting? Why can we not talk about the production of vessels entirely by welding as is being developed in Japan, why can we not hear what prospects there are of producing engineering items by powder metallurgy? Why can no one show us if these are unsuitable subjects? They are the possible long term developments arising from all the money being spent on metallurgical R. and D. Metallurgy is after all only a means to an end – the production of engineering hardware – and it may not always be necessary to study in enormous detail some aspects of it and neglect as a result the ultimate objective of advancing the technology of producing better hardware.

G. G. Saunders (*Whessoe Ltd*)

May I at this juncture raise two points of engineering significance with regard to the development of improved steels.

In the pressure vessel industry it is necessary to convert the nominally flat plates supplied by steel makers into cylinders and petal plates for spheres and dished ends. This can be achieved either by pressing cold which introduces plastic strain into the steel, or by pressing hot in the austenite region.

As steel strength increases, the loading capacity required for the former increases and problems of elastic spring back increase. If the latter is used, then many of the beneficial effects on strength and toughness induced for instance by controlled rolling procedures are lost.

Furthermore if composite plates are hot formed, the weld metal used to join them loses much of its strength on normalizing, and considerably undermatches the plate steel.

Is anyone involved in the development of higher strength steels prepared to comment on these aspects?

W. J. Murphy (*U.S. Steel Corporation*)

Steels containing bainite. With respect to the subject of bainitic steels, I should like to point out that there are steels being considered for Arctic gas-transmission line-pipe applications which have a microstructure consisting of a small amount of bainite (or even martensite) on patches in a matrix of polygonal ferrite. One such steel being proposed by U.S. Steel has a nominal composition of 0.1 C, 1.3 Mn, 0.25 Mo, and 0.04 Nb; controlled rolling is used to achieve a fine grain size for strength and toughness, rolling is finished just below the Ac_3 to provide a small amount of deformed ferrite for added strength, precipitation hardening is provided by niobium carbide, and the molybdenum may be varied somewhat to provide a range of yield strength through variations in the amount of bainite or martensite (relative to pearlite) in the microstructure. In the as-rolled condition, plates of the steel exhibit the absence of yield-point behaviour (i.e. exhibit a 'round-nose' stress–strain curve) and a high rate of work-hardening such that when the line pipe fabricated from the plate is expanded, minimum yield strengths up to 550 MN/m^2 (80 $klbf/in^2$), depending upon composition, result. These strength levels are accompanied by excellent low temperature toughness properties.

Line pipe fittings. With respect to the question concerning the availability of steels for gas-transmission line-pipe fittings for Arctic applications, in the U.S.A. low carbon quenched and tempered alloy steels are being considered to achieve the required strength and low-temperature toughness. Relatively heavy plates of a suitable alloy steel composition are fabricated by welding and forming into the fittings which are then heat treated by quenching and tempering. The major problem relates to achieving the required toughness, for the most severe low temperature applications, in the heat-treated weld metal. This problem is under investigation with available weld metals being evaluated and new weld metals being developed.

K. Sachs (*Guest, Keene & Nettleford*)

In connection with the improved pipe-line fittings that Mercer (this discussion) has called for, and looking further ahead attention should be drawn to the possibility of pressure die casting of ferrous alloys. This can now be done commercially. At present it is limited to fairly small castings. Its extension to larger sizes is an important development step which will be taken in due course.

M. Cohen (*Massachusetts Institute Technology*)

The concept of structure-property relations embrances many levels of structure. This hierarchy is not necessarily put there by nature but is, in large measure, a construct of the human mind. It constitutes our attempt to understand the inner machinery of materials and to connect this microworld to the outer manifestations which we will call 'properties'. But there is a dualism or reciprocity here; in a real sense, structure is determined by properties just as fundamentally as the properties are determined by structure. This interplay is a powerful one, and has actually been the central theme of physical metallurgy over the past fifty years. It has proved to be an endeavour of much intellectual excitement and satisfaction.

However, physical metallurgy also bridges over into society through the *performance* of metals in the service of man. This linkage involves processing as well, and can be represented succinctly as:

STRUCTURE/PROPERTY/PERFORMANCE
/PROCESSING/

Thus, the impact of physical metallurgy arises not only because of its appeal as a way of probing the nature of the metallic world, but also because the knowledge derived can be harnessed directly in response to the needs of society.

As it happens, much of the attention of physical metallurgists has been occupied with the structure/property link of the chain, and the stimulation has been very gratifying. But society will not pay for only that piece of the system in the long run; the field must become more fully populated toward the engineering end of the spectrum. It will not do to leave the applications, and the resulting feed-back of experienced knowledge, to a separate group of professionals labelled 'the engineers'.

In fact, the development of microalloyed steels, as nicely portrayed in this session, provides an excellent model of the intermingling of scientific and engineering metallurgy, and the accompanying advances are truly impressive. This success may well herald a new unity in metallurgy.

J. E. Roberts (*British Steel Corporation*)

Burdekin (this discussion) raised the question of residual stress levels in plates produced by various methods. While there may be differences inherent in the various metallurgical procedures, it is likely that these are outweighed by the fact that the final operation for most, if not all, plates and sections, is a levelling or straightening operation. For this to be successful there must be a residual stress pattern introduced into the product, the level of the stresses being proportional to yield stress, rather than how that yield stress was achieved. It is perhaps an area where the engineer has to come to the rescue of the metallurgist by developing processing machinery which would avoid the need for final shape correction. Whether this would really reduce performance problems, after cutting, forming, welding and perhaps heat treatment (stress relief) has been involved in producing a completed structure is open to further debate.

L. M. Wyatt (*Central Electricity Generating Board*)

No one admires more than I do the progress made in the development of higher tensile steels, but I must bring to the attention of the meeting a serious limitation in their application.

Most of our more advanced designs are limited by fatigue and the law governing the propagation of a crack in fatigue varies very little between materials. If, therefore, we change from a carbon-manganese steel to a higher tensile steel (it is immaterial whether the higher tensile properties are produced by quenching and tempering controlled rolling or alloying) we must limit either the dimensions of flaws existing in the component as manufactured or the number of cycles to which it is subjected during operation. If we do the former we must increase the standard of inspection and, therefore, the fabrication cost, if the latter we decrease the usefulness of the plant. Either procedure tends to offset the economic advantages gained by the use of the higher tensile steel, and in a recent example we have had to revert to carbon-magnanese steel because the increased fatigue life of the plant gave no essential operational flexibility.

We are now beginning to understand and to some extent reduce creep crack propagation by controlling the structure of the material. The control of fatigue crack propagation by changing material structure would be more difficult but not I believe absolutely impossible, and the benefits to be gained would be very great.

J. Savas (*Republic Steel Corporation*)

I would like to propose that engineers start looking at some new concepts in a field that has already been investigated but hardly touched, and that is the cryogenic field where they may want to consider moving liquid natural gas for long distances in much smaller pipes. Many of the problems that have been encountered in looking at these micro-alloy steels might very easily become redundant by looking at more highly alloyed steels which would require less material, for example 9 % nickel or possibly 10 % nickel – steels that could operate at − 162 °C. At this point the economics would have to be looked at together with the technology from the engineers point of view so that a whole new concept with respect to the transfer of materials can be examined.

J. Barford (*Central Electricity Generating Board*)

I recognize an underlying feeling in the discussions so far that the engineering requirements have and are being satisfactorily met, so that the contributions from physical metallurgy are necessary and sufficient. I am not at all sure this is true. Undoubtedly engineering practice got along quite well before the term physical metallurgy was coined to describe a newly emerging discipline, as evidenced by the presence today of many large and complex structures made in the last century.

It is debatable whether, in the absence of physical metallurgy, we would have the range of engineering alloys presently available. However, I believe we would have had a great many of them. Obviously this would have been without an understanding of how they work, but nevertheless I feel most of them would have been developed on the well established principle of 'suck it and see' – as indeed most of them were. And here engineering advances allowed by the availability of better materials owe at least as much and probably more to the improvements in metal winning and production processes than to physical metallurgy *per se*.

There are three questions to which, as metallurgists and engineers, we might address ourselves: are the metallurgists supplying the right materials, are they supplying the right information on those materials, and does the engineer really know what he wants? It is far from clear to me that the answers to all these questions is an unqualified yes.

It seems that metallurgists have been fairly successful in developing materials that perform well in certain tests, but it is not always clear what relevance these tests have to component behaviour. For example, in the first session we discussed the relevance of the most common of these tests, a simple, uni-axial tensile test. However, it is when material properties change with time that metallurgists have been least successful in predicting long term behaviour, except in developing empirical predictive techniques by using simple tests that are believed to have some relation to what the material experiences in service. This covers the whole range of predictive behaviour requirements from creep, fatigue (and worse, a combination of the two) to corrosion, stress corrosion and a whole range of time dependent, environmentally affected behavioural phenomena. It is in the last field where the established science of physical metallurgy could and should make its future impact. What is the response of a material in a real engineering situation and what effects do, perhaps small, changes in fabrication methods or service conditions have on material behaviour?

W. G. Ferguson (*University of Auckland*)

During this conference we have heard of the application of special steels to pipe-line construction. These pipe-lines are essentially subjected to static loading during their life. In situations where such pipe-lines are subjected to earthquake loading, is this taken into account during the materials selection and design stage?

J. H. Van der Veen (*Hoogovens, Netherlands*)

In response to Cohen (this discussion) I would state that we make steel to supply a certain desired 'performance'. Nowadays many 'performances' are successfully supplied by composite materials, such as reinforced concrete, prestressed concrete, reinforced plastics, etc. Let us not forget that steel, in a way, can be considered to be a composite material. Lately, in order to achieve a desired dispersion of the components in the steel, such as grains, precipitates, grain boundary particles, etc., we have taken advantage of the transformation and recrystallization characteristics of the steel, as these can be influenced by alloying elements and brought into effect by low temperature, high reduction rolling and various heat-treatments. Perhaps more attention could successfully be paid to the possibilities of achieving the desired components, and their dispersion, during controlled solidification. Also, it might be worthwhile, for the same reason, to study the possibilities of powder metallurgy.

M. Cohen (*Massachusetts Institute Technology*)

The future for physical metallurgy will probably lie in becoming a more integral part of the overall science/engineering spectrum which I discussed earlier. In other words, there appears to be much greater potential for physical metallurgy in responding to the pull of society than in relying unduly on the technological push of its own creations. In this framework, we can expect to see more physical metallurgists at work on metals processing and performance.

The new thrust, in turn, will be greatly affected by energy requirements, first because materials production and fabrication are energy-intensive, and secondly because all advances in energy-conversion schemes tend to become materials-limited. Resistance to corrosion and oxidation are then likely to be just as critical as the more familiar mechanical properties. Nondestructive testing will also become more crucial, and deserves special attention on the part of physical metallurgists. And, of course, cost factors will inevitably have to be weighed in the balance. The list of important constraints and roadblocks is long, indeed, but the essential point for our purposes here is that such issues not only have to be dealt with in physical metallurgy, but they are more demanding than the well-known game of optimizing only strength and fracture toughness.

The following three contributions were received after the meeting:

T. N. Baker (*University of Strathclyde*)

I would like to refer to Morrison & Chapman (3.2). The first point concerns the fragmenting of the Hall–Petch equation to obtain a number of separate contributions to the yield stress σ_y. Frequently when rolling is finished in the ferrite phase, for a steel containing microalloying additions, dislocations and particles are in intimate association. Particles present before the final rolling will tend to restrict dislocation movement while particles which nucleate later, often do so on dislocations. The final dislocation density is therefore a function of the volume fraction, morphology and size of the precipitates, some of which are in turn dependent on the dislocations themselves. Under such circumstances it is unrealistic to separate the dislocation and precipitation hardening contributions to yield stress. The low σ_{ppt} contribution in the example quoted by Morrison & Chapman may be partially explained in this manner. In a steel studied at the University of Strathclyde with the composition 0.04 % C, 1.6 % Mn, and 0.063 % Nb, controlled rolled to between 800 and 600 °C, σ_{ppt} was found to be in the region of only 100 N mm^{-2}.

Another factor which can give a misleading picture of the importance of the various yield stress components is the presence of sub-grains in steel finish rolled in ferrite. In this case the K_y value is reduced from about 20 N/mm$^{\frac{3}{2}}$ for polygonal ferrite to about 10 N/mm$^{\frac{3}{2}}$ for sub-grains, thereby giving a different contribution (dependent on the ratio of subgrain size to grain size) of the grain size to the σ_y term.

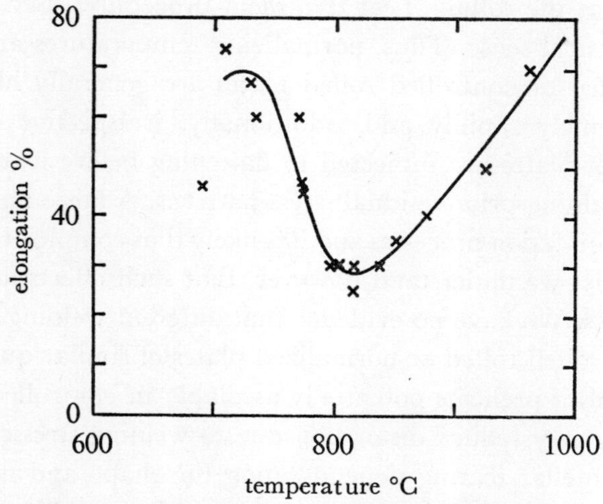

FIGURE 1. Elongation % plotted against testing temperature for a mild steel.

The second point I wish to raise concerns the discussion on the influence of austenite grain size on the austenite/ferrite transformation temperatures and the subsequent possibility of producing a softer ferrite through controlled rolling. Evidence for a softer or more ductile ferrite has not received a great deal of attention in the literature. Robbins *et al.* (1961) have reported an increase in the ductility of pure irons at about 800 °C as determined by torsion tests carried out at a strain rate of 47 %/s. We have also found a similar increase in the ductility of aluminium-killed mild steel at about 730 °C, figure 1; the no. 13 Hounsfield specimens being tested at temperature under a vacuum of better than 0.01 Pa (10^{-4} mmHg). In this case the minimum in ductility is associated with temperatures in the two phase $\alpha+\gamma$ region. The ductility then increased by more than 100 % as the tensile-testing temperature was decreased, and reached a maximum at about 730 °C. It is possible that this temperature may be more than 50 °C below the Ac_1 due to the influence of deformation on the transformation temperature.

Reference

Robbins, J. L., Shepard, O. C. & Sherby, O. D. 1961 *J. Iron Steel Inst.* **199**, 175.

R. McCallum (*National Engineering Laboratory*)

Referring to Cohen's structure-property-performance-processing link (this discussion), in many instances components are not specified with demands for particular properties if a wrought material, freedom from porosity if a casting or of theoretical density if made from powder.

All components contain defects, and detailed information is lacking on the degree of imperfection which can be tolerated even though still satisfying functional performance specifications, especially if the component is manufactured from powder, is as-cast, or made from reconstituted material. This involves the philosophy of fitness for purpose for many of today's engineering components.

J. E. Roberts, J. A. Chapman and W. B. Morrison (*British Steel Corporation*)

The direct answer to Burdekin's question (this discussion) on residual stress is that we have no evidence suggesting that the rolling, heat treatment procedures used materially affect the residual stress in any practical sense. Thus, normalizing temperatures and finish rolling temperatures in either as-rolled or controlled rolled plates are generally high enough for stress relaxation to occur extremely rapidly and, additionally, irrespective of production route, a substantial proportion of plates are subjected to flattening before leaving the works which entirely removes any remaining prior residual stress patterns. A large number of plates moreover, are formed during fabrication processes and it is likely that complex residual stress patterns will arise as a result of this; we understand however, that such effects are catered for during normal welding procedures. We have no evidence that different welding practices need to be applied for as rolled, controlled rolled or normalized plates of similar quality, although it has been suggested that the lower preheats potentially available in controlled rolled steels due to the lower potential c.e.v., may reduce distorttion due to welding stresses.

The susceptibility to lamellar tearing depends upon the shape and volume fraction of inclusions. The elongation of silicates and sulphides varies with hot rolling temperature and the degree of deformation at any particular temperature. In Si-killed steels it is likely that there would be little difference in inclusion shape between as rolled, normalized and controlled rolled steels, though as rolled steels could show greater silicate elongation; similar considerations would apply to semi-killed steels. While it is true that in Al-treated steels, low temperature rolling tends to cause the type II sulphides to become elongated, there is an increasing tendency for S level to decrease, with the increasing use of continuously cast steels, and the presence of such elongated sulphides would assume less importance. For all normal applications, controlled rolled steels have been fully successful, and show no detrimental characteristics compared with similar steels normalized.

Where there is a particularly high risk of lamellar tearing, special precautions may be needed for any type of steel, such as the use of very low S levels (typically around 0.006 % S) or the use of rare earth metal treatments to modify sulphide form and to increase the refractory nature of the sulphides which minimizes any shape change from hot rolling.

The fissures referred to by Burdekin occur as a result of structual and crystallographic texture and are not directly related to the presence of inclusions. A large number of results from tests on both commercially and laboratory rolled steels with fissures have shown no correlation between through thickness reduction in area and the presence of fissures.

With regard to the influence of S on cold cracking susceptibility, some early work[1, 2] did indeed suggest that as the S level decreased below about 0.020 %, there was an increased susceptibility to hydrogen induced cold cracking arising from welding. However, B.S.C.[3] and Japanese experience[4] has been that welding procedures recommended originally for steels of normal S levels (0.02 %–0.04 %) have proved satisfactory for steels of very low S levels (< 0.01 %).

References

(1) Smith, N. & Bagnall, B. I. 1968 *Br. Welding J.* **15**, 63.
(2) Hewitt, J. & Murray, J. D. 1968 *Br. Welding J.* **15**, 151.
(3) Haddrill, D. M. *The Welder* **4**, 23.
(4) Gondob, H. & Nakasugi, H. 1973 Processing and properties of low carbon steels *A.I.M.E.* **1973**, 133.

3.6 Development of Al–Zn–Mg–Cu alloys for aircraft

By H. Y. Hunsicker
Aluminium Company of America, Alcoa Center, Pennsylvania 15069, *U.S.A.*

[Plate 23]

Historical development of the high strength Al–Zn–Mg–Cu alloys during the last 50 years reflects the influence of progressive changes in engineering and functional requirements related to the evolution of aircraft size, design concepts and manufacturing methods with consequent changes in dimensions and configurations of structure components. If the only criterion of merit for such application were high strength uncomplicated by stress-concentrating geometric factors or environmental influences, compositions available at least 40 years ago would have been highly suitable, since strength of the early materials fully equalled that of the alloys in use today.

These early alloys were, however, not adequately resistant to stress-corrosion cracking even in the form of thin-section products such as sheet, thin extrusions and tube prevalent in aircraft of the earlier portion of the period. Development of the chromium-containing alloy 7075–T6 in the 1940s provided a solution to this problem for the thin-section products. As experience accumulated in extending the use of 7075–T6 to larger and thicker components produced from plate, thick-section extrusions or forgings, transverse direction stress-corrosion cracking was encountered along with decreasing strength and an undesired degree of mechanical anisotropy, particularly ductility. Alloy 7079, which had more uniform mechanical properties, enjoyed a transient period of popularity but proved to be even more deficient in resistance to stress-corrosion. The latter problem was eliminated with development of 7075–T73 involving use of a two-stage precipitation treatment with an accompanying significant strength penalty. Despite the strength trade-off, this material is extensively used in medium section thickness components of current large American aircraft and has an eminent success record.

More highly alloyed compositions of this alloy family, 7178 and 7001, capable of developing higher strength than 7075–T6, appeared in the 1950s. The former is used in compressively stressed portions of wing structure but is not highly favoured for newer designs because of somewhat lower fracture toughness. The latter was considered deficient in this attribute and consequently not adopted for aircraft usage.

Further developments based on 7075 include the T76 temper with high resistance to exfoliation corrosion and mechanical properties intermediate to those of the T6 and T73 temper products. Alclad 7075 employing a high strength cladding is used in certain large aircraft. More closely controlled variants of 7075, i.e. 7175 and 7475, coupled with special fabricating practices developed within the last 10 years are characterized by higher fracture toughness leading to increasing application. The newer alloys 7049 and 7050, introduced within the last 5 years, provide further advance in combinations of mechanical properties and environmental resistance resulting in adoption for retrofit programmes and inclusion in newly designed aircraft.

Development and successful application of these materials in today's aircraft is a result of the persistent and systematic efforts of many scientists and engineers employing practical experimental and analytical methodology consistent with the Rosenhain tradition. The good record of reliability which these vehicles and materials have accumulated, and may confidently be expected to continue to accumulate, is a tribute to these people and their talents and dedication.

The invitation to participate in the 1975 Rosenhain Memorial Conference is appreciated greatly by your speaker and the company he represents. The Conference planning committee suggested a presentation on high strength aluminum-base alloys for aircraft structures application with emphasis on those of the Al–Zn–Mg–Cu type, in particular the American alloys and viewpoint. The appropriateness of the subject derives from the involvement of Dr Rosenhain and the National Physical Laboratory in the early and continuing development of aluminum alloys for aircraft. The general concurrence of the birth and growth of the aluminum industry with the age of powered flight has been noted many times. In the maturing period of these industries, over the last 30 years, development and application of the high strength Al–Zn–Mg–Cu alloys has played an important role in making possible the increased size and high reliability of today's commercial aircraft.

Alloys of the Al–Cu–Mg type, with lineage from the 1906 duralumin and age hardening discoveries of Wilm, continue to play a prominent part in not only supersonic aircraft, where mechanical properties at elevated temperature are the basis for their selection, but also in components of subsonic commercial aircraft of latest design. Alloy and temper selection, whether naturally aged or precipitation heat treated, depends upon service requirements, whether emphasis is on fracture toughness or on higher strength and stability at elevated operating temperatures. The well-known 4 Cu, 2 Ni, 1.5 Mg 'Y' alloy (2218)†, developed at the N.P.L. under Rosenhain, as well as the more readily fabricated variant 2018 have been used for many years principally in engine applications and can be regarded as precursors of RR58 (2618), which was selected for major Concorde structural applications.

While lineage of the currently-used Al–Zn–Mg–Cu alloys cannot be traced with continuity to the N.P.L. activity under Rosenhain's direction, the information published in 1921 (Rosenhain, Archbutt & Hanson 1921) undoubtedly influenced and stimulated the later developments. As recounted in this publication, Al–Zn–Mg and Al–Zn–Cu–Mg alloys were made as early as 1911 in the N.P.L. programmes as an extension of work with Al–Zn–Cu. The addition of Mg greatly increased fabricating difficulties but by 1917, compositions such as alloy 'E' (20 Zn, 2.5 Cu, 0.5 Mg, 0.5 Mn) were successfully extruded and rolled after extrusion. Ultimate tensile strength values ranging to 580 MPa were reported for this material naturally aged after water quenching. This compared with about 420 MPa for duralumin. Evaluation of alloy 'E' and the other experimental alloys included a broad scope of mechanical tests, including notch-impact and fatigue as well as corrosion and stress-corrosion tests. Results of these tests raised questions concerning suitability of the stronger alloys for structural use. The observations concerning compositional effects on metallurgical structures, extent of recrystallization, grain shape, etc., and their correlation with properties were astute and consistent with current views.

It would not be appropriate to recount the subsequent developments without recognizing the influence of Rosenhain's many publications advocating observance of a scientifically based, physical metallurgy approach in alloy development and of those which exemplify his keen observations and interpretations of relations between metallurgical structures and the functional properties of alloys. Specifically, his insistence on the necessity for accurate information on equilibrium relations, his concern with the nature of grain boundaries and plastic deformation as well as with mechanisms of corrosion and stress-corrosion cracking have had considerable

† The four-digit numbers cited throughout this paper are those of the current Aluminum Association designation system adopted in 1952.

impact on subsequent research and on development of many aluminum alloys. His later interest in gas removal, impurity effects, grain refinement and ingot structures has influenced both their compositions and processing.

Earliest references generally cited as direct ancestry for the modern 7XXX alloys are those resulting from work of the German investigators Sander and Guertler, who demonstrated the exceptionally high tensile properties possible with fully heat treated ternary Al–Zn–Mg alloys based on the Al–$MgZn_2$ section (Sander 1926). Their work was followed up in Germany, and concurrent development was proceeding in Great Britain, France and Japan during the 1930–40 decade. Space limitations do not permit tracing these developments, and they had only modest direct influence on the American activity, which is the primary focus of this review.

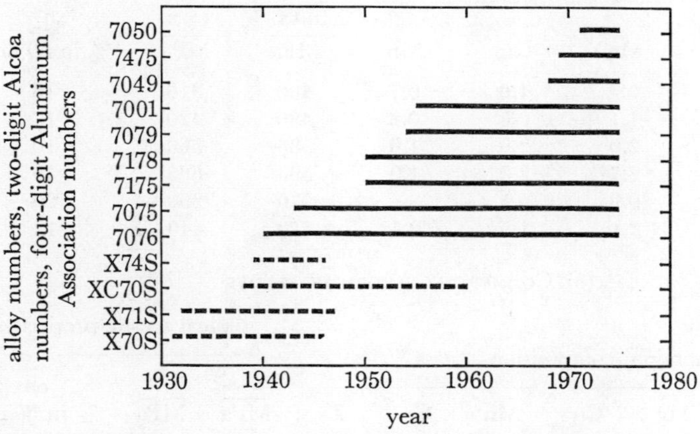

FIGURE 1. Chronology of American Al–Zn–Mg–Cu alloys. Arranged by year number assigned and showing period of listing.

1930–1940: early American Al–Zn–Mg–Cu alloys

Development activity on Al–Zn–Mg–Cu alloys, not on an urgent or intensive basis, was undertaken in the late 1920s by Alcoa's laboratories. Following preliminary production and testing in the laboratories and trials in the plants during 1930–31, two compositions were given the first of the two-digit Alcoa numbers for wrought alloys starting with 7, which denoted zinc as the alloying element present in greatest mass percentage. The first two alloys (table 1*a*, figure 1) were identified by the designations X70S and X71S, the 'X' indicating in both cases the experimental status. The former was proposed for producing forgings and had moderately high strength. For forging, emphasis was to achieve ease of hot working because of then-existing plant equipment limitations. Development activity continued with numerous variants of X70S, including higher strength compositions, such as XC70S, which were predecessors of 76S (now 7076), and more forgeable, less quench-sensitive alloys such as X73S, which was used for a short time to produce radial aircraft engine crankcases.

X71S, intended for producing sheet and extrusions, had higher alloy content and considerably higher strength (U.S. patent 1924729, 1933). Alloys of this series, of which only two are listed in table 1*a*, were considered for construction of dirigibles for the U.S. Navy, but their relative susceptibility to stress-corrosion cracking (s.c.c.), the high ratios of yield strength to tensile strength, lower elongations than for naturally-aged Al–Cu–Mg alloys, poor forming characteristics, and difficulty of fabrication were deterrents to acceptance. If high static strength were the only criterion of engineering value, it is evident that X71S, or alloys based on the ternary Al–Zn–Mg system that readily attain equal or higher strengths, would have been accepted and used much earlier than was actually the case.

In 1937 information was published on equilibrium phase relations in the ternary Al–Zn–Mg system with Al solid solution limits from 200 to 400 °C (Fink & Willey 1937). Extension of this work to the quaternary system was not undertaken for several years, but availability of the ternary data provided a basis for some fresh approaches.

TABLE 1. AMERICAN Al–Zn–Mg–Cu ALLOYS

(*a*) PARTIAL LIST OF EARLY ALCOA EXPERIMENTAL ALLOYS

alloy number	year listed	nominal composition, % Zn	Mg	Cu	Mn	typical tensile properties† u.t.s. MPa	t.y.s. MPa	el., % in 50 mm	product
X70S	1931	10.0	0.4	1.0	0.7	400	310	18	forgings
XC70S‡	1938	8.0	1.1	0.5	0.8	480	420	17	forgings
X71S	1932	10.0	2.0	2.0	1.0	585	560	10	sheet
XB71S‡	1932	9.0	2.7	2.0	1.0	595	565	10	sheet
X73S	1935	5.2	0.9	0.5	—	370	290	20	forgings
X74S	1939	5.2	2.1	1.5	0.4	510	440	12	sheet

(*b*) COMMERCIAL AMERICAN ALLOYS

alloy number	year listed	nominal composition – mass % Zn	Mg	Cu	Mn	Cr	Zr	typical tensile properties[1] u.t.s. MPa	t.y.s. MPa	el. % in 50 mm	products[2]
7076[3]	1940	7.5	1.6	0.7	0.6	—	—	530	490	14	F[3]
7075[4]	1943	5.6	2.5	1.6	—	0.25	—	570	500	11	S P, F, E
7175[5]	1950	5.6	2.5	1.6	—	0.25	—	620	560	11	F
7178	1950	6.8	2.7	2.0	—	0.25	—	600	540	10	S, P, E
7079	1954	4.3	3.3	0.6	0.2	0.2	—	540	470	14	S, P, F, E
7001	1955	7.4	3.0	2.1	—	0.25	—	675	625	9	E
7049[6]	1968	7.7	2.5	1.6	—	0.16	—	540	470	11	F, E
7475[4]	1969	5.7	2.3	1.6	—	0.22	—	560	490	12	S, P, F
7050[7]	1971	6.2	2.3	2.25	—	—	0.12	530	460	11	P, F, E

† Different heat treating conditions were employed for different compositions and products; some alloys not heat treated to maximum strength temper.

‡ Letter prefixes used to distinguish related compositions established sequentially. These would not qualify as modifications of the original under current Aluminum Association rules.

1, Properties for forged or rolled products. Extruded products have higher strengths. 2, S-sheet, P-plate, F-forgings, E-extrusions. 3, Propeller blade forgings only. 4, See text concerning T73-type and T76 tempers. 5, 'Premium Strength' forgings. 6, Properties listed for T73-type temper. 7, Properties listed for T736-type temper.

The alloy development programmes were spurred to some degree by the appearance of RR77 early in 1937 and by emerging military needs for higher strength materials. Considerable emphasis was placed on resistance to s.c.c., and many hundreds of tests were performed on a large number of compositions fabricated in the laboratories. By mid-1938 a composition had been selected for sheet that was considered to have a combination of the major solute elements providing the most favourable balance of strength, resistance to s.c.c. and other characteristics including fabricating properties. This Al–Zn–Mg–Cu–Mn alloy was designated X74S, and a programme was initiated to produce sheet in the production mill for evaluation by potential users.

The account by Dix (1950) discloses the disappointment encountered in trials of X74S parts which had been subjected to plastic deformation and fit-up stresses. Under these conditions early s.c.c. was encountered, and this effectively sent the development 'back to the laboratory' for several years.

1940–1950: alloys 7076 and 7075

Introduced in 1940, alloy 7076 (table 1*b*) is the oldest continuously-produced American alloy of the 7XXX type, figure 1. The composition was selected to produce forged aircraft propeller blades meeting urgent military needs for a more abrasion- and fatigue-resistant material than then-standard 2025. The amounts produced have never been great because of its specialized application, but this alloy is still produced today.

After the late-1939 observations of s.c.c. in formed sheet components, X74S was eliminated from further consideration and laboratory work intensified to develop more predictive s.c.c. testing procedures and to improve the alloy performance. Among many other compositional and heat treating factors evaluated was addition of small amounts of several sparsely-soluble, high-melting elements, Zr, Cr, V, Mo, W. These were added individually in amounts of 0.15 to 0.25% to X74S, which in all cases contained 0.4% Mn. As results from boiling 6% NaCl solution tests of deformed sheet specimens of these alloys were completed, it was apparent that outstandingly better s.c.c. performance was observed with the Cr-containing composition. Immediately a series of compositions was prepared with 0.05 to 0.45% Cr, and from these it was concluded that 'the presence of 0.2 to 0.35% Cr definitely prevents cracking' (U.S. patent 2240940, 1941).

Some time was required to gain further assurance by confirming and repeated tests, to establish that Mn could be omitted when Cr was present, and to ascertain that d.c. ingots could be produced and fabricated in the mill. By March 1941 the first sheet conforming to a low alloy content within the 7075 specification range had been made in the mill and tested. This composition, corresponding to X74S but with Cr in place of Mn, was identified as XA75S. When higher strength was requested, the Zn, Mg and Cu contents were all adjusted upward. This composition was initially designated XB75S. By early 1943 all laboratory work with XB75S was highly favourable, and it was then established as 75S (now 7075).

Broad evaluation programmes were conducted by the aircraft industry, some involving substitution of the higher strength alloy in existing designs, others involving new designs, all at this time for military purposes. The alloy was developed first as a sheet material and soon found suitable for extrusions. First production of hand forgings was in mid-1945, and by early 1946 thick (to 90 mm) plate had been rolled from 200 mm × 1000 mm d.c. ingots and tested. The adverse effect of the Cr addition on development of strength with low quenching rates was noted early. This tended to limit application of the alloy to products having section thicknesses of 75 mm or less. Early production, fabrication and forming problems were solved, and industry acceptance resulted in rapid growth of production so that by 1950 this alloy accounted for 22% of Alcoa's shipments of strong, heat treatable aircraft products (Nock 1953).

Among the alloys of table 1*b*, 7075 has been produced and used in the greatest quantity by a very wide margin, and high mechanical reliability has been experienced in modern aircraft employing this basic composition in a variety of structural components and fabricated forms. The engineering trade-offs considered in establishing the proportions of the major solute elements, Zn, Mg and Cu were those of continuing importance today, and this basic composition has been found particularly adaptable to new temper and compositional controls to meet more stringent requirements for fracture toughness and resistance to exfoliation and s.c.c. For about 15 years from its original introduction, the only tempers applied were those of the T6 type. Resistance to s.c.c. in moderately thick products in the T6 temper was inadequate, but the alloy was found to have highly favourable response to the higher temperature T7X type precipitation treatments employed later to greatly increase resistance to s.c.c. and exfoliation corrosion. Increasing base metal purity coupled with special processing provided substantial gains in fracture toughness. This basic alloy and the variants of composition and temper developed since its introduction are expected to retain a continuing role in aircraft applications for some time.

Although only incidental in relation to the development story, it seems opportune to report in connexion with this conference† some previously unpublished information from Alcoa Laboratories on the properties and structure of 7075-W (as-quenched, naturally aged) sheet that was tested and examined after 25 years natural ageing. The changes in tensile properties of this early production lot were monitored over the 25 year period, figure 2, and additional changes are apparently continuing at a very low rate at the present time.

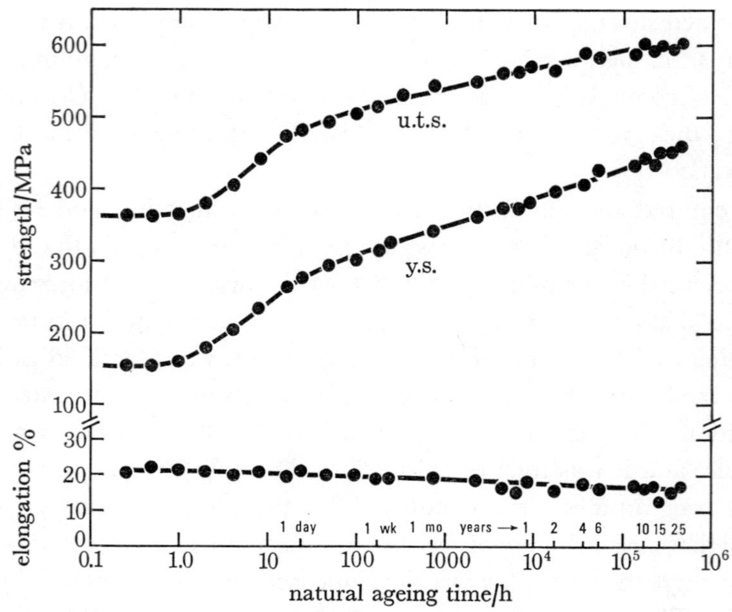

FIGURE 2. Tensile properties transverse to rolling direction of 1.6 mm thick 7075–W sheet as a function of natural (room temperature) ageing time. Solution heat treated at 466 °C, cold water quenched, flattened by stretching, and naturally aged.

The transmission electron microstructure (figure 3, plate 23) of the material examined after 25 years gave unmistakable evidence of the quasi-spherical g.p. zones normally observed in the precipitation heat treated T6 temper. Zones formed during the stages of natural ageing in which the greatest strength changes occur have not been resolved by t.e.m. nor are those formed in long-time naturally aged 2XXX alloys. Estimated mean zone diameter in the W (25 years) material was 1.2 nm and approximate range < 1.0 to about 1.5 nm. The estimated zone density was $4 \times 10^{18}/cm^3$. A brief experiment showed that a zone structure of similar apparent size and

† The 1921 Eleventh Report to A.R.C. (Rosenhain *et al.* 1921) contains comparative short time and 10-year natural ageing data for some German-produced duralumin.

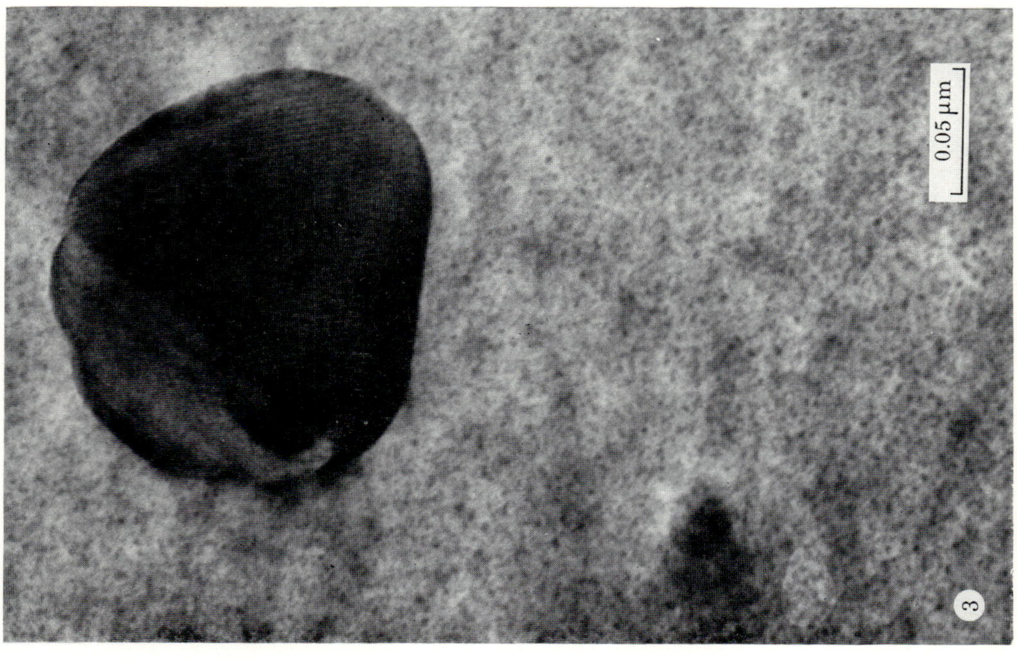

FIGURE 3. Transmission electron microstructure of 7075–W sheet after 30 years natural (room temperature) ageing. Estimated mean size of g.p. zones about 1.5 nm. Large dark area is $Al_{12}Mg_2Cr$ precipitate.

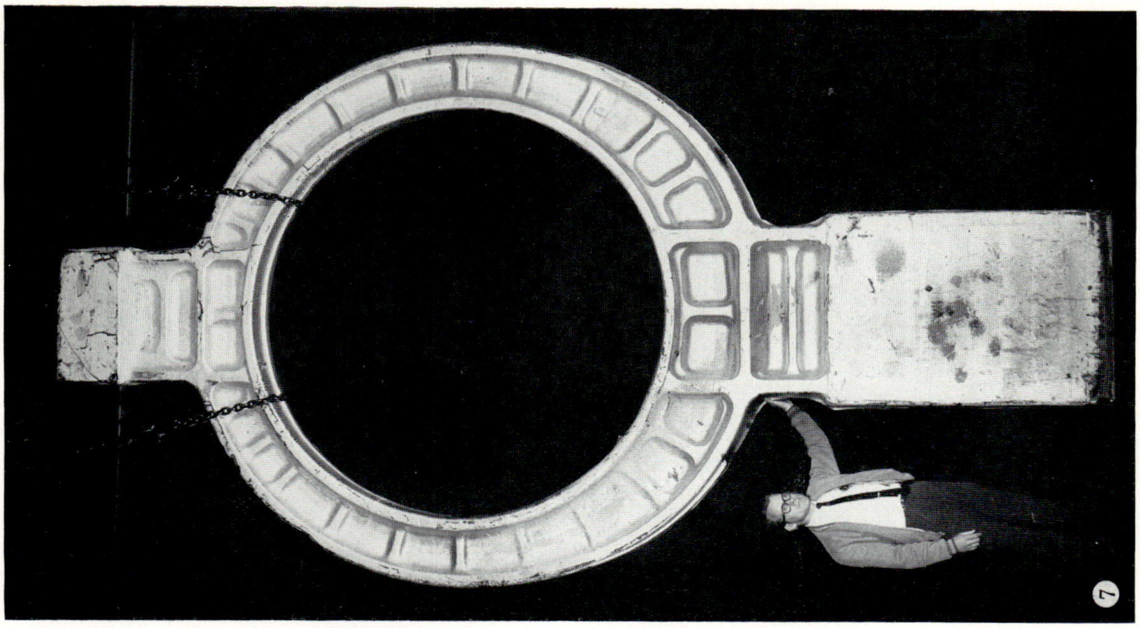

FIGURE 7. Die-forged 7075–T73 integral centre engine support and vertical stabilizer spar for McDonnel-Douglas DC-10. Four similar forgings are used in each plane.

density was produced by a precipitation treatment on freshly quenched material of 20 min at 120 °C, but the yield strength of the material so treated was only 265 MPa compared with the 25 year r.t. material's 460 MPa.

1950–1960: *7178*, *7001*, *7175*, *7079*, *7075–T73*

7178. In response to insistent industry and military demand for an alloy with higher strength than 7075–T6 (target yield strength ~ 10% higher), the 7178 composition was introduced in 1950 (see Nock 1953). This alloy has been used on a selective basis principally for compressively stressed members of both military† and commercial‡ planes but has fallen into general disfavour for newer designs as consideration for damage tolerance and fracture toughness assumed greater prominence.

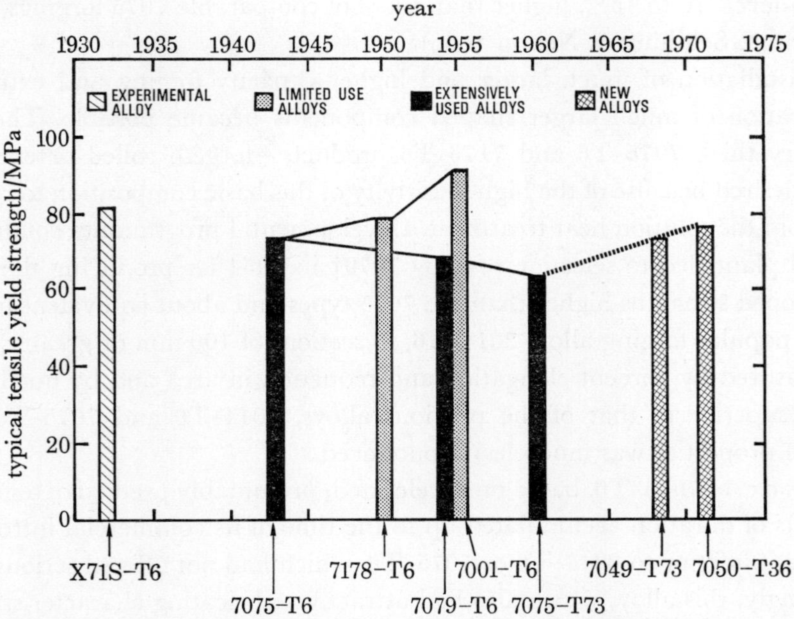

FIGURE 4. Typical tensile yield strengths (0.2 p.s.) of American Al–Zn–Mg–Cu alloys related to year of introduction.

7001. In 1955 the 7001 composition was introduced providing another step on the ladder of increasing static strength capability (Harvey Aluminum 1955). The even more limited ductility and toughness of this alloy in T6 type tempers were considered inadequate to permit its use in major aircraft structure components. The T75 temper developed later offered significantly improved resistance to s.c.c. and exfoliation corrosion. However, 7001–T75 has also not been accepted, mainly because of its fracture toughness characteristics, which are intermediate to those of 7075–T6 and 7178–T6 (ATC 1966).

As evidenced in figure 4, following the appearance of these two higher strength alloys, primary effort was directed towards improving properties other than static strength in the absence of stress-concentrating features. Later alloys are characterized by lower yield and ultimate strengths, representing trade-off of strength to upgrade tolerance for environmental, geometric and manufacturing factors, fatigue or other damage in service.

† Examples: McDonnell-Douglas F4 – alclad fuselage sheet (originally T6, now T76); Boeing B52 extruded wing stiffeners (upper surfaces) and original production upper wing skins; Lockheed C130 main wing plates.

‡ Examples: Lockheed Electra main wing plates; Boeing 727 upper wing skins.

7175. Concern with the degree of mechanical anisotropy of moderately large and thick-section 7075–T6 products in the late 1940s led to establishment in 1950 of the 7175 composition modification. This differed from 7075 only in specifying lower maximum limits on the impurity elements Fe and Si coupled with slight adjustments in maximum Mn and Cr contents in recognition of beneficial ductility effects from reduced volume fraction of undissolved phases. The 7175 specification was applied generally in producing parts of large cross section to meet more stringent elongation requirements in directions perpendicular to that of major extension in working.

With even tighter composition controls and special fabricating procedures 7175 is used to produce Alcoa's 'Premium Strength' forgings, which originated considerably later. These products, available today in two tempers, T66 and T736, provide superior fracture toughness at yield strength levels 10 to 15% higher than those of comparable 7075 forgings in equivalent tempers (Kauffman, Schilling & Nelson 1969).

7079. With installation of much larger and higher capacity forging and extrusion presses (1952–4), fabrication of much larger shaped components became possible. The mechanical properties of very thick 7075–T6 and 7175–T6 products – forged, rolled or extruded – were lower than was desired because of the high sensitivity of this basic composition to rate of cooling in quenching from the solution heat treatment. Developmental programmes conducted in both laboratories and plants led to selection of alloy 7079† in 1954 for producing the larger parts. This alloy developed strengths higher than the 7075 types and about equivalent to those of the previously most popular forging alloy, 2014–T6, in sections of 100 mm or greater thickness. Its ductility, as measured by percent elongation and reduction in area and by notched specimen properties, was superior to that of the previous alloys, 2014–T6 and 7075–T6, and directionality of these properties was much less pronounced.

Resistance to s.c.c. of 7079–T6, based on accelerated, presumably predictive tests and natural environment tests of duration accumulated up to the time of its commercial introduction, was indicated to be not inferior to 2014–T6 or 7075–T6, which had not offered serious problems in service. Accordingly, this alloy, which also had attractive fabricating characteristics, appeared to meet the requirements considered important at that time exceptionally well. Through the latter 1950s and into the early 1960s 7079 was specified and produced in high volume for a variety of new and older aircraft designs in a wide range of product forms for primary structure elements as well as some skin sheet (which exhibited higher fracture toughness than equivalent 7075–T6 sheet).

By the latter part of the 1950s, it became apparent that this alloy was susceptible to s.c.c. under environmental conditions of service, and a disturbing frequency of cracking was encountered with products of thick and non-uniform section such as die forgings, some extrusions and parts produced by programme-machining thick plate. This was undoubtedly influenced by part size and configuration and in some cases by new service parameters applicable to such parts as hydraulic landing gear cylinders. The greater susceptibility relative to the previously-used alloys had not been indicated by laboratory tests. This apparent divergence between 'accelerated' tests and service experience was found to be associated with alloys containing either no Cu or less than about 1% of this element, whereas good correlation was found with alloys having Cu contents equal that of 7075 (1.6% Cu) or higher (Lifka & Sprowls 1967).

† Alloy 7079 is essentially the same as the German alloy identified originally as Hy43 (ISO type AlZn4Mg3Cu; DIN 3.4345).

As a consequence of the service experience with 7079–T6 and inadequacy of 7075–T6 in large, thick-section products, work previously underway was intensified to find a basic materials solution to the s.c.c. problem.

7075–T73. Highest strengths are developed in alloys of the type under discussion by precipitation heat treatment at temperatures lower than 150 °C, a conventional T6 treatment being 24 h at 120 °C. Alternate two-stage treatments, such as 4 h at 96 °C followed by 8 h at 157 °C, were developed and used starting in the 1940s to reduce the total cycle time in producing 7075–T6 sheet (Nock & Wooll 1947).

A broad experimental programme was undertaken with 7075 in which the effects on resistance to s.c.c. of many variants of isothermal and two-stage ageing were evaluated. This work culminated in development of 7075–T73 products, which were introduced commercially in 1960. It was established that precipitation treatment at temperatures higher than 160 °C progressively improved s.c.c. resistance. Higher strength (figure 5) was obtained when the treatment at these temperatures was preceded by a lower temperature stage to nucleate a finer, higher-density precipitate dispersion, and the resulting combinations of strength and s.c.c. resistance were superior. The T73 temper is therefore produced by applying a two-stage precipitation treatment.

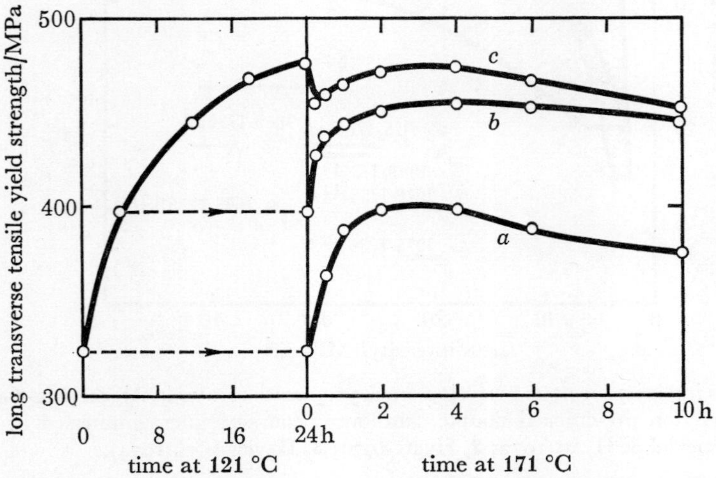

FIGURE 5. Comparison of yield strengths (0.2 p.s.) of 7075 plate resulting from *a* isothermal (171 °C) and *b*, *c* two-stage (121/171 °C) precipitation heat treatments.

Based on many hundreds of accelerated (chloride solutions) and natural environment tests employing smooth surface specimens stressed in the short-transverse direction either in a bending mode or in direct tension, properly processed 7075–T73 material resists s.c.c. at tension stresses of at least 300 MPa. By comparison, the same materials in T6 temper would not be expected to resist s.c.c. consistently in the accelerated tests at stresses over about 50 MPa (Brown *et al.* 1972). Although most of the accumulated information involved in alloy development and proof testing has been obtained from smooth specimen tests, there has been recent interest in measurements of stress-corrosion crack velocity as a function of the crack tip stress intensity. Such data have been obtained from fatigue-precracked double cantilever beam specimens, and comparative values for short transverse direction tests of several of the alloys and tempers discussed here are illustrated in figure 6 (Speidel & Hyatt 1972; Hyatt 1970; Davies, Nordmark & Walsh 1974).

The 7075–T73 development is considered a highly significant contribution to the technology of the Al–Zn–Mg–Cu alloys, enhancing their service reliability and engineering value in the form of large aerospace structural components. Several million kilograms of products in this alloy/temper have now been in service for several years with no reported s.c.c., and it is anticipated that there will be continued large volume usage of this material. One of the largest and most interesting die forgings currently being produced in 7075–T73 is the integral engine mount and vertical stabilizer spar for the McDonnell-Douglas DC-10†, figure 7, plate 23 (Turley & Gassner 1972).

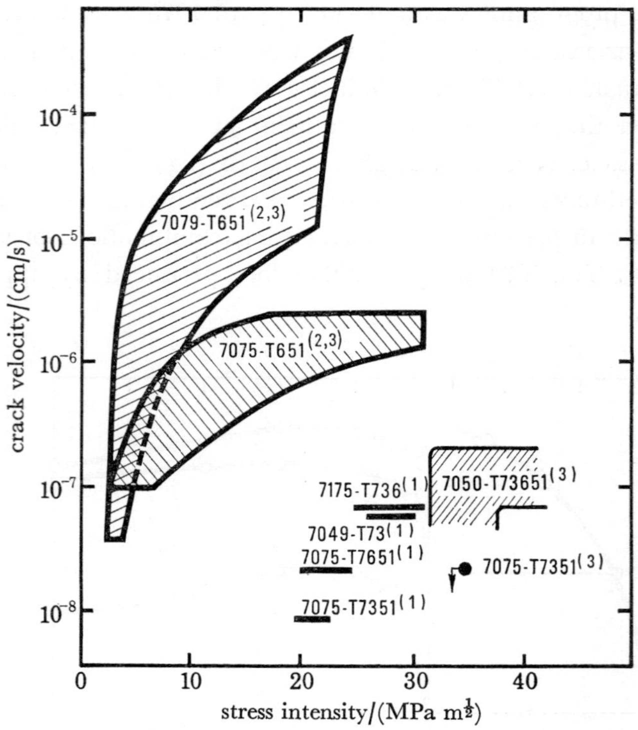

FIGURE 6. Effect of stress intensity on stress-corrosion crack propagation rates in Al–Zn–Mg–Cu alloys tested with short-transverse direction pre-cracked double cantilever beam specimens immersed in saturated aqueous NaCl solution (1, Speidel & Hyatt 1972; 2, Hyatt 1970; 3, Davies *et al.* 1974).

Various hypotheses have been proposed to explain the decisive effect of overageing on s.c.c. resistance and to relate it to microstructural features including the presence of precipitates in grain boundaries; precipitate-free zones (p.f.z.); electrochemical relationships between grains, precipitate-containing boundaries and p.f.z.; dislocation pile-up at grain boundaries; and matrix dislocation-precipitate interactions. The theory attributing s.c.c. in T6 temper to selective dissolution of an anodic p.f.z. accelerated by tension stress, and the higher resistance of T73 temper to equalization of the electrochemical potentials is consistent with some of the experimental observations (Sprowls & Brown 1969; Lifka & Sprowls 1972). The principal microstructural difference between the T6 and T73 tempers, however, is in the intragranular precipitate, which changes in structure from predominantly coherent g.p. zones of T6 to the metastable η',‡ of T73. The change in precipitate structure is the basis for the dislocation-

† As-forged length 5.33 m diameter 2.77 m mass 2360 kg.

‡ Equilibrium phase field extends across quaternary diagram joining $MgZn_2$ and AlCuMg. Evidence from electrochemical solution potential changes indicates Cu precipitates in the transition form of this phase during the higher temperature stage of the T73 treatment.

matrix precipitate interaction hypothesis (Speidel 1971). Although a variety of evidence can be cited as supporting one or the other of the mechanisms suggested, the question must be regarded as incompletely resolved at the present time.

1960–1975: 7075–T76, 7175–T736, 7049, 7475, 7050

Exfoliation corrosion – 7075–T76

Exfoliation corrosion had been a problem with 7075–T6 in certain aircraft components in which the resistance to s.c.c. had been entirely adequate because of part configuration and the nature and direction of stresses applied. Although 7075–T73 provided very high resistance to exfoliation, designers were reluctant to accept the strength reduction, about 15% for this temper. It was determined that less-drastic overageing than required for T73 produced high resistance to exfoliation. Based on this, the exfoliation-resistant T76 temper was evolved in mid-1960s providing strengths about midway between those of the T6 and T73 tempers. In some cases, it has been possible to apply T76 temper material in place of T6 without redesign or increase in weight. 7075–T76 products are highly resistant to s.c.c. at short-transverse direction stress levels of 170 MPa.

Higher strength with high s.c.c. resistance – 7175, 7049, 7050

Application of 7075–T73 in place of the higher strength T6 temper material in some cases required part redesign with an associated weight penalty. Reaction from the aircraft industry was a request to restore the T6 temper strength or higher combined with very high s.c.c. resistance. This motivated new research efforts on a wide geographic scale with more intensive evaluations of Zn, Mg and Cu amounts and ratios; variations in the supplementary elements for grain control and s.c.c. improvement – Cr, Mn and Zr; other additions, most prominently Ag; and numerous processing variants. Silver additions (0.3–0.4 mass % Ag) were widely investigated and generally considered to offer improved resistance to s.c.c. In the U.S. the benefits were not regarded as commensurate with the added cost, and the limited supply of Ag was considered a strong deterrent to extensive commercial use in an aluminum alloy.

7175–T736 Forgings. In 1968 forged products having significantly higher strength than 7075–T73 were developed employing special composition controls within the 7175 specification and non-conventional thermal processing and fabricating processes. The static strengths of these 'Premium Strength Forgings' are comparable with those of 7075–T6, and the resistance to s.c.c. is high although not considered equal to that of 7075–T73 products. Alloy 7175–T736 forgings have been used for a variety of specialized applications.

7049–T73. Alloy 7049 (Luhan & Summerson 1970) also introduced in 1968, differs from 7075 principally in its higher Zn and lower Cr contents, with somewhat more restrictive limits on the ranges of major and impurity elements. The higher Zn:Mg ratio was found to provide higher strength/s.c.c. resistance combinations, the lower Cr reduced quench sensitivity, and Cu, Fe and Si were controlled to improve ductility and fracture toughness at the higher strength level.† This alloy in T73 temper provides high resistance to s.c.c. coupled with strengths comparable with those of 7079–T6 in parts having a section thickness of 75 mm or over. Alloy 7049–T73 forgings have been in service for several years, prominently in landing gear retrofit programmes replacing 7079–T6 in F4, F111, B52 and C141 aircraft. Forgings of this alloy are

† A modification of 7049 with lower maximum limits on Fe (0.20% max.) and Si (0.15% max.) has been established as 7149.

also being specified for selected applications on newer aircraft such as DC10 and F5E (Anon 1974). Extrusions in T73511 and T76511 tempers are available, and design data are being developed for their application.

7050. An alloy of more recent vintage, 7050 (Staley, Hunsicker & Schmidt 1971), represents a greater departure in composition. Designed with particular reference to production of thick section parts in which lower quenching rates are unavoidable and for application of the T73 and T76-type precipitation treatments, 7050 has higher Zn and Cu contents relative to 7075, Zr in place of Cr, and more restrictive limits on the major elements as well as impurity elements. Zr, functioning as the high-temperature precipitate-forming addition to suppress recrystallization, reduces the strength loss with decreasing quench cooling rate characteristic of the Cr- or Mn-containing alloys. The $ZrAl_3$ phase formed by Zr avoids combination with principal alloying element species, in contrast with Cr which forms $Al_{12}Mg_2Cr$ or Mn which forms $Al_{20}Cu_2Mn_3$. The Zr-containing phase is present in smaller volume fraction as compact, nearly spherical particles. These factors were considered as favouring higher fracture toughness.

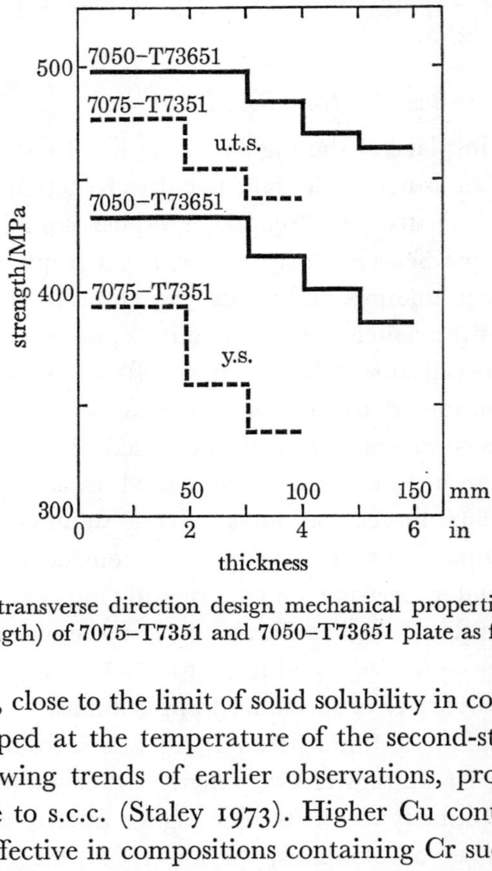

FIGURE 8. Comparative long-transverse direction design mechanical properties (ultimate tensile strength and tensile yield strength) of 7075–T7351 and 7050–T73651 plate as function of thickness.

The higher Cu content, close to the limit of solid solubility in combination with Zn and Mg, increases strength developed at the temperature of the second-stage precipitation treatment (Staley 1974), and, following trends of earlier observations, provides higher strength for a specific level of resistance to s.c.c. (Staley 1973). Higher Cu contents than those of the 7075 range are not similarly effective in compositions containing Cr such as 7178.

Selection of second-stage precipitation conditions for 7050, as with 7075 and 7049, involves a trade-off compromise between strength and resistance to s.c.c. Tempers designated by types T76, T736 and T73 are established. The resistance to s.c.c. of 7050–T736 is considered to be at least equivalent to that of 7175–T736, which has been used for several years with satisfactory experience, but does not equal the 'virtual immunity' of 7075–T73 or 7050–T73.

The strength advantage of 7050–T73651 compared with 7075–T7351, increasing with thickness, is apparent in the design yield strength values for plate charted as a function of thickness in figure 8. Beyond the thickness shown for 7075–T7351, the strengths of this alloy are generally too low to be of important structural value, while up to 150 mm the Zr-containing alloy has nearly as high strength as the thinnest 7075–T7351 product.

Evaluation of resistance to s.c.c. of 7050 involved hundreds of tests principally with small tensile specimens, 3.2 mm diameter in the stressed section, machined from a variety of products. Emphasis was placed on the performance of specimens oriented to allow stress application in the short transverse direction relative to the wrought structure. Exposures included industrial atmosphere and seacoast environment as well as recognized accelerated laboratory tests.

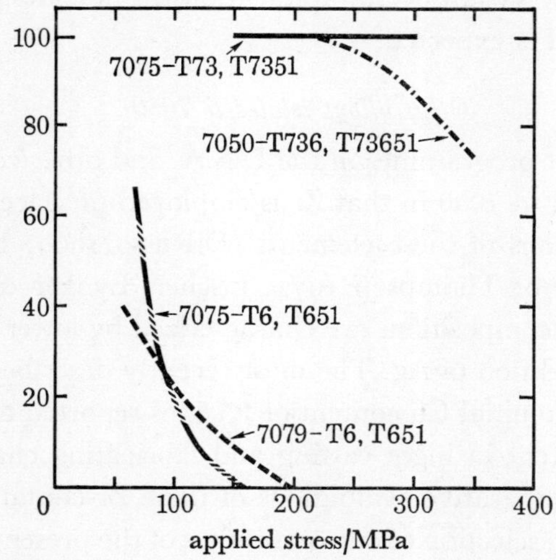

FIGURE 9. Results of industrial atmosphere stress-corrosion tests of 50 to 63 mm thick plate during four years exposure at New Kensington, Pa., U.S.A.

Results from four years exposure in the industrial atmosphere of specimens from development-stage plate complying with the 7050–T736 specification and from similar tests of the earlier alloys provided the information for figure 9. Performance in the seacoast exposure was comparable to that in the industrial environment. Specimens removed from plate produced in a commercial run have been exposed for over 16 months at stresses up to 310 MPa with no s.c.c. failures in either industrial or seacoast environment. These observations, coupled with those from numerous accelerated exposure (NaCl – alternate immersion) tests support a high ranking of resistance to s.c.c. at stress levels to at least 240 MPa for plate and other 7050–T736 products.

TABLE 2. REPRESENTATIVE (MEAN) VALUES OF PLANE-STRAIN STRESS INTENSITY FACTOR, K_{Ic}, FOR PLATE IN 25 TO 50 MM THICKNESS RANGE†

	$K_{Ic}/(\text{MPa m}^{\frac{1}{2}})$, in direction indicated		
alloy–temper	l–t	t–l	s–l‡
7075–T651	29.7	25.3	19.8
7475–T651	44.0	38.5	29.7
7075–T7651	28.6	26.4	20.9
7475–T7651	47.3	41.8	31.9
7075–T7351	34.1	29.7	22.0
7475–T7351	55.0	49.5	36.3
7050–T73651	38.2	31.7	26.9

† Compact-tension specimens per ASTM E399.
‡ Values for 50 mm thickness only.

Fracture toughness tests were performed on a range of commercially-fabricated 7050–T736 products to determine applicable values of plane-strain stress intensity factor, K_{Ic}. The values obtained exceeded those determined for the same types of products in 7075–T73 (e.g. plate data, table 2). This is significant because the normal accompaniment of higher strength, if derived through change in precipitation treatment or increased alloy content maintaining the same purity, is reduced fracture toughness.

Alloy 7050 plate and forgings are being produced currently for several retrofit programmes replacing the originally specified material which was considered inadequate in resistance to either exfoliation corrosion or s.c.c. Several applications in new designs have been made and increased use of this material is expected.

Other alloys related to 7050

Research and development programmes in the U.S.A. and other countries have led to other compositions that are related to 7050 in that Zr is employed in place of Cr or Mn, or in combination with smaller additions of these elements (DiRusso 1969; DiRusso & Buratti 1974; Hyatt & Schimmelbusch 1970; Thompson 1972; Fischer, Lynker & Markworth 1972; Reynolds & Harris 1974). These compositions are characterized by lower Cu content than 7050, in one case coupled with an addition of Ag. The most recently described composition, Alcan GB X3066 (DTD 5120) with a nominal Cu content of 1.7 % is reported to have an advantage over 7050 from its lower Cu content in ingot casting and fabricating characteristics (Reynolds & Harris 1974). Additional comparative evaluations of these Zr-containing alloys will assuredly be made resulting in eventual selection of one or another of the present specifications or perhaps an evolutionary compromise.

Higher fracture toughness alloy 7475

Although 7075–T6 aircraft structure components are considered as having an excellent performance record from the viewpoint of mechanical reliability and damage tolerance, additional assurance against the probability of critical stress failure in flawed, poorly machined, service damaged or fatigue cracked parts is regarded as highly desirable. About the same time that more quantitative techniques for measuring fracture toughness were evolving from the basic concepts of linear elastic fracture mechanics, programmes were initiated to achieve superior fracture toughness at high strength levels.

From early work (Nock & Wooll 1963) it was concluded that the Al–Zn–Mg–Cu alloys exhibited an inherent fracture toughness/strength combination advantage over Al–Cu–Mg types in the elevated temperature precipitation strengthened tempers. This is considered as related in part to differences in deformation mechanisms associated with the coherent type precipitate of the former contrasted with the non-deforming precipitate of the latter although other factors are also contributory. For either type alloy, changes in content of major alloying elements or in state of precipitation produce inversely related changes in strength and fracture toughness, a trade-off relation not uncommon to other metal-base alloys. The improvement desired, however, was increased toughness with no sacrifice in strength.

Our early investigations, since confirmed by published studies of many other investigators, demonstrated that highly significant improvements in fracture toughness could be obtained by reducing the volume fraction of insoluble (and undissolved soluble) intermetallic-phase particles present in the alloy structure. Work was concentrated on the 7075 composition since gains were modest in higher strength alloys, such as 7178.

The 7475 composition specification differs from 7075 in the more restrictive limits placed on the impurity elements, principally Fe and Si (0.12% and 0.10% max., respectively) which form sparsely soluble phases, and includes minor but significant adjustments in Mg and Cu ranges. The composition changes were coupled with modified process and mill controls found to provide supplemental toughness improvement through changes in grain size, homogeneity, and size and interparticle spacing of the $Al_{12}Mg_2Cr$ precipitate.

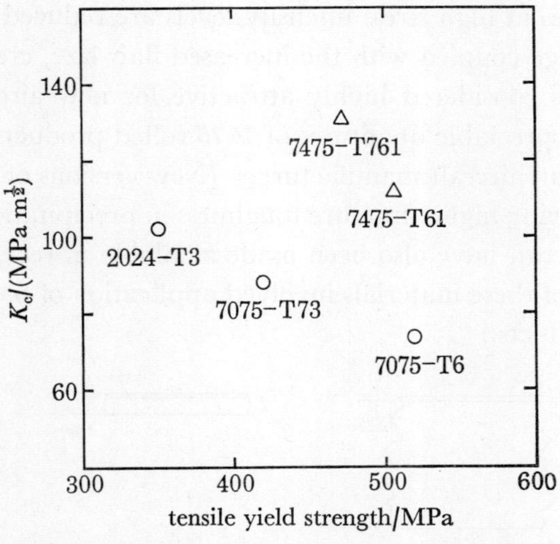

FIGURE 10. Fracture toughness (stress intensity factor, K_c, values) in relation to tensile yield strengths for sheet of five alloy/temper combinations. Mean values from 400 mm wide centre-cracked panels tested without anti-buckling guides.

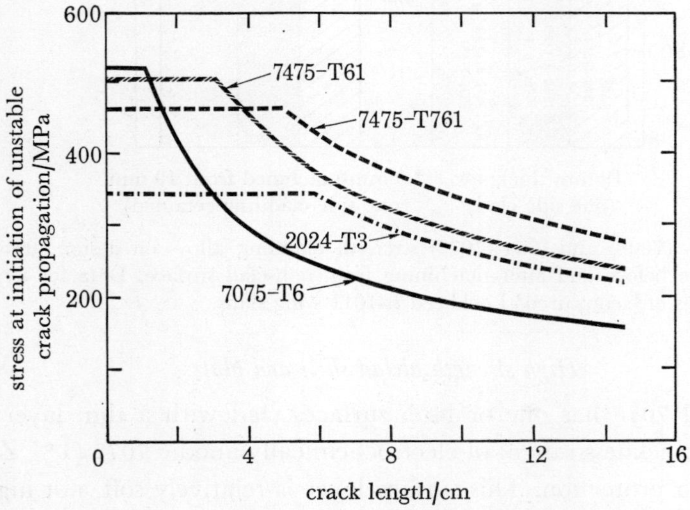

FIGURE 11. Gross section stress at initiation of unstable crack propagation plotted against crack length for wide sheet panels of four alloy/temper combinations.

Combinations of fracture toughness and tensile yield strength for 7475 sheet in two tempers produced in accordance with the process specifications are compared with those of sheet of conventional commercial grade alloys in figure 10. Alloy 2024–T3 (naturally aged temper) has been the American aircraft designers historical preference for those structural members in which high fracture toughness is most essential. The aim with 7475 was to couple a level of toughness at least equal to that of 2024–T3 with higher strength, and this was achieved with 7475. This is illustrated further by curves of residual strength of large cracked sheet panels as a function of crack size in figure 11. It is apparent that cracks which develop by fatigue, for example, in 7475–T61 or 2024–T3 may grow to about double the size of those in 7075–T6 before the stress becomes critical and unstable fracture occurs. The exfoliation-resistant 7475–T761 has higher crack-length tolerance. Comparative plane-strain fracture toughness data for plate of alloys 7075, 7475 and 7050 in applicable tempers are listed in table 2.

Fatigue crack growth rates at high stress intensity levels are reduced with the higher toughness material. This advantage coupled with the increased flaw size, crack length and damage tolerance are characteristics considered highly attractive for new aircraft designs as well as retrofit programmes, and appreciable quantities of 7475 rolled products have been supplied to both American and European aircraft manufacturers. (New versions of Al–Cu–Mg and Al–Cu alloys (2124, 2048, 2419) having higher fracture toughness in precipitation heat treated tempers preferred for supersonic aircraft have also been made available in recent years. The improvement in fracture toughness of these materials involved application of basic principles similar to those outlined for 7475 products.)

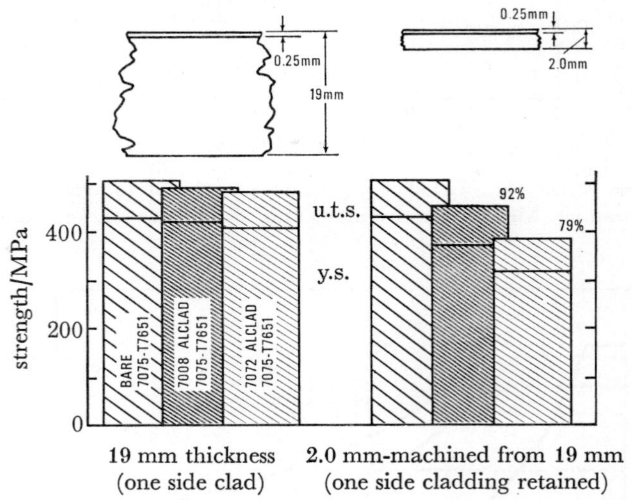

FIGURE 12. Effect of high- (7008) and low- (7072) strength cladding alloys on design strengths of composite (one-side alclad) plate before and after machining from non-clad surface. Data for 2 mm final thickness applies to thin portions of 'sculptured' Lockheed L-1011 wing skins.

High strength alclad sheet and plate

Conventional alclad 7075 has one or both surfaces clad with a thin layer (1.5 to 4% of thickness specified by thickness range) of electrochemically anodic 7072 (1% Zn alloy) to provide galvanic corrosion protection. This surface layer is relatively soft, not highly resistant to abrasion, and measurably reduces strength and fatigue resistance.

The high strength alclad sheet and plate of this alloy family, exemplified by 7008 alclad 7075, was conceived, produced in both laboratory and plant, and evaluated in the late 1940s. However, it was not used commercially for about twenty years until development of the Lockheed L1011 (Frisbee 1974). The cladding alloy is of the Cu-free Al–Zn–Mg type, develops strength by precipitation and is anodic to the 7075 core affording galvanic protection.

In sheet gauges the strength advantage amounts to only about 5% but this is coupled with higher resistance to abrasion and to fatigue. In the wing panels of the L1011 produced from plate, the higher strength cladding provides a static strength advantage of as much as 15% (figure 12) because of the greater proportionate thickness of cladding to core alloy in the thin webs formed by removal of metal from one surface. The harder surface layer also improves contour forming by shot peening (Brandel & Klass 1971).

Prospectus

Considerable advance has been made in overcoming early disadvantages of the Al–Zn–Mg–Cu alloy products – inadequate resistance to s.c.c. and exfoliation corrosion. Fracture toughness has been improved at high static strength levels. At this time these properties are not considered as restricting the design and application of these alloys for subsonic aircraft. On the other hand, the structures and components in many cases are regarded as fatigue-critical from a design stress viewpoint, and increased resistance to fatigue crack initiation and decreased growth rate would enhance the value of aluminum alloys for these applications. Many mechanistic studies of fatigue have been made, and these efforts are being intensified. However, they have not yet provided unequivocal guidelines to the alloy designer concerning direction or paths to seek increased fatigue performance. The higher toughness materials have shown some advantage in reduced growth rates at high stress intensities, but there is evidence that delayed initiation and growth retardation in earlier stages may occur in compositions and structures characterized by features that are quite in opposition to those favouring high toughness. These matters deserve additional study and further definition of basic principles that can be applied towards improvement.

Unique metallurgical structures developed by unconventional thermomechanical treatments have been reported to increase ductility, fracture toughness and resistance to fatigue. Treatments applied in final stages of processing (f.t.m.t.) generally consist of a sequence of precipitation treatments interspersed with cold or warm working operations. These procedures are intended to develop precipitate-stabilized, dense, uniform dislocation distributions.

More recently another type of thermomechanical treatment (i.t.m.t.) has been proposed to be applied much earlier during fabrication and intended for application to hot-worked products that are normally characterized by unrecrystallized structures. The i.t.m.t. processes are designed to develop either a final recrystallized structure of unusually fine grain size or a fine-grained recrystallized structure that is modified by a moderate amount of final hot working.

Evaluation of materials processed by these treatments has not been sufficiently complete to demonstrate that they fulfill all engineering property requirements or that the possible advantage in one characteristic is not offset by an untenable loss in another. Some of the treatments involve considerable additional cost and are difficult to control on a commercial scale. These treatments are not now applied commercially, but investigations are continuing to determine their possible merit, feasibility and cost effectiveness.

The development and successful application of these materials based on the quaternary Al–Zn–Mg–Cu system is the result of the efforts of many scientists and engineers presently and formerly associated with industry, government agencies and academia. The good record of service and reliability which these present-day materials have accumulated is a tribute to these people and their efforts. Although further advances in engineering properties will undoubtedly be made with this family of materials, the products available today should very satisfactorily fill the needs of the aircraft industry for the next several years.

References (Hunsicker)

Anon 1974 Advances in aerospace materials – processing technology. *Met. Prog.* **105**, 41–75.

ATC 1966 Evaluation of Aluminum Alloy 7001-T75. Aerospace Technical Council Report No. 101 (December 1966).

Brandel, W. W. & Klass, L. S. 1971 Ball forming solves contouring problems. *Met. Prog.* **100**, 56–57.

Brown, R. H., Sprowls, D. O. & Shumaker, M. B. 1972 The resistance of wrought high strength aluminum alloys to stress corrosion cracking. Stress corrosion cracking of metals – A state of the art. *A.S.T.M., STP* 518, pp. 87–118.

Davies, R. E., Nordmark, G. E. & Walsh J. D. 1974 Design mechanical properties, fracture toughness fatigue properties, exfoliation and stress corrosion resistance of 7050 sheet, plate, hand forgings, die forgings and extrusions. NASC Contract N00019-72-C00512, Final Report November 1974.

DiRusso, E. 1969 New aluminum alloys with high mechanical resistance for production of forgings. *Allum. Nuova Met.* **38**, 409–13.

DiRusso, E. & Buratti M. 1974 The properties of two new aluminum forging alloys Zergal 3 and Zergal 4. *Alluminio* **43**, 31–41.

Dix, E. H., Jr 1950 24th Campbell Memorial Lecture, Aluminum-zinc-magnesium alloys – their development and commercial production. *Trans. Am. Soc. Metals* **42**, 1057–1127.

Fink, W. L. & Willey, L. A. 1937 *Trans. A.I.M.E.* **124**, 78–109.

Fischer, G., Lynker, F. W. & Markworth, M. 1972 A non-quench sensitive Al–Zn–Mg–Cu alloy with additions of chromium, zirconium and silver *Aluminium* **48**, 413–8.

Frisbee, L. E. 1974 The Lockheed Tristar – an operational overview, *Aeronaut. J.* **78** (September 1974) (765), 389–402.

Harvey Aluminum 1955 Research Results at Harvey Aluminum. *Mod. Metals* **11**, 54.

Hyatt, M. V. 1970 Use of precracked specimens in stress corrosion testing of high strength aluminum alloys. *Corrosion* **26**, 487–503.

Hyatt, M. V. & Schimmelbusch, H. W. 1970 Development of a high strength stress-corrosion resistant aluminum alloy for use in thick sections. AFML-TR-70-109.

Kaufman, J. G. Schilling P. E. & Nelson F. G. 1969 Fracture toughness of aluminum alloys. *ASM, Metals Eng. Q.* **9**, 39–47.

Lifka, B. W. & Sprowls D. O. 1967 Stress corrosion testing of aluminum alloy 7079-T6 in various environments. *A.S.T.M., STP* 425, pp. 342–362.

Lifka, B. W. & Sprowls, D. O. 1972 Significance of intergranular corrosion in high-strength aluminum alloy products. *A.S.T.M., STP* 516, pp. 120–144.

Luhan, J. V. & Summerson, T. J. 1970 Development of 7049-T73 High Strength, Stress-Corrosion Resistant Aluminum Alloy Forgings. *ASM, Metals Eng. Q.* **10**, No. 4, 35–42 (November 1970).

Nock J. A., Jr 1953 Today's Aluminum Aircraft Alloys. *Trans. S.A.E.* **61**, 209–220.

Nock, J. A., Jr & Hunsicker, H. Y. 1963 High strength aluminum alloys. *Metals* **15**, 216–224.

Nock, J. A., Jr & Wooll, A. C. 1947 New short time aging practice for 75S aluminum alloy sheet. *Iron Age* **159**, 57–59.

Reynolds, M. A. & Harris J. G. 1974 Development of a tough high-strength aluminum alloy with improved stress-corrosion resistance. *Aluminium* **50**, 592–9.

Rosenhain, W., Archbutt, S. L. & Hanson, D. 1921 Some alloys of aluminum. *Alloys Res. Comm. Rep.* No. 11, London: Inst. of Mech. Eng.

Sander, W. 1926 Aluminum–magnesium–zinc alloys. *Z. anorg. Chem.* **154**, 144–151; also U.S. Patent 1629699 (1927).

Speidel, M. O. 1971 Current understanding of stress corrosion crack growth in aluminum alloys. *The theory of stress corrosion cracking in alloys* pp. 289–344. Brussels: NATO Scientific Affairs Div. Pub.

Speidel, M. O. & Hyatt, M. V. 1972 Stress-corrosion cracking of high strength aluminum alloys. *Adv. Corrosion Sci. Technol.* **2**, 115–335 (Plenum Press).

Sprowls, D. O. & Brown, R. H. 1969 Stress corrosion mechanisms for aluminum alloys. *Proceedings of International Conference on Fundamental Aspects of Stress Corrosion Cracking, NACE* 466–506.

Staley, J. T. 1973 Stress-corrosion cracking in aluminum alloys. *ASM, Metals Eng. Q.* **13**, 52–7.

Staley, J. T. 1974 Aging kinetics of aluminum alloy 7050. *Met. Trans.* **5**, 929–32.

Staley, J. T., Hunsicker, H. Y. & Schmidt R. 1971 New aluminum alloy X7050. A.I.M.E., TMS Paper F71-7.

Thompson, D. S. 1972 RX720 alloy, an aluminum alloy for high strength thick section applications. Reynolds Metals Publication.

Turley, R. V. & Gassner R. H. 1972 Large aluminum alloy forgings fly high in the DC10. *Met. Prog.* **101** 48–50.

3.7. Recent developments in ultra high strength steels and their applications

By M. J. May, T. Gladman and E. F. Walker

British Steel Corporation Physical Metallurgy Centre,
Swinden Laboratories, Moorgate, Rotherham S60 3AR

[Plate 24]

As far as the development of new classes of ultra high strength steels is concerned the past decade has seen little progress apart from a few notable exceptions such as the trip steels. However, the introduction of fracture mechanics concepts to the evaluation of material properties has been responsible for an enormous breakthrough in the understanding of the factors which influence the fracture processes in conventional high strength steels. This approach has concerned not only the resistance of these materials to unstable fracture under rising load but also the stable crack propagation which can occur in aggressive environments under static load and in both aggressive and inert environments under cyclic loading conditions. Utilization of fracture mechanics has demonstrated the improvement which can be obtained in the properties of high strength steels through the use of thermomechanical treatments, hardening by intermetallic compounds and control of inclusions. Though much still remains to be learned in each of these fields, this paper highlights some aspects of the progress which has been made in the understanding of the fracture, fatigue and stress corrosion cracking of high strength steels.

1. Introduction

The strength levels attainable in the so-called ultra high strength steels, i.e. with yield strengths greater than 1240 N/mm^2 (180 klbf/in^2) have not increased significantly over the last decade. The strength levels are still ranked as shown in figure 1, with cold drawn pearlitic steels (Embury & Fisher 1966) having the highest strength levels, followed by ausformed martensitic steels (Schmatz *et al.* 1963), maraging steels, quenched and tempered martensitic steels, and precipitation-hardened austenitic steels (Peckner 1963). The application of such steels is rarely limited by inadequate strength but rather by other factors such as fatigue resistance, toughness, environmental fracture behaviour and modulus. Consequently, over recent years considerable effort has been devoted to the improvement of these properties and an appreciation of the metallurgical and mechanical factors that are involved in the control of these properties.

Improvements in strength have occurred in the lower strength structural steels, an important feature of these developments being the maintenance of adequate levels of toughness and formability. Recent developments in the structural steels, for example inclusion control to improve toughness, have obvious implications for the higher strength steels, and are reflected in current interests in the toughness, fatigue properties, etc., of the ultra high strength steels. However, it is important that these studies should be directed at the factors which limit the application of these steels and due account must be taken of the service environments and modes of failure.

2. Ultra high strength steels

There are many forms of high strength steel, including the high strength low alloy steels, where a yield strength of 350 N/mm² (50 klbf/in²) is properly regarded as a high strength steel, in that a conscious attempt is made to attain the level of strength, either by the application of grain refinement or precipitation strengthening techniques. High strength low alloy steels have been developed to give yield strengths of up to 750 N/mm² (110 klbf/in²) with relatively simple microstructures and inexpensive manufacturing process routes.

However, ultra high strength steels may be defined as those having yield strengths greater than 1240 N/mm² (180 klbf/in²) and involving specific thermal, mechanical, or thermo-mechanical treatments. Of these, the quenched and tempered through-hardening alloy steels are the older and perhaps better known.

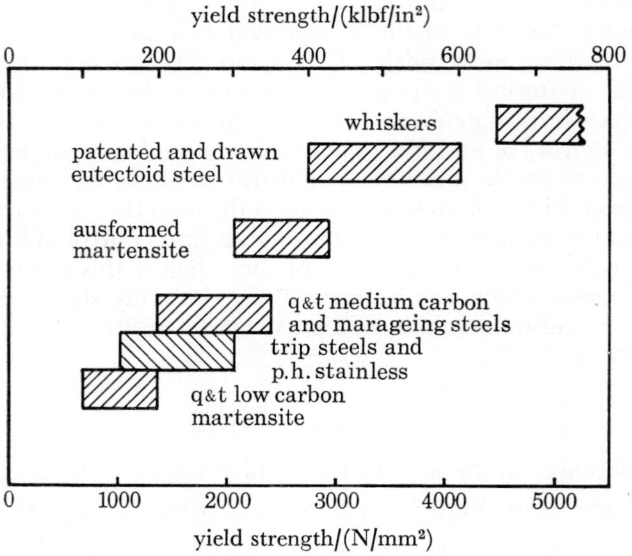

Figure 1. High yield strength steels.

2.1. *Low alloy quenched and tempered steels*

A large number of steels have been developed over the years, some of which have been used more extensively than others. Typical of these steels are 817M40 (BS. En.24), 835M30 (BS. En.30B) which are nickel–chromium–molybdenum steels, and 897M39 (BS. En.40C) a 3.5 % Cr–Mo steel, which was developed in the 1940s during a time of nickel shortage.

As indicated in figure 1, this type of steel can be used to produce a wide range of strength levels, depending upon the carbon content and tempering temperatures employed. The effect of tempering temperature on the strength of some typical low alloy quenched and tempered steels is shown in figure 2*a*. Increasing the tempering temperature up to 150–200 °C results in little loss of strength due to the fine precipitation of carbide. Tempering at progressively higher temperatures reduces the strength but increases the ductility. The corresponding changes in fracture toughness, K_{Ic}, are also shown in figure 2*a*. While the strength decreases with increasing tempering temperature, the K_{Ic} values are increased. The importance of these changes on the design stresses based on either the yield stress or on fracture resistance are shown in figure 2*b*. In the lightly tempered condition, a fracture mechanics analysis

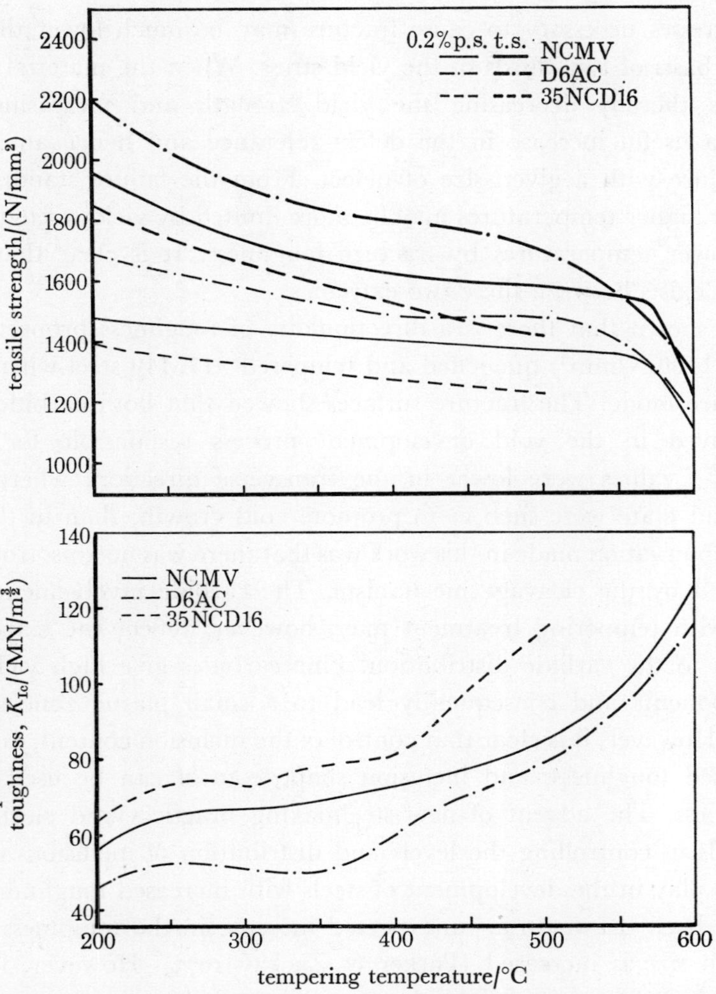

FIGURE 2*a*. Tensile and fracture properties of NCMV, DGAC and 35NCD16 steels.

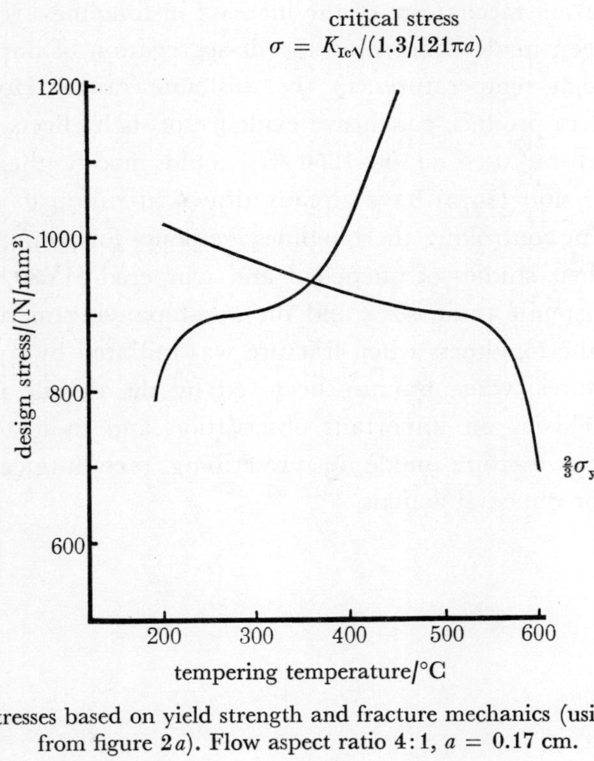

FIGURE 2*b*. Design stresses based on yield strength and fracture mechanics (using aggregate data from figure 2*a*). Flow aspect ratio 4:1, a = 0.17 cm.

indicates that the stresses necessary to cause fracture may be much lower than those given on the usual design basis of two thirds of the yield stress. When the material is tempered at higher temperatures thereby decreasing the yield strength and increasing the fracture toughness, there is a useful increase in the defect tolerance and hence an increase in the critical stress for failure with a given size of defect. From the failure standpoint, however, the steel tempered at higher temperatures may be more limited by yield strength than was the steel tempered at lower temperatures by fracture toughness. It is clear that an optimum tempering treatment exists between these two extremes.

Dulieu (1974) has shown that there is a directionality of toughness properties in an ultra high yield strength (1400 N/mm^2) quenched and tempered 817M40 steel when failure occurs by the ductile fracture mode. The fracture surfaces showed that both carbide and inclusion contents were involved in the void development process responsible for ductile crack propagation. The K_{Ic} values were lower in the transverse direction, where the inclusion orientation in bar and plate were such as to promote void growth, than in the longitudinal direction. A further observation made in this work was that there was no anisotropy of toughness when failure occurred by the cleavage mechanism. The carbide involvement implied from the change in K_{Ic} with tempering treatment may, however, reflect the extreme sensitivity of the yield strength to the carbide distribution. Fine carbides give high yield stresses, low work hardening exponents and consequently lead to a small plastic zone size and rapid cavity development. However, it is clear that control of the inclusion content, and distribution, will produce increased toughness, and inclusion shape control can be used to reduce the anisotropy of toughness. The advent of new steelmaking practices and electroslag refining of high strength steels in controlling the levels and distribution of inclusions obviously have an important part to play in the development of steels with increased toughness.

Attention has also been drawn to an unexpected increase in the toughness of martensitic steels when the grain size is increased (Parker & Zackay 1973). However, it is significant to note that such expectancy arises from the sensitivity of cleavage fracture to the grain size, and fractures involving the ductile failure mode would not be expected to conform to such concepts. The exact mechanism of the increase in toughness is not precisely defined, and suggestions have been made concerning the de-segregation of impurities from austenite grain boundaries at high temperatures in the austenite range. However, re-segregation experiments have failed to produce conclusive evidence of such effects and it is possible that the very high temperatures used (1100–1250 °C) could modify the sulphide distribution significantly. Dolby & Knott (1972) have already drawn attention to the importance of the operative failure mode in controlling the toughness response following changes in the metallurgical variables. In their studies of quenched and tempered HY80, they showed that an increase in the prior austenite grain size (and the accompanying increase in the martensite colony size) decreased the toughness when fracture was initiated by quasi-cleavage, whereas at higher test temperatures, when fracture occurred by the ductile mechanism, the grain size had little effect. This is an important observation and indicates the significance of identifying the operative fracture mode in prescribing metallurgical structures for steel development and also for remedial actions.

2.2. *Trip steels*

Trip steels are essentially warm worked metastable austenites. The compositions are designed to give a stable austenite at ambient temperature. The austenite is then deformed at a temperature above its M_d temperature to give a high strength deformed austenite substructure. The temperature of the prior deformation of the austenite controls the metastability of the resultant structure. The objective is to produce a degree of metastability that will permit substantial strain induced transformation of the austenite at small plastic strains at the operative service temperature. Thus any plastic strain at the crack tip will result in strain induced transformation, extensive work hardening and enlargement of the plastic zone size. The transformation produces a volume increase of some 3 % and this in itself causes extensive modification of the state of triaxiality at the crack tip (Antolovitch & Singh 1971).

A typical composition of a steel showing this behaviour is 0.25 % C, 2.0 % Mn, 2.0 % Si, 8.0 % Ni, 9.0 % Cr and 4.0 % Mo. When such a steel is solution treated, deformed by 35–80 % reduction in thickness at 450 °C, yield strengths of 1030–1520 N/mm² are produced. Plane stress fracture toughness values of more than 200 MN/m$^{\frac{3}{2}}$ have been obtained, but stress corrosion testing has shown that these materials may not offer any improvement over low alloy steels of the same strength level in hydrogen bearing environments. This is attributed to the formation of fresh hard martensite at the tip of the crack which is susceptible to the environmental attack and also has a low tolerance to hydrogen induced cracking.

2.3. *Cold worked austenitic steels*

Cold worked austenitic steels can attain high yield strengths, particularly if the stacking fault energy is low, as for example in low nickel high manganese austenites. Deformation of the low stacking fault energy austenite results in the formation of deformation twins, high dislocation densities and ϵ-martensite (cph). The strength levels are in excess of 1380 N/mm² following 50 % cold work. Because of their basic structure, these steels are not susceptible to hydrogen cracking.

The cold deformation aspects of these steels, however, indicate that the applications may be limited to thin plate or strip products.

2.4. *Maraging steels*

Maraging steels containing 18 % Ni with 7–9.5 % Co, 3–5 % Mo and 0.1–0.8 % Ti can be used to provide yield strengths in the range 1350–2050 N/mm² depending on the specific alloy content. The heat treatment is simple, consisting of an austenitising treatment, air cooling to form an essentially carbon-free martensite, and ageing to produce a fine distribution of intermetallic precipitates. A 12 % Ni steel with 3–5 % Cr, 3 % Mo, aluminium and titanium can give yield strengths of 1000–1350 N/mm² after a similar treatment. Provided that the carbon content is maintained at a very low level and that the formation of deleterious titanium compounds, e.g. nitrides, sulphides, carbo-sulphides, are avoided, high levels of toughness and stress corrosion resistance can be achieved. Yield strength levels of 1700 N/mm² and 2050 N/mm² are associated with K_{Ic} values of 85 MN/m$^{\frac{3}{2}}$ and 70 MN/m$^{\frac{3}{2}}$, respectively. Reported $K_{Is.c.c.}$ values vary widely, ranging from 12 MN/m$^{\frac{3}{2}}$ to 132 MN/m$^{\frac{3}{2}}$, depending on the compositional factors, impurity contents and processing procedures and probably testing conditions. Steels of this type are costly, due to their high alloy compositions and manufacturing practices, but in general possess good combinations of strength and toughness.

2.5. *Ausformed steels*

The basic strengthening mechanisms involved in ausformed steels may include refinement of the martensitic plate size, inherited cold work from the austenite, and strain induced precipitation in the austenite prior to the transformation to martensite. Directionality of toughness may be expected in materials failing by the ductile fracture mode, as outlined for the case of conventional quenched and tempered steels, §2.1. However, there is also the possibility of directionality of toughness in materials failing by the cleavage mode arising from the elongated austenite grain structure which imposes differences in the cleavage facet sizes according to the test direction in the transformed martensite.

3. Factors limiting the use of high strength steels

3.1. *Current applications*

The uses of bulk structural steels result in a range of fairly common requirements such as weldability, machinability, formability, etc., whereas the applications of higher strength steels are usually of a more specialized nature. It is convenient therefore to consider briefly some typical applications where strength is a prime factor and thereafter discuss in more detail the important considerations which limit the utilization of higher strength steels.

Undoubtedly the development of higher strength alloys has been motivated by the necessity to design efficient lightweight structures for aircraft, aerospace vehicles and military projectiles. Many low alloy quenched and tempered and maraging steels have been used in structures of this type and more recently the cold worked austenitic stainless alloys have been used in the fabrication of aerofoils.

High quality quenched and tempered steels are used for armour plating, often in combination with a lower strength more ductile alloy in order to achieve the required resistance to piercing and spalling. Military bridges, both fixed and transportable, use appreciable quantities of high strength materials, particularly maraging steels.

The increased loading of marine gears has promoted the more widespread use of high strength steels although the most sophisticated are usually gas carburized. The motor industry, however, is a bulk user of high strength steels for many moving parts, including gears and valves. Transport in general uses large quantities of high strength steels for such components as linkages and road springs and small quantities of such steels are to be found in earth moving equipment and shear pins in tracked vehicles.

3.2. *Fracture toughness*

Current structural design codes stipulate the maximum allowable membrane stress within a structure in terms of some fixed proportion of the relevant tensile properties. Indeed it is most common to find that the applied tensile stresses are limited to a maximum of two thirds of the appropriate yield stress of the alloy used. It is widely recognized, however, that an increase in strength will usually result in a decrease in fracture resistance.

The safety of structures and components designed and fabricated in such a way that unacceptable tensile stresses are not produced has indeed proved, over the years, to be a satisfactory approach as is demonstrated by the low incidence of structural and component failures resulting from gross overload effects. However, it is equally obvious that conventional design procedures do not adequately control the incidence of low stress, low energy fractures.

Although the risks of low stress fracture have long since been recognized it is only within relatively recent times that, through fracture mechanics concepts, it has been possible to introduce procedures that permit a quantitative analysis of such failures. Previously the integrity of the structure was preserved through the selection of materials on the basis of experience and reference to the ductile-brittle transition characteristics as observed in an

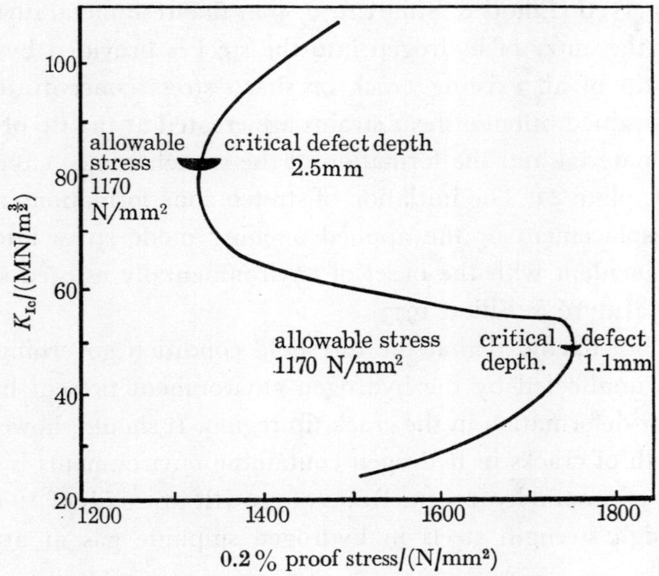

FIGURE 3. Comparison of critical crack sizes. Defect aspect ratio 4:1.

impact test. With the advent of higher strength steels the risks of low stress fracture were increased and in addition the ductile-brittle characteristics of these steels were less well defined and as such the Charpy test became less reliable. An analysis of the significance of defects based on the theory of linear elastic fracture mechanics can clearly demonstrate the controlling importance of fracture toughness. For example, in figure 3, it is seen that the critical stress for fracture from a crack of fixed size and geometry is highly dependent upon the fracture toughness and applied stress. The defect tolerance of the steel tempered at the higher temperature and subjected to stresses of almost yield point magnitude is still more than twice that demonstrated by the higher strength alloy under the influence of an equal stress equivalent to two-thirds of the pertinent proof stress.

The development of high strength steels is therefore severely limited by fracture toughness considerations. The required toughness must be related to the specific design and careful consideration must be given to the consequences of fracture and the possible use of fail-safe design procedures together with an appraisal of the capabilities of non-destructive testing procedures. The latter consideration is necessary to identify the size of defect that can be reliably detected and hence the fracture resistance of the material necessary to tolerate appreciably larger defects.

It is useful to remember that an approximate measure of the critical depth of a surface flaw or half breadth of an embedded flaw is obtained from the multiple $0.4(K_{Ic}/\sigma_y)^2$.

3.3. *Stress corrosion and hydrogen cracking*

Recent work (McIntyre 1973 *a*) has established that stress corrosion cracking of high strength steels in aqueous environments can generally be attributed to the effects of atomic hydrogen. This is produced by the catalytic dissociation of water at the metal surface. The hydrogen atoms occluded by a process of adsorption and diffusion reduce the fracture stress of the steel by assisting the formation of stable microcracks which link up and cause macroscopic crack growth.

It has also been observed (Elliott & Stuart 1970) that the fresh metal surface that is required in order to facilitate the entry of hydrogen into the steel is provided by the formation of a stretch zone at the tip of an existing crack or sharp stress concentration. As a structure containing a crack is strained, intense shear strains are created at the tip of the crack and lead to the separation of material and the formation of the stretch zone. A typical stretch zone is illustrated in figure 4, plate 24. The initiation of stretch zone formation can be related to the crack tip opening displacement or the applied opening mode stress intensity, K_{I}, and has been shown to be coincident with the onset of environmentally assisted stable crack growth defined by $K_{\mathrm{Is.c.c.}}$ (McIntyre & Elliott 1973).

It may, therefore, be concluded that the threshold condition governing the onset of stress corrosion cracking is unaffected by the hydrogen environment present but solely related to the mechanics of shear deformation in the crack tip region. It should, however, be emphasized that the rate of growth of cracks in hydrogen containing environments is strongly dependent upon the availability of atomic hydrogen. Rates of growth in excess of 10 mm/min have been recorded for some high strength steels in hydrogen sulphide gas at atmospheric pressure (McIntyre 1973 *b*).

The large amount of data now available on the susceptibility of steels to environmental fracture supports the opinion that the susceptibility is markedly increased at higher strength levels. Indeed, this is confirmed by fractographic studies that have indicated that stretch zone formation occurs more readily in higher strength steels, although its subsequent development is retarded. That is to say much larger stretch zones are easily observed in lower strength steels. The reasons for this are presently unclear but it must be recognized that environmental considerations will restrict the use of certain high strength steels for many applications. It would seem that the control of the stretch zone is a fundamental requirement that may lead to improved resistance. However, it is suggested that the formation of the stretch zone is primarily dependent upon the attainment of a critical shear strain that appears to be more readily produced in high strength steels having low strain hardening exponents.

These observations suggest that the development of ultra high strength steels with improved resistance to environmental fracture may best be explored by a study of the more highly alloyed steels where an adequate measure of fresh surface repassivation may be achieved by the creation of protective films. Such steels would probably be expensive and their use limited to special applications.

Finally, it must also be remembered that the above discussion is centralized around the observation that the stretch zone provides the fresh metal surface to initiate the primary reaction. It must be appreciated that a number of alternative mechanisms may be equally effective in introducing the hydrogen into the steel. These may include fatigue cracks, the creation of defects as a result of anodic dissolution or active path corrosion mechanisms, the

Figure 4. Typical illustration of a stretch zone formed between the fatigue and fast fracture regions.

(*Facing p.* 384)

emergence of slip steps at free surfaces particularly in regions of stress concentration, or mechanical damage.

3.4. *Fatigue*

It is convenient to consider the importance of strength on both the fatigue life measured in a conventional test and on the rate of propagation of fatigue cracks. Many publications (Harris 1961; Forrest 1962; Barnby 1972) have expressed a belief that steels exhibit a fatigue limit that is easily related to their tensile strength. Indeed it is suggested for high strength steels that the fatigue limit can range between 30 and 60% of the tensile strength.

This fact, however, implies that fatigue would be no more serious a problem in higher strength steels. This is contrary to the observations (Razim 1968) that retained austenite, present in increasing volumes in many higher strength steels, has a highly deleterious effect on fatigue life.

It must be appreciated, however, that the fatigue life of a structure is controlled by both the initiation and the subsequent propagation of a fatigue crack. In recent years a great deal of research effort has been directed towards the study of fatigue crack propagation and the general conclusions appear to suggest that the fatigue crack growth rates are independent of tensile strength and directly related to the alternating stress intensity ΔK. A limiting factor on fatigue life is the critical crack size for unstable fracture which, being dependent upon the fracture toughness, K_{Ic}, is reduced as the tensile strength is increased.

In summary, therefore, it is clear that by virtue of the high design stresses which will reduce the initiation endurance and the lower fracture toughness associated with higher strength steels, the risks of fracture resulting from fatigue are significantly increased.

3.5. *Corrosion fatigue*

In considering the influence of strength level on the resistance of steels to corrosion fatigue, it is firstly necessary to define the nature of the conjoint action (McEvily & Wei 1971, Austen & Walker 1975) in order to avoid confusion. In a previous section on stress corrosion it was suggested that the hydrogen atom could be identified as the agent responsible for the subsequent fracture processes. When this mechanism is supported by fatigue damage, the synergistic effect is said to be a result of 'stress corrosion fatigue' (s.c.f.). More conventional is the active path or anodic dissolution mechanism which for clarity is termed 'true corrosion'. In this instance the conjoint action should be similarly referred to as 'true corrosion fatigue' (t.c.f.).

The corrosion fatigue strength of steels resulting from the latter process has been shown to be significantly reduced (Shalon & Raphael 1959). While it is possible to produce corrosion resistant stainless grades in order to overcome this problem, the use of the usual forms of high strength carbon martensitic steels could be somewhat limited.

On considering the markedly increased susceptibility of the higher strength low alloy steels to hydrogen induced crack growth mechanisms, it is perhaps to be expected that these steels are also extremely prone to accelerated failure in corrosion fatigue. Indeed, certain ultra high strength steels show pronounced degradation when subjected to fatigue cycling in a normal 'damp' air environment.

3.6. *Other factors*

The application of high strength steels may entail features other than an appreciation of fracture mechanics. The steel may have to be welded, and the availability of suitable electrodes and procedures to eliminate the risks of weld cracking on heat treatment or subsequent failure at the weld is an important consideration. One of the major features that limits the application of high strength steels is the elastic modulus. The increases in strength that have been attained in steels have generally arisen from a genuine desire for weight saving. The reduced section that can be achieved as a result of the use of higher yield strength steels, however, could result in failure by elastic collapse, because the modulus remains constant while the section size decreases. The modulus cannot be changed to any significant extent in steels, and it is important to ensure that any weight savings made on the basis of design against plastic distortion do not result in elastic collapse. Design considerations have a major part to play in the application of ultra high strength steels from this point of view.

4. Application of high strength steels

From the foregoing remarks it might appear that the applications for ultra high strength steels are somewhat limited. This would not only be a disconcerting conclusion but equally a mistaken one. The factors which can limit the use of these steels must be fully appreciated if the metallurgist is to be able to contribute gainfully to the development of the high strength materials. Progress will only be made if, on the one hand the engineer appreciates the folly of designing sophisticated components solely on the basis of strength. Meanwhile, on the other hand, the metallurgists must use their expertise in collaboration with others to identify the relationships between the microstructure, the operative fracture mechanisms and the properties that are indeed required.

In terms of the ductile fracture process, considerable effort has been directed at the control of inclusion content and of inclusion shape. Further work should be directed towards an understanding of the effects of inclusion size and the spatial distribution of inclusions. Other second phase particles can also influence the ductile fracture process, a typical example being cementite. A clearer understanding of the role of cementite, alloy carbides, and intermetallic second phase particles could prove very useful in developing high strength steels with high levels of toughness. In terms of the cleavage fracture process, obviously the application of grain refinement principles is indicated, but second phase particles may be the important trigger for general cleavage failure if particle cracking occurs. Studies of these aspects may not provide strikingly new alloys, but at least would enable a rationalization of the many existing ultra high strength steel compositions.

This implies not so much a sweeping away of the conventional methods, codes, practices which are the cornerstone of design, but much more the recognition that some flexibility is required in order to accommodate the relatively new but now proven principles of fracture control.

The authors wish to thank Dr K. J. Irvine, Head of Research, Special Steels Division, British Steel Corporation, for permission to publish this paper.

REFERENCES (May *et al.*)

Antolovich, S. D. & Singh, B. 1971 *Metall. Trans.* **2**, 2135–2141.
Austen, I. M. & Walker, E. F. 1975 *Phys. Metall. Centre Rep.* PMC/7422/–/75/A.
Barnby, J. T. 1972 *Fatigue*, M & B Technical Library, TL/ME/4.
Dolby, R. E. & Knott, J. F. 1972 *J. Iron & Steel Inst.* **210**, 857–865.
Dulieu, D. 1974 B.S.C. Conference 'Inclusions and their effects on properties', Leeds.
Elliott, D. & Stuart, H. 1970 Fractographic features associated with general yielding in low strength steels. *Proc. 3rd Annual SEM Symp., Chicago.*
Embury, J. D. & Fisher, R. M. 1966 *Acta Met.* **14**, 147–159.
Forrest, P. G. 1962 *Fatigue of metals.* London: Pergamon Press.
Harris, W. J. 1961 *Metallic fatigue.* London: Pergamon Press.
McEvily, A. J. & Wei, R. P. 1971 *Proc. Int. Conf. on Corrosion Fatigue – Chemistry, Mechanics and Microstructure, Connecticut*, 1971 (NACE), p. 324.
McIntyre, P. 1973*a* The relationships between stress corrosion cracking and sub-critical flaw growth in hydrogen and hydrogen sulphide gases. *Int. Conf. on Stress Corrosion Cracking and Hydrogen Embrittlement of Iron-Base Alloys, Unieux-Firminy, France*, 14–16 June 1973.
McIntyre, P. 1973*b* Influence of environment on crack growth in high strength steels. *1st Physical Metallurgy Centre Conf. on Mechanics and Mechanisms of Crack Growth, Churchill College, Cambridge*, 4–6 April 1973.
McIntyre, P. & Elliott, D. 1973 The role of stretch zone formation in environmentally activated crack growth in steels. *3rd Int. Conf. on the Strength of Metals and Alloys, Cambridge* 20–25 August 1973.
Parker, E. R. & Zackay, V. F. 1973 *Engng Fract. Mech.* **5**, 147–165.
Peckner, D. 1963 *Mater. Design Engng*, March, 101–112.
Razim, C. 1968 *Haterei Technische, Mitterlungen* **23**, 1–8.
Schmatz, D. J., Schaller, F. W. & Zackay, V. F. 1963 *Relation between structure and strength in metals and alloys.* London: National Physical Laboratory, January 1963.
Shalon, R. & Raphael, M. 1959 *Am. Conc. Inst. J.* 1251.

3.8. Nickel base materials developments for high temperatures

By R. B. Nicholson
European Research and Development Centre, International Nickel Limited

[Plate 25]

1. Introduction

The purpose of this paper is to review the development of nickel base alloys for service at high temperatures with particular reference to their use in gas turbines. Although the development has included a broad range of alloys designed for a variety of service conditions, it is characterized by the family of age-hardening compositions known as the 'superalloys'. Other papers in the conference will be devoted to the contribution which these alloys have made to the gas turbine industry. Nonetheless, it is worth remarking here that improvements in the physical metallurgy and processing of the superalloys have made possible an average increase of 10 K per year in operating temperature over the last 35 years; a record unmatched by any other alloy development. When added to the increased gas temperature made possible by integral cooling of components in the turbine, itself greatly assisted by developments in the metallurgy of fabrication and casting, it is not surprising that the engineer has been able to transform the gas turbine from a curiosity into one of the fundamental power generating machines in a period of about 30 years.

2. Early history

Although the phenomenon of age hardening was discovered in an aluminium alloy, it was the classic paper of Merica, Waltenburg & Scott (1920) in which the basic condition for age hardening, a decreasing solid solubility with decreasing temperature, was first recognized. It was therefore logical that Merica should apply the then novel concept to alloys of interest to the nickel industry and the result was K-Monel (Merica 1932) (a precipitation hardened nickel–copper alloy now known as 'Monel' K-500) and some of the earliest patents on the precipitation hardening of nickel base alloys with aluminium (Pilling & Merica 1929). Another early recognition of the possibility of age hardening nickel–chromium alloys was due to Chevenard (1929).

At that time the alloy now known as Inconel† 600 (based on 77 % nickel, 16 % chromium, 7 % iron) - see table 1, was being produced for ambient temperature applications where corrosion resistance was required. Thus it was natural that age hardening modifications of this alloy should be made. Two of the earliest were Inconel W (now known as Inconel 722) hardened by additions of titanium & aluminium and Inconel X (still widely used as Inconel X-750) with a further addition of niobium.

Meanwhile in England, the early development of age hardening nickel alloys was influenced by work on the nickel–chromium heat and oxidation resistant alloys which showed the outstanding characteristics of the 80 % nickel, 20 % chromium composition (Tapsell & Bradley 1925). Thus when Pfeil (Pfeil 1940; Pfeil, Allen & Conway 1952) had the task of developing superior alloys for the Whittle gas turbine, it was again logical that he should turn to age

† Trademark.

hardening versions of the 80/20 Ni–Cr base. Just a few years earlier, Jenkins & Bucknall (1935) working at the National Physical Laboratory under Walter Rosenhain's direction had shown that the creep properties of copper alloys could be improved by age hardening and had made the vital observation that the conditions for optimum hardness did not correlate in any simple way with the conditions for optimum creep properties. The immediate result of Pfeil's work was the development of a modified non-age hardening 80/20 alloy Nimonic† 75 and an age hardening version Nimonic 80, later modified to become Nimonic 80A.

Thus were the distinctive compositions of the American and British families of nickel base superalloys born, the former soon to be overtaken, albeit temporarily, by cast and wrought cobalt alloys, and the latter to form the basis of gas turbine construction in British engines and many foreign engines ever since.

In the context of the present conference, it is worth noting that the origin of superalloys was thus firmly based on a sound knowledge of the physical metallurgy of age hardening alloy constitution and creep deformation. Since that time, the pace of development has often been hectic under the pressure of great national crises and it would be wrong to deny that inspired empiricism has played a major part in the invention of new alloys. However, a careful reading of the authoritative accounts of these developments (Betteridge 1959; Betteridge & Heslop 1974; Sims & Hagel 1972) will show that the physical metallurgy of even the most complex alloys has been well documented and, for the most part, well understood and used to stimulate further progress.

3. Alloy requirements

The requirements for alloys to be used in the turbine section of a gas turbine are among the most stringent met by materials in any environment. As well as the obvious and basic requirement of resistance to rupture under conditions of high temperature and high stress, other mechanical properties such as room temperature ductility, intermediate temperature strength and fatigue resistance, environmental properties such as resistance to oxidation and hot corrosion, physical properties such as low density and compatible thermal expansion and production characteristics such as fabricability and/or castability are required. Every alloy developed is inevitably a compromise between these and other requirements and is usually designed for service in a particular part of a gas turbine, e.g. blades, vanes, disks, or combustion chamber. For a long time wrought alloys led the way, but cast alloys have been highly successful as blade and vane materials, while powder metallurgy techniques are now showing promise in several parts.

The sheer variety of service requirements and alloys developed to meet them makes it impossible to deal with the subject in anything but a superficial way and the reader is referred to other accounts for details (Betteridge 1959; Betteridge & Heslop 1974; Sims & Hagel 1972; Glenny, Northwood & Burwood-Smith 1975).

4. Alloy constitution

The basis of superalloys is a Ni–Fe–Cr austenite (γ). A high nickel content confers improved basic oxidation resistance, improved solubility of alloying elements and better phase stability. The chromium addition gives excellent resistance to oxidation and hot corrosion. Chromium

† Trademark of Henry Wiggin and Co. Ltd.

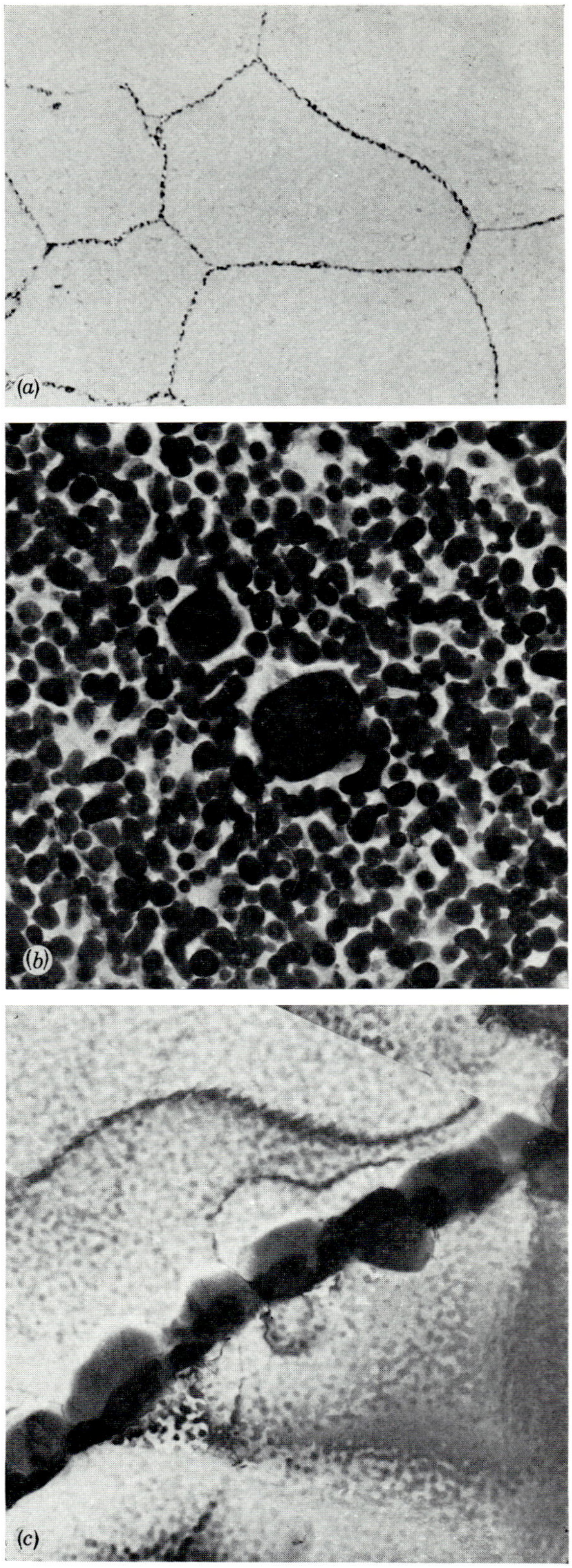

FIGURE 1. Typical microstructures of Nimonic 105 after the following heat-treatment: 4 h at 1150 °C, air cool + 16 h at 1050 °C, air cool + 16 h at 850 °C, air cool: (*a*) optical micrograph X375 showing carbide precipitation at grain boundaries, (*b*) electron micrograph (extraction replica) × 20000 showing γ', (*c*) electron micrograph (thin foil) × 6500 showing grain boundary carbides enveloped in γ' (Betteridge & Heslop 1974).

(*Facing p.* 391)

contents as high as 50 % are beneficial for the latter (Beltram & Shores 1972) but additions in the range 8–20 % are common for the former. Aluminium, as well as being an important hardening element, confers good oxidation and corrosion resistance at higher concentrations than can normally be tolerated in the bulk of the alloy. However, aluminide coatings formed by the diffusion of aluminium contents into the surface of a superalloy component are widely used (Grisaffe 1972). The introduction of a high iron content, even to the extent of making the alloy iron base is generally done for reasons of cost.

The principal hardening phase is γ' Ni_3Al. Despite its nomenclature this is an equilibrium second phase in the binary Ni–Al or ternary Ni–Cr–Al system. It is also formed as a metastable phase in the Ni–Ti and the Ni–Cr–Ti system and is the first formed phase in a wide variety of Ti, Al and Nb containing superalloys. Many of the most important properties of superalloys are due to the γ' phase and its relation with the austenitic matrix. The two phases are almost exactly matching in lattice parameter so that the precipitate particles are normally coherent with the matrix. This confers a uniquely low particle coarsening rate on the γ'/γ system so that the alloy overages very slowly even at temperatures of the order of 0.7 of the melting point. Yet the system is strongly age hardening and has excellent creep resistance because of the coherency strains set up between the precipitates and the matrix and because of the high temperature strength of the precipitate conferred by its ordered crystal structure.

A second metastable phase which occurs in some superalloys is γ^* based on Ni_3Nb. Although this phase is also coherent and ordered, it is less effective than γ' at higher temperatures because it is less stable.

The other major source of high temperature properties in superalloys is solid solution strengthening. The principle elements involved are chromium itself, cobalt (which may be present up to 20 %) and molybdenum, tungsten and tantalum which may be added up to a total of 15 %. These elements also dissolve in the γ' phase and may thereby have a twofold hardening effect.

The major limitations in the addition of all these elements to the austenitic base and the origin of much of the complexity of superalloy development is the balancing of improved high temperature properties with problems of oxidation resistance, fabricability and phase stability – some examples of this balance are considered below.

Finally, creep resistance and creep ductility are determined as much by the structure of the grain boundaries of the alloy as by the structure of the grains (figure 1, plate 25). Broadly speaking insufficient strengthening of the grain boundaries will cause high creep rates and too much hardening will lead to poor creep ductility. The control of grain boundary structure in superalloys is largely practised through precipitation of carbides, principally chromium carbides, but substituted with iron and the refractory metals. The elements boron and zirconium also play an important role in some alloys.

Chemical composition is, of course, only part of the story and the proper development of the whole microstructure can only be achieved by choice of the correct casting, fabricating and heat treatment conditions.

The interrelationship of some of these factors is considered below in the context of the development of a selection of superalloys.

5. Wrought alloys

(a) *Chemistry and heat-treatment*

The development of wrought superalloys is well illustrated by the Nimonic series. Figure 2 shows the improvement in stress rupture properties bought about over a period of 20 years. This improvement was gained by increasing the solid solution strengthening and γ' strengthening of the alloys. Table 1 shows that by comparison with the early alloys Nimonic 75 and 80A, the later alloys have a substantial cobalt content, an increased aluminium and titanium content and, finally, an addition of molybdenum. At the same time the chromium content was approximately halved. The constitutional effect of the changes in cobalt and chromium is to increase the γ' solvus and increase the amount of titanium and aluminium which can be precipitated as γ' by an age hardening treatment. Therefore the alloy retains its strength to a higher temperature and is stronger in creep at any given temperature (figure 3).

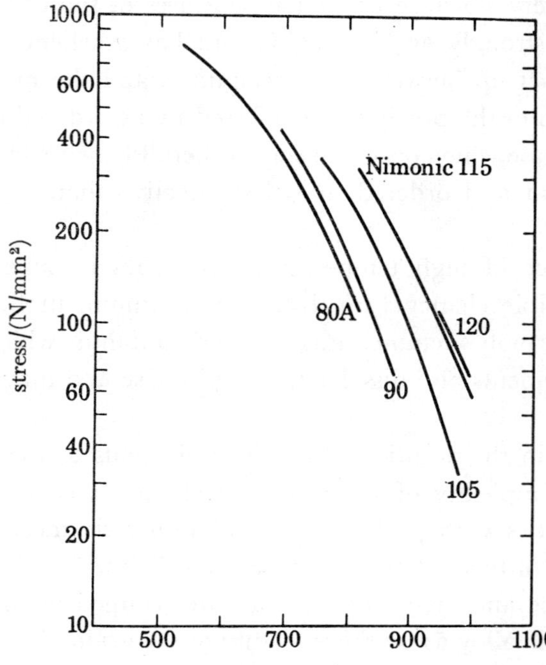

FIGURE 2. Illustrating progress in wrought Nimonic alloys: stress to produce rupture in 100 h as a function of temperature (Betteridge & Heslop 1974).

The penalty in increasing the alloy additions is well illustrated by table 2: a reduction in melting temperature and a drastic reduction in the hot working range. It was this rapid convergence of the γ' solvus temperature and the solidus temperature which prevented any further major development in this line of alloys. More recently a second line of related alloys, Nimonic 81, EPK 55 and EPK 57, has been developed to meet the need for alloys of similar strength to the original Nimonic series but with greater hot corrosion resistance. The latter is achieved by increasing the chromium content by some 10 % over the corresponding lower chromium alloy. Because of the reduced solubility for aluminium and titanium, niobium is added as an additional γ' former to maintain strength.

TABLE 1. TYPICAL COMPOSITIONS OF SOME WROUGHT AND CAST NICKEL BASE SUPERALLOYS

	Cr	Co	Al	Ti	Nb	Mo	W	Ta	Fe	C	B	Zr	others
Inconel 600	15.5	—	—	—	—	—	—	—	8.0	0.08	—	—	—
Inconel 722	15.0	—	0.6	2.5	—	—	—	—	7.0	0.04	—	—	—
Inconel X-750	15.0	0.5 max	0.9	2.5	1.0	—	—	—	7.0	0.04	—	0.03	—
Nimonic 75	19.5	—	—	0.4	—	—	—	—	5.0 max	0.10	0.001 max	0.05 max	—
Nimonic 80A	19.5	2.0 max	1.4	2.5	—	—	—	—	1.0 max	0.10 max	0.005 max	0.10 max	—
Nimonic 90	19.5	18.0	1.5	2.4	—	—	—	—	1.0 max	0.13 max	0.005 max	0.10 max	—
Nimonic 105	14.8	20.0	4.7	1.2	—	5.0	—	—	1.0 max	0.16 max	0.008 max	0.14 max	—
Nimonic 115	15.0	14.8	5.0	4.0	—	4.0	—	—	1.0 max	0.15	0.06 max	0.20 max	—
Nimonic 120	12.5	10.0	4.5	3.5	—	5.7	—	—	1.0 max	0.08 max	0.025	0.05 max	—
Nimonic 81	30.0	2.0 max	0.9	1.8	—	—	—	—	1.0 max	0.05	0.002	0.06	—
EPK 55	28.5	20.0	1.2	2.3	0.75	—	—	—	0.5 max	0.05	0.006	0.07	—
EPK 57	24.3	19.7	1.4	3.0	0.95	1.5	—	—	0.5 max	0.05	0.012	0.05	—
Udimet 500	17.5	16.5	2.9	2.9	—	4.0	—	—	4.0 max	0.15 max	0.010 max	—	—
Udimet 700	15.0	17.0	4.3	3.4	—	5.3	—	—	4.0 max	0.15 max	0.05 max	—	—
IN 713 LC	12.5	—	6.1	0.8	2.2†	4.2	—	—†	0.5 max	0.12 max	0.012	0.10	—
IN 100	10.0	15.0	5.5	4.7	—	3.0	—	—	0.5 max	0.18	0.014	0.06	1.0 V
Mar-M 200	9.0	10.0	5.0	2.0	1.0	—	12.5	—	—	0.15	0.015	0.05	—
Mar-M 246	9.0	10.0	5.5	1.5	—	2.5	10.0	1.5	—	0.15	0.015	0.05	—
IN 738	16.0	8.5	3.4	3.4	0.9	1.8	2.6	1.8	0.5 max	0.17	0.010	0.10	—
MA 753	19.5	2.0 max	1.4	2.5	—	—	—	—	1.0 max	0.10 max	—	—	1.3 Y_2O_3

† Nb+Ta.

Heat-treatment of superalloys is a major subject in itself. As well as a standard heat-treatment which is designed to give the best combination of the desired properties, special heat-treatments to fully develop a specific property are also available. The principal microstructural parameters are grain size, carbide precipitation at grain boundaries and γ' precipitation in the matrix. Whilst the early alloys had relatively simple heat-treatments, e.g. Nimonic 90, 8 h at 1080 °C, air cool plus 16 h at 700 °C air cool, later alloys have three stage or multi-stage treatments with the intermediate stages being designed to give the optimum grain boundary microstructure of carbide particles enveloped in a film of γ' (figure 1). Other stages are designed to give two size ranges of the γ' precipitate for the best combination of strength at both high and intermediate temperatures.

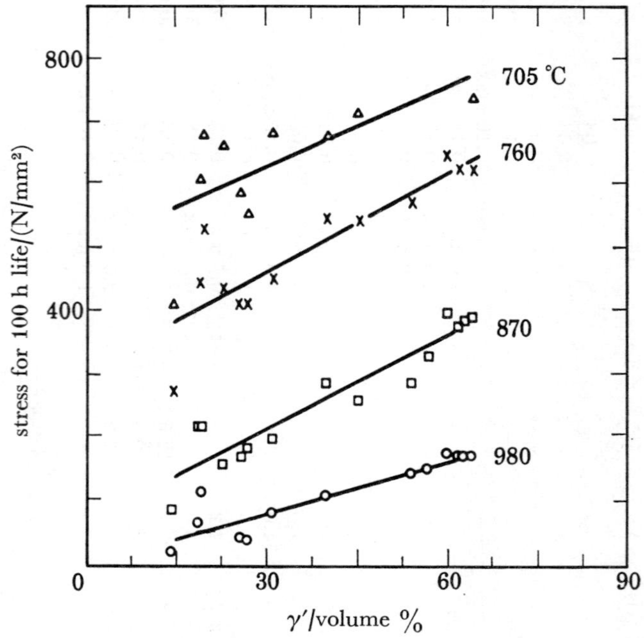

FIGURE 3. Illustrating the relation between stress rupture properties and volume fraction of γ' (Decker 1969) for a variety of nickel base superalloys.

TABLE 2. THE EFFECT OF ALLOY DEVELOPMENT ON MELTING POINT AND HOT WORKING RANGE (BETTERIDGE & HESLOP 1974)

	solidus/°C	maximum pre-heating temperature/°C	minimum working temperature/°C
Nimonic 75	1340	1220	950
Nimonic 80A	1320	1170	950
Nimonic 90	1310	1170	950
Nimonic 105	1290	1220	1000
Nimonic 115	1260	1150	1100

(b) *Composition control via electron theory*

Superalloys are unique among engineering materials in that modern electron theory is regularly used both as an aid to the design of the alloy and for routine production control.

The highly alloyed materials developed in the 1960s sometimes suffered from a drastic embrittlement or loss of creep properties due to the formation of σ phase. This phase, based on the composition FeCr, is one of a group of *topologically* close packed phases (as distinct from the beneficial *geometrically* close packed phases γ and γ^*) which are highly detrimental to superalloy properties. A particular problem with their formation is that they often appear after long times in service although absent in the as-heat-treated alloy.

A technique, since become known as Phacomp (an acronym for phase computation) was developed to predict the propensity of an alloy to form σ (Woodyatt, Sims & Beattie 1966). The technique is based on Pauling's (1938) theory of the disposition of the 3d electron orbitals in the first long period of transition elements. A fraction of the orbitals hybridize with the p and s orbitals to create the metallic bond while the remainder form non-binding orbitals which partly fill the available electron holes in the d shell. The number of non-binding orbitals increases through the series Cr, Mn, Fe, Co, Ni, giving 'electron hole numbers', N_v as follows:

Cr 4.66 Mn 3.66 Fe 3.66 Co 1.71 Ni 0.66

Das, Rideout & Beck (1952) showed that the γ/σ phase relation depended on the average electron hole number $\bar{N}_v$ given by

$$N_v = \sum_{i=1}^{n} m_i (N_v)_i$$

where m_i is the atomic fraction of the ith element of electron hole number N_v and n is the number of elements in the alloy. The limit of stability of the γ phase is reached at $\bar{N}_v \approx 2.50$ depending on precise composition. Alloys such as Nimonic 115 and Udimet† 700 were among the first to be designed using Phacomp and quality control in their production uses this principle.

In making Phacomp calculations, it is important to recognize that it is the *matrix* composition which determines the propensity for σ formation so that the removal of σ-resisting elements in γ' such as Ni and σ-promoting elements such as Cr, Mo and W in carbides must be taken into account (Sims 1972). Compositional changes may continue to occur over long periods of service – hence the possibility of σ forming after, say, 3000 h service in an alloy which is initially σ free.

(*c*) *Property optimization via trace element control*

Superalloys are particularly susceptible to property changes from the introduction of elements present in minute quantities, typically 1–100 parts/10^6 compared with the additions of 0.1–30 % of the elements so far considered. As with most trace element phenomena, the reason for the remarkable effect of such small additions is that the elements are concentrated at the grain boundaries where, for example, they affect the morphology and stability of the grain boundary carbides (Decker & Freeman 1960). Table 3 shows the effect of boron and zirconium additions to Udimet 500 – similar results are obtained in most superalloys so that boron and zirconium have become standard additions in all alloys in quantities which have to be tailored to meet the main chemistry of each alloy.

Other trace elements are highly detrimental to the properties of superalloys. For example, lead and bismuth greatly reduce the hot workability of the alloy (figure 4) and also affect the stress rupture life and ductility (Wood & Cook 1963). Control of harmful trace additions

† Trademark of Special Metals Inc.

is best obtained by careful selection of raw materials or the use of vacuum refining techniques; more recently both methods have been employed for optimum results.

(*d*) *Cobalt alloys*

The negligible use of wrought cobalt alloys in gas turbines is due to the fact that, despite the 70 K higher melting point, they are invariably weaker than their nickel base counterparts. There is a simple metallurgical reason for this: cobalt alloys can be strengthened by solid solution strengthening and carbide precipitation (Morral *et al.* 1969) but to date nobody has been successful in finding a hardening precipitate in cobalt base alloys which has the many charms and few vices of γ' (Sullivan 1970).

TABLE 3. EFFECT OF BORON AND ZIRCONIUM ON CREEP OF UDIMET 500 AT 870 °C (SIMS & HAGEL 1972)

alloy	life at stress of 170 N/mm² (h)	elongation (%)	stress for minimum secondary creep rate of 0.004 %/h (N/mm²)
base	50	2	120
+0.19% Zr	140	6	157
+0.009% B	400	8	190
+0.009% B+0.01% Zr	647	14	232

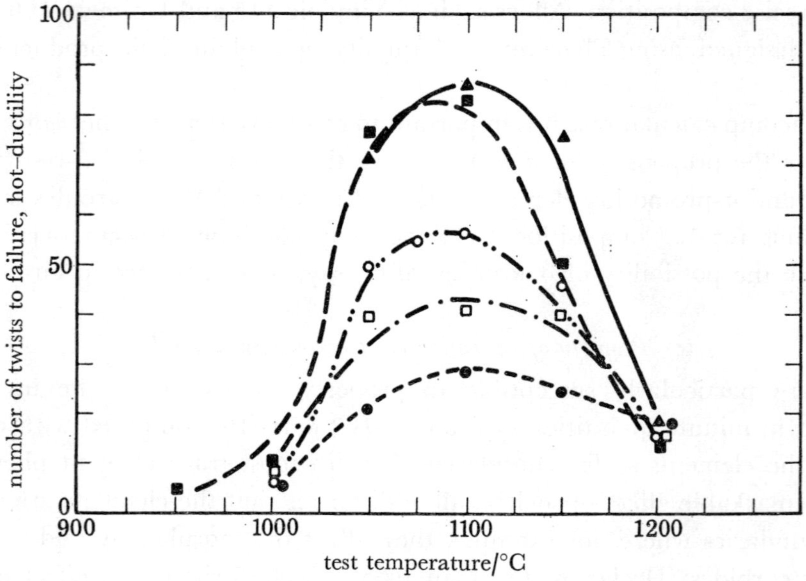

FIGURE 4. Illustrating the impurity problem in superalloys: the effect of lead content on torsion ductility of Nimonic 105 (Betteridge & Heslop 1974). Actual Pb content (% by mass): ▲——▲, < 0.0005; ■ — — ■, 0.0005; ○—·—○, 0.0014; □—··—□, 0.0029; ● - - - ●, 0.0110.

6. CAST ALLOYS

In contrast to the situation with wrought alloys, cobalt base superalloys held an early advantage in comparison with nickel base superalloys. Historically this seems to have been due to the existence of alloys based on the cobalt–chromium–molybdenum system (which had been developed for dental and medical purposes) at a time when the first American gas turbines were being constructed. Cast alloys also offered the possibility of high production rates and low conversion costs. Later these alloys were replaced by wrought nickel base alloys as a result of the superior properties developed in the latter.

In the last decade there has been a resurgence of interest in cast alloys and a large number of new nickel base compositions have been developed and successfully produced in large tonnages. The main reason for this change has been the apparent lack of further development potential in the wrought alloys. The movement to cast components at once removes problems of hot workability (most cast alloys cannot be heat-treated to give a single phase structure). Some γ/γ' eutectic is formed on solidification and the volume fraction of γ' can be as high as 60% giving a corresponding improvement in high temperature properties (figure 3). Table 1 shows the compositions of some typical cast alloys such as IN-713, IN-100, MAR-M†200, MAR-M246 and IN-738. Their chemical compositions show significantly increased titanium and aluminium contents compared with the wrought alloys, plus major additions of tungsten and other refractory metals.

In terms of properties, the cast nickel base alloys have allowed a further 50–75 K advance in metal temperature compared with the best wrought alloys. The advance in operating temperature has been greater than this, however, because of the opportunities offered by improved casting technology for the incorporation of cooling passages in cast components. While this feature was first developed in wrought alloys more than 20 years ago, it has reached its greatest sophistication in the modern cast blade or vane.

7. New developments

Despite the high level of technology reached in wrought and cast nickel base superalloys, the pace of development shows few signs of slackening. Several recent advances deserve mention.

(*a*) *Grain size control*

The importance of grain size control in both wrought and cast alloys cannot be overemphasized – in nickel base superalloys the whole range of grain sizes from ultrafine 10 μm grains to single crystals is encountered.

In wrought alloys, it is grain refinement which is important particularly for components such as disks which operate at intermediate temperatures. Grain refinement leads to improved tensile and fatigue properties and is usually achieved by some type of thermomechanical processing. For example the 'Minigran'‡ process (U.A.C. 1972) involves mechanical working below a precipitate solvus temperature but above the matrix recrystallization temperature so that grain growth after recrystallization is impeded by the precipitates.

Grain refinement is also important in conventional cast alloys where it again leads to improved tensile properties. Grain size control is achieved by mould coatings or, more recently, by inoculation techniques (Miles 1974). At the other end of the spectrum Ver Snyder and co-workers (Piearcey & Ver Snyder 1966; Kear & Piearcey 1967) have obtained large columnar grains and even single crystals in a variety of alloys. This is an astonishing achievement in alloys as complicated as superalloys and in components as irregular in shape as turbine

† Trademark of the Martin Metals Company.
‡ Trademark of United Aircraft Corporation.

blades and vanes. Current results suggest that a 30–80 K advantage is obtained over conventionally cast alloys with a major improvement in rupture ductility (Ver Snyder 1970). Clearly the absence of transverse grain boundaries in directionally solidified alloys and any grain boundaries at all in single crystal alloys offers a means of escaping from the compromise between matrix strength and grain boundary ductility which is inevitably faced in all modern superalloys.

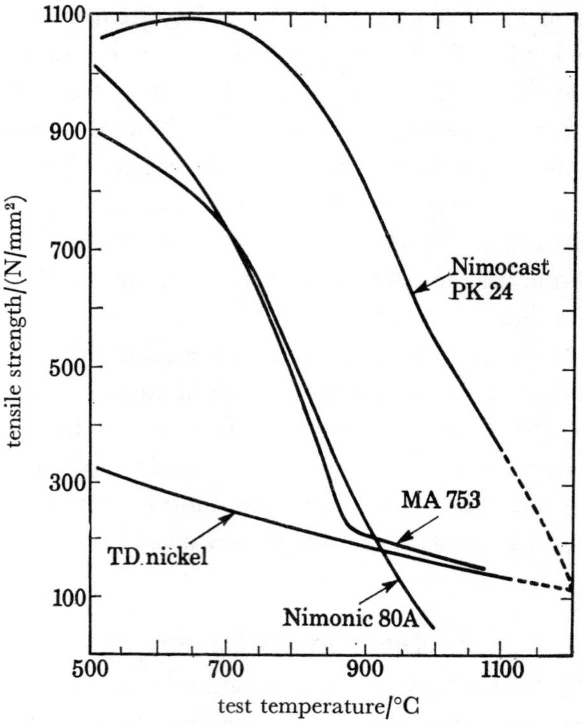

FIGURE 5. Progress in superalloys: tensile strength versus temperature for a typical wrought alloy (Nimonic 80A), a typical cast alloy (Nimocast PK 24 or IN–100), a simple dispersion hardened alloy (TD Nickel) and a combined precipitation hardened/dispersion hardened alloy (MA 753) (Betteridge & Heslop 1974).

(*b*) *Dispersion hardening*

Another development is the use of dispersion hardening to supplement precipitation hardening by γ' in wrought alloys. All γ' hardened alloys show reduced strength at high temperatures because of the solution of the γ' precipitate, whereas alloys hardened by an insoluble dispersion of, say, an oxide, do not have this characteristic. Early dispersion hardened alloys such as thoria-dispersed nickel ('TD nickel') showed good high temperature properties but poor intermediate temperature properties, because the γ' forming elements could not be introduced through the chemical reduction techniques used for making the powders which are the feedstock of these dispersion containing alloys (Anders, Alexander & Wartel 1962). Recently a high energy impact milling process known as mechanical alloying (Benjamin 1970) has allowed the production of alloys which combine the best features of γ' hardened and dispersion hardened materials (figure 5). This powerful combination offers great opportunities for the designer.

(*c*) *Powder metallurgy*

Dispersion hardening is a special example of powder metallurgy techniques which are already generally important and look like becoming more widely used in the future. The use of inert gas atomized powder of conventional alloys gives freedom from macrosegregation which is a major problem in large superalloy ingots (Hopkins 1974). The technique allows improved workability and better intermediate temperature properties in ingots made by the compaction of atomized powder. Advances in hot isostatic pressing have allowed ingots of great structural integrity to be made from powder. Currently the powder production route is used primarily for large components such as disks but it is possible that it will also allow new developments in blade and vane alloys thus improving the competitiveness of wrought products against cast products in this area. Interestingly, cast products may themselves be improved by using one step in this powder route, hot isostatic pressing of the casting, which could allow their structural integrity to approach that of wrought products.

Obviously it is too early to say which of these many novel approaches will become the standard practice of the future but it is clear that the metallurgy of superalloys will continue to advance and hence provide new opportunities for the gas turbine engineer to design more efficient and powerful turbines.

References (Nicholson)

Anders, F. J., Alexander, G. B. & Wartel, W. S. 1962 *Metal Prog.* **84**, 88.
Beltram, A. M. & Shores, D. A. 1972 In Sims & Hagel (1972).
Benjamin, J. S. 1970 *Met. Trans.* **1**, 2943.
Betteridge, W. 1959 *The Nimonic Alloys*. London: Edward Arnold.
Betteridge, W. & Heslop, J. (ed.) 1974 *The Nimonic Alloys*. London: Edward Arnold.
Chevenard, P. 1929 *Comptes Rendus* **189**, 846.
Das, D. K., Rideout, S. P. & Beck, P. A. 1952 *Trans A.I.M.E.* **194**, 1071.
Decker, R. F. 1969 *Steel strengthening mechanisms*. Zurich: Climax Molybdenum Co.
Decker, R. F. & Freeman, J. W. 1960 *Trans. A.I.M.E.* **218**, 277.
Glenny, R. J. E., Northwood, J. E. & Burwood-Smith, A. 1975 *Met. Rev.* **20**, 1.
Grisaffe, S. J. 1972 In Sims & Hagel (1972).
Hopkins, D. J. 1974 In Betteridge & Heslop (1974).
Jenkins, C. H. M. & Bucknall, E. H. 1935 *J. Inst. Metals* **57**, 141.
Kear, B. H. & Piearcey, B. J. 1967 *Trans. A.I.M.E.* **239**, 1209.
Merica, P. D. 1932 *Trans. A.I.M.E.* **99**, 13.
Merica, P. D., Waltenburg, R. G. & Scott, H. 1920 *Trans. A.I.M.E.* **64**, 41.
Miles, A. G. 1974 In Betteridge & Heslop (1974).
Morral, F. M., Harbraken, C., Coutsouradis, D., Drapier, J. M. & Urbain, M. 1969 *Met. Eng. Q.* **9**, 1.
Pauling, L. 1938 *Phys. Rev.* **54**, 899.
Pfeil, L. B. 1940 U.K. Patent 583162.
Pfeil, L. B., Allen, N. P. & Conway, C. G. 1952 *Iron and Steel Inst. Spec. Rep.* **43**, 37.
Piearcey, B. J. & Ver Snyder, F. L. 1966 *J. Aircraft* **3**, 390.
Pilling, N. B. & Merica, P. D. 1929 U.S. Patents 2048163–2048167.
Sims, C. T. 1972 In Sims & Hagel (1972).
Sims, C. T. & Hagel, W. C. 1972 (ed.). *The superalloys*. New York: John Wiley and Sons.
Sullivan, C. P. 1970 (ed.) *Cobalt-based superalloys*. Brussels: Cobalt Information Centre.
Tapsell, H. J. & Bradley, J. 1925 *Engineering* **120**, 614, 648, 746.
U.A.C. 1972 United Aircraft Corporation Patent U.S. 3. 660, 177.
Ver Snyder, F. L. 1970 2nd International Powder Conference on Strength of metals and alloys, p. 1011, *American Society for Metals*.
Wood, D. R. & Cook, R. M. 1963 *Metallurgia* **67**, 109.
Woodyatt, L. R., Sims, C. T. & Beattie, H. J. 1966 *Trans. A.I.M.E.* **236**, 519.

3.9 Titanium alloys for engineering structures

By P. H. Morton

Imperial Metals Industries (Kynoch) Ltd, P.O. Box 216, Kynoch Works, Birmingham B6 7BA

[Plates 26 and 27]

The range of properties obtainable in titanium alloys derives from the use which is made of the β–α phase transformation, and alloying elements are classified according to their effect on the transformation temperature. The relation between composition and heat treatment on the one hand and the resultant microstructure and mechanical properties on the other hand are considered. In addition to the commonly used 'general purpose' alloy Ti–6Al–4V, more advanced alloys have been developed for three main applications, namely high strength forging alloys, creep resistant alloys and sheet alloys. For each type of alloy a different balance of material properties is required and the process of optimizing the alloy composition and heat treatment to give the best balance in each case is discussed. Factors affecting the cost of titanium alloys are outlined and consideration is given to the likely trends of titanium alloy development in the future.

Alloy development has produced a wide range of titanium alloys noted for their high strength/density ratio, elevated temperature properties up to 550 °C and corrosion resistance. However, as titanium alloy manufacture involves sophisticated melting and fabrication techniques and because metal scrap generated during manufacturing is not readily recycled, titanium alloys are expensive when compared with other engineering materials. As a result, titanium alloys tend only to be used in applications where high performance is required and high strength to weight ratio is of prime importance – for example, in gas turbine aero engines, airframe structures and in general engineering applications such as centrifuges where higher rotational speeds can be obtained by minimizing centrifugal stresses.

The relatively high cost of titanium alloy manufacture arises principally as a result of the metal's high reactivity and affinity for the elements, O, N, C and H. The solubility of these interstitial elements in titanium is greater by several orders of magnitude than in other commonly used metals, as is shown in table 1, and poor ductility arises if the interstitial content rises too high. Because titanium will reduce any known carbide or oxide refractory, melting has to be carried out in a copper crucible, cooled by water or liquid metal (Na/K). Likewise because titanium will react with any gaseous environment containing O, N, C or H, melting and hot processing have to be carried out in vacuum or inert gas if contamination is to be avoided. In practice, hot processing is normally carried out in air and surface contamination is subsequently removed by descaling and pickling; the oxide film that forms during processing dissolves progressively into the underlying metal, giving it a hard and brittle surface up to around 0.2 mm thick, which has to be pickled or machined off before further fabrication. Any hydrogen that is absorbed diffuses very rapidly through the metal and can only be removed by subsequent vacuum degassing so that it is advisable to use an oxidizing atmosphere and to avoid gas fired furnaces. The melting and fabrication of titanium alloys is thus relatively expensive. Clean scrap can be recycled, but there is no simple way of recycling oxygen contaminated material and, as a consequence, only a proportion of the titanium scrap arising from its processing is in practice recycled into titanium or titanium alloys.

TABLE 1. SOLUBILITY AT ROOM TEMPERATURE OF O, N, C AND H IN Ti, Fe, Cu AND Al

	interstitial element			
metal	oxygen	nitrogen	carbon	hydrogen
titanium	14.5 mass %	~20 mass %	0.5 mass %	100 parts/10^6
iron	~1 part/10^6	<5 parts/10^6	100 parts/10^6	<1 part/10^6
copper	<5 parts/10^6	<1 part/10^6	<1 part/10^6	<1 part/10^6
aluminium	<1 part/10^6	<1 part/10^6	<1 part/10^6	<1 part/10^6

Principles of heat treatment

The wide range of mechanical properties attainable in titanium alloys stems from the use which is made of the allotropic transformation from the high temperature β (b.c.c.) phase to the lower temperature α (h.c.p.) phase. The fundamental effect of alloy additions is to change the shape and extent of the two phase α plus β field. Elements can therefore be categorized as: α stabilizers such as aluminium, tin and oxygen which raise the transformation temperature; β isomorphous additions such as vanadium and molybdenum which depress the transformation temperature; β stabilizing eutectoid formers such as chromium and iron; and the neutral element zirconium.

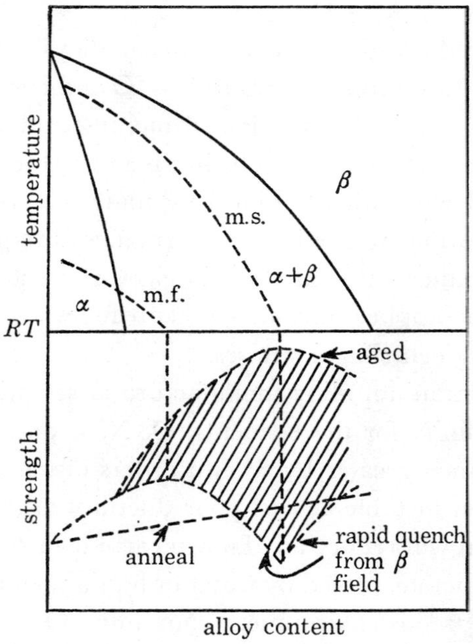

FIGURE 1. Schematic diagram of heat treatment of binary β-isomorphous alloy system.

The principles of heat treatment of titanium alloys can be understood by considering the simplified case of a binary alloy containing a β stabilizer addition (Goosey 1969). The effect of such an element on the β to α transformation is shown in figure 1, which also includes a schematic diagram of the trends in tensile strength that result from different heat treatment procedures. From figure 1 it can be seen that the strength of annealed material will increase as the alloy content or percentage β phase increases.

However, for material rapidly cooled from the β phase field, a more complex strength–composition relation exists. This relation is dependent on the transformation of β to α that can occur in α–β alloys, producing a supersaturated solution of α, sometimes referred to as α'.

When the alloy addition is slight, some strengthening can be obtained by the martensitic transformation, but the increase in strength is very much less than that found in ferrous materials. The maximum strength obtainable from the martensitic transformation occurs at the composition having M_f at room temperature. Increasing alloy content above this level results in a progressive increase in the amount of metastable β that is retained on quenching from the β field, a change in structure that is accompanied by a gradual reduction in as-quenched strength. The latter decreases to a minimum at the composition when M_s is at room temperature, i.e. the lowest alloy content at which 100 % metastable β can be retained by rapid cooling. This composition leads to a maximum strength on ageing, the reaction being decomposition of metastable retained β to form α plus enriched β.

General purpose 'α plus β' alloy, Ti–6Al–4V

The most commonly used titanium alloy is Ti–6Al–4V (Ti 318) which is used extensively as an 'all purpose' alloy in both the U.S.A. and Europe for aerospace applications. The alloy contains modest quantities of both α stabilizer (aluminium) and β stabilizer (vanadium) and combines a minimum tensile strength of 960 N mm^{-2} with good creep resistance up to about 380 °C. It finds use in forgings and in sheet, rod and wire form.

The as cast ingot contains coarse primary alpha and, after initial breakdown, further hot working in the α plus β phase field is necessary in order to break up the structure and distribute the α phase in finely divided form; if insufficient work is put into the structure it contains long primary α particles which tend to delineate the prior β grain boundaries and lead to reduced ductility. In most applications the hot working is followed by annealing at 700 °C which produces a structure of α phase containing particles of retained β as shown in figure 2, plate 26. Typical properties are given in table 2.

For certain specific applications, such as fasteners (see figure 4, plate 26) and compressor disks, additional strength is obtained by heat treatment. This consists of a solution treatment

TABLE 2. COMPOSITION AND TYPICAL ROOM TEMPERATURE TENSILE PROPERTIES OF Ti–6Al–4V AND HIGH STRENGTH FORGING ALLOYS

composition	heat treatment	0.2 % proof stress N/mm^2	u.t.s. N/mm^2	elongation on 5D %	reduction in area %
(Ti 318) Ti–6Al–4V	annealed	930	990	15	40
(Ti 318) Ti–6Al–4V	solution treated + annealed	950	1030	14	37
(Ti 550) Ti–4Al–2Sn–4Mo–0.5Si	solution treated + aged	1000	1100	14	43
(6–6–2) Ti–6Al–6V–2Sn–0.7(Fe, Cu)	solution treated	1044	1134	16	48
(6–2–4–6) Ti–6Al–2Sn–4Zr–6Mo	solution treated	1120	1240	12	34
(Ti 551) Ti–4Al–4Sn–4Mo–0.5Si	solution treated + aged	1200	1310	13	40
(Ti 680) Ti–11Sn–2¼Al–4Mo–0.2Si	solution treated + aged	1190	1310	15	48

high in the α plus β phase field and rapid quenching, followed by an annealing or ageing treatment. The resultant structure consists of equiaxed α grains separated by areas of fully transformed β phase. Typical microstructure is shown in figure 3, plate 24 and properties in table 2.

High-strength forging alloys

For applications such as engine mounting brackets (see figure 5, plate 24), undercarriage components, flap tracks and engine compressor disks, alloys with greater strength are required and a number of alloys have been developed for the purpose. In these alloys greater use must be made of the heat treatment potential of titanium alloys. In terms of the simplified binary system shown in figure 1, it is necessary to increase the β stabilizer content to that at which M_f is substantially below room temperature. It might be thought that the optimum could be achieved if sufficient β stabilizer is added to bring M_s down to room temperature. Such alloys, the so-called metastable β alloys, are indeed capable of being heat treated to the greatest strength of known titanium alloys, but in practice their tensile ductility is poor except in thin sections and their use is consequently restricted to sheet applications and fasteners. For normal high strength forging applications, 'α plus β' alloys are therefore used in which the amount of β stabilizer is chosen to be intermediate between that of Ti–6Al–4V and that of the metastable β alloys.

The high strength α plus β alloys therefore all contain a sufficient quantity of β stabilizer to provide the heat treatment capability which results from ageing of transformed β, as well as substantial quantities of α stabilizers aluminium (and/or tin) to provide solid solution strengthening of the α phase. A typical example of this development is given by the I.M.I. alloy Ti 550 (Ti–4Al–2Sn–4Mo–0.5Si) and other examples include 6–6–2, 6–2–4–6, Ti 551 and Ti 680 (see table 2).

The principal properties which had to be optimized in developing an alloy of this type were, in order of importance:

High specific strength (i.e. strength/density ratio) at room temperature.

Retention of this strength to moderate temperatures (400 °C).

Adequate ductility.

Forgeability.

Depth hardenability.

In order to achieve high specific strength, molybdenum has nearly always been chosen as the β stabilizer element. Tungsten is equally potent but suffers from high density; vanadium is a less potent strengthener and its effect drops off rapidly with increase in temperature; and other β stabilizers either lose their strengthening effect at elevated temperatures or form eutectoids which reduce heat treatability and ductility.

Further strengthening is obtained in Ti 550 by addition of a small quantity of silicon. Addition of silicon to this and other alloys was originally made on the basis that fine particles of silicide might give a measure of precipitation strengthening and improve creep resistance. In practice the addition was found to be beneficial, but subsequent studies have shown that although the solubility in the α phase is quite low, it was nevertheless acting principally as an α solid solution strengthener. Excess silicon not in solution forms small silicide particles which have relatively little effect.

Forgeability is difficult to predict and in the alloy development stage is generally optimized empirically, but it is recognized to be primarily a function of the extent of the α plus β range. The upper transus temperature should be sufficiently high to enable extensive forging to be

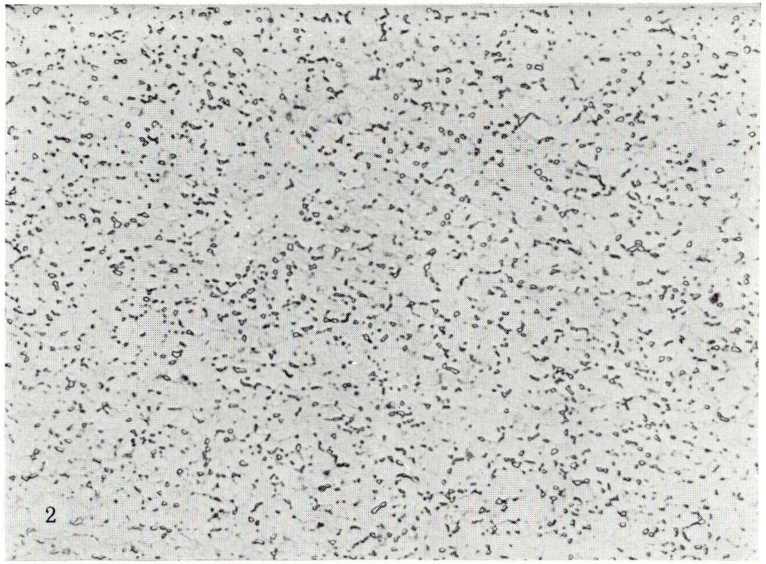

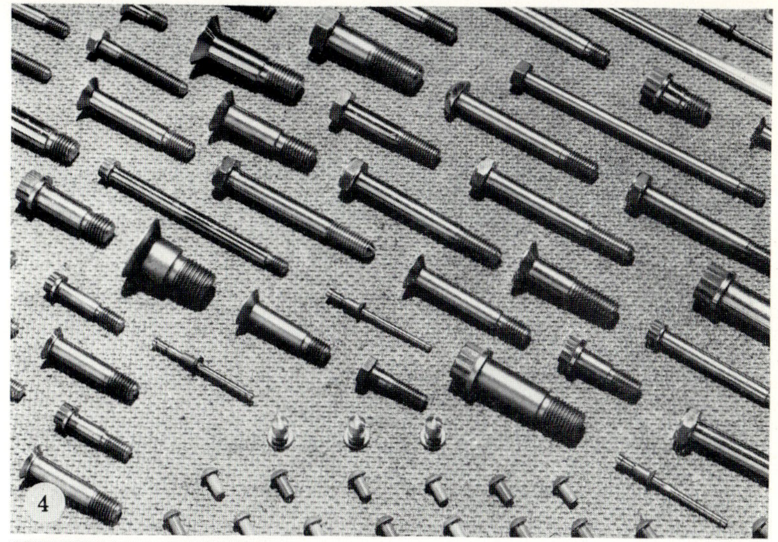

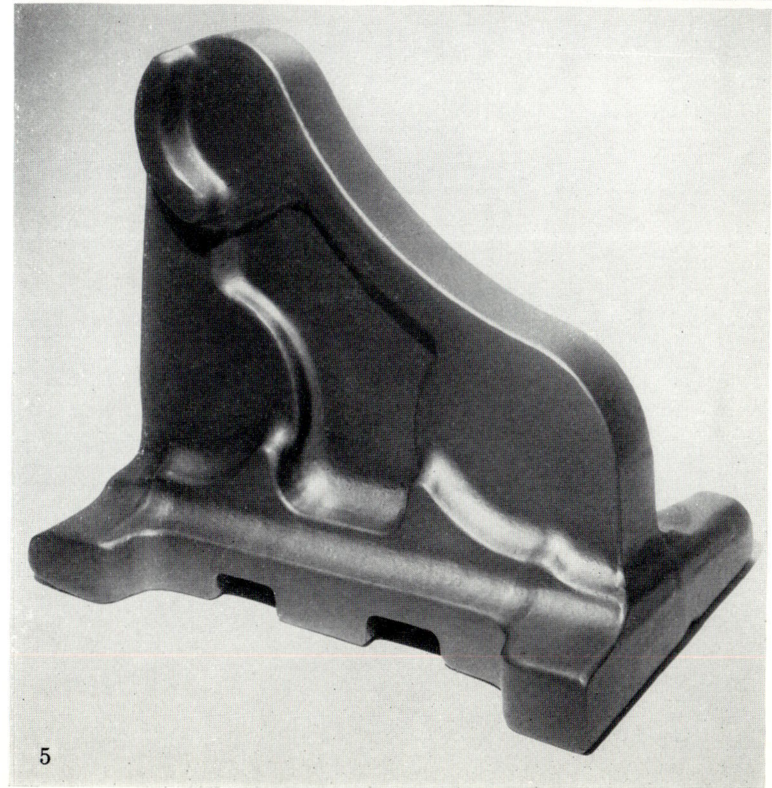

FIGURE 2. Microstructure of annealed Ti–6Al–4V (Magn. ×400).

FIGURE 3. Microstructure of solution heat treated and annealed Ti–6Al–4V (Magn. ×200).

FIGURE 4. A selection of fasteners made in Ti–6Al–4V.

FIGURE 5. Engine-mounting bracket forged by Cameron Iron Works for Rolls-Royce RB 211 engine in Ti 550. Piece weighs about 40 kg. (Courtesy Rolls-Royce (1971) Ltd.)

(Facing p. 404)

FIGURE 6. Compressor disk for Rolls Royce 'Spey' engine in Ti 679. (Courtesy Rolls-Royce (1971) Ltd.)

FIGURE 7. Microstructure of Ti 685 after β heat treatment and ageing (Magn. $\times 50$).

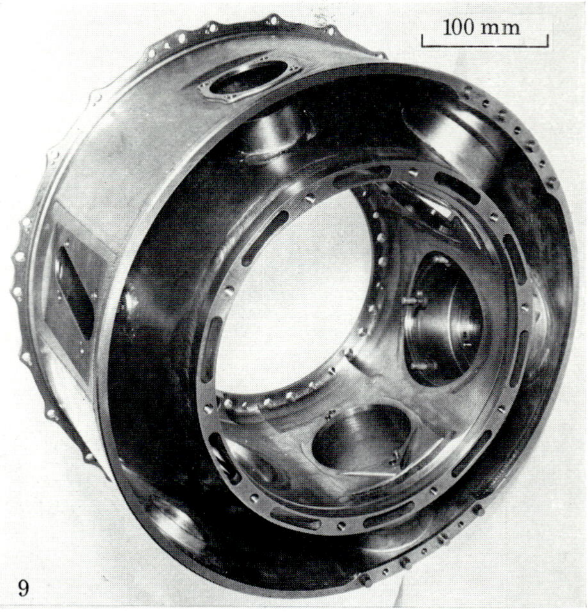

FIGURE 9. Inner unit of intermediate casing of the Rolls-Royce/SNECMA Olympus engine, made in Ti 230. (Courtesy Rolls-Royce (1971) Ltd.)

carried out in the α plus β field, because heating into the β range causes β grain growth and lower ductility.† The α plus β range must be wide enough to enable forging to continue to lower temperatures without any martensitic transformation.

The property of depth hardenability, i.e. the thickness of section in which high strength can be obtained by solution treating and ageing, is largely a function of the β stabilizer content. In general, the greater the amount of molybdenum the better the depth hardenability.

Two other properties of great importance to the designer of high performance structures are fatigue resistance and fracture toughness. Both are dependent on detailed microstructure and can be significantly improved by choice of a suitable thermomechanical treatment. Numerous specialized heat treatments have been reported and/or used which improve these properties but as yet insufficient is known to define in any detail the mechanisms involved. In general the best fatigue properties are obtained in fine grained α plus β structures with a relatively high proportion of primary α. Fracture toughness, on the other hand, is improved by solution treatment high in the α plus β field to give a relatively low proportion (10–25 %) of primary α and further improvements are possible by detailed attention to the form of the surrounding transformed β.

Increasing attention is being given to the decrease in fracture toughness which occurs when salt solution is present at the base of a fatigue crack, the so called $K_{1s.c.c.}$. As yet, however, there is no known example of a titanium alloy failure caused by cracking under these circumstances, and there is some doubt as to the significance of the laboratory tests.

Looking to the future, alloys of still higher strength can certainly be expected to be used. Several such alloys have already been developed, for example Ti 551 (Ti–4Al–4Sn–4Mo–0.5Si), and 6–2–4–6 (Ti–6Al–2Sn–4Zr–6Mo), etc., but they are not widely used because of rather low fracture toughness. Greater understanding of the influence of optical and electron-optical microstructure and more detailed study of the micro segregation of elements within these structures will lead to improved fracture and fatigue properties and this should open the way to more widespread use of very high strength alloys. Basic studies which could usefully contribute to this end include measurement of continuous cooling curves, which have been very little used so far in the study of titanium metallurgy, and clarification of the martensite transformation temperatures and transformation mechanisms in complex alloys. Studies of the various possible martensite transformations in titanium alloys have been almost entirely confined to simple binary alloys and remarkably little is known about the commercially important compositions. Totally new compositions of alloys seem unlikely to be needed although minor changes may well be made and greater use may perhaps be made of β eutectoid formers not used in current alloys.

Creep resistant alloys

In order to obtain maximum performance and fuel economy in gas turbine engines, the operating temperatures of the h.p. compressor has to be raised to as high a temperature as possible. At these temperatures creep resistance becomes paramount, whereas room and elevated temperature tensile strength become less important. The α plus β alloys so far described do not retain creep resistance much above 400 °C. To attain higher temperatures it is necessary to use one of the alloys with lower β stabilizer content, the so-called 'near α' alloys (see table 3). These can be divided into two families according to the heat treatment used.

† An important exception is that β forging is found to improve one specific aspect of fatigue behaviour (a flatter curve in the Goodman diagram) which is of importance in helicopter rotor applications.

TABLE 3. COMPOSITIONS AND TYPICAL ROOM TEMPERATURE TENSILE PROPERTIES OF CREEP RESISTANT, NEAR α ALLOYS

composition	heat treatment	0.2 % proof stress N/mm²	u.t.s. N/mm²	elongation on 5D %	reduction in area %
(Ti 679) Ti–11Sn–2¼Al–5Zr–1Mo–0.25Si	α+β solution treated+aged	1120	1160	15	40
(6–2–4–2) Ti–6Al–2Sn–4Zr–2Mo–0.1Si	α+β solution treated+aged	1020	1100	13	35
(8–1–1) Ti–8Al–1Mo–1V	α+β solution treated+aged	900	1000	17	38
(Ti 685) Ti–6Al–5Zr–0.5Mo–0.25Si	β solution treated+aged	900	1020	11	21

(i) *α plus β heat treated*

The first commercial alloy specifically designed to meet creep requirements was the I.M.I. alloy Ti 679 (Ti–11Sn–2¼Al–5Zr–1Mo–0.25Si), which has been used for many years for the construction of h.p. compressor disks (see figure 6, plate 25), and compressor blades operating at temperatures up to 450 °C.

Development of this alloy took place in three well defined stages. The first stage was determination of the maximum amount of α stabilizer (and of the neutral element zirconium) which could be added without severe loss of ductility. The second was addition of sufficient β stabilizer to make the alloy heat treatable and forgeable, without reducing the creep resistance at maximum temperatures by too much. Thirdly, sufficient silicon was added to further boost the strength and creep resistance, without excessive precipitation of silicide phases. At this stage of titanium alloy metallurgy it was taken for granted that the heat treatment should be similar to that for the forging alloys already described, namely solution treating in the α plus β phase field followed by an ageing, or more strictly a stabilizing anneal, at lower temperatures.

It had been well known for some time that if a titanium alloy contained more than about 8 % aluminium – or a corresponding amount of tin – an ordering reaction took place with the formation of Ti_3Al or Ti_3Sn and a consequent loss in ductility. It was subsequently discovered, however, that even alloys with slightly lower α stabilizer content became embrittled after exposure at temperatures around 500 °C. The tensile ductility of material after being stressed at elevated temperatures (the post creep ductility) was found to be significantly lower than before creep testing. No ordering could be detected from X-ray diffraction patterns, but the advent of thin foil transmission electron microscopy showed that this form of instability was the result of short range ordering which took place during creep testing. Establishment of the upper limit of α stabilizer content to avoid this phenomenon led to the now well known empirical formula for the stability/ordering parameter, namely:

$$\mathrm{Al}+\frac{\mathrm{Sn}}{3}+\frac{\mathrm{Zr}}{6}+10\times \mathrm{O_2}\ (\text{mass }\%)$$

and it was concluded that this parameter must be less than about 8 for complete freedom from ordering. In Ti 679 (containing 0.15 % O_2) the value of this parameter is 7.9.

The β stabilizer content of Ti 679 was chosen as 1 % Mo, at which level the increase in elevated temperature tensile strength outweighs the slight inherent reduction in creep resistance but the amount of β stabilizer is still sufficient to make the alloy forgeable and heat treatable. The silicon addition was optimized at 0.2 %; since the maximum solubility of silicon in the α phase is approximately 0.15 %, this leaves only a small amount out of solution.

Although this alloy achieved the maximum creep resistance possible at that time, it nevertheless forfeited to a certain extent its forgeability and density, and later alloys of the same general type have attached more importance to these two properties. Low density is, of course, particularly important in a rotating part. The best known example of such an alloy is the

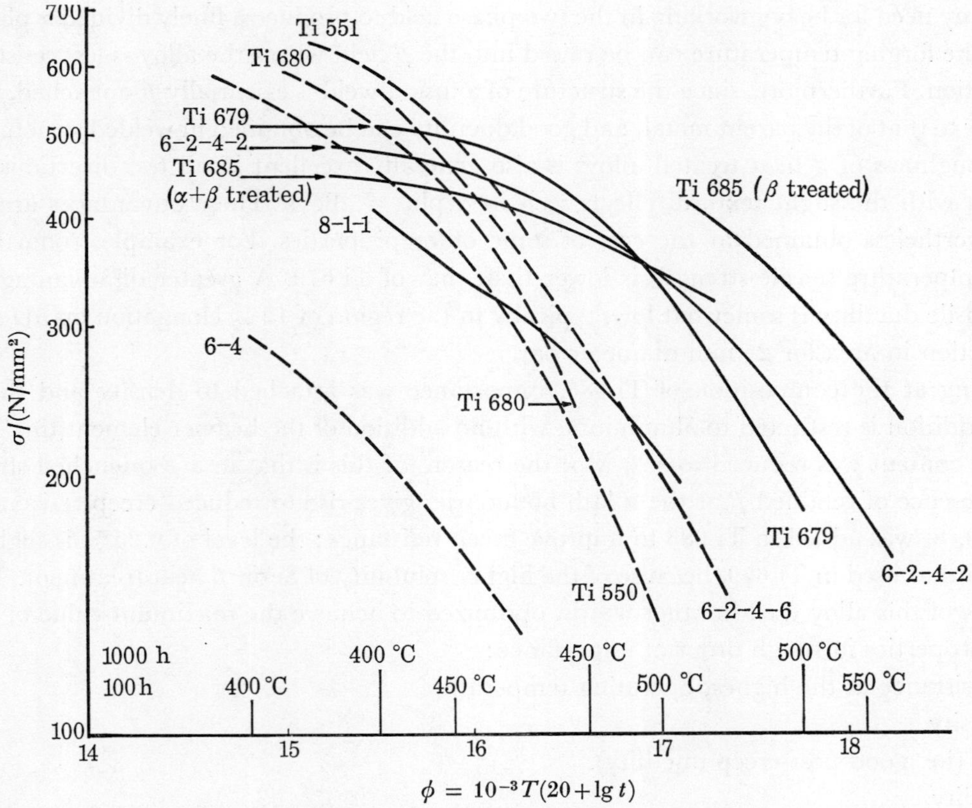

FIGURE 8. Larson–Miller plot showing the creep strength as a function of temperature and time for high-strength $\alpha+\beta$ forging alloys (dashed lines) and creep-resistant near α alloys (solid lines).

American alloy 6–2–4–2 (Ti–6Al–2Sn–4Zr–2Mo–0.1Si). Substitution of more aluminium for some of the tin significantly reduced its density with only a slight reduction in creep strength. Raising the molybdenum content to 2 % significantly improved the forgeability, although at the expense of some creep performance.

Another creep resistant 'near α' alloy of the same type is 8–1–1 (Ti–8Al–1V–1Mo). This alloy has been used in the U.S.A. for compressor disks and blades, but its applications have been curtailed because of metallurgical stability problems. The ordering parameter (greater than nine even with low oxygen content) is greater than for any other commercial alloy.

(ii) *β heat treated*

The alloys Ti 679 and 6–2–4–2 probably represent something close to the ultimate achievable in alpha plus beta treated creep resistant titanium alloys, but a significant step forward was taken with the introduction of I.M.I. Ti 685 which is heat treated in the β field and therefore represents an entirely new family of alloys. Quenching of this type of alloy from the solution treatment temperature produces a structure with no primary α; it is entirely composed of 'transformed β' (see figure 7, plate 25) consisting of a fine martensitic type structure or Widmanstätten α (dependent on the rate of cooling) separated by very small quantities of retained β phase. Ageing of this structure at about 550 °C stabilizes and stress relieves it.†

The introduction of β heat treatment brings with it certain well established advantages. Of principal importance is the improved creep resistance as shown in figure 8. Also, since there is no longer any need for heavy working in the two phase field to produce a finely divided α plus β structure, the forging temperature can be raised into the β field where the alloy is less resistant to deformation. Furthermore, since the structure of a fusion weld is essentially β quenched, it is very similar to that of the parent metal, and good ductility can be obtained in welded structures. Fracture toughness of β heat treated alloys is also generally excellent in all test directions, in comparison with the slight textural effects seen in α plus β alloys. These advantages are, of course, nevertheless obtained at the cost of some other properties. For example, room and elevated temperature tensile strength is lower than that of Ti 679. A greater disadvantage is that the tensile ductility is somewhat low, typically in the region of 12 % elongation on 5D and 23 % reduction in area for 25 mm diameter bar.

In arriving at the composition of Ti 685, importance was attached to density and the α stabilizer addition is restricted to aluminium, with no addition of the heavier element tin. The β stabilizer content was reduced to $\frac{1}{2}$ % Mo; the reason for this is that in a β quenched structure the presence of retained β at the α lath boundaries gives rise to reduced creep resistance. Like Ti 679, Si was added in Ti 685 to improve creep resistance; the level of 0.25 % is slightly higher than that used in Ti 679, because of the higher solubility of Si on β heat treatment. The composition of this alloy was, in other words, optimized to achieve the maximum value of the following properties in rough order of importance:

Creep resistance at the highest operating temperature.
Low density.
Stability (i.e. good post-creep ductility).
Forgeability.
Weldability.
Fracture toughness.

Experience with the alloy has shown that if it is to be of maximum usefulness in compressor disks, two other properties should now be included in the list, namely low cycle fatigue strength and crack initiation and propagation characteristics. Normally Ti 685 has very good crack propagation characteristics under low cycle fatigue conditions below 150 °C. Recent work has, however, shown that when it is used in the thickest sections the resultant coarse transformed β structure reduces the crack propagation rate to that of other titanium alloys. What is needed is a fine Widmanstätten β transformation product, and several experimental alloys are being developed both in the U.K. and in the U.S.A. which show promise in this direction.

Undoubtedly there is room for still further improvement in creep resistance of titanium alloys, and the upper temperature limit for use of titanium alloys will in the future be raised significantly above the present level. A new problem, however, begins to arise in that we are approach-

† The structural effects of ageing involve changes in the morphology and structure of minor phases and precipitates and are too complex to be discussed in this review.

ing the temperature at which significant oxidation of the titanium alloy surface begins to occur. If operating temperatures are raised above about 550 °C, it may well be necessary either to select a composition having greater inherent oxidation resistance or, more likely, to develop some form of surface treatment which enhances oxidation resistance without diminishing fatigue resistance. Beta treated near α alloys (i.e. of the Ti 685 type) will probably be developed which lift the maximum operating temperature by a further 30 °C or so, but beyond that stage some new type of titanium alloy will be needed. Several years ago a number of 'stable β' alloys were developed which contained sufficient beta stabilizer to give a structure entirely of stable β phase with no α present at all. Such alloys needed to contain, however, such large quantities of β stabilizer that they became very much denser and, like the refractory metals that they began to resemble, processing became much more difficult usually requiring a high temperature extrusion operation. A more likely development would be the introduction of a new precipitation or dispersion hardening phase. The problem of finding a suitable dispersion for titanium alloys will, however, be more difficult than in other alloy systems. As stated previously, the high affinity for oxygen, nitrogen and carbon and the high solubility of these elements in titanium means that very few of the compounds which might be considered are likely to be stable when in contact with titanium at elevated temperatures.

TABLE 4. COMPOSITION AND TYPICAL ROOM TEMPERATURE TENSILE PROPERTIES OF SHEET ALLOYS

composition	heat treatment	0.2% proof stress N/mm²	u.t.s. N/mm²	elongation on 5D %
(Ti 125) Ti–0.13 O_2	annealed	360	460	28
(Ti 155) Ti–0.28 O_2	annealed	540	640	24
(Ti 230) Ti–2.5Cu	solution treated	524	604	26
(Ti 317) Ti–5Al–2.5Sn	annealed	750	825	15

Sheet alloys

In the majority of applications where titanium is required in sheet form, e.g. engine fire walls or ducting, one or other grade of commercially pure material is used. These are in fact alloys of titanium and oxygen. Typical tensile properties are given in table 4.

For many applications requiring somewhat stronger, weldable sheet, the I.M.I. alloy Ti 230 (Ti–2½% Cu) is used (see figure 9, plate 25). This alloy was developed specifically as a heat treatable sheet alloy and is metallurgically interesting in that it is the only titanium alloy which is strengthened by a classical age-hardening reaction. The solubility of copper in α titanium diminishes with decreasing temperature, and heat treatment consists of solution treating at the temperature of maximum solubility (eutectoid temperature) followed by quenching and ageing to give a structure of α plus compound. The commercial ageing treatment is a two stage process of nucleation at a lower temperature followed by growth of the precipitate at a higher temperature. Typical properties of the alloy in both the solution treated and the fully heat treated condition are shown in table 4.

Where titanium alloy sheet is required with a greater room temperature strength than Ti 230, the more highly alloyed all-α alloy Ti–5Al–2½Sn (Ti 317) has found some applications in the past. However, it is not an easy alloy to fabricate and is no stronger than Ti 230 at slightly elevated temperatures; it has in general been superseded by the potentially more attractive α plus β alloy Ti–6Al–4V. Neither alloy is really well suited to sheet applications and in both cases pressing to shape has frequently to be carried out warm. No viable alternative high strength sheet alloy has, however, been developed to date. The best high strength heat treatable sheet alloys to emerge from intensive development programmes have been the meta-stable β alloys such as Ti–15Mo (Ti 205), Ti–3Al–8V–6Cr–4Mo–4Zr (Beta C), Ti–13V–11Cr–3Al (B120 VCA) and Ti–11$\frac{1}{2}$Mo–6Zr–4$\frac{1}{2}$Sn (Beta III). These alloys are readily cold formable in the β solution treated condition and can be aged to very high strength, although with correspondingly low ductility (see table 5). However, it is inherent in alloys of this type that they are not weldable, have relatively high density, and suffer from instability at relatively low temperature. Moreover in large sections they suffer from problems of chemical segregation and grain size control which lead to poor mechanical properties, particularly tensile ductility and fatigue performance. Consequently no meta-stable β alloy is extensively used at present.

Table 5. Composition and typical room temperature properties of meta-stable β titanium sheet alloys

composition	heat treatment	0.2 % proof stress N/mm²	u.t.s. N/mm²	elongation on 5D %
(Ti 205) Ti–15Mo	solution treated	583	760	25
	solution treated + aged (550 °C)	970	1050	11
	solution treated + aged (475 °C)	1250	1390	5
(Beta III) Ti–11.5Mo–6Zr–4.5Sn	solution treated	730	850	20
	solution treated + aged (530 °C)	1060	1130	8
	solution treated + aged (480 °C)	1230	1300	6

It is difficult to predict what development is likely to take place in titanium sheet alloys for the future. Improved *medium* strength sheet alloys (somewhat stronger than Ti 230) could probably be developed, although at present there is insufficient demand to justify the R & D effort. Improved *high* strength sheet alloys will require an entirely new approach, which is hard to visualize at present.

The way ahead

Greater knowledge of the interrelation between heat treatment, microstructure and fracture properties is likely to lead to gradually increasing use of the higher strength forging alloys, possibly accompanied by some changes in composition. New alloys will be developed which will further raise the temperature of operation, probably in conjunction with surface treatment to improve oxidation resistance. The demand for improved medium strength alloys and sheet alloys is probably insufficient to bring about significant improvements in this area, though studies of textural strengthening effects may in time extend the range of applications of existing alloys.

Because of the inherent cost of processing titanium, greater output and/or increased recycling of scrap are unlikely to reduce dramatically the cost of its use in any particular application. A more fruitful area for cost reduction is undoubtedly improvement in yield. Isothermal forging is now being seriously studied in both the U.S.A. and U.K. and this process could significantly improve process yield because most titanium alloys can be rendered superplastic at about 900 °C. The superplastic deformation characteristics of Ti–6Al–4V sheet are already being studied on a development scale, where use is being made of both vacuum forging and blow moulding techniques. The success of this work could herald the introduction of sheet alloys designed specifically for superplastic processing.

Reference (Morton)

Goosey, R. E. 1969 Heat treatment of engineering components, *Iron and Steel Inst. Sp. Pub.* 124, p. 75.

3.10 Future developments in non-ferrous metallurgy

By G. B. Brook and W. E. Duckworth
Fulmer Research Institute, Stoke Poges, Buckinghamshire, SL2 4QD

[Plate 28]

A major advance in the physical metallurgy of non-ferrous metals in recent years has been the use of knowledge, built on the foundation laid by Rosenhain, to predict unusual properties which can have considerable value in engineering applications. Much of the paper is therefore concerned with discussing these new discoveries and showing how they provide a possible escape from the straight-jacket of conventional ways of improving non-ferrous metal properties.

Another development, still in its early days of understanding and application, is the modification of grain boundary properties by trace elements. The use of these both at grain boundaries and in the matrix will increasingly produce new combinations of properties, particularly in age-hardenable alloys.

Many new advances in metallurgical processes are envisaged, based upon an extension of physical metallurgical knowledge and techniques into new areas. New structures, with improved properties produced by novel casting and powder metallurgy methods are possible.

It is concluded that general progress will be made in response to the demands of society for new materials or better use of existing ones. Metals like aluminium and copper are most likely to benefit in this way.

Introduction

It is not inappropriate for there to be a contribution from the Fulmer Research Institute at this conference. Although the Institute was founded more than ten years after the death of Rosenhain, its founder, Colonel Devereux, was concerned with the commercial development of one of Rosenhain's major research programmes at the National Physical Laboratory. This was the aluminium copper age hardening alloy known during the 1914–18 war as Y alloy. This was originally developed as a cast alloy but its composition was adapted by metallurgists at Rolls Royce to produce one of the RR series of forging alloys which was subsequently exploited commercially by Devereux. These are wrought alloys from which the alloy used in Concorde has been developed.

Devereux maintained his interest in age hardening aluminium alloys throughout his life, and it was not coincidental that much of the Fulmer Research Institute's early reputation was gained in further studies on the aluminium copper age hardening system (Brook 1963). The significance of this work in future developments of non-ferrous alloys will be discussed later.

Throughout the major part of his working life Rosenhain was concerned with the relation between the internal structure of metals and their properties. In his later life he was very much involved with studies of the transformations occurring in steels. It is, perhaps, relevant therefore, on the occasion of his centenary, to point out how much in the last 25 years attention has been shifting from the martensite transformation which was so exclusively studied in steels to the more interesting transformations occurring in non-ferrous metals. It is in studying these transformations, in understanding their behaviour, and in exploiting them for the purposes of mankind that some of the next advances in non-ferrous metals will occur.

It is also pertinent to comment that at the last notable metallurgical centenary, that of Sorby (1964), a great deal of discussion took place on the relations between structure and properties of metals. All this discussion might be said to have been of the *post hoc* and not of the *propter hoc* type. That is to say it was concentrated on how known properties could be understood in terms of their structure and very little was said on how unknown properties could be predicted as a result of a study of structure.

Rosenhain, too, was of the school which concerned itself with explaining events rather than predicting them but one finds it hard to predict events until the basis on which they may be predicted has been thoroughly understood. So, it is appropriate to take the diagnostic work of Sorby, Rosenhain and their successors to provide the basis for the prediction of the properties which will be produced in non-ferrous metals over the next quarter century, based on the foundations that they laid.

Exploitation of metallurgical phenomena

Rosenhain's earliest work was concerned with the phenomena of plastic flow and he made some of the first observations relating deformation and microstructural changes (see Rosenhain 1934). Although it is 75 years since this work was done there is still much that can be learned and put to use from a knowledge of these relationships.

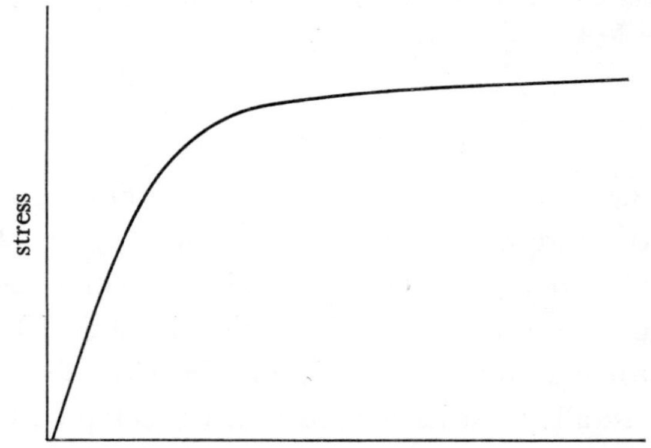

Figure 1. The conventional stress–strain diagram.

Stress–strain relations

The conventional stress strain relationship which describes plastic flow in metal is shown in figure 1. This form of stress strain diagram has dominated metallurgical thinking since the time of Hooke (1678) and the limitations of systems which obey this type of relation are now fully understood. One of the major ones is that it is very difficult to increase strength without sacrificing ductility and the type of strength–ductility relation shown in figure 2 is very common for a wide variety of non-ferrous metals and alloys.

One way out of this dilemma which has been studied extensively in recent years has been the composite type of structure. This has been fully reviewed in recent publications by Kelly and his associates (Kelly 1973). Gas turbine blade applications of non-ferrous composite structures have recently been developed (Lemkey & Thompson 1971).

Superelasticity and shape memory

Another way forward has been indicated in recent years by the discovery of a different shape of stress–strain curve in various systems. This is shown in figure 3 (Rachinger 1960; Pops 1970). It will be seen that before the first onset of permanent set or strain hardening arises at point B there is a very long period of rapid extension under fairly low load. This is the so called superelasticity era.

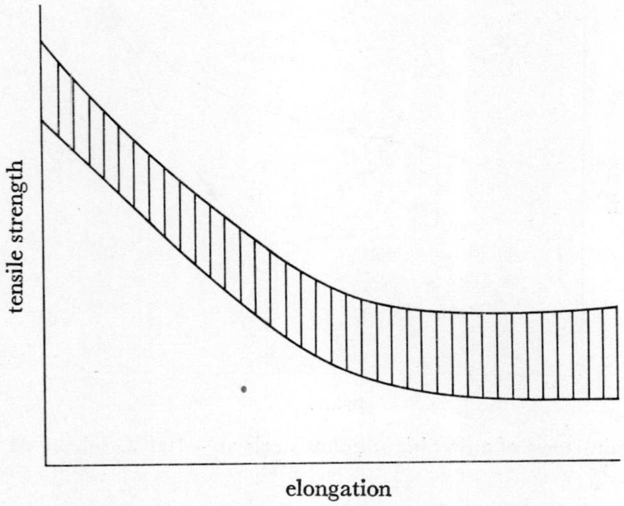

FIGURE 2. The inverse relation of strength and ductility.

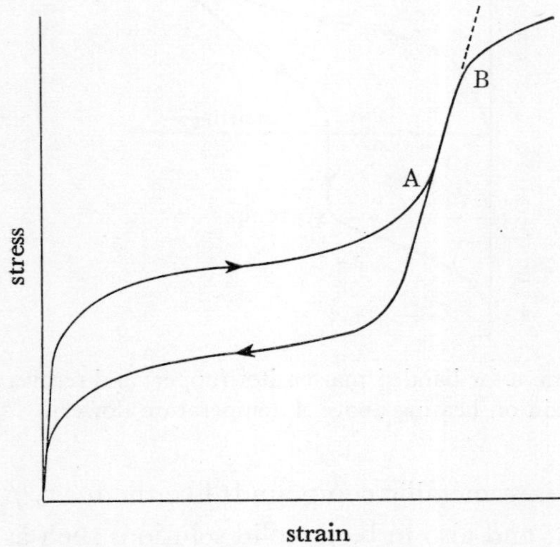

FIGURE 3. Stress–strain curve for superelastic alloys (above M_s temperature).

Superelasticity, otherwise known as rubber-like behaviour, was first observed by Chang & Read (1951) in single crystals of the AuCd intermetallic compound in which the martensitic transformation can be induced by deformation. This anomalously high, fully reversible, apparently elastic, strain was associated with the stress induced growth of martensitic plates deformed at a temperature between the M_s and M_d temperatures. These plates then shrank and finally disappeared as the complete recovery of the strain took place when the stress was reduced (Rachinger 1960; Pops 1970). This is not true (Hookean) elasticity as figure 3 shows.

True elastic deformation of the structure is replaced at a low stress by deformation arising from the stress induced transformation to a martensitic structure. By changing composition this can be made to occur at any stress below that required to cause slip in the structure. There is little strain hardening associated with the martensitic transformation. After strains of up to 15% (point A), the rate of hardening increases as the martensite is deformed elastically until the plastic yield point of the martensite is reached (point B).

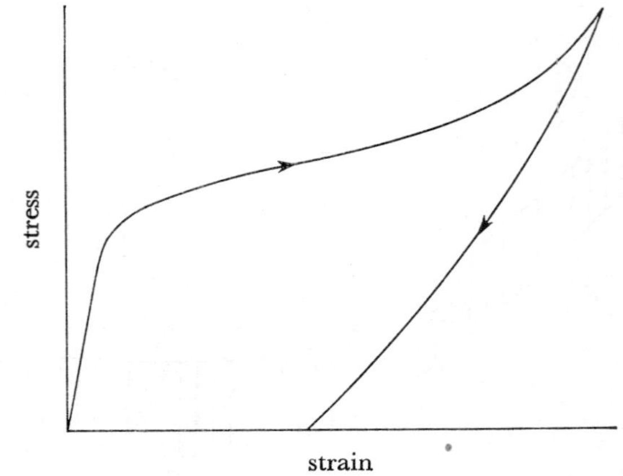

FIGURE 4. Stress–strain curve of austenitic stainless steels at −196 °C (above M_s temperature).

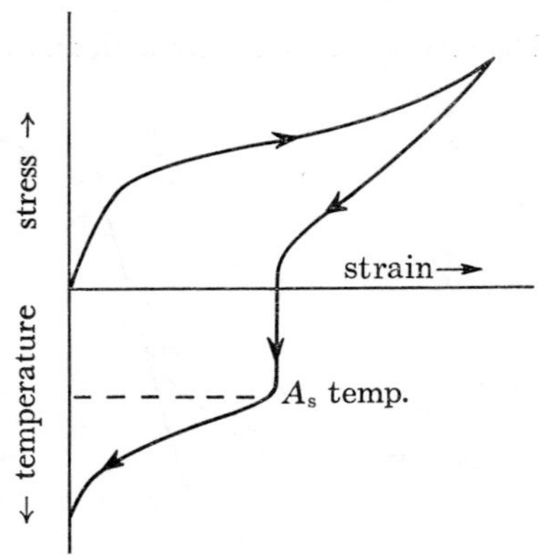

FIGURE 5. Stress–strain curve for banded martensites (upper) and recovery of pseudo-plastic strain on heating above A_s temperature (lower).

This behaviour is found in intermetallic compounds like the b.c.c. β phases of the group 1B metal alloy systems and TiNi; and also in b.c.c. solid solutions such as the γ phase of uranium alloys and the austenitic stainless steels. At temperatures as low as −196 °C, large transformation ductilities are possible in austenitic steels of appropriate composition which have stress strain curves like those in figure 4 in which only part of the initial strain is superelastic.

Unlike those in carbon steel martensite, the interfaces of these plates are mobile, hence their ductility. For those metallurgical students brought up in Rosenhain's belief that the typical type of martensite is a steel martensite, i.e. a hard brittle form of martensite, it perhaps is still unusual for them to realize that a steel martensite is most untypical of the martensite family and that martensites generally are soft and ductile.

When parallel-banded martensites are deformed the stress–strain curve shown in figure 5 (upper) is obtained. In favourably oriented grains, alternate plates are reoriented by the stress-induced movement of the plate boundaries. When this motion is complete the reoriented martensite deforms elastically until plastic yielding occurs. On unloading after the reorientation of the martensite, a permanent extension is retained because the reverse movement of the martensitic plate boundaries is not complete when the deformation is carried out below the M_f temperature.

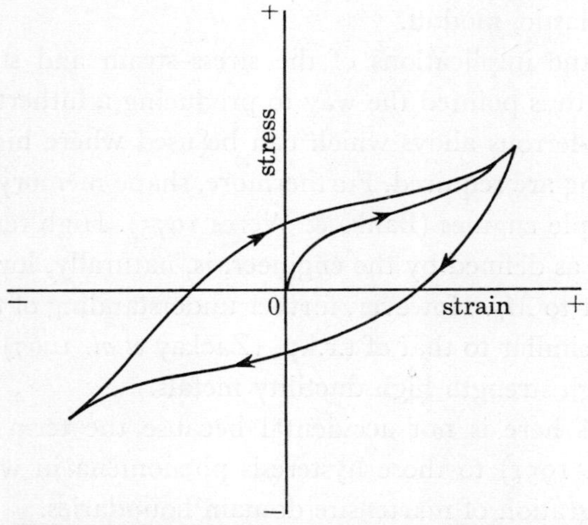

FIGURE 6. Complete mechanical hysteresis curve for banded martensite.

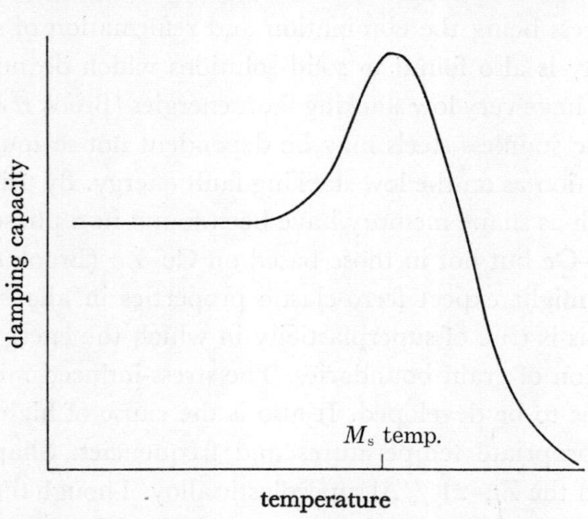

FIGURE 7. Variation of damping capacity of banded martensites with temperature.

The residual strain can then only be reversed by (1) heating above the A_s temperature to revert the structure to the high temperature form by which the original shape is restored, figure 5 (lower), or (2) by applying a reverse stress by which the complete hysteresis loop can be generated as shown in figure 6. The whole of figure 5 thus demonstrates the well known shape memory effect (Buehler 1963).

By mechanically cycling the material through the hysteresis loop at or below the M_s temperature, a high energy absorption or damping capacity can be demonstrated (figure 7).

As reversible deformation of these alloys is accommodated by the cyclic motion of martensitic plate boundaries, the slip processes which normally lead to the development of a fatigue crack do not take place. Thus alloys with the banded microstructure have extremely good fatigue properties, e.g. a TiNi alloy has withstood a stress of 240 N/mm^2 (35000 lbf/in^2) for more than 10^8 cycles. Fracture has been reported at a stress of 480 N/mm^2 (70000 lbf/in^2) after 10^7 cycles (Buehler 1963). High duty service components made in TiNi with shape memory are shown in figure 8, plate 28.

Above the M_s temperature, the superelastic state of these alloys demonstrates not only good fatigue and damping properties but also effective elastic strains of up to 15–25% at a range of different, apparently elastic, moduli.

An understanding of the implications of the stress–strain and strain–temperature curves shown in figures 3–6 has thus pointed the way to producing a hitherto unattainable combination of properties in non-ferrous alloys which can be used where high ductility, high fatigue strength and high damping are required. Furthermore, shape memory alloys can be used as the operating element in simple engines (Banks & Weres 1975). High tensile strength can also be achieved but yield stress, as defined by the engineer, is, naturally, low unless the materials are used at temperatures near to M_d. However, further understanding of the phenomena described may enable mechanisms similar to that of t.r.i.p. (Zackay *et al.* 1967) in steels to be developed to produce a range of high strength high ductility metals.

The reference to steels here is not accidental because the term ferro-elasticity has been applied (Chang & Read 1951) to those hysteresis phenomena in which the hysteresis arises from the drag on the migration of martensite domain boundaries.

The cause of the hysteresis is not known with certainty but the evidence points increasingly to the energy absorbing process being the elimination and reformation of stacking faults (Brook *et al.* 1975). Ferro-elasticity is also found in solid solutions which do not undergo martensite transformations but which have very low stacking fault energies (Brook *et al.* 1975). For example ferro-elasticity in austenitic stainless steels may be dependent not so much on the presence of the martensitic transformation as on the low stacking fault energy. By using this understanding ferro-elastic properties such as shape memory have been found in α phase solid solutions based on Cu–Al, Cu–Si and Cu–Ge but not in those based on Cu–Zn (Brook *et al.* 1975).

On these grounds one might expect ferro-elastic properties in alloys demonstrating other hysteresis phenomena. This is true of superplasticity in which the energy absorbing process is the vacancy aided migration of grain boundaries. The stress-induced migration of boundaries allows high uniform strains to be developed. It also is the cause of high damping capacity in superplastic alloys at appropriate temperatures and frequencies. Shape memory has been demonstrated at Fulmer in the Zn–21% Al superplastic alloy. Though the shape memory strain is too small to be of practical use it is sufficient to cause lack of shape stability, and hence is of importance in the commercial application of the alloy.

Grain boundary phenomena

The role of grain boundaries, their structure and their migration is understood more clearly than in Rosenhain's day but there is a need for still greater knowledge of the role of trace elements and their interactions with both grain boundaries and vacancies. New developments in the control of structure and therefore properties will result from this.

It is known from measurements of internal friction on pure metals (Weinig & Machlin 1957) that less than 1% of solute is sufficient to completely saturate grain boundaries with solute atoms and that one solute can be displaced by another depending on the relative sizes, solubilities and chemical nature of the atoms. Thus in many alloys, the nature of the boundaries is

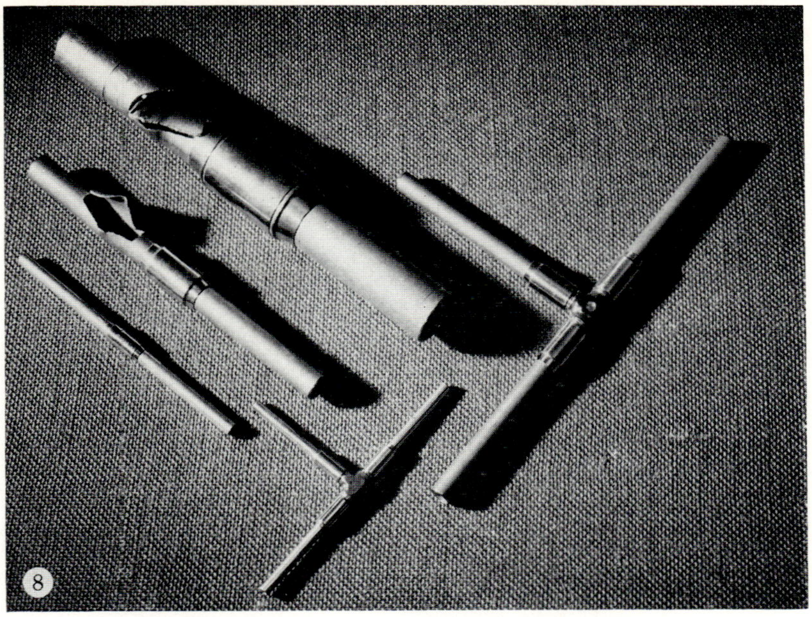

FIGURE 8. Some parts made in TiNi shape memory alloy.

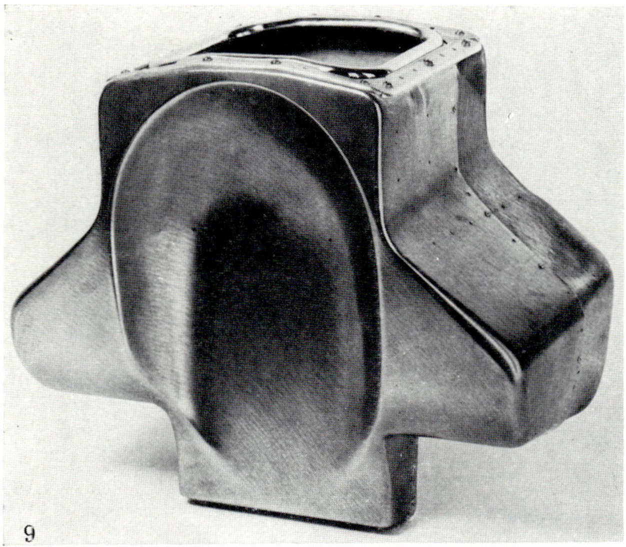

FIGURE 9. Superplastically formed componet,

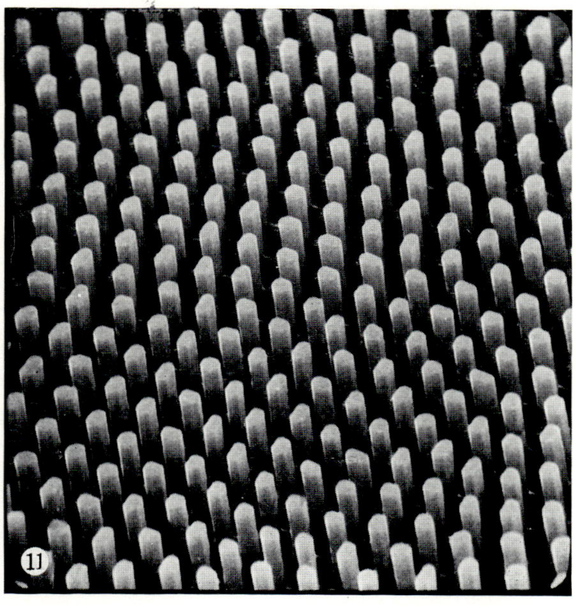

FIGURE 11. Deep etched transverse section of metal/carbide directionally solidified eutectic (Magn. × 1666).

(*Facing p.* 418)

dependent also on a minor solute. This will control their lateral rate of migration, and also cohesive strength.

For example, certain elements will reduce the rate of migration of boundaries in copper giving a fine stable grain size whereas others will cause rapid boundary migration especially during slow deformation, inducing a condition very similar to superplasticity. Similarly, trace elements are used in aluminium copper alloys to convey superplastic properties (Superform 1975).

Superplasticity deserves an honourable mention in any non-ferrous paper concerned with Rosenhain's work. In 1920 he observed that a sheet of Zn 7 % Al 4 % Cu on cutting with shears 'flew to pieces unless the cutting was done very slowly'. On slow bending extensive deformation was possible. 'The material behaved, indeed, as though its properties lay somewhere between pitch and glass, being more viscous than the former, but much less so than the latter.' During discussion of the relevance of these findings Rosenhain did not envisage any industrial exploitation. 'Alloys described in the present paper appear to be disappointing from the point of view of practical utilisation.' Figure 9, plate 26, shows a commercial component now being made from superplastic alloys (Superform 1975).

It is well known that additives of zirconium to tin bronzes can reduce or prevent hot shortness (Eborall 1954); zirconium, beryllium and many other elements can prevent the grain boundary migration which gives rise to discontinuous precipitation and embrittlement in many age-hardenable alloys in gold, copper and aluminium base systems (Brandes 1975*a*).

Many solid solutions of low stacking fault may display a ductility trough at elevated temperature just below $0.5T_m$, e.g. at about 450 °C in copper alloys, which may be a consequence of the inability of the grain boundary to migrate and remove stress concentrations developing at the boundary during deformation. This is the type of problem which might be alleviated by the use of trace elements which can preferentially segregate to the grain boundary. The modification of grain boundary behaviour by means of trace elements is another technique which will increasingly be exploited in the development of improved and unusual properties in non-ferrous alloys.

Trace elements in solution

The interaction of small quantities of certain elements with lattice defects is another aspect of metallurgy which has been little studied or used. The interaction of trace elements with vacant lattice sites in aluminium alloys has allowed the diffusion of certain atomic species to be controlled. By this, the nucleation of fine and stable precipitates has been achieved (Brook & Hatt 1969) which have the practical advantage of producing a high strength, tough and creep resistant alloy (Brook 1975).

It has been known for some time that cadmium, indium and tin can interact with vacancies in aluminium–copper alloys and thus reduce the free vacancy concentration available for aiding diffusion in copper (Brook 1963).

A consequence of this is a reduced rate of g.p. zone formation and of hardening at ambient temperature. However, at higher temperatures, e.g. about 150–200 °C, this interaction is less stable and the abnormally high vacancy concentration produced allows a very rapid nucleation of the precipitates to take place. The trace element atoms can then be identified at the semi-coherent precipitate/matrix interface where they reduce the rate of precipitate coarsening. There is recent evidence that these atoms are located at the growth dislocations at the interface (Sankaran & Laird 1974).

More effective use of trace elements had to await the realization and confirmation that pairs of elements would interact with each other and with vacancies in solid solution and produce a much more marked effect than the elements cadmium, indium and tin. This has been a consequence of the realization that small amounts of magnesium and germanium or magnesium and silicon associate in solid solution in aluminium and also interact with free vacancies. These elements together have effects which the individual elements do not have separately (Brook & Hatt 1969).

Most precipitation-hardenable alloys need to be heated to temperatures close to their melting point at which the alloy has relatively low strength and can readily distort. This is followed by a rapid cool of severity varying from water or oil, to air jets which again can cause distortion. In large sections, cooling rates are often inadequate to retain all the solute in solution. Strength is developed by ageing at a much lower temperature at which a fine dispersion of precipitate can be achieved. This method of achieving high strength has the disadvantages already mentioned; it has others, notably the difficulty of restoring strength once the material has been welded.

There is a need for new strengthening methods which overcome these difficulties. One such method can be applied where the precipitate is normally disordered but which can be ordered by thermal treatment. In such circumstances, the disordered precipitate is usually coherent with the matrix and provides little hardening until it is ordered. Thus on cooling from the solution treatment temperature fast cooling is not needed as precipitation can be allowed to take place. This means that large castings can be made in such alloys without the need for further solution treatment. By heating at the appropriate temperature to order the precipitate, a high strength can then be achieved without the need for a separate solution treatment. Equally, after a welding operation, local heating only is needed to reorder the precipitate. This method of heat treatment has been developed for copper alloys containing nickel and manganese but it can be applied to other systems in which an orderable precipitate can be engineered by devising an appropriate structure from an understanding of the alloying behaviour of the system.

Thus trace elements in solution can be used, again, to confer unusual combinations of properties and will increasingly be so used in future, particularly in age hardenable alloys.

Thus, whereas in his day Rosenhain sought to understand alloys developed by empirical methods, it is becoming possible nowadays to use our knowledge of atomic processes to predict specific properties. From an understanding of martensite structures with mobile interfaces, the ferro-elastic phenomena of shape memory, superelasticity and high damping capacity have been predicted. By studying the hysteresis process ferro-elastic phenomena occurring in solid solutions with low stacking fault energy not undergoing martensitic transformations have also been predicted. A better knowledge of the interactions between grain boundaries, solute atoms and vacancies will no doubt lead to new means of achieving superplastic properties and other properties which depend on the hysteresis of boundary migration. The work done on the effect of trace elements in Al–Cu and Cu–Ni–Mn alloys enabled improved alloys to be developed.

Exploitation of metallurgical processes

Casting technology

Clearly, no review of the prospects for non-ferrous metallurgy would be complete without reference to future processing developments. Since Rosenhain's death there have been many major changes in non-ferrous processing technology. Continuous casting has come to be established as a routine production process for aluminium and copper alloys, and as this is a process for quantity production, it is unlikely in the near future that the production demands for the other non-ferrous alloys will be sufficient to warrant an extension of continuous casting techniques to these other metals.

The process is being taken up by steel-makers and the non-ferrous industry can be proud to have pointed the way to it.

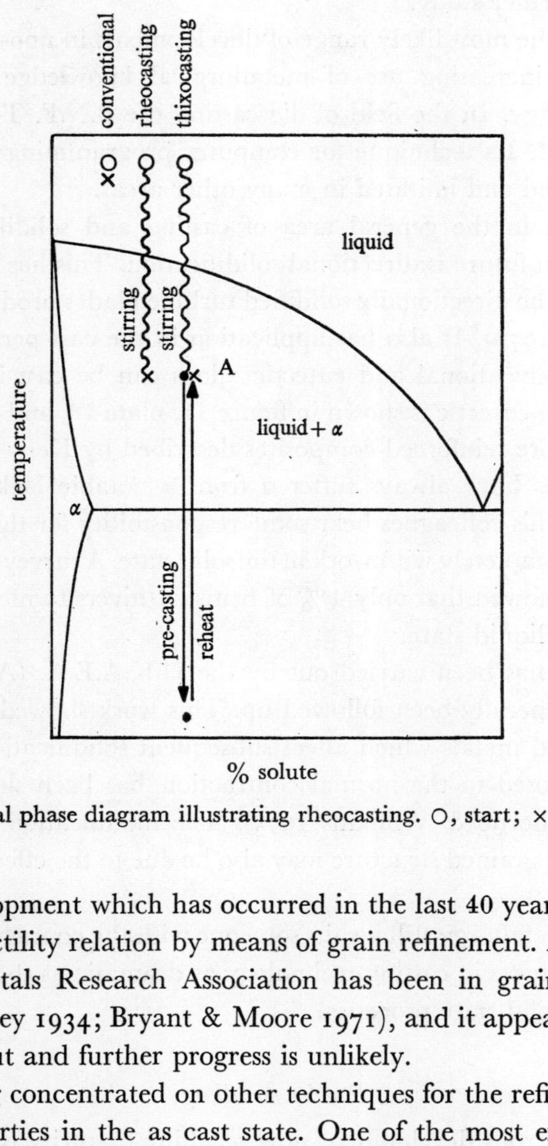

FIGURE 10. Hypothetical phase diagram illustrating rheocasting. ○, start; ×, casting temperature.

Another casting development which has occurred in the last 40 years has been the improvement of the strength–ductility relation by means of grain refinement. A major advance by the British Non-Ferrous Metals Research Association has been in grain refinement by mould dressing (Genders & Bailey 1934; Bryant & Moore 1971), and it appears now that this method has been fully worked out and further progress is unlikely.

Attention is thus being concentrated on other techniques for the refinement of structure and for improving the properties in the as cast state. One of the most exciting developments in recent years, which will certainly be given much attention in future, is the process of rheocasting being developed by Flemings and his colleagues at M.I.T. (Flemings & Mehrabian

1973). The process has been thoroughly described in recent papers but a brief account can be given by reference to figure 10.

The alloy is allowed to cool from the initial fully-liquid state into the middle of the liquidus solidus region, shown as point A in the diagram, during which the melt is stirred vigorously, the primary dendrites are fractured and the melt becomes thixotropic. It can be used in this form or quickly quenched. If the quenched material is reheated to point A the thixotropic properties are restored and the liquid can then be injected into suitable moulds to produce as cast material of desirable properties. A production advantage of this process is that the thermal energy required to reheat to point A is obviously lower than that required to completely re-melt the material and the time taken for the heat to disperse from the moulds is also less. Therefore the production cycle is shortened, thermal energy saved and less thermal fatigue and other wear and tear occurs in the dies. In these times of energy saving this is a major advance which deserves considerable further study.

This points to one of the most likely range of developments in non-ferrous metallurgy in the foreseeable future, the increasing use of metallurgical knowledge to improve production economics and save energy. In the field of die casting the B.N.F. Technology Centre (1974) has again taken the lead. Its technique for computer programming of die casting operations will no doubt be extended and imitated in many other areas.

Another development in the general area of casting and solidification which will have specialized application in future is directional solidification. This has found first application in the aerospace field with the directionally solidified turbine blades produced by Pratt & Whitney (Lemkey & Thompson 1971). It also has application in the cast permanent magnet material field (Nesbitt 1969). Conventional and eutectic alloys can be cast in this way. The type of structure obtained with a eutectic is shown in figure 11, plate 26, and the resemblance between this structure and the fibre reinforced composites described by Kelly (1973) is obvious.

Solidification processes have always suffered from a notable lack of theoretical research. Possibly Rosenhain and his colleagues bear some responsibility for this because they identified physical metallurgy so completely with work in the solid state. A survey carried out by B.I.S.R.A. some years ago (1964) showed that only 4% of British University metallurgical research work was concerned with the liquid state.

Much valuable work has been carried out by the U.K.A.E.A. (Anon 1973) on the liquid state but this has not generally been followed up. This work showed the existence of ordered structures in many liquid metals which affect subsequent solidification behaviour. Expansion on solidification, as opposed to the normal contraction, has been shown to be due to liquid structures (Rivlin, Waghorne & Williams 1975). The modification of aluminium silicon by sodium to produce a fine grained structure may also be due to the effect of sodium on the liquid structure.

Possibly liquid state metallurgy still awaits someone with the courage and vision of Rosenhain to point the way to advances in casting technology as dramatic as those which physical metallurgy has produced in solid state processes.

Powder metallurgy

Another major non-ferrous development now reaching maturity is powder metallurgy, and in particular two aspects; the manufacture of specific shapes and the rolling of strip from powder. Isostatic forming and powder forging are now established production techniques which only need the impetus of full commercial exploitation to take them further and new metallurgical developments in these areas are unlikely until commercial use has been fully established.

In the power rolling field the Sherritt-Gordon process (Blore *et al.* 1966) for the rolling of nickel foil is well established and the technique is being followed in the stainless steel industry (Sturgeon & Taylor 1972). Again, the production demands in other non-ferrous alloys are unlikely to encourage further widespread commercial exploitation of this technique in other metals.

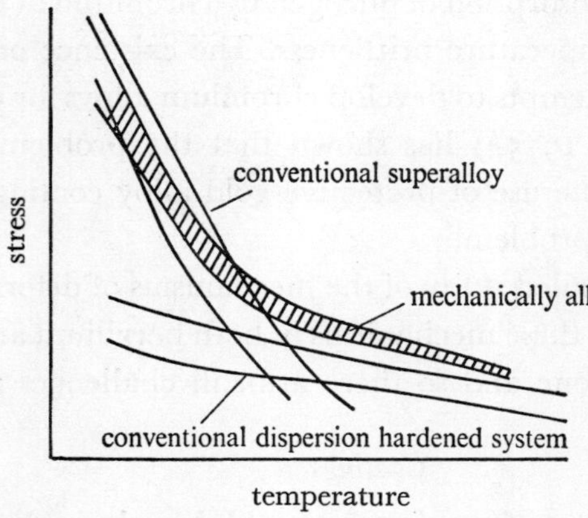

FIGURE 12. Comparison of stress/temperature properties for conventional and mechanically alloyed superalloys.

However, one development of powder metallurgy, mechanical alloying, which also started in non-ferrous applications and has been imitated in the ferrous metals industry is likely to be the major development in non-ferrous powder metallurgy in the future. The process has been fully described (Benjamin & Cairns 1970). It is sufficient to say here that the initial stages of the process are high energy ball milling of pre-alloyed or elemental powders with small amounts of suitable dispersoids. During this high energy milling the metal powders ingest the dispersoids in such a form as to produce individual particles which can then be compacted by extrusion or isostatic forming at a suitable temperature into very strong ductile components. Some indication of the achievements of the process are shown in figure 12.

The technique thus provides another way of raising the ductility–strength relation, and also of conferring other properties such as oxidation resistance. As well as its obvious advantage in improving the properties of existing materials for high temperature performance and other arduous conditions, the process may offer a way out of the two remaining most intractable problems in non-ferrous metallurgy, that of producing chromium and beryllium alloys in a strong, tough, ductile form.

As all other previously promising techniques have failed, it is perhaps too soon to expect success but this is one of the few remaining techniques which is yet to be explored. If this could overcome the problems with these two metals then both could have a bright future in many applications, especially aerospace. Chromium because of its high temperature characteristics, beryllium because of its extreme lightness and high modulus. Rosenhain would have been particularly pleased at the latter because of his own early work with beryllium.

The problem with lack of ductility in this metal has been the high critical resolved shear stress for non-basal slip. Thus twinning has been the dominant mode of failure and the interactions of twins with each other and with grain boundaries initiates cracks which spread catastrophically. Attempts to reduce the critical resolved shear stress for pyramidal slip by purification of beryllium have not been successful in improving ductility.

Chromium alloys have been standing in the wings for many years awaiting their opportunity to replace nickel alloys in gas turbine engines. For three reasons they have made no progress. The first is the ingenuity by which the effective upper temperature at which nickel alloys can be used has been raised, first by internal cooling and secondly by the use of castings.

The other two reasons are defects in chromium; firstly the lack of ductility at low temperatures which is due in part to the effect of dissolved nitrogen in raising ductile/brittle transition temperatures and secondly, the absorption of nitrogen by chromium at operating temperatures, thereby increasing the lower temperature brittleness. The existence of two interrelated problems has militated against any attempts to develop chromium alloys for useful service. However, recent work at Fulmer (Brandes 1975*b*) has shown that the problem of nitridation at high temperatures can be solved by the use of protective gold alloy coatings and this justifies new attempts to solve the brittleness problem.

Success will come from the detailed study of the mechanisms of deformation and of the effect of solutes and insoluble phases on these mechanisms in both beryllium and chromium. Sufficient of this work has not yet been done and so there are still challenges awaiting the solid state physical metallurgist.

Coatings

One major application for the use of non-ferrous metals has always been in corrosion resistant applications either as bulk metals or as coatings. With the increasing emphasis being placed on conservation, this role is likely to increase, especially in the form of coatings where the advantages of the relatively higher price non-ferrous metals can be combined with the low price and ready availability of the appropriate substrate.

The technology of so many of the coating methods, electroplating, spraying, dipping, etc. is so well established that major advances in any of these are unlikely.

Plasma spraying and chemical vapour deposition are among the latest techniques for which there is still development scope. Certainly there is still much to be learnt about the intelligent application of coatings, not only for corrosion resistance, but for resistance against other processes of attrition such as wear and erosion.

Conclusions

The extent to which non-ferrous metallurgical phenomena and processes will continue to be further exploited at the pace which was characteristic of the years 1945–1970 will depend upon the priorities set by society.

There are those who argue that emphasis should be given to the better use of existing materials rather than the development of new ones (Maddock 1975). This view will encourage process development as will increasing concern with energy utilization. Sir Alan Cottrell (1975) has recently pointed out how energy intensive metal manufacture is. Thus it is to be expected that some of the new low energy processes discussed in this paper will make rapid advance.

New material developments, involving more exploitation of metallurgical phenomena, will most likely be stimulated by the needs of material conservation and new energy forms. For example, the use of new battery types is being delayed by materials problems (Coe 1975). Indeed, all engineering advance is ultimately dependent upon advances in materials technology and wherever the engineer finds his greatest challenges in future so will the non-ferrous metallurgist be found.

Which metals are likely to be there? Copper and aluminium undoubtedly. Their present production volume and advantageous engineering properties almost guarantee this. Fears about a possible copper shortage should not be exaggerated. There have only been 12–15 years supply of copper since the time of Alexander the Great! (Dunham 1974). Nickel's main future lies in the steel industry but it will remain pre-eminent in high temperature applications until the outstanding problems with chromium and cobalt are solved or until it is replaced by improved iron alloys or ceramic components. Manganese, a much neglected metal, may well find greater uses in view of its ready availability and low energy content. Tin will continue in its traditional role until its price rivals that of the previous metals, whose use will also remain relatively stable. Titanium will depend for its future on developments in aerospace technology unless advances in scrap recovery encourage its wider use in the chemical industry. Zirconium is very much dependent upon the future of nuclear power. The traditional use of lead will continue but zinc may find itself increasingly challenged in its die casting role by plastics and aluminium and as a coating by other alternatives. These are the two common metals whose future supplies are most uncertain (Dunham 1974).

Some of the less common non-ferrous metals may well come to the fore during the rest of the century. Sodium has long been on the threshold of many major applications from the cooling medium in fast reactors to the replacement for copper as an electrical conductor and now in its role in the sodium/sulphur battery. The later derivatives of the battery may well find us learning a great deal more in the future about alkaline metals, and others in the first group of the periodic table which have been traditionally rather neglected.

Vanadium, niobium, tantalum and the rare earths will find increasing applications in the steel industry and may have individual uses such as tantalum as a coating in the chemical industry and in the electro-ceramics.

To conclude finally it would have been hard to predict at the time of Rosenhain's death in 1934, at a time when both nuclear power and space exploration were dreams of science fiction writers, and where a flight across the Atlantic would only be undertaken by the adventurous hero, that within eleven years the first nuclear device would be exploded, that in 20 years trans-Atlantic crossing at near the speed of sound would be common place, and that within 35 years man would have walked on the Moon's surface and returned safely to Earth.

We have less than 25 years of this century left. Mankind is beginning to learn the lesson of its profligacy in materials and energy. It is therefore the common place prediction that the next 25 years will see much of the vast knowledge about non-ferrous metals gained since the death of Rosenhain applied to making much better use of the materials we have, and to using them with less consumption of energy. The metallurgist has a significant role to play here, and it is to be hoped that the present generation plays its part as well as Rosenhain played his.

The authors are most grateful to their many colleagues who have contributed ideas and suggestions for inclusion in the paper.

References (Brook & Duckworth)

Anon 1973 *Proc. 2nd Int. Conf. Props. Liquid Metals.* Tokyo 1972.
Banks, R. & Weres, O. 1975 *Proc. 1st Int. Symp. on Shape Memory Effects and Applications.* Toronto: A.I.M.M.E. (In the Press.)
Benjamin, J. S. & Cairns, R. L. 1970 *Int. Powder Met. Conf.* New York: Plenum Press.
Blore, M. H. D. *et al.* 1966 *ASM, Metals Engng* **6**, 54.
B.N.F. Tech. Centre. 1974 *Am. Mach. Spec. Report* 670, Computer Diecasting.
Brandes, E. A. 1975*a* *Gold Bull.* **8** (2), 41–47. Johannesburg: Chamber of Mines of South Africa.
Brandes, E. A. 1975*b* *Gold Bull.* Johannesburg: Chamber of Mines of South Africa. (In the Press.)
British Iron & Steel R.A. 1964 Report MG 67/64.
Brook, G. B. 1963 *Precipitation in Metals.* SR3. Fulmer Research Institute, Stoke Poges.
Brook, G. B. 1975 *Design Engng*, April 65–67.
Brook, G. B. *et al.* 1975 *Proc. 1st Int. Symp. on Shape Memory Effects and Applications.* Toronto: A.I.M.M.E. (In the Press.)
Brook, G. B. & Hatt, B. A. 1969 Inst. Met. Proc. Int. Symp. *Mechanism of Phase Transformations in Crystalline Solids* 1968, pp. 82–84. London: Inst. Met.
Buehler, W. J. 1963 *Proc. 7th Navy Sci. Symp.* Report ONR-16 **1** (AD 42178) 1–7 (also Raychem Corporation 1970 Cryofit Tube Fitting System).
Bryant, M. D. & Moore, A. 1971 *Br. Foundrym.* **64**, 215–229.
Chang, I. C. & Read, T. A. 1951 *Trans. A.I.M.M.E.* **191**, 47–52.
Coe, R. E. 1975 *The Metall. & Matls Technologist* **7**, 145–148.
Cottrell, A. 1975 *Metals & Mater.* March 32–33.
Dunham, K. 1974 *New Scient.* **64**, 129–130.
Eborall, R. J. 1954 Proc. N.P.L. Symp. Fracture of Metals at High Temperature, 229–241.
Flemings, M. C. & Mehrabian, R. 1973 *Trans. Am. Foundrym. Soc.* **81**, 81–88.
Genders, R. & Bailey, G. L. 1934 Research Monograph No. 3, BNFMRA.
Hooke, R. 1678 *De Potentia Restitutiva.*
Kelly, A. 1973 *Strong Solids.* 2nd edn, pp. 157–220. Oxford: Clarendon Press.
Lemkey, F. D. & Thompson, E. R. 1971 *Met. Trans.* **2**, 1537–1544.
Maddock, I. 1975 *New Scient.* **65**, 375–378.
Nesbitt, E. A. 1969 *J. appl. Phys.* **40**, 1259–1265; *J. Metals* **21**, 20A.
Pops, H. 1970 *Met. Trans.* **1**, 251–258.
Rachinger, W. A. 1960 *J. Aust. Inst. Metals* **5**, 114–117.
Rivlin, V. G., Waghorne, R. M. & Williams, G. I. 1975 *Gold Bull.* Johannesburg: Chamber of Mines of South Africa. (In the Press.)
Rosenhain, W. 1934 Obituary. *J. Inst. Met.* **54**, 313–315.
Rosenhain, W. *et al.* 1920 *J. Inst. Metals* **23**, 261–317.
Sankaran, R. & Laird, C. 1974 *Mater. Sci. & Eng.* **14**, 271–279.
Sorby 1964 Iron and Steel Institute. *Metallography* 1963. Proceedings of the Sorby Meeting, Sheffield 1963. London: I.S.I. SR80.
Sturgeon, C. M. & Taylor, R. L. S. 1972 *Metal strip from powder.* Monograph Series ME/14. London: Mills & Boon.
Superform Ltd 1975 Literature on Supral 150.
Weinig, S. & Machlin, E. S. 1957 *Trans. A.I.M.M.E.* **209**, 32–41.
Zackay, V. F. *et al.* 1967 *Trans. A.S.M.* **60**, 252–259.

KEYNOTE SPEECH

By R. W. K. Honeycombe (*Department of Metallurgy and Materials Science, University of Cambridge*)

1. Introduction

The papers in this session cover large areas of ferrous and non-ferrous physical metallurgy, so it would be inappropriate in the time available even to attempt to summarize the contents. Instead I shall be very selective, and in doing so try to establish some of the contributions of physical metallurgy already made to successful alloy development and some which can be anticipated. Physical metallurgy, while it is a coherent discipline, can be said to exist at a number of different levels which interact with each other. The original impetus came from the study of microstructure in the optical microscope pioneered by Sorby, combined with the concept of equilibrium diagrams which defined relations between chemical composition and structure, the latter being more precisely defined at a later stage by X-ray diffraction studies.

2. High strength aluminium alloys

The gradual development of aluminium alloys over the period 1920–1940, and in particular the Al–Zn–Mg–Cu alloys discussed by Hunsicker (3.6), owed much to the approach referred to above, which by 1939 had produced alloys with strengths up to 450–550 N mm^{-2}. However, while these alloys relied on precipitation hardening for much of their strength, detailed study of the dispersed phases responsible was not yet possible, indeed the high resolution electron microscopy needed was not available until after 1945. This illustrates that alloy development can go a substantial distance using only the 'surface layers' of physical metallurgy as a guide. The word 'guide' is used advisedly, because although by this time most relevant binary and ternary equilibrium diagrams were known at least in broad outline, they could not be precisely applied to complex alloys with four or five deliberate alloying additions, nor did they describe the metastable structures obtained by most strengthening heat treatments. In 1940 stress corrosion in these alloys was already a serious problem which could only be dealt with by an empirical search for the alloy with maximum resistance to stress corrosion cracking, together with the desirable strength level and reasonable workability. Precise scientific analysis of this phenomenon had to wait until well after World War II. In the decade immediately following World War II, deeper layers of physical metallurgy began to be penetrated; in particular the basic principles underlying the strength of metals and alloys were established by extensive development of dislocation theory and related experiments. At the same time the increased resolution of electron microscopes led to the detailed examination of the various stages of precipitation in alloys and eventually to the direct observation of dislocations in thin foils. These major developments in basic physical metallurgy made it much easier to understand the frequently complex behaviour of aluminium alloys, but unfortunately the basic work which was usually done for good reason on simple binary aluminium alloys, was not always extended with the same zeal to commercial alloys. The consequence was that, while the fine microstructure of commercial alloys became easier to study, the complexities of behaviour, for example on ageing, ensured that empirical development would continue. However it is, I believe, fair to say that empirical development carried out with an increased background knowledge is a great improvement on unaided empiricism.

It is clear from Hunsicker's paper that the improvement of the Al–Zn–Mg–Cu alloys in the period 1960 until the present day had not led to increased strength levels for the optimum strengths had almost been reached in the previous decade. Indeed, the emphasis in recent years has been in the combination of the appropriate strength level with resistance to corrosion, in particular stress corrosion cracking. Alloys have now been developed with virtual immunity to stress corrosion, but it must be said that the basic principles behind the empirical solutions have not yet been fully spelt out. This arises from the fact that it has not been possible to study grain boundaries, frequently the sites of stress corrosion cracking, in sufficient detail until comparatively recently. It is now possible, for example by use of electron energy analysis cobined with electron microscopy, to study the solute depletion profiles in aluminium alloys

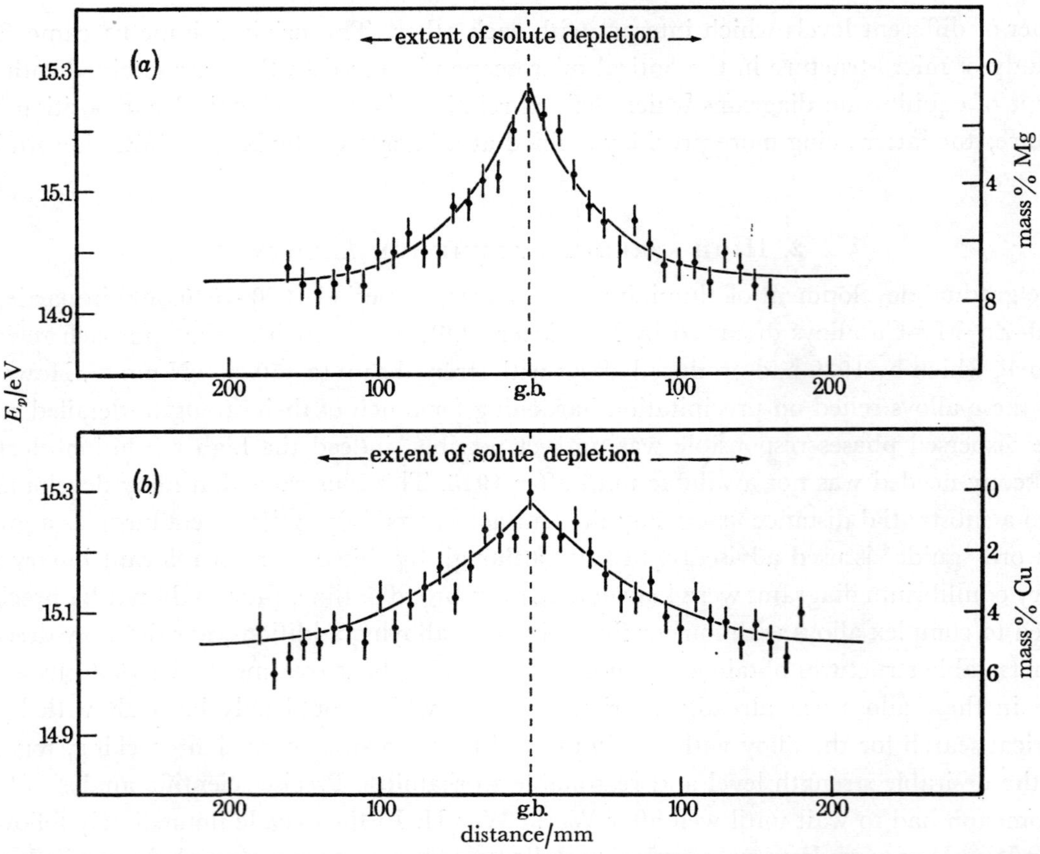

FIGURE 2. Typical energy loss/composition profiles across a grain boundary in (*a*) Al–7.2 mass % Mg aged 200 h at 125 °C, (*b*) Al–4.4 mass % Cu aged 100 min. at 220 °C (Doig & Edington).

sensitive to stress corrosion (Doig & Edington 1974). Figure 1, plate 29 shows the grain boundary precipitation of β' in an Al–7 mass % Mg alloy which leads to s.c.c. sensitivity, while figure 2 shows the electron energy loss profiles in two alloys which can be directly related to solute depletion. So far only binary and ternary aluminium alloys have been studied in this way, but further work on complex commercial alloys should be done, and in my view would be very useful.

The other over-riding requirement of many modern high strength aluminium alloys is that they should have good fracture toughness at high static stress levels – this applies in the

---

DESCRIPTION OF PLATE 29

FIGURE 1. A typical grain boundary in an aged Al–7.2 mass % Mg alloy showing β' Al_3Mg_2 precipitates (Doig & Edington).

FIGURE 5. A unidirectionally grown lead-silver eutectic (*a*) as grown, (*b*) isothermal treatment for 120 h at 558 K, (*c*) 94 hours in gradient of 4 K mm^{-1} at less than 554 K. Optical micrograph transverse sections (magn. × 500), (*d*) etched, same as (*c*) scanning electron micrograph (magn. × 1000) (Jones).

FIGURE 6. The dispersion of vanadium carbide in an Fe–V–C alloy after isothermal transformation and ageing at 725 °C for (*a*) 5 min, (*b*) 15 min showing coarsening on dislocation network. (Dunlop & Honeycombe.)

FIGURE 7. An iron–0.04 mass % carbon alloy with 680 ppm antimony. Cracking at the iron carbide/ferrite interface. (Rellick and McMahon.)

FIGURE 8. Beginning of isothermal transformation of a Ti–7 mass % Ni alloy at 640 °C. Optical micrograph × 300. (Chatfield.)

FIGURE 10. Intergranular cavities in a failed Calder Hall fuel element. (Magn. × 45 (Eldred).)

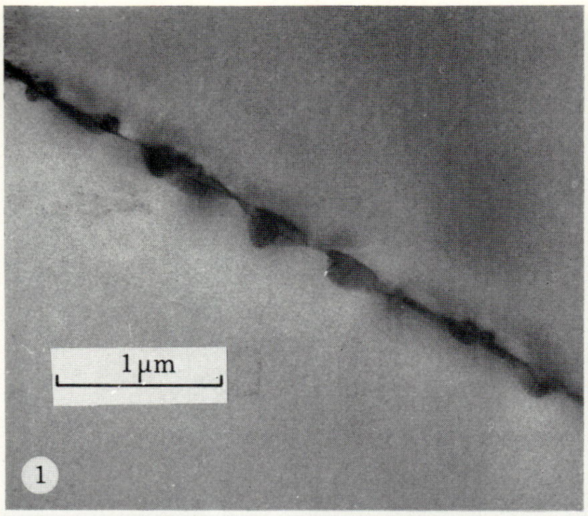

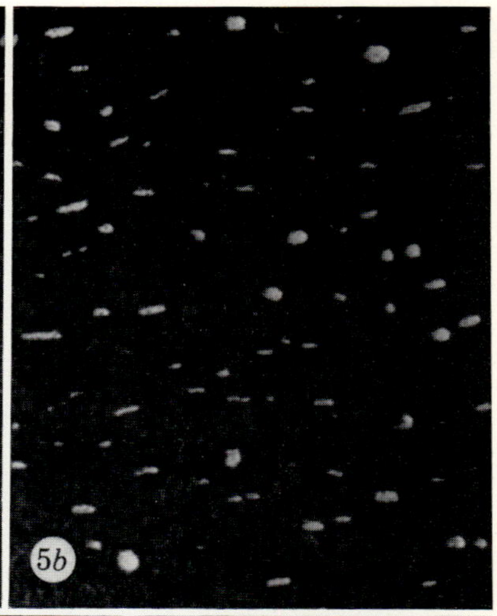

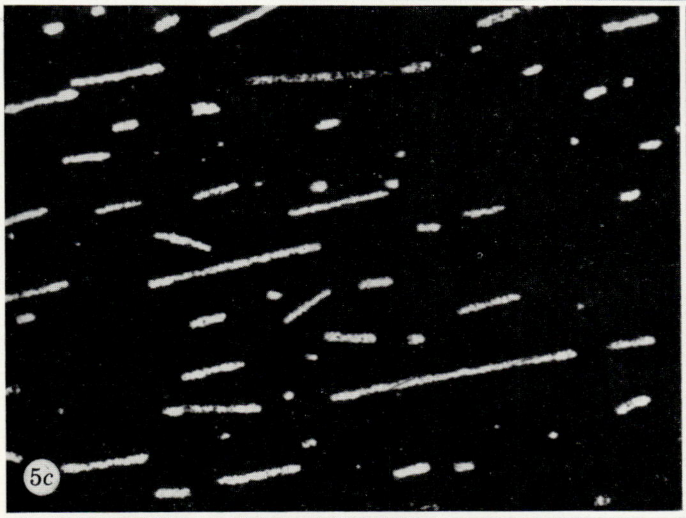

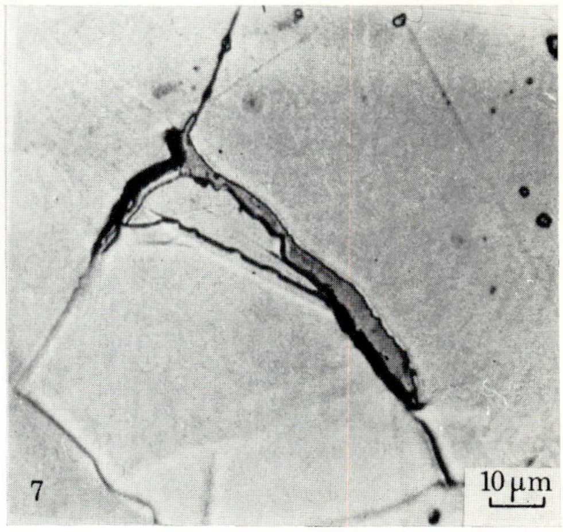

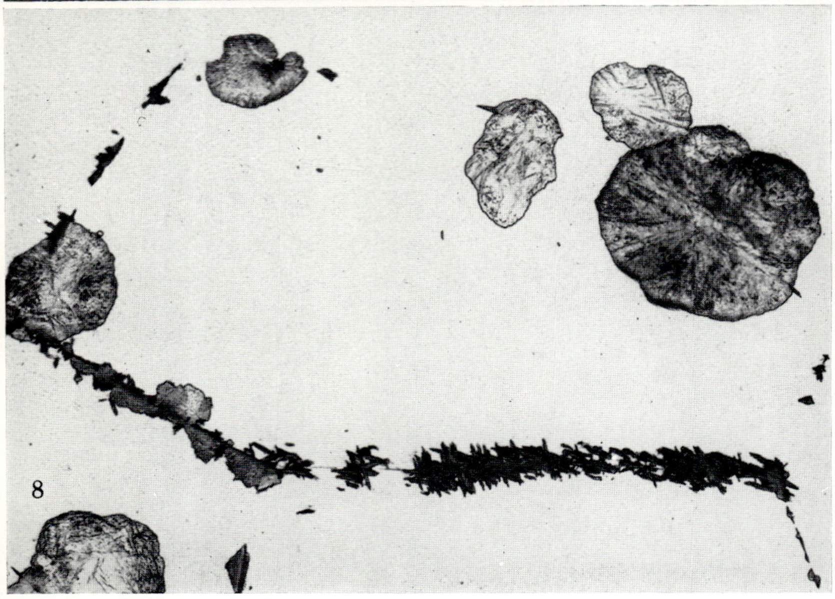

FIGURES 1, 5*a*–*d*, 6*a*, *b*, 7, 8 and 10. For description see opposite.

presence of stress corrosion cracking which is essentially a brittle crack propagation, and even in the absence of this phenomenon when frequently inadequate resistance to crack propagation occurs. In the latter case, the harmful role of coarse particles of intermetallic compounds is now fully understood, and a positive effort is made both by control of composition and heat treatment to minimize their occurrence. The 7474 alloy referred to by Hunsicker has thus close limits imposed on impurity elements such as iron and silicon, and also much attention is paid to control both of chemical homogeneity and of grain size, together with careful particle size control of the strengthening precipitate phase ($Al_{12}Mg_2Cr$).

Finally the aluminium alloy scene would not be complete without some reference to fatigue failure. While it might be thought intuitively that high fracture toughness and fatigue resistance should go hand in hand, Hunsicker throws some doubt on this view, and clearly there is room for more work in this area. Nevertheless the use of thermomechanical treatments (pioneered in the alloy steel field) have shown promise of simultaneously increasing ductility, fracture toughness and fatigue resistance.

3. Nickel-base alloys – high temperature properties

While many basic investigations carried out on aluminium alloys have provided a good scientific framework for future alloy development, it is, I think, fair to say that the use of scientific principles has found fullest scope in the development of precipitation-strengthened nickel-base alloys for use under stress at high temperatures. Moreover it is in this type of application that the metallurgist and engineer work in closest contact, for the technical rewards to be gained by raising the operating temperatures of a gas turbine by only a few degrees are easily worth an intensive programme of alloy development to meet the challenge.

The use of age hardened alloys at high temperatures brings up the important question of stability of microstructure, because the particular creep properties of an alloy can only be maintained at the operating temperature if the microstructure does not alter markedly during the service life of the part. Put the other way around, the service life of the part will be determined to a substantial degree by the ability of the microstructure to resist significant changes. Nicholson (3.8) has discussed the γ' hardening phase in nickel-base alloys and pointed out that its stability at high temperatures is dependent on the close lattice matching between γ' and the nickel-base solid solution matrix. The resulting low energy interfaces do not provide a high driving force for coarsening. The idea of deliberately matching crystal lattices to each other arises from well-defined concepts concerning phase transformations in alloys, and it is indeed satisfying that such concepts can be used along with more empirical methods to improve, for example, the creep properties of an important group of high temperature alloys.

Stability of microstructure is important in another related field, that of the directional solidification of alloys in particular eutectic alloys for possible applications as turbine blades. As Nicholson has pointed out, the growth of single crystal turbine blades has now been shown to be technically achievable, with obvious advantages in the elimination of grain boundaries which play such an important role in creep failure. Brook & Duckworth (3.10) have also emphasized the use of directionally solidified eutectics for similar applications (their figure 11). There is however some reason for concern about the thermal stability of some eutectic structures, not so much at constant high temperatures, but in particular, when the alloy is subjected to a temperature gradient which is the normal circumstance in practice.

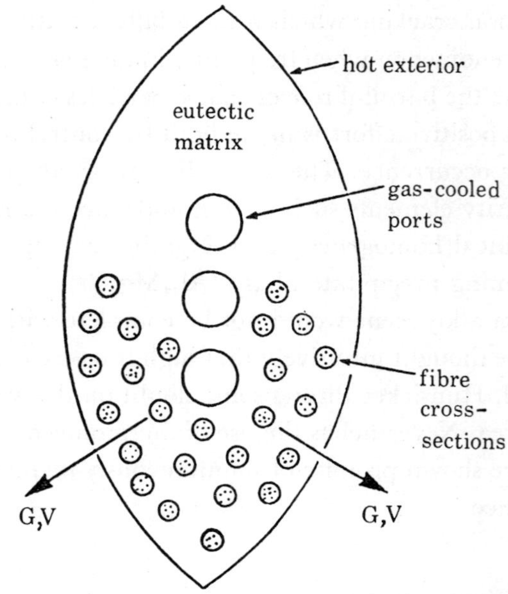

FIGURE 3. Schematic cross-section of a unidirectionally solidified fibrous turbine blade in service (Jones).

It is now well established (Barnes & Mazey 1963; Shewman 1964; Ho 1970) that inclusions can migrate along a temperature gradient as a result of the driving force provided by thermal diffusion. In the case of two mutually insoluble metals A and B, A forming the inclusions and B the matrix, the motions of the inclusions are limited entirely by diffusivity of B atoms; this has been fully described theoretically (Ho 1970). Jones (1974*a*) has applied the theory to the turbine blade situation where the temperature gradient G is normal to the fibre axis (figure 3), and the rate controlling process is the diffusivity D_i. The velocity of migration of the fibre V_{fibre}, can be written as

$$V_{(\text{fibre})} = D_i \delta Q G / 2RT^2 r, \tag{1}$$

where δ is the thickness of high diffusivity layer, D_i the diffusivity of matrix metal B in the interphase boundary, Q the heat of transport of B, r the fibre radius, R the gas constant, and T the absolute temperature.

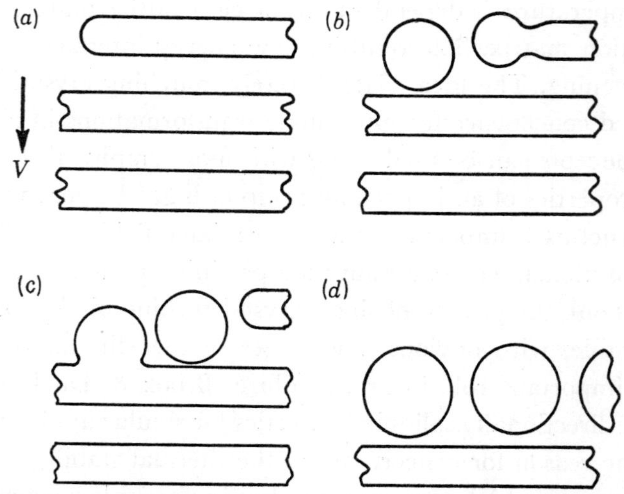

FIGURE 4. Sequence illustrating how in a temperature gradient a spheroidized eutectic rod can catalyse the spheroidization of its neighbours (Jones).

Jones shows, by use of typical metallic values that the fibres will migrate a distance of 1–5 μm in 10 h at 1000 K in a gradient of 100 K mm^{-1}, which is comparable to the inter-rod or lamellar spacing of a typical eutectic alloy. Such an effect would be expected to lead to a rapid break-down of the aligned structure. This break-down is accelerated by irregularities in r, the fibre radius, but even if it is constant, variation in interfacial kinetics e.g. due to anisotropy, impurities, etc., can readily lead to the degeneration of eutectic rods. The process of progressive spheroidization is shown schematically in figure 4*a–d*. Experiments have been carried out on directionally solidified Al–Al_3Ni (Jones & May 1975) and lead–silver eutectics (Jones 1974*b*) and in both cases a dramatic effect of temperature gradient on structural stability was shown. In the former case, increases in coarsening rate of up to twenty times were measured in the temperature range 868–883 K. The coarsening of the lead–silver eutectic is illustrated in figure 5*a–d*, plate 29, in which a transverse section of the original directionally grown eutectic (figure 5*d*) is compared with an isothermally heated specimen and one in a gradient as low as 4 K mm^{-1}. Such data would appear to call into question the use of directionally solidified eutectics for turbine blade use, and give some comfort to those who have produced sophisticated nickel-base alloys with very stable dispersed precipitates. Even in these circumstances the stability of finely dispersed phases needs close attention, a problem not only in nickel-base alloys, but also alloy steels and other alloys employed under stress at elevated temperatures.

Particle coarsening at elevated temperatures is defined to some degree by the Lifshitz–Wagner theory which assumes diffusion control:

$$(r_t^3 - r_0^3) = (K/RT)\,V^2\,CD\gamma t, \tag{2}$$

where r_0 and r_t are the mean particle radii at zero time and time t, D is the diffusivity of solute in matrix, γ the particle interface interfacial energy, C the conc. of solution in equilibrium with a particle of infinite size, V the molar volume of precipitate, and K a constant.

However, this theory gives only rough agreement with experiments for various reasons, of which the most relevant are connected with the acceleration of growth beyond that predicted. At least two types of sites in crystals provide for this.

(*a*) Grain boundaries and interphase boundaries: Precipitate particles nucleate at grain and interphase boundaries because they are energetically favourable sites, but having done so, the accelerated diffusion of solute along such interfaces leads to more rapid coarsening than within the grains.

(*b*) Matrix dislocations: Recent work (Dunlop & Honeycombe 1975) on low alloy steels has shown that very fine carbide dispersions within grains coarsen predominantly by selective growth of those particles lying on dislocation lines (figure 6, plate 29). It has been found that a relatively low dislocation density can significantly increase coarsening rates, because of the ability of dislocations to climb, and for their strain fields to interact with those of coherent precipitates. The result is a much broader size distribution of particles than predicted by the Lifshitz–Wagner theory. This theory gives a linear relation between r^3, the mean particle radius, and time for volume diffusion controlled growth, and between r^2 and time for interface reaction control. However a linear relation between r^5 and time was obtained in the case of vanadium carbide coarsening in ferrite, as predicted by Kreye (1970) for coarsening of particles lying on dislocations:

$$(r_t^5 - r_0^5) = (K'/RT)\,V^2\,CD_d\,q\delta Nt, \tag{3}$$

where D_d is the diffusion coefficient for dislocation pipe diffusion, q the effective cross-sectional area of the dislocation pipe diffusion path, N the no. of dislocations intersecting each particle, and K′ constant.

So we are now aware of at least some of the important principles determining the stability of microstructure. The application of these principles to industrial alloys is often very difficult, consequently the empirical approach is still necessary; however it now has substantial backing from basic physical metallurgy.

4. High strength steels

As May and colleagues have pointed out in their paper, achievement of very high strength in steels is no longer the real challenge. Today the interest is in high strength steels with great resistance to crack propagation under static load, under cyclic load and often in corrosive environments. This goal of a satisfactory combination of diverse properties is much more difficult to achieve, and calls for extensive basic knowledge often in new fields, in combination with a sensible empirical approach. Steel metallurgists have perhaps clung rather too long to the traditional 'quench and temper' approach to high strength alloy steels. The martensite transformation which occurs on quenching is a valuable way of raising the dislocation density (as high as 10^{11}–10^{12} cm^{-2}), but the high interstitial hardening due to carbon in solid solution brings with it brittleness. The tempering treatment precipitates most of the carbon as iron carbides or alloy carbides, to a large extent on to what is left of the quenched-in dislocations which now form an effective strengthening network in a ferrite matrix. The result is a strong and frequently (but not always) tough material.

One of the dangers of this route to high strength is temper embrittlement which is induced in some steels by tempering in a critical temperature range (500–600 °C), or by cooling at critical rates through this temperature range. It is known that impurity elements, in particular the elements Sn, Sb, As and P are responsible for the embrittlement, and recent work by Rellick & McMahon (1974) has shown that this can be due to the rejection of these elements during carbide formation and growth. Figure 7, plate 29, shows an example of intergranular cracking along a carbide-ferrite interface in a simple iron–0.04 % carbon alloy doped with antimony (680 parts/10^6). If the temperature is low enough to avoid much diffusion, then the elements remain at or near the carbide matrix interface and nearby ferrite/ferrite interfaces. In plain carbon steels such embrittlement is transient, being removed by tempering for long times, however in certain alloy steels, e.g. containing nickel and chromium, the effect is not readily removed in this way. Auger spectroscopy of intergranular fracture surfaces leaves no doubt that the embrittling agents are elements such as antimony and arsenic. In this connection it is interesting to note that a recent survey by the C.E.G.B. has revealed that elements such as As, P, Sn, Pb and Sb are the main contributors to cracking in heat affected zones around welds in steels commonly used in power plant (Townsend 1975). The best solution to such problems in the long run is the use of purer steels with the harmful impurities below certain very low limits. However this trend at present runs contrary to efforts to increase use of scrap with the consequential build up in the resulting steels of harmful impurity elements. The risks in allowing this trend to continue should be fully realized and prompt action taken.

As May and co-authors (3.7) have pointed out, several alternative ways to high strength involve modification either of the austenitic structure or of the martensite resulting from low temperature transformation. The austenitic condition can be retained to room temperature by suitable alloying, and then high strength with good ductility is obtained simply by heavy cold work. A refinement of this approach involves the warm working of austenite of carefully balanced composition to give a high strength deformation structure. Cooling to room temperature retains the austenite, but this phase is now metastable; under the action of plastic deformation it undergoes strain-induced transformation to martensite leading to high strength levels, good ductility and resistance to crack propagation. However these methods involve very close control of composition, working and heat treatment. The maraging steels provide an alternative solution where carbon is eliminated from the steel so that a low carbon martensite can be formed by air cooling; at the same time the composition is adjusted to allow the precipitation of intermetallic compound dispersions on subsequent ageing. Again high strength and toughness are achieved, but high concentrations of expensive alloying elements are needed. The ausforming process provides yet another alternative route in which metastable austenite (in the TTT curve 'bay') is heavily deformed prior to transformation to martensite followed by tempering in the normal way. The undoubted success of this process in achieving high strength, together with toughness in a range of alloy steels is not yet fully explained, and the method places severe limitations on the applications for which it can be used.

All the above processes are tedious, difficult to control and/or use high concentrations of expensive alloying elements. They provide solutions to the strength with toughness problem but at a price that cannot always be paid, except in critical applications. There are clearly needed easier and cheaper routes to strong, tough steels. An obvious approach which should be further explored is the direct decomposition of austenite to a fine-grained dispersion strengthened ferrite without going via martensite. The change in carbide solubility as the γ–α transformation takes place can lead to very fine dispersions of alloy carbides formed as the ferrite grains form in the austenite (Davenport & Honeycombe 1971). This direct transformation approach has been shown to be successful in low alloy steels where, when it is combined with controlled rolling which achieves extremely fine grain sizes, is capable of reaching yield stresses in excess of 500 N mm^{-2}. It seems likely that yield stresses very substantially in excess of this will be achieved when the physical metallurgy of direct transformation of austenite is better understood.

It is tempting at this stage to define the ideal type of structure of a high strength steel, but as yet we can only do this in part. It is of course impossible to avoid sulphide and oxide inclusions in commercial steels and their control is of great importance, however we shall concentrate on the basic steel structure. In normal circumstances the matrix should consist of ferrite with as fine a grain size as possible (around 1 μm), equi-axed, containing a fine uniform dispersion of carbide or intermetallic phases with no large grain boundary or other carbide particles. In addition a dislocation-network should exist within the grains stabilized by some of the fine particles to provide sub-grain strengthening. Finally the grain boundaries should be as free as possible of certain impurity elements already referred to. It is clear that such structures can be approached in several different ways, but the ideal approach does not yet exist.

In this section the alternative methods to high strength and toughness in steels have been briefly referred to. Several of these approaches could well find effective application in non-ferrous alloys, but it must be admitted that research in thermo-mechanical treatment of alloys has concentrated mainly on steels.

5. Titanium alloys

Titanium alloys share with many steels the occurrence of a phase change which allows the structure of the alloys to be varied over wide limits, though it is clear that the ramifications of the γ–α transformation in iron have been studied in very much greater detail than the β–α transformation in titanium. This is not surprising since the developments in titanium alloys have occurred primarily since 1950, but it has meant that the titanium alloys have not yet exploited fully the available strengthening routes. For example, there are very few precipitation hardened titanium alloys available, and only one is made commercially. It is a little disappointing that the development of titanium alloys has contained a large empirical element, but this is still inevitable today when most alloys finally chosen for applications are complex in composition, and represent delicate balances between desirable properties. However a close examination of the development to date shows that broad physical metallurgical principles have been widely used, but these relate mainly to the surface levels of understanding rather than the more profound details e.g. of alloy theory in alloy design, or of dislocation theory in the optimization of mechanical properties.

Indeed it would appear to be a valid point that progress may well have been more rapid had a little more attention been given to the fine microstructure and its relationship to mechanical properties, in particular fatigue resistance and fracture toughness. It is surprising that more work has not been done on TTT diagrams or on the structures resulting from isothermal decomposition of β titanium. Recent work on the titanium–nickel alloys indicates that there are many analogies with ferrous structures, but the structures are extremely complex. Figure 8, plate 29, shows the beginning of transformation from metastable β at 640 °C in a Ti–7 % Ni alloy; both pearlitic and bainitic-type structures are already evident in the microstructure.

6. The future in non-ferrous metallurgy

In their paper, Brook & Duckworth have examined some of the more exciting recent developments in physical metallurgy such as memory alloys and superplasticity, and have also bravely tried to predict future trends in such fields as well as the modes of processing which may be extensively used. While it is impossible to review all of their paper it is perhaps worth emphasizing the criteria which may become more commonplace in deciding development trends.

We have all rather suddenly become more energy conscious and appreciative of many other aspects of conservation. One result of this increased sensitivity is that more positive efforts will be made to use expensive metals sparingly, and to replace increasingly rare metals by other materials. In the past and still today we are often profligate with our metallic resources. In ferrous metallurgy clear steps have been taken towards the use of steels leaner in alloying elements, but just as strong as the older alloys they have replaced. To do this has meant an awareness of several important principles in physical metallurgy such as the grain size dependence of the yield stress and the effect of the γ/α phase transformation on the dispersion of carbides. Without limiting consideration to non-ferrous metallurgy I think that several approaches to alloy development and improvement can be mentioned as compatible with the new technological climate, and thus we would expect them to be increasingly employed in future.

(*a*) *Grain refinement*

Physical metallurgical research has clearly underlined the importance of grain refinement as a strengthening mechanism, which has the added advantages of also increasing toughness in alloys which show ductile brittle transitions. While the non-ferrous industry has in some areas used grain refinement very effectively, it is probably in the steel field, in particular in control-rolled micro-alloyed steel, that the most spectacular advances have been made. Yet there is a big challenge still in both ferrous and non-ferrous metals and alloys where grain refinement could be much more widely practiced with resultant improvement in mechanical properties at relatively small cost.

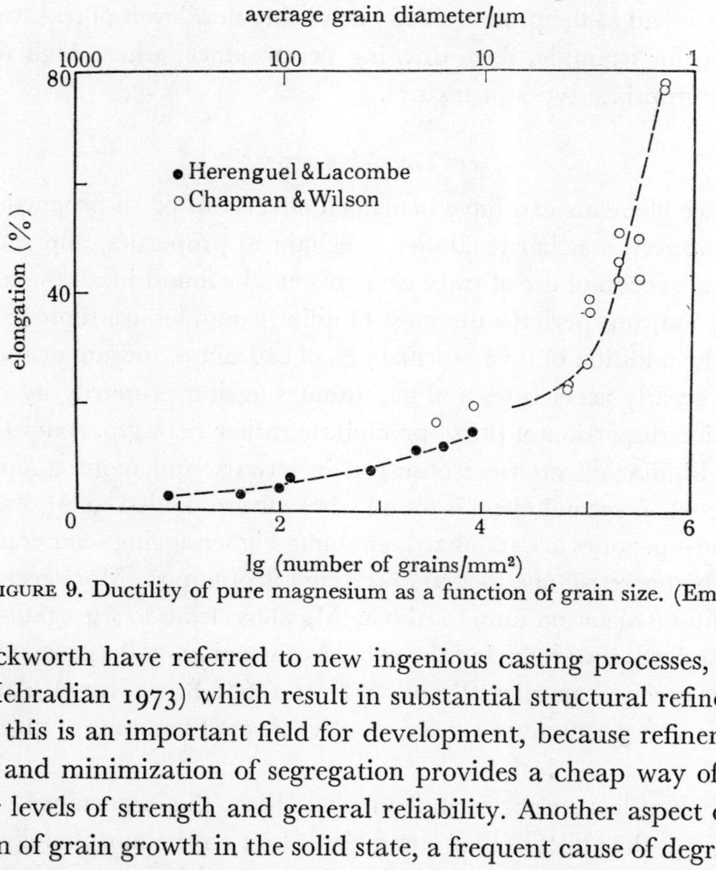

FIGURE 9. Ductility of pure magnesium as a function of grain size. (Emley.)

Brook & Duckworth have referred to new ingenious casting processes, such as rheocasting (Flemings & Mehradian 1973) which result in substantial structural refinement. I share their conviction that this is an important field for development, because refinement of the primary grain structure and minimization of segregation provides a cheap way of upgrading existing alloys to higher levels of strength and general reliability. Another aspect of grain size control is the prevention of grain growth in the solid state, a frequent cause of degradation of mechanical properties. The great benefit to room temperature ductility in refining the grain size of pure magnesium is illustrated in figure 9. In nuclear reactors for power generation, serious problems have arisen in Magnox fuel cans due to excessive grain growth of the magnesium solid solution (Pugh *et al.*). At the operating temperatures, marked cavitation can take place at the boundaries (figure 10, plate 29) and eventually lead to intergranular failure with resultant leaking of radioactive material into the cooling gas. This problem has now been largely overcome by semi-empirical means, but the challenge of maintaining in many alloys a fine grain size under stress at elevated temperatures still remains. Some basic theory exists, for example Gladman (1966), but each alloy still needs a separate detailed approach to provide a satisfactory practical solution to the problem.

(*b*) *Texture control*

Preferred orientation of grains in alloys can have both detrimental and advantageous effects on mechanical properties. In non-ferrous metallurgy, particularly in copper and aluminium-base alloys, control of texture is now commonplace, but there are important aspects yet to be fully exploited, e.g. texture strengthening which is particularly effective in hexagonal materials such as titanium and magnesium alloys. Turning to the ferrous field, I think it is fair to say that it is only recently that deliberate control of texture has been accepted as a method of upgrading mechanical properties (Mintz, Morrison, Morris & Davies 1975). This belated exploitation arises partly from the complexity of textural changes where a phase change is superimposed on the effects of deformation and annealing. This is clearly a field where physical metallurgy must give a lead to the process side in achieving new levels of textural control which will bring benefits in, for example, deep drawing performance, where high R values can be obtained with the appropriate types of texture.

(*c*) *Trace elements*

In some alloys, trace elements can have beneficial effects out of all proportion to their concentration on such properties as hardenability, mechanical properties, and stability of microstructure. Examples of successful use of trace elements can be found in all the areas covered by the papers in this session, but perhaps the most familiar examples are those in the aluminium alloys in particular the addition of 0.05–0.1 mass % of cadmium, indium or tin to aluminium–copper alloys which greatly accelerates and accentuates ageing primarily by encouraging the formation of a very fine dispersion of the θ' precipitate rather than g.p. zones 1 and 2. (Hardy, Silcock *et al.* 1956). Similar effects were observed in ternary and more complex commercial alloys (Polmear 1960–1). A second effect is found when silver is added to Al–Zn–Mg–Cu alloys when, in contrast, the g.p. zones are stabilized, enabling higher ageing temperatures to be used, with the result that better resistance to stress corrosion is obtained. More recently it has been found that small additions of germanium to Al–Cu–Mg alloys leads to segregation of germanium at the precipitate interfaces thus reducing the rate of coarsening, and as a result improving the elevated temperature creep properties (Brook & Hatt 1969; Brook 1975). An important but complicating factor is that germanium is only effective if small amounts of magnesium are also present.

In his paper, Nicholson has emphasized, for nickel alloys, the importance of trace elements at grain boundaries, e.g. the effect of boron and zirconium on the morphology and stability of grain boundary carbides. Similar effects result from boron additions to stainless steels. However trace element segregation occurs also to interphase boundaries where the impurity atoms alter interfacial energies and also interact with interface dislocations. These effects play a most important part in controlling the scale of the microstructure, its stability at elevated temperatures, and hence in determining the mechanical properties both at room and high temperatures.

While some of the above phenomena have been known for many years, new trace element effects continue to be found in both high purity and in commercial alloys. Fortunately we are now just at the stage when the techniques capable of resolving the various trace element phenomena are becoming available, e.g. Auger spectroscopy, atom probe, electron energy analysis and scanning transmission electron microscopy. We can thus look forward to a better understanding of trace element phenomena, and to hopefully a less empirical approach to their exploitation in industrial alloys.

The author is grateful to several of his colleagues whose work is referred to in this paper, and who have provided illustrations.

References

Barnes, R. S. & Mazey, D. J. 1963 *Proc. R. Soc. Lond.* A **275**, 47.
Davenport, A. T. & Honeycombe, R. W. K. 1971 *Proc. R. Soc. Lond.* A **322**, 191.
Doig, P. & Edington, J. W. 1974 *Proc. R. Soc. Lond.* A **339**, 37.
Dunlop, G. L. & Honeycombe, R. W. K. 1974 *Phil. Mag.* **32**, 61.
Flemings, M. C. & Mehrabian, R. 1973 *Trans. Am. Found. Soc.* **81**, 81–88.
Gladman, T. 1966 *Proc. R. Soc. Lond.* A **294**, 298.
Hardy, H. K. 1956 *Inst. Metals* **84**, 429.
Ho, P. S. 1970 *J. appl. Phys.* **41**, 64.
Jones, D. R. H. 1974*a* *Metal Sci. J.* **8**, 37.
Jones, D. R. H. 1974*b* *Mat. Sci and Eng.* **15**, 203.
Jones, D. R. H. & May, G. J. 1975 *Acta Met.* **23**, 29.
Kreye, H. 1970 *Z. Metall.* **61**, 108.
Mintz, B., Morrison, W. B., Morris, P. P. & Davies, G. J. 1975 *Texture and the properties of materials* (ed. G. J. Davies *et al.*). London: The Metals Society.
Polmear, I. J. 1960–1 *J. Inst. Metals* **89**, 51, 193.
Pugh, S. F. *et al.* 1972 *J. Br. Nuclear Energy Soc.* **11**, No. 4.
Rellick, J. R. & McMahon, C. J. 1974 *Met. Trans.* **5**, 2439.
Silcock, J. M., Heal, I. J. & Hardy, H. K. 1956 *J. Inst. Metals* **84**, 23.
Townsend, R. D. 1975 Meeting on Materials in Power Plant, Institution of Metallurgists, Edinburgh.

Discussion (Chairman I. J. Polmear (*University of Monash*))

G. B. Brook (*Fulmer Research Institute*)

I should like to enlarge on the part of our paper (3.10) in which we considered the effect of trace additions of specific elements on precipitation and on methods of controlling precipitate size. By the term 'trace elements' I differ from those working with steel and mean deliberate additions of specific elements in amounts from a few hundreths to about one tenth of an atomic percentage. The atoms of these elements tend to be different in size from those of the matrix and so it is natural that they should seek some site in the lattice at which they can reduce their energy. This they do by segregating preferentially to defects and interfaces and this is how their effects can be utilized. There are four typical sites:

1. Grain boundaries which have recieved most attention. Here trace elements can be used to control the rate at which boundaries can migrate. This can be at higher or lower rates than the normal migration rate of the boundaries in a pure alloy depending on the nature of the trace element. Thus they can cause grain refinement or at the other extreme, they can introduce a property akin to superplasticity for which only about 0.1 atomic percentage is needed. Unwanted trace additions can cause embrittlement but intentional additions can replace detrimental elements at boundaries and thus increase ductility. Embrittlement is often caused by discontinuous precipitation which is a consequence of grain boundary migration. The use of boron in steel to increase the depth of hardening is an example of a trace element effect at grain boundaries. By reducing boundary migration, the formation of pearlite is inhibited.

2. Dislocations which have been well treated especially in steels and aluminium alloys.
3. The matrix/precipitate interface which is now thought to be a special case of segregation to dislocations, in this case to growth dislocations at the matrix/precipitate interface. Trace elements have been detected at this interface and as long as they remain there, growth of the precipitate is prevented. This avoids the coarsening to which Professor Honeycombe has referred.
4. Vacant lattice sites where they can control diffusion rates. This is extremely important as less than 0.1 % of one atomic species can control the whole diffusion process over a wide range of temperature. If the trace element forms part of the precipitate or structure which is produced by diffusion, then the trace element will accelerate precipitation; whereas if the trace element plays no part in the precipitate structure, it denies vacancies to the diffusing species. We have cases of elements added to aluminium–copper alloys delaying g.p. zone formation for more than three years, when in the absence of the trace element, g.p. zone formation would be complete in about 48 h.

I would like to concentrate on the last two processes and illustrate them by an example which enabled us to develop a new engineering material from first principles using our knowledge of trace element effects.

In aluminium–copper alloys, the precipitation of the θ' phase can be controlled by trace elements. As little as 0.05 % of certain elements will preferentially interact with vacancies and will thereby prevent the diffusion of copper. At the same time abnormally high concentrations of vacancies are retained in solid solution. However the interaction with vacancies and trace elements is not stable at all temperatures. If the alloy is heated to about 160–170 °C, the complex decomposes and an abnormally high transient vacancy concentration results in rapid diffusion of copper and dense nucleation of a fine distribution of θ'. This produces an alloy of higher strength than is normally achieved. Furthermore, trace elements themselves diffuse to the dislocation on the semi-coherent matrix/θ' phase interface and by preventing further growth retain a fine precipitate size. This has been deduced from X-ray diffraction experiments and has recently been confirmed by electron microscopy. Some trace elements such as tin, indium or cadmium are only sparingly soluble in aluminium and eventually precipitate in their own lattices. In this event, the atoms at the interface dislocations redissolve in the matrix and the precipitate coarsens rapidly.

Thus it became necessary to seek more stable trace element configurations. From their effect on diffusion of copper to form g.p. zones, we discovered that some elements, which had little or no effect when added alone, had considerable effect when added together in the correct proportion. Such pairs of elements are magnesium plus germanium and, less effectively, magnesium plus silicon. There is considerable evidence that these pairs of atoms exist in association in solid solution in aluminium. Mg + Ge prevent g.p. zone formation after quenching aluminium–copper alloys from solution treatment temperatures. The interaction between Mg + Ge and vacancies is so strong at 30 °C that no zones formed in three years. On heating to 160–170 °C, very fine stable precipitates were obtained. The same X-ray diffractions were found for combined Mg + Ge additions as were found for single additions of Cd and In; these were absent for single additions of Mg or Ge.

This suggested that combined additions of Mg and Ge to an aluminium–copper alloy would produce an alloy of high strength and stability. We found that when made on a commercial scale, this alloy not only had the high strength and creep resistance we predicted but also high fracture toughness because we had few inclusions and good stress corrosion resistance as trace elements had nucleated the precipitates uniformly so avoiding depleted zones adjacent to coarse precipitates at grain boundaries. Competing alloys use 1 % Fe + 1 % Ni, which form fine inclusions, to achieve creep resistance. By using trace elements to control precipitate coarsening, the need for iron and nickel disappears as does the lack of toughness associated with the inclusions.

This is an example of what can be achieved by making use of what is known about trace elements. By adding selected trace elements to a well known alloy system, it has been possible to design a new engineering alloy with a combination of properties which could not be achieved by any other means.

I would also like to refer to the question raised in an earlier session about increasing the elastic modulus of aluminium alloys. This can be increased quite readily by incorporating insoluble intermetallic compounds of high modulus. Such an approach was used many years ago by Dudzinski and his colleagues† who achieved a Young modulus of 83 GN/m^2 (12×10^6 lbf/in^2). However this is achieved at the expense of ductility and fracture toughness. A high specific modulus can be achieved in Al–Mg–Li alloys again by introducing an insoluble high modulus intermetallic compound. However I think that the most effective way of producing high modulus materials is through fibre reinforced composites rather than conventionally alloyed materials.

E. D. Hondros (*National Physical Laboratory*)

I would like to comment on the important question of residual impurity levels, their effects on mechanical properties and their control.

Recent work is showing that residual species affect both high temperature properties (creep resistance, weld re-heat cracking etc.) and low temperature properties (temper brittleness etc.) and although similar elements are involved, the hierarchy of surface activity is different according to whether it is a high or a low temperature phenomenon: thus, copper and antimony appear to be more active for the high temperature rather than low temperature phenomena.

The question of controlling the residual levels by chemical fixation processes (as in the manner of content of sulphur surface activity by manganese additions) is worthy of future research effort. The amount of residuals at the grain boundaries depends on the bulk level and on temperature – below a certain bulk level, the grain boundary enrichment level will not promote embrittlement. With agents such as phosphorus and tin, there does not exist a thermochemically equivalent additive as manganese does for sulphur. Our work suggests that certain additives can reduce the bulk residual levels to tolerable amounts.

L. M. Wyatt (*Central Electricity Generating Board*)

I would like to refer to Honeycomb's allegation (Keynote) that the forgemasters like to use scrap by which I assume he means melted down car bodies and tin cans. I personally do not believe that the forgemasters themselves prefer this because they appreciate as well as we do that it leads to the accumulation of undesirable tramp elements in their steel. The reasons why some of them are forced to use inferior scrap arises from the quite illogical siting of most of the alloy steel plants at places like, to mention only two, Sheffield and Terni, miles from any blast furnace. This difficulty is compounded by the apparent fact that even though they may be

† N. Dudzinski, J. R. Murray, B. W. Mott & B. Chalmers. 1948 *J. Inst. Met.* **74**, 291–314.

part of an organization which itself makes tonnage steel direct from hot metal they may be denied low residual scrap which arises from the structural steel works. The result is that the cheaper tonnage steel gets better quality scrap than the more expensive alloy steel. If the forging plant could be located near the tonnage steel plant and closely integrated with it, it should be possible for them to be supplied with high purity hot metal which they can further refine or failing this with high purity scrap arising from the tonnage operation. I would have thought that it would have been purely a matter of administration to ensure that the high purity scrap for the tonnage steel produced direct from ore be supplied to the forgemasters and if this could be arranged it should be possible to produce high quality forgings with a low residual content very much more economically than at present.

I must protest at Honeycomb's statement that a serious hazard could arise from the grain boundary cavitation of Magnox cans. This does cause considerable operational inconvenience and an increase in cost because fuel elements have to be removed before they have completed their economic life. However, the onset of cavitation is readily detectable and it would take months or even years after it is originally detected before any conceivable hazard could arise either to personnel or to the plant. The worst that can arise from delay in discharging fuel elements would be a slight increase in the radioactivity of the reactor coolant which might lead to a very slight decrease in the sensitivity of the equipment which detects failed fuel elements. There is no difficulty in discharging such elements long before any noticeable effect on the reactor could occur.

H. Y. Hunsicker (*Alcoa*)

An engineering viewpoint expressed during the earlier sessions was that the aluminium-based aircraft materials would be considerably more satisfactory if they had better corrosion resistance, better stress corrosion resistance, higher fracture toughness and higher fatigue properties. Indeed, advances which would provide superior performance in these characteristics presumably would enable the designers and engineers to build more efficient aircraft having greater safety and reliability, longer life and requiring less frequent inspection and maintenance. Actually, the materials that are flying today are giving good performance and the improved materials developed recently that will be used in the new generation of aircraft are expected to overcome certain of the past deficiencies. The physical metallurgists and industry have consistently strived to offer materials that provide an effective compromise between the various engineering property requirements and sound material economics.

The matter of corrosion and stress corrosion resistance of the strong aluminium alloys has been of concern to the producers for a number of years. The early (1930s) development of alclad products was a significant advance in reducing corrosion problems. This well-established technique in which a thin cladding layer is applied in the rolling process for flat-rolled products or by co-extrusion has its basis in metallurgical and electrochemical concepts. The cladding compositions are designed with electrochemical characteristics that provide galvanic corrosion protection to the underlying core material. The traditional cladding compositions have generally been of low strength, and as pointed out by Forsyth (2.3) lead to earlier fatigue crack initiation. As described in the paper, a higher strength alclad product is used in one of the new wide bodied jets providing better resistance to fatigue crack initiation along with manufacturing and strength advantages.

Some of the earlier discussions and comments have intimated that basic understanding of stress corrosion cracking mechanisms is in a rather poor state of resolution for metals generally. This is partly because of the recognition of greater varieties of metallurgical structure features that can be observed with advanced techniques and can participate in this process as well as the increasing number of suggested interactive effects of environment and stress with the metallurgical structures. Cohen (discussion session 3, part 1) commented on the breadth of the dimensional scale on which we must deal with metallurgical structure features, which range from those of macrodimensions to those of atomic and lattice defect dimensions, and suggested that alloy properties and characteristics must be regarded as an integration of the effects of all such elements of structure. We must continue to be concerned not only with the effects on the submicroscopic scale but also with those of the much grosser features as, for example, the grain boundaries, which were of such great interest to Dr Rosenhain. These thoughts are quite relevant to the matter of stress corrosion behaviour of the high strength aluminum alloys. As related in my paper, stress corrosion resistance of the Al–Zn–Mg–Cu alloys is in a much better state of affairs with use of the overaged tempers than had been the case with the full strength tempers. The classic stess corrosion failure mode of the latter had always been observed as intergranular. This tendency has been overcome with the overaged tempers. However, we have observed that under conditions of very high tensile stress and highly aggressive corrosion environment, these materials now fail by a transgranular fracture mode. This is not expected to be a problem in using these materials, but illustrates how as we make progress in overcoming one situation, increase the severity of testing or service factors and push materials closer to performance limits, an entirely new situation may be encountered. It further emphasizes the need for greater depth of understanding of mechanisms.

In possibly a reverse sense with respect to the directional change of the limiting factor of structure, we have observed with this same alloy system that a change in predominant fracture mode limits fracture toughness with increasing strength. As we increase alloy content to develop higher strength than we now use (yield strengths of 700 MPa are readily developed) a transition occurs from transgranular to predominantly intergranular fracture mode. This is accompanied by a precipitous loss in fracture toughness. Thus, again the grain boundaries, their nature and morphology, and all the local modulations associated with them become matters of considerable importance and present challenges to us to acquire further understanding which should permit achieving new advances in useful engineering properties.

In the course of the discussions of the last two days there were frequent references to the need for tests that are reliably predictive of real-life service performance. To facilitate material development, optimization and selection, short-time or accelerated methods that can be correlated with long-time behaviour in time-dependent characteristics are critical needs. With respect to stress corrosion performance, as related in my paper, we had some years ago an unfortunate experience with poor predictions from accelerated laboratory testing of certain alloy compositions, specifically the lower copper content alloys of the Al–Zn–Mg–Cu type. However, the same testing procedures employed in qualifying material in a different composition range (higher copper contents) of the same alloy system provided short-time data which correlate very closely with long-time service information. Thus, details of the methods used to achieve valid predictive testing can be strongly influenced by the particular material being tested.

Dependable, meaningful quality assurance tests are also essential. In this respect it is interesting to point out that ostensibly indirect testing procedures are employed for quality control and assurance with 7075–T73 alloy products. To insure that the manufacturing operations have been properly carried out to provide very high resistance to stress corrosion cracking, a specified combination of two other properties not obviously related to stress corrosion resistance must be met. These are a maximum value of tensile yield strength and a minimum value of electrical conductivity. These properties are determined on samples from each production lot and are a basis for release of acceptable material. To maintain the established statistical correlative basis for this specification, frequent stress corrosion tests of both accelerated and long-time types are conducted. The information accumulated for over 15 years with properly processed 7075–T73 material meeting these specified mechanical and physical property requirements has provided a very high confidence level that it will have excellent resistance to stress corrosion cracking through long time service.

P. G. Partridge (*Royal Aircraft Establishment*)

Following Cohen's comments (discussion to session 3, part 1) I feel justified in dwelling for a moment on one of the most important factors affecting materials selection - that is the cost of the material. Hunsicker (3.6) has given us some idea of the enormous R. and D. effort which has been devoted to improving these alloys over the last 40 years. The strength improvements currently obtained amount to 5–15 % and the more complex heat treatments involve extra cost. But presumably a 15 % improvement is considered worthwhile. Imrie (1.5) however reported that an increase in strength of less than 15 % would not be sufficient to justify changing from a 300 M steel in undercarriages because of the increased ancillary costs. James (1.4) states that a significant improvement in airline economics could be achieved if the resistance to corrosion in Al–Zn–Mg alloys was improved. In view of the improvements described by Hunsicker it would have been nice to have had some numbers from James. For example how much improvement does he need and what cost is he prepared to pay?

Morton (3.9) says greater output/increased recycling is unlikely to reduce the cost of titanium. Cost reduction is more likely to arise by improving the yield, as in isothermal forging. He believes new alloys are unlikely to be needed in the foreseeable future. Similarly Glenny & Hopkins (1.6) conclude that the need for new materials is tempered by the desire to use proven materials at minimum cost. Forsyth (2.3) also suggested that the fatigue problem could be tackled more profitably by improving our knowledge of defects and N.D.T. rather than continuing alloy development.

It is clear from these papers that cost rather than mechanical properties is often responsible for the selection of a particular material. I feel that this has not been emphasized sufficiently. I wished for more data on how the manufacturers maximum permitted material cost is obtained and in the aerospace industry for example on the cost of the material + processing cost as a percentage of the total aircraft cost and as a percentage of the total operating cost throughout the life of the aircraft over 10–20 years.

It was therefore refreshing to have Cohen & Owen (3.6) report the simple but striking fact that for a typical US steel company the ratio of R. and D. investment/cash flow was about 3 % compared with the ratio of technology capital investment/cash flow of about 60 %. They emphasize too that the total engineering system should be considered. This attitude of mind seems to be exactly what is required, as much in the non-ferrous as in the ferrous field.

The papers suggest that a significant increase in profitability is unlikely to be achieved by any improvement in existing alloys. However airline operators calculate their break-even figures on the assumption that the aircraft operate with only 50 % payload. Would not a change in airline operations to increase payload result in a much greater and immediate increase in revenue than could possibly be achieved by a 15 % improvement in material properties?

Following that – would it not be possible to feed back some of the increased revenue into the materials area to enable the designer/material engineer to use more expensive materials with greater reliability/performance? This would benefit the operator still further over long periods. I think this is what Cohen & Owen meant by the total engineering system. This approach suggests to me that profitability and metal cost should be looked at more closely in the future, perhaps by a multidisciplinary team analogous to that recommended by Brown (2.7) for tackling the stress corrosion problem.

G. Cooper (*Institut Cerac S.A.*)

A word in defence of maraging steels, which have been said to be of minor interest because of their high costs. In our laboratory we make a large number of one-off parts for special machines. Here, maraging steels are very valuable, since they are easily machined into complex shapes, very often to final dimensions, and simply heat-treated. We thus avoid the expensive learning process involved in the development of heat treatments for complex parts, and to a large extent, we avoid the costs of finish machining. The overall cost of the maraging component is thus less in real terms than those made from cheaper competitive steels.

The modest properties of maraging steels do not normally concern us, as experimental machines usually have a very low life expectancy for other reasons.

W. M. Imrie (*Dowty*)

The prime goal in undercarriage design is to save weight. As many of the major components are of tubular shape, weight can often be saved by increasing the bending modulus of rupture. In general, this property increases as the tensile stress increases. Figure 1 illustrates this relationship for three typical D/t ratios as well as indicating how an increase in tensile strength from the normal high strength alloys (S 99, S 132, etc.) to the ultra-high strength alloys (4340, 300 M, etc.) can effect a considerable percentage reduction in weight of a typical cylindrical part. Any significant increase in strength beyond that of 300 M may well not be cost-effective because of machining problems etc.

With regard to the concept of applying fracture-mechanics to undercarriage steels, figure 2 indicates the relation of critical crack length (worst case, part-through crack for thick wall) and fracture toughness at various stress levels. The minimum crack length in large components considered to be reasonably certain of detection under field or production conditions and by standard N.D.T. techniques, is around 0.25 in (6 mm.) It can be seen from figure 2 that this figure (which is appreciably greater than that quoted by other contributors to this conference) requires a fracture toughness at the 225 klbf/in^2 (1560 N/mm^2) level, i.e. the minimum proof stress of 300 M, exceeding 100 klbf/in$^{\frac{1}{2}}$ (690 N/mm^2). Thus a 100 % improvement in the current figure would be required. Figure 3 illustrates the fracture toughness required for various critical crack lengths at the specified minimum proof stress of many of the high and ultra-high tensile steels. The horizontal lines on the graph represent the scatter of fracture-toughness results, obtained from figure 5 of my paper (1.5) and elsewhere. The similarity between this method of presentation and that of figure 4 of Brown (2.7) relative to the stress corrosion resistance of the titanium alloys is interesting. The general trend for the low alloy steel is shown

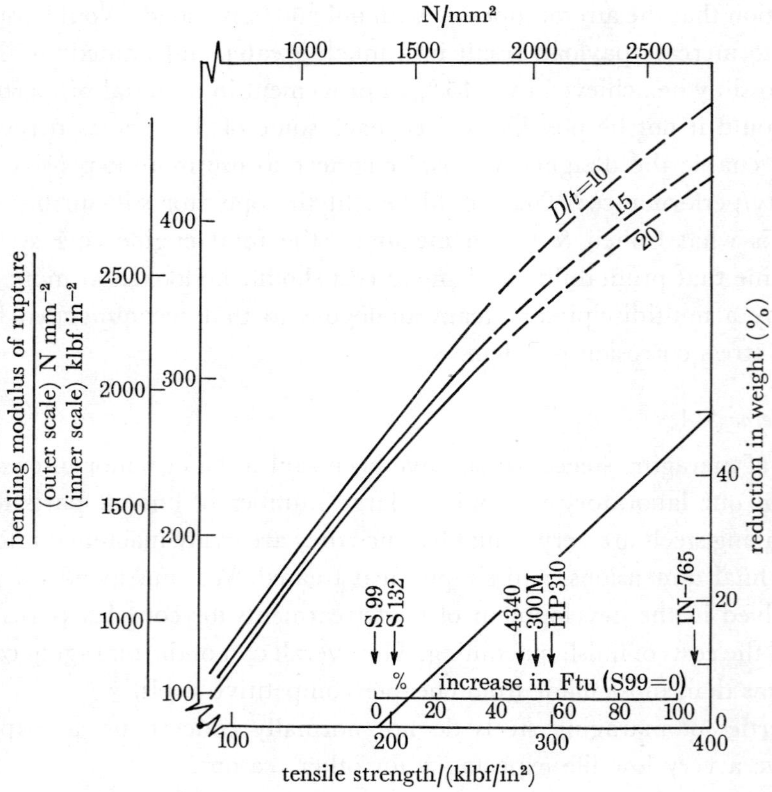

FIGURE 1. Relation of Ftu and Fb for steel tubular members, and effect of increase in Ftu on weight reduction (42 in o.d.).

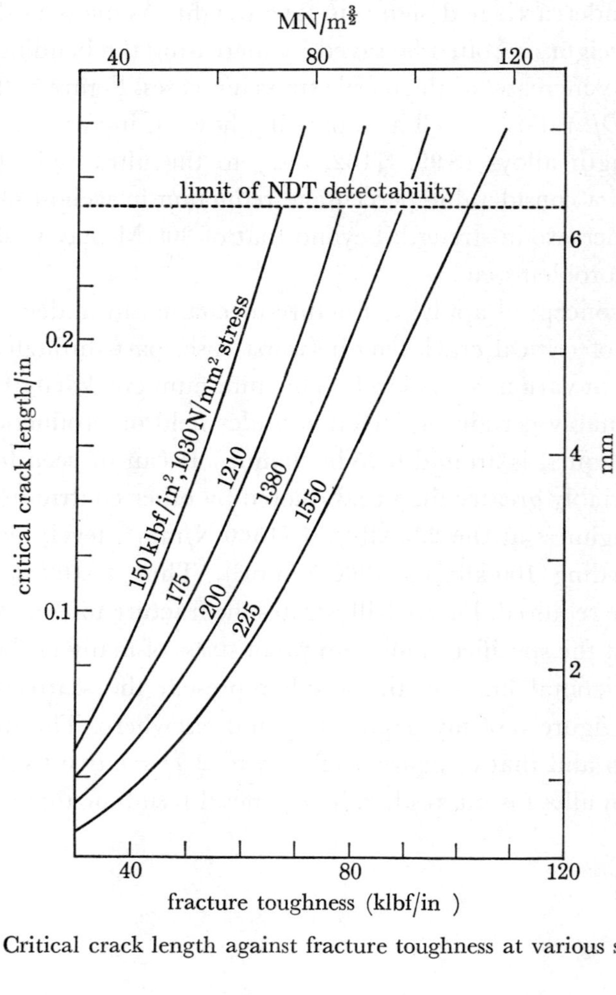

FIGURE 2. Critical crack length against fracture toughness at various stress levels.

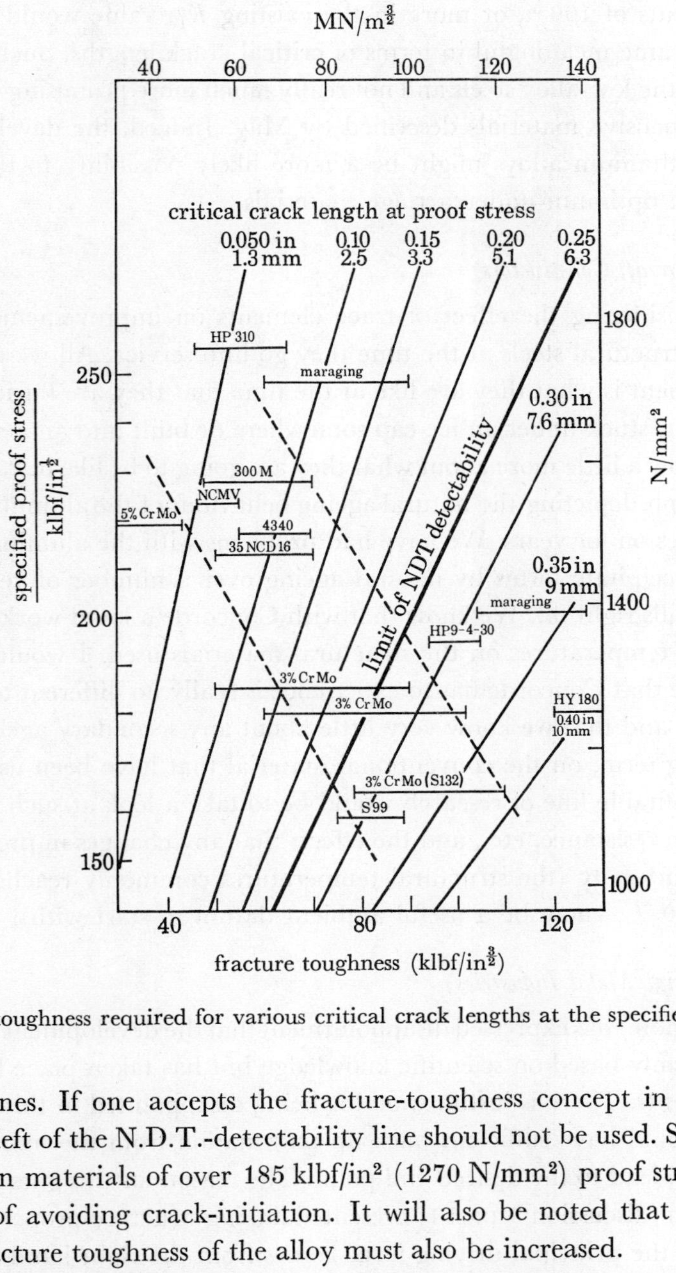

FIGURE 3. Fracture toughness required for various critical crack lengths at the specified minimum proof stress.

by the broken lines. If one accepts the fracture-toughness concept in its entirety, then all materials on the left of the N.D.T.-detectability line should not be used. Such a decision would exclude all known materials of over 185 klbf/in^2 (1270 N/mm^2) proof stress, which highlights the importance of avoiding crack-initiation. It will also be noted that as the proof stress is increased, the fracture toughness of the alloy must also be increased.

The relatively highly alloyed HY 180 possesses the highest fracture-toughness of the alloys shown in figure 3, but its low proof stress would involve a considerable weight penalty. It is understood that a higher strength version of this alloy has now been developed, claimed to retain high toughness, and it would be interesting to hear details from any of the United States Steel Corporation representatives present. The Trip steel described by May (3.7) possesses even lower strength than HY 180 and this fact, together with the high alloy content and expensive processing, is such as to make this steel unattractive. The strongest steel on the right of the N.D.T. line is the 90 ton (200 klbf/in^2, 1400 N/mm^2) grade maraging steel and it is interesting to note that this material was chosen for the light-weight military bridge designed, I believe, on fracture-mechanics principles by the M.V.E.E.

As already explained, and as again indicated by figure 3, if the strength of 300 M is to be retained, improvements of 100 % or more in the existing K_{Ic} value would have to be made before the results became meaningful in terms of critical crack lengths. Such an improvement is most unlikely with the low-alloy steels and not really much more promising with the relatively sophisticated and expensive materials described by May. Indeed, the development of higher strength, lower-cost titanium alloys might be a more likely possibility to the replacement of 300 M as the current optimum undercarriage materials.

H. Tyrer (*British Aircraft Constructors*)

We have been considering the effect of trace elements on improvements in properties in trying to get better structural steels at the time they go into service. All we really know about materials at the moment is what they are like at the time that they are buried in the ground, thrown into the water, stuck under an ice cap somewhere or built into an aeroplane structure. Surely we need to know a little more about what they are going to be like in 25 years time. Hunsicker (3.6) has a graph depicting the natural ageing behaviour of the aluminium–zinc–magnesium alloys which goes on for years. We have had problems with the aluminium 10 % magnesium alloys where precipitate forms by natural ageing over a number of years and the stress corrosion resistance falls right off. We know that with Concorde a lot of work has gone into the study of the effect of temperatures on the structural materials used. I would suggest that the temperature envelope that Concorde has to experience is really no different to that of ordinary sub-sonic aeroplanes, and that we know very little about any secondary ageing characteristics and their effects, long term, on the conventional material that have been used for years. It is considered that a profitable line of research would be to take a look at such things as changes in K_{Ic}, stress corrosion resistance, etc., and the effects that any changes in properties may have on future structural integrity (the structural temperatures commonly reached on the ground in the tropics – say 60 °C – may be a useful ambient datum to start with).

P. H. Morton (*Imperial Metal Industries*)

Honeycombe (Keynote) has expressed disappointment that the development of titanium alloys has not been more firmly based on scientific knowledge but has taken place largely as a result of empirical experiments. There is justification for being disappointed in this respect; it would be very nice if one could have sufficient back up information from the more basic studies of physical metallurgy to lead to the desired end point. The reason for this goes back to the chain of relations that Cohen (discussion 3, part 1) presented linking structure-properties-performance. As a metal producer, the principal driving force comes from the right hand end of the chain. We experience a very strong 'pull' there from the customers - from the society of British Aircraft Constructors if not from society in general. There is always a paramount need to meet performance requirements; there has not been *time* available to develop alloys purely from the metal physics end of the chain. You may call this an empirical approach if you like, but I venture to suggest that if we had not taken this approach we could not have reached the present high level of performance and general sophistication in titanium alloys in the short period of 29 years that has lapsed since titanium alloys were first studied.

That is not to say that work on metal physics or physical metallurgy is not urgently needed – it certainly is. The empirical approach, to be successful, has always to be based on a background of sound physical metallurgical knowledge and principles. Much has been done but certainly the fundamentals of titanium alloy behaviour need much more studying. I agree entirely that more information is needed on factors affecting the structure of titanium alloys – including time–temperature–transformation diagrams – and on the relation between structure and properties. I hope this is a challenge that the Universities will take up.

I was interested to hear Partridge (this discussion) on the importance of total cost in determining what alloy is used in any particular application, a sentiment with which I entirely agree. In my reference to the cost of titanium, the point I was making was not that greater output and increased recycling would not reduce material cost, but that it would not reduce it dramatically, by which I meant a reduction of 50 % or so, whereas improvements of utilization (or yield) in the fabricated component *could* in many instances have a much bigger effect on the total cost of a component. New alloys will certainly be required for use at high temperatures, but in other applications more attention will in future be devoted to alloys for which fabrication costs can be reduced to a minimum. Reduction in total cost is certainly now one of the main driving forces for innovation in this field.

R. B. Nicholson (*International Nickel Ltd*)

I would like to make some comments, in the context of the papers under discussion, relating to Barford's lament (discussion 3, part 1) that physical metallurgy is too often used in an explanatory role rather than a predictive one. Physical metallurgy has made a unique contribution to what Cohen & Owen (discussion 3, part 1) call the 'bank of available knowledge' on the major properties of alloys based on aluminium, iron, nickel and titanium without which the advances recorded in the papers on these materials could not have been made. The bank of knowledge is much less when it comes to properties such as fatigue, fracture toughness and localized corrosion and the resulting problems in commercial materials are clearly recorded in the same papers.

Therefore there is no need to be apologetic about the explanatory role of physical metallurgy as long as it leads to the application of the knowledge to improve materials in the ways recorded in this Conference. Of course, for this pattern to be effective, there needs to be close contact between the materials producer, the component manufacturer and the end-user. Honeycombe (Keynote) suggested that the nickel superalloy producers and the gas turbine industry have been a good example of this collaboration.

If the contact is less close, problems can arise. After all, any textbook on metallurgy will 'explain' that iron oxidizes in hot CO_2. Presumably someone omitted to 'predict' this to the designer of a certain type of nuclear reactor!

W. E. Duckworth (*Fulmer Research Institute*)

We have been discussing the prospects and promises of physical metallurgical advance but if we are to keep the trust of the engineers in our ability to improve his materials we must try to predict the disasters. We have not been very good at this in the past and have been caught napping. This is why I welcomed Jones' work on the lead–silver eutectic metioned by Honeycombe (Keynote). The stereoscan picture was especially prophetic. It looked just like a graveyard.

The following five contributions were written after the meeting:

D. E. Miles (*National Physical Laboratory*)

With reference to Honecombe's remarks (Keynote) on the degradation of directionally solidified eutectics, I feel he is painting a pessimistic view which I do not feel is justified when considering turbine blade applications. Work by M. McLean at N.P.L. on a more realistic high temperature eutectic alloy duly shows significant degradation of the composite microstructure in the presence of very high temperature gradients (10^2 K mm^{-1}) close to the melting point ($0.9T_m$). Directionally solidified eutectics are obviously thermodynamically unstable when held in a temperature gradient but so are conventional superalloys. What is most important are the kinetics at the service temperature and until the experiments have been carried out on the actual alloy under consideration or the theories have been made more exact the true significance of degradation in a temperature gradient can not be realized. I feel of more immediate interest to the engine designer are the anisotropic mechanical properties of the directionally solidified eutectics – all of which have different properties. For example one material I have been looking at having a cobalt, chromium matrix reinforced with chromium carbide fibres has creep rupture properties along the axis more than 2 orders of magnitude greater than the 60 or 90° off axis properties. On the other hand an axial ductility of 3 or 4 % increases to up to 40 % when 60 or 90° off axis.

J. E. Restall (*National Gas Turbine Establishment*)

With reference to composition control via electron theory Nicholson (3.8), the assumption made in the Phacomp calculation procedure used for predicting alloy instability is that it is the composition of the austenitic matrix phase which determines the propensity for sigma phase formation. A suggestion was originally made by Mihalisin and Pasquine that the alloyed γ' phase in commercial nickel superalloys is at least as likely to precipitate σ, or other topologically-close-packed-phase, during thermal exposure as a high N_v matrix phase. This suggestion has been confirmed at N.G.T.E. by making bulk experimental alloys based on the composition of γ' found in superalloys which have shown a propensity for sigma phase formation. Despite this comment it still appears that Phacomp is a very powerful technique for controlling the microstructural stability of nickel-base alloys.

On the subject of the beneficial effects of unidirectional solidification (u.d.s.) on the mechanical properties of cast nickel superalloys, many years of work at N.G.T.E. involving a number of nickel-base superalloys allow us to confirm that the creep-ductility and rupture-lifetime are improved by the u.d.s. process, but the level of improvement depends very markedly on the alloy composition and the temperature and stress at which the creep tests are done. Our results indicate that it is possible to attain up to a maximum of about 60 °C improvement for a 100 h rupture life in the temperature range 750/850 °C, but the more general level of improvement is in the range 15–30 °C. A temperature advantage of 80 °C at test temperatures in excess of 1000 °C is unlikely to be realized by u.d.s. of a conventional superalloy; such improvements appear to be more relevant to the strongest directionally-solidified eutectic materials. An important reason for the current inerest in u.d.s. arises from the significant improvements in thermal fatigue resistance and structural soundness of nickel-base superalloys.

D. R. Holmes (*Central Electricity Generating Board*)

I should like to strongly support Partridge's remarks (this discussion) about the importance of material costs as we have heard too little on this subject. There are three important parameters which determine whether a material should be used for a given application namely mechanical properties, corrosion resistance and cost. Cost, the most important parameter, is placed last because the true cost of a material for use in a particular application can only be determined when the other two parameters are known; both are required to determine the thickness and expected life of a component.

In the C.E.G.B. we endeavour to take account of all three parameters in the light of our statutory obligation to provide a secure supply of electricity as economically as possible (not, I must emphasize, to provide total energy supplies as efficiently as possible). As an example in 1968 we went through an exercise to see whether our generation could be made more efficient and therefore provide electricity at lower cost to the consumer by the introduction of more advanced materials and more advanced steam conditions into the power stations being built. To a first approximation, neglecting labour and various other secondary costs, the generation cost of a unit of electricity (U) is given by

$$U = \frac{c}{\eta} + 100\,\frac{Pr}{L \times 365 \times 24},$$

where c is the cost in pence/kW h of the chemical or nuclear energy of the fuel, η is the generation efficiency, P is the capital cost in £/kW of the plant, r is the return on capital required and L is the load factor, defined as the electrical energy actually produced over the plant lifetime, divided by the electricity that could be produced if the plant worked non-stop to full capacity.

Differentiating and rearranging, we find that

$$\Delta P = 87.6\,\frac{cL\Delta\eta}{r\eta^2}.$$

When current values for the parameters were substituted, we found that the permissible increase in capital costs (ΔP) for a substantial increase in steam temperature of 100 °C, corresponding to an efficiency increase ($\Delta\eta$) of 2 %, was only £4/kW. No materials were available with the necessary combination of mechanical properties, corrosion resistance and cost to permit this increase. Note also that because of the η^2 term in the denominator, the more efficient we become, the less money is available for further improvements.

With the recent trebling of fuel costs, the position is currently under review, but our tentative conclusion is that because of the large increase in r, due to the necessity to introduce inflation accounting, the scope for the introduction of more expensive materials is still very limited. Finally it should be noted that if the load factor of the plant becomes low ($L \rightarrow 0$) due to the inability of the materials to withstand the uprated conditions, then the cost of the units generated approaches infinity ($U \rightarrow \infty$).

A second point, in defence of those responsible for the choice of mild steels for use in Magnox reactors, criticized by Nicholson, it must be pointed out that the protective oxidation of mild steel in high pressure CO_2 was well known and studied, but the phenomenon of *breakaway* corrosion was not. The dramatic increase in rate by factors up to 10^2, which occurs after several years exposure, is still not entirely understood in spite of many years of research.

C. A. Stubbington (*Royal Aircraft Establishment*)

I should like to discuss aspects of texture in titanium alloys. One facet of this topic concerns property anisotropy – a feature of titanium alloys since they were introduced. It is now clear that this anisotropy is due, at least in part, to textual variations within, and between, components resulting from differences in composition, working temperature, amount of work and heat treatment. Macro and microstructural differences, again a function of the parameters listed above, also contribute to property anisotropy.

Another aspect of texture concerns its use in a controlled manner. The biaxial strengthening of correctly textured titanium sheet is well documented, but recently thought has been given to the exploitation of some of the properties conferred by preferred orientation, in components other than those fabricated from sheet. Properties which are significantly improved if the preferred orientation is controlled so that basal planes are normal to the stressing direction include Young's modulus, yield and tensile strength, and high cycle fatigue strength. To obtain and use material with known preferred orientation will require not only close control of alloy composition, thermo-mechanical processing and heat treatment, but also design procedures which take account of the resulting property anisotropy. It will be necessary to require pole figure determinations in procurement specifications and to ensure that in practice material is always used in the correct orientation. Additionally, it must be pointed out that while the properties mentioned earlier are maximized if the stresses are normal to the basal plane, this stressing mode is likely to result in minimum stress corrosion and sustained load properties, and very high crack growth rates with fatigue stressing involving dwell at maximum load.

Thus, further opportunities exist for metallurgists and engineers to cooperate to develop thermo-mechanical processing routes which ensure that the properties of titanium alloy components are both optimum and isotropic, and to investigate the circumstances in which titanium alloy components, fabricated to have a strong preferred orientation and maximum unidirectional properties, can be used to full advantage.

E. W. Skerrey (*British Aluminium Co. Ltd*)

Hunsicker (3.6) presented an account of a valuable alloy field where heat treatment of some of the alloys mentioned is running to five digits, e.g. 7050–T73651. This implies a complex metallurgical situation at the end of the heat treatment and we must know how well the advantages such as good resistance to stress corrosion and exfoliation with good mechanical properties, can be maintained in subsequent fabrication, e.g. cold working, hot working, curing of adhesives, paint storing.

4. THE FUTURE ROLE OF PHYSICAL METALLURGY IN RELATION TO ENGINEERING PRACTICE

By Sir James Menter, Treas.R.S. (*Tube Investments Ltd*)

Engineering practice is all about designing, constructing and maintaining structures for the comfort and use of mankind.

In order to achieve his purpose the engineer takes materials from the world around him and assembles them into an artefact to meet a defined need. In earlier times need could be satisfied quite simply by gathering materials as they are found in nature and assembling them into useful forms with a minimum amount of modification, such as chiselling rough stones into simple geometrical forms to make walls. The early engineer was his own materials scientist and the great pyramids, cathedrals, aqueducts and the like remain as testimonies to his understanding of the properties of natural materials. The progress of science and technology has changed all that. The improvement of properties made possible by separating metals from their ores opened up entirely new vistas for structural design. But the intricacy and complexity of all the intervening stages from ore in the ground to a useful modern engineering structure are so diverse that engineering practice has to be seen now as only the last step in a long sequence of professional activities encompassing geological prospecting, through mining to smelting, refining and fabrication. Thus engineering practice has come to depend for its successful prosecution on many sciences and technologies among which metallurgy is paramount. But metallurgy itself is now subdivided, as the subtitle of our conference shows, and throughout the papers you will find reference to a variety of sub-subspecies including 'solid state physical metallurgy'. In reflecting on the scope of my contribution to the conference I was led to seek the origin of this adjectival division of the discipline because I believe it has an important bearing on the future.

What better authority could there be than the progenitor of the subject whose name we honour today. A father should know his own child. Chapter 1 of Rosenhain's famous book *An introduction to the study of physical metallurgy* published in 1914 gives us the definition of physical metallurgy, although one must read beyond the first line to find it:

> 'The term "Physical Metallurgy" speaks largely for itself, since it ovbiously connotes that branch of science which deals with metals in their physical aspect. The term is justified by the need for a distinguishing name for that branch of the knowledge of metals which has to a large extent grown up in the last fifty years—a branch which concerns itself with the nature, properties and behaviour of metals and of alloys, as distinct from the far older branch of metallurgy which deals with the reduction of metals from their ores. The term "Metallurgy" has formerly been almost entirely confined to this latter meaning, and those who have grown old in this idea are a little apt to resent an innovation which gives to the old term "Metallurgy" a wider and more general meaning than it formerly bore, so as to include the newer knowledge of metals. This inevitable widening of the old term, however, demands a subdivision, so that the department of metallurgy which relates to the reduction of metals and their refining may well be termed "process" or "chemical" metallurgy, leaving to the younger branch of the science the newer term "Physical Metallurgy".'

This passage brings out two important points which I wish to develop. The first, and more obvious, is the bifurcation of the discipline into chemical and physical. The exact territorial domain staked out for physical metallurgy is made precise in another passage from his book:

> 'Every process which the metals undergo, from the moment they first leave the refining furnaces as metals, until they reach their ultimate user, forms the direct subject matter of our science—for in each and all of such processes, are involved those phenomena associated with them which it is the object of this science to study and to elucidate.'

The metallurgy, physics and chemistry of the liquid state and of raw materials and the chemical deterioration of metals and alloys in use are thus excluded from the domain of physical metallurgy, although it should be said in fairness to Rosenhain that he admits later in the book, in a brief reference to corrosion, that the subject is so extensive as to call for a further separate book.

The second point I wish to draw your attention to in Rosenhain's definition of physical metallurgy is that he gives the subject fifty years of existence before 1914, yielding pride of place as the true progenitor of the subject presumably to Sorby who pioneered the use of the optical microscope for the study of metals in 1864. We have here another important clue to the scope of the subject as seen by Rosenhain. Not only was he inclined to exclude chemistry but saw the main thrust of his attack in terms of the use of physical instrumentation for quantitative studies of microstructure of well characterized materials.

In his time this rather narrow focus was undoubtedly correct. Writing as he was in 1914 within two years of the demonstration by von Laue of the diffraction of X-rays by crystals and with the insight he had himself gained into the crystalline nature of plastic deformation through his studies with the light microscope, it was inevitable that he should set rolling a bandwagon on which the players, with their newly acquired instruments were to perform a vast repertoire of tunes.

With the hand of Rosenhain on the horse's rein there was no immediate danger of this concentration of effort on metals in 'their physical aspect', as he put it, leading to neglect of other important fields. His breadth of vision and wide experience outside the research laboratory ensured that this work was properly related not only to the broader field of metallurgy as a whole but even more importantly to real engineering practice. For this he no doubt owed much to his upbringing and education as a civil engineer. Sir Alan Cottrell may have something to say about that in his contribution to this conference on the future tasks for the educator. However, after Rosenhain and particularly in the post-war years, the subject became increasingly identified as the study of microstructure by advanced physical techniques. The bandwagon gained ever greater momentum from the weight of electronic equipment taken on board. For all the astonishing wealth of new insights obtained in this way one cannot help reflecting that, as with musical instruments, the background roar of pop music made possible by electronic sophistication sometimes drowns the purity of melodic line of a Bach fugue played on the well tempered clavier. By this I mean that it is by no means easy for the practical man who wishes to put new knowledge to use to distinguish what Crussard has called the master ideas which carry industrial practice forward.

The predominance of the physical approach was intensified in the 1940s following the emergence of the concept of the dislocated crystal lattice which stimulated widespread work on the construction of theoretical models of lattice defect dynamics and interactions to 'explain' the observed macroscopic behaviour of metals under stress.

These combined influences of physical technique and physical theory served for a time only to widen the gulf with studies of the chemical aspects of metals on the one hand and with metallurgical industry and engineering practice on the other. There are signs in some of the papers of our meeting that these separatist tendencies are now diminishing, even reversing. If our conference has done no more than give a modest push in the direction of reunifying the study of materials of construction and of orienting it towards meeting the pressing needs of the engineering designer, it will have been very well worth while. It is gratifying to hear, for example, from Cohen & Owen that already M.I.T. has recently recombined its graduate programmes in chemical process and physical metallurgy after more than twenty years of separation.

Since some of my best friends are physical metallurgists or physicists and lest these observations be mistakenly interpreted as a rejection of their endeavours I must briefly sketch out the position we seem to have reached in characterizing the microstructure of engineering materials, before turning to the materials needs of engineering design and practice and expressing some personal views on what appear to be the most pressing.

In terms of technique the research metallurgist now has at his disposal a veritable arsenal of instrumental weapons with which to determine quantitatively (albeit with some inaccuracy at the limits of observation) the spatial relationships of each atomic species in a solid microstructure. As to the bulk microstructure this is done with the electron microscope and the electron probe analyser and X-ray diffraction equipments, all now designed and operated in a great variety of forms according to the nature of the information sought. Lattice spacings are resolvable down to atomic dimensions, line and planar defects can be seen as well as small aggregates of point defects, precipitated phases can be seen and characterized both as to composition and orientation with respect to the matrix and any strain field which they may cause. Surfaces, either free or internal (fracture faces) may be characterized with equal refinement by high and low energy electron diffraction or by observations on the inelastic scattering of electrons or ions. Monolayers, even partial monolayers of foreign atoms can be detected. Some variants of these techniques permit the stripping off of atoms, layer by layer and identifying them individually in one way or another.

In singling out these observational techniques for characterizing microstructures I am not for a moment suggesting they are the beginning and end of modern physical metallurgy. Many other techniques are used in conjunction with the microstructure approach to measure macroscopic properties. The supreme importance of microstructural observation lies in the facility it provides for characterizing quite precisely the unit mechanisms governing the macroscopic behaviour thereby providing a secure foundation on which to build models of macroscopic behaviour which can help both to diagnose shortcomings in desired properties and suggest new approaches to improving properties.

The most reliable information from these studies relates to essentially static physical situations in solids. Indicative information may be gained in some instances on physical dynamics or chemical changes of microstructures.

Thus from the point of view of technique development as applied to the unravelling of microstructure the current situation may be said to be quite satisfactory, except that more might be done towards better means of characterizing unit dynamical processes. Perhaps the high voltage electron microscope will help in this respect.

What contribution then has all of this microstructural work made to solving problems of interest to the engineer? The engineer's first requirement of his material is that it should retain its shape. To this end he designs to a yield stress. Yield is undoubtedly the best characterized material property in terms of microstructure. A variety of strengthening mechanisms are clearly defined – work hardening, precipitation hardening, dispersion hardening, substructure hardening, martensitic hardening, etc. All depend in one way or another on a refined microstructure whose yield behaviour in some simple structures can be described quantitatively (Petch & Orowan), although more work is required to secure quantitative understanding of more complex microstructures where more than one hardening mechanism may be operating simultaneously.

As a broad generalization I think it is reasonable to say that physical metallurgy has now provided the means of specifying the microstructural design requirements to secure a high yield stress at ambient temperatures in a given metal matrix. There is also considerable flexibility in the choice of microstructure to secure the desired stress level, thus leaving some room for manoeuvre in modifying the microstructure design to secure other desired properties.

For engineering structures requiring rigidity at high temperatures, however, the situation is less satisfactory. The unit lattice processes involved in creep are complex, involving intricate interactions between matrix lattice defects, precipitated phases and grain boundaries. Engineering designs are frequently based on the assumption that some change in shape is permissible over the life time of the component. We still lack the certain knowledge of how large this change will be unless we conduct tests of comparable duration to the lifetime. It must therefore remain an important goal of the physical metallurgist to develop reliable predictive guidelines for the design of creep resistant alloys.

Subject to the achievement of consistency in bulk production, which I shall be discussing later, there is little difficulty now in realizing adequate intrinsic ambient temperature strength (yield stress) in metals and alloys but of course the engineering designer is concerned with specific strength, since this determines the dead weight of his structure and also with elastic stiffness or specific modulus which must be large enough for the structure to resist collapse by elastic buckling. The efforts of the physical metallurgist to develop ever higher performance materials in respect of these two properties have been constrained by the strait jacket of the density and modulus of metals. Nothing can be done about density and it appears that the modulus can be affected only marginally by alloying or texture modification. An escape route has been found, however, in moving to non-metallic materials which present a new challenge to the physical metallurgist. Besides the advantages arising from specific strength and stiffness of an order of magnitude greater than most metals, non-metallic materials offer new opportunities for escape from the poor high temperature strength and high chemical reactivity of many metals and, in their composite forms, an escape from the limitations of the phase chemistry of alloys. It is not surprising that there are offsetting disadvantages. Ductility is frequently poor and there are joining problems to be overcome.

Nevertheless, the potential of this new range of materials opens up new perspectives for the engineering designer but before they can be brought into general use it is essential to characterize the critical aspects of their microstructures, to develop adequately controlled manufacturing procedures and establish property standards on well characterized material in a manner analogous to that initiated and pursued for metals by Rosenhain and his successors.

Perhaps some metallurgists have seen these developments as a threat to the entrenched position of metals and alloys as the preferred materials for engineering structures. It is timely for the metallurgist to come to terms with the exciting opportunities presented by these new trends and to embrace the problems of these new domains in order to create a sound basis for material specification as an aid to design. It is gratifying to note that our National Physical Laboratory has taken up and followed, in respect of non-metallic materials, the tradition and example set by Rosenhain for metals so many years ago, with programmes of work aimed at establishing reliable engineering data on the properties of well characterized materials. Similar responses can be identified in our educational institutions but more needs to be done within our professional institutions to come to terms with these new trends.

Moving on from the problems of intrinsic and specific strength, the next problem for the designer is to join individual components into an engineering structure. The development of welding science and technology has perhaps done more than any other single field of endeavour to advance engineering practice, particularly on large structures. However, to a greater or lesser extent all welding techniques involve potentially deleterious metallurgical changes, create the risk of dangerous notch effects if improperly executed, and generate potentially dangerous stress concentrations by thermal effects. All of these problems have been the subject of intensive studies by physical metallurgists and must continue to be. The metallurgical and stress problems are frequently particular to joint geometry, making it difficult to formulate generalized standard specifications. Since the joints in a structure are so critical to its integrity, it is of the utmost importance for the engineering designer, and particularly the detail draughtsmen, to appreciate just what is implied by that little shaded area he puts on his drawing to indicate a welded joint. Physical metallurgical knowledge of welding will be increasingly essential *within* the drawing office. Apart from questions of joint integrity, there is the question of post-welding stress relief procedures mentioned by Cotton, where for practical economic reasons much greater care is required and no doubt further research required towards specifying for stress relief only those parts of a structure which need it. In concluding my comments on joining, I should mention the need to develop joining techniques for non-metallics and composites and to couple that with a challenge to the physical metallurgist to develop techniques for metallic materials which obviate the difficulties inherent in current practice.

For all the care taken by the designer to ensure that his structure is mechanically sound, failure by fracture may still occur. Several authors have reminded us yet again of some of the more spectacular failures of large engineering structures which have occurred both in service and in the course of construction. Some engineers have expressed the view that one cannot expect engineering practice to advance without some failures. As a professional man I find that view hard to accept and there is growing evidence that the lay public will find it increasingly unacceptable. We have certainly advanced a very long way from Bouche's failure to design his bridge against the wind forces in the Firth of Tay and his negligence in remitting to a carpenter responsibility for supervision of the cast iron work. The design fault was considered then quite inexcusable by the enquiry into the disaster and would be unthinkable today, but disasters still do occur as a result of human error in the manufacture of materials and during their assembly. We frequently find that these errors occur due to lack of adequate professional supervision on the job. There is much that needs to be done to enhance mutual understanding between those engaged in laboratory studies of fracture and the engineering designer – as Cotton puts it in his paper[1.3] 'not only are specific answers required to specific questions but general principles need to be revealed and communicated to engineers in language which they can understand'.

There is equally a pressing need for this understanding to be disseminated into the supervisory hierarchy of production plants and construction sites. These problems require the sustained combined attention of managers, educators and professional bodies over a long time.

Fracture results from stress concentration which in turn may have a variety of origins. Most important among these are the heterogeneities in the microstructure, some of which, of course, are designed into the microstructure to secure desired yield or creep properties. Their stress concentrating tendencies must be balanced against their virtues in other directions. I want to refer more particularly to those adventitious heterogeneities such as non-metallic inclusions which originate from undesirable impurities present in the raw materials or picked up from the atmosphere or from refractories during refinement. Their absolute incidence may be diminished by more sophisticated refining techniques but at present only with additional cost to the end product. Their effect can be mitigated to a degree by modification of their shape and interfacial relationship with the metal matrix by minor additions to the melt. Their impact on the properties of the matrix is more or less irreversibly predestined by the manner in which they separate from the melt in solidification. In discussing the problem of sulphide inclusions in structural steels van der Veen (3.4) remarks that 'as the accent of this conference should be on physical metallurgy a detailed discussion of steelmaking alternatives (for achieving lower inclusion levels) should not be attempted'. I find this remark interesting because it substantiates my earlier plea for the reunification of chemical and physical metallurgy. If this problem of non-metallic inclusions were treated more as a total problem of raw materials, liquid metal and solidification by those researching in the field, faster progress might be made towards rendering them less harmful.

In parallel with continuing studies of the sources of heterogeneity in materials there is a need for the development of more discriminating techniques for ensuring that the uniformity of the material is maintained at the desired level in the course of production. A good works practice with appropriate production control procedures is an essential base on which to build but it is equally essential to monitor these procedures by the application of a relevant non destructive testing technique on the end product. The first requirement for a relevant NDT technique is that the information it yields should ideally be uniquely related to that shortcoming in the material property which is known to be responsible for material failure in assembly or use. Thus there will be a continuing important role for physical metallurgy in pinpointing the precise origins of material failures and in developing our understanding of the exact mechanisms of fracture initiation. In order to use this knowledge as a means of monitoring production, much more work is needed in collaboration with physicists and electronics engineers to relate the precise character of material heterogeneities to their signal response to test probes. Even before we can tackle this problem in the context of real heterogeneities in the materials of construction there is much that is still unknown about, for example, the modes of propagation and the effect of test piece macrogeometry on the transmission of ultrasound. Model studies such as Schlieren techniques in which the progress of ultrasonic pulses through a transparent solid can be directly observed provide valuable help in getting to grips with real situations and computer simulation techniques have an important supporting role in such studies.

Model work, in parallel with painstaking analysis of the signal response of precisely characterized heterogeneities should eventually supply the basis on which material can be selected or rejected for a specific application. Test procedures must be capable of operation 'on line' at a speed commensurate with that of the flow of material through the production process. This requirement poses a major challenge to the electronic engineer. His skills in fast signal processing have been successfully bent to the needs of other disciplines such as computery and medical diagnosis. There is an urgent need to bring them to bear in support of the physical metallurgist to serve the need of the engineer for homogeneous materials of construction.

Before leaving the topic of fracture I must make brief reference to the special problem of failure under cyclic loading. Forsyth's paper (2.3) shows very clearly that great progress has been made in controlling failure under these conditions by close attention to details of engineering design, material manufacture and assembly. He also makes it plain that our understanding of the atomic basis of fatigue is, as yet, inadequate as a basis for developing intrinsically fatigue resistant alloys. This must remain an important challenge for the physical metallurgist.

James in his paper reminds us that the definition of the design engineer's responsibility given by the Fielden Committee on Engineering Design includes 'his interest throughout the designed life of the product in service'. This implies among other things the need to design protection against deterioration of the structure by corrosion.

In commenting earlier on the tendency within physical metallurgy to confine its scope to the application of the methods of physics to the study of metals I noted the implicit exclusion of corrosion from its domain. The many references in this conference to corrosion and particularly its more insidious form of stress corrosion leave no doubt as to its practical importance. Corrosion phenomena seem about as diverse as the infinite variety of combinations of material and environment and have yet to be comprehendable with a limited catalogue of principles such as govern our understanding of the tensile strength of materials. What greater challenge could the physical metallurgist want than figure 1 of Brown's paper (2.7) in this conference, which summarizes the microstructural features believed possibly involved in stress – corrosion cracking – what a rogues' gallery! Brown suggests that the problem can only be understood and stress corrosion controlled through the combined endeavours of the metallurgist, the applied mechanics specialist and the corrosion scientist-engineer. In enumerating these 'professions' he is of course doing no more than personifying those branches of science which are relevant to his figure 1. Stress corrosion is a materials science problem, *par excellence*. Corrosion as a whole tends to be the Cinderella of each of the component disciplines. If there is a Prince Charming in the audience, would he please step forward and take Cinderella to the ball. Aside from stress corrosion and corrosion fatigue, a related but distinct phenomenon, both of which are of crucial importance to the engineer because they may lead to quite sudden and unexpected structure failures, the question of general corrosion is of continuing and increasing importance as maintenance costs rise and as more massive structures are built in environments where access for maintenance is difficult.

The offshore structures described in this conference form one such class. Other notable examples are to be found in the nuclear power industry where general surface oxidation has been a source of major difficulty in inaccessible locations leading in one case to substantial and permanent downrating of reactor temperature and power output. In this sphere of engineering one observes a sharp contrast between the brilliant successes of physical metallurgy in the development of nuclear materials capable of withstanding the hitherto totally unknown hazards of nuclear radiation and the failure to recognize and solve corrosion problems of the conventional materials for containment and cooling.

General surface protection is still mostly achieved by the applications of coatings, frequently requiring repeated applications through the lifetime of the structure. There seems to be scope for more work towards the development of materials with inherently corrosion resistant surfaces or alternatively of techniques of surface treatment which render the surface resistant without the application of a coating. Here I have in mind the remarkable effect of trace elements on the protective properties of surface films and the possibility of creating such films directly by such techniques as ion implantation borrowed from semiconductor technology.

My discussion so far has centred on the technical problems of materials inherent to the creation of an engineering structure. I would now like to refer to a range of new technical challenges and opportunities arising from broad economic factors which will increasingly impinge on engineering design and through that on the role of the metallurgist. Cost has always been a very important factor in the choice of engineering materials and in engineering design, and the cost-effective approach to meeting a perceived human need has in recent years led to the emergence of the relatively new discipline of value engineering. The main driving forces for this development have been the rising costs of raw materials and labour, and more recently of energy, all of which costs have recently taken a sharp upward turn throughout the world. While one or other of these cost increases may be temporarily contained, even diminished by trade recessions such as that we currently experience, there is little doubt that the general trend will continue upwards. Furthermore there will be an increasing cost impact from the anticipated exhaustion of a limited number of critical elements on which the metallurgist and engineer have come to rely.

The problem of raw material exhaustion will lead to the growth of what might be called substitution or contingency metallurgy (although I am loath to invent yet another adjectival branch of the discipline). It is interesting to note in passing that this is no new problem. Rosenhain refers to it in his 1923 Sorby Lecture on 'Present and future problems of metallurgy', where he expressed particular concern about the potential availability of platinum, tungsten and beryllium. Incidentally his hopes for beryllium as an intrinsically attractive structural material have yet to be realized as indicated by Brook & Duckworth (3.10) in this conference. Rosenhain was able to reassure himself and his audience in that lecture by throwing a challenge to the engineers to open up wider areas and greater depths of exploring and exploiting the resources of the Earth's crust – a challenge which they have met, as our conference has certainly shown in relation to the raw materials for energy production. But there is another escape open to us, arising indeed from Rosenhain's own influence on metallurgical science, which for that reason would have given him great satisfaction. From our vastly increased understanding of the microstructure of materials and how their properties can be controlled by chemical composition, I believe we can be confident that this knowledge and more yet to come will point the directions for the substitution metallurgist to overcome these resource exhaustion problems.

It should be mentioned here that potential resource exhaustion together with the general rise in material costs and the recognition of the value of stored energy and labour in scrap is likely to give increasing prominence to the metallurgy of recycled materials.

Turning now to the other cost problems, labour and materials, these pose a challenge to develop shorter routes from liquid to the desired finished shape of an engineering component, to develop forming techniques less wasteful of discard material and to conserve energy not only in extraction and refining but also in heat treatment to achieve enhanced mechanical properties.

This is a very broad field and I have time only to mention a few examples, which particularly illustrate the growing interpenetration of metallurgical science and technology with engineering design.

Perhaps the most spectacular achievement in cost-effectiveness of engineering design is the remarkable transformation of electronic technology over the last thirty years. I suppose the critical breakthrough, aside from the invention of the transistor, was the availability of ultra pure materials made possible by the parallel invention of zone refining, an offspring of physical metallurgy. It is unlikely that equally spectacular economies in the use of materials are possible in mechanical engineering but there is one significant analogy between the two fields. The micro-miniaturized electronic component or chip, now embraces the functions formerly performed by a whole circuit of separate interconnected components. In other words, circuit design, formerly a separate activity from component design, has now been subsumed within the component. The material and its function have become a unified design problem calling for expertise both in materials design and electronics. The same trend can be observed in mechanical engineering design, for example, in the turbine blade manufactured by unidirectional solidification referred to by Brook & Duckworth. The extension of this general design concept wherein the finished or near finished component is formed directly from liquid (or for that matter powder) with a highly sophisticated microstructure designed to give enhanced properties calls for a new breed of designer.

Powder metallurgy offers the advantage of economy in material and shortening of process route from raw material to finished component and in so called mechanical alloying there is the additional advantage of escape from the phase limitations of liquid metallurgy. Powders manufactured by ultra-high speed quenching again show potentially interesting enhanced properties resulting from the metastable microstructures obtainable. Fleming's rheocasting promises yet further prospects of a highly economic route for component manufacture.

In sheet forming the advent of more or less conventional alloys such as aluminium 6 % copper, with superplastic properties permitting elongations of 1000 % or more without plastic instability, opens prospects for the designer to create complex shapes with a single piece of material which hitherto needed to be machined from plate or fabricated from a number of separate pieces. My own company has been particularly concerned with developing such a material and promoting its use. An example is shown in figure 9 of Brook & Duckworth's paper. Our experience in trying to extend the use of the material reveals a need for continuing debate and mutual education not only between the material developers and engineering designers but for both of these groups to be educated by the press operators, who through practical experience gain an intimate knowledge of the boundaries of component design possibilities.

Thus with all these new technologies which offer the prospect of more cost effective designs we may expect, especially in relation to component design, a shift from traditional methods based on the machining and fabrication of semis like rod, bar, section, plate and sheet whose properties are codified in data handbooks towards what might be called metallurgical designing in which the component emerges from a creative synthesis of knowledge about process route, microstructure and engineering function.

In conclusion I return to Rosenhain's definition of physical metallurgy. No one can doubt that the emphasis he placed on the study of metals 'in their physical aspect' was timely and fruitful beyond measure. In the limited time of this conference we have been able only to review its impact on mechanical engineering and even in this field we have scarcely had time to do full justice to the great leap forward that has occurred since his time in our understanding of the atomistic basis of the engineering properties of materials. Rosenhain justified the separate identity of physical metallurgy on the basis of a need to distinguish it from what he called the 'old metallurgy'. It seems to me timely for it to be fully assimilated into a 'new metallurgy' which in turn will increasingly assume an even broader aspect as materials science and technology.

Within this new metallurgy there is a pressing need for the distillation of knowledge gained from atomistic studies into 'general principles expressed in language which engineers can understand' – to quote Cotton again. In approaching that task it is important to distinguish those problems where the unit atomic processes have a direct bearing on engineering properties from those in which the unit processes can be adequately treated on a statistical basis and described by the equations of continuum mechanics.

The new metallurgist will increasingly approach materials development as a total problem of liquid physics and chemistry, solidification control and shape control. He may well encounter, coming from the other direction an engineer who will be increasingly knowledgeable about microstructures and materials manufacturing processes, wishing to spell out detailed requirements, hitherto regarded as the preserve of the metallurgist. This is not to imply that all practitioners of metallurgy and engineering will be expected to be competent in all aspects of the other's trade. Of course the individual subdivisions of both will continue to be essential for the advance of knowledge and of practice. What I am really calling for is an increasing number of designers who can deploy the accumulated wealth of materials knowledge to the creation of cost effective and energy conservative artefacts. How are we to breed these designers? That is a question which I must leave to Sir Alan Cottrell in the next session of the conference.

Discussion (Chairman R. W. Cahn (*Sussex University*))

J. G. Ball (*Imperial College London*)

I was naturally extremely interested in the points which Sir James (Keynote) raised on the re-unification of the field of metallurgy, and particularly as he started from the point of view of the definition given by Rosenhain in his 1914 textbook. We have been trying to do this for the past 15 or 16 years in Imperial College. Appropriately, it was in the forerunner of Imperial College, the Government School of Science Applied to the Arts, that the teaching of metallurgy in this country really started, initially to exploit the mineral resources of the top 23 cm of the surface of this country. At that stage, it was the winning of the primary metal from the mineral that dominated, and it was Percy's task to organize teaching of this subject when he came from the practice of medicine in Birmingham, to take up the chair of metallurgy in the Government School. This Government School, as a result of a wrangel with T. H. Huxley, became the Royal School of Mines. Percy left in rather a huff because of an argument with Huxley, but his successors such as Sir William Roberts Austin and Sir Harold Carpenter, in fact devoted their energies to what was essentially physical metallurgy. While this was going on, the teaching of chemical or extraction metallurgy extended into the teaching of chemical analysis, which was practiced near the mine where the primary metal was extracted. Subsequently a lot of the teaching spread from the Royal School of Mines to the growing number of academic departments in the country. The chemical side of metallurgy tended to be rather arid, concentrating as it did on chemical analyses which were pursued by ancient methods bearing no relation to the interesting physical techniques which we have at the moment. Thus, in the minds of many students, physical metallurgy burgeoned forth from the exploitation of X-ray and microscope techniques, and the development of dislocation and alloy theory to become an interesting and dominant topic in metallurgy teaching.

With the more recent development of chemical metallurgy in terms of thermodynamics, a much more interesting scientific basis was achieved, and it is the exploitation of that which, led by Professor Richardson, we have been attempting for the past 15 or so years. I do not think Sir James means to confine chemical metallurgy solely within the extraction context; he means something much wider. The chemistry of materials can, and should, embrace a much wider range of science than it does at present. But when we attempt to include this in a teaching curriculum, we are in danger of forcing the student in too many directions. It is essential, and I think this also comes from what Sir James has said, that you not only talk about chemistry and physics, you must include someting about engineering properties, fluid flow, heat transfer and so on. The big danger is that the student finds such diverse matter indigestible. We found this continuously during our attempts to integrate this very wide range of science and technology into a single undergraduate course. The alternative is to go to a general type of degree and there are many aruguments in favour of this, but where joint honours degrees have been tried in the past, they have not been noticeably successful in attracting the best sort of students. We have consequently been very wary in moving in this direction. However, the climate is moving again in this way, and it might be possible to exploit the build up of modular degrees. Nevertheless, the chemist and the physicist have distinctive ways of thinking and they do not always easily marry together. This is a serious problem which should not in any way be swept aside. If you try to develop this integration where there is an intense and large research activity, you are almost inevitably having to do it with specialists, who have their own special interests to pursue. I personally think that we might get the breakthrough by giving more recognition and status to devoted teachers who are not married particularly to one branch of the subject.

C. Edeleanu (*Imperial Chemical Industries*)

If Sir James (Keynote) had added to the list of subjects he has mentioned as being related to each other, 'tribology', 'soil mechanics' and 'noise and vibration' he would have described a service which goes by the name of Engineering Technical Services in that part of I.C.I. in which I work. Since we try to provide an integrated service to our engineering colleagues I cannot but agree with Sir James that none of us can afford to take a narrow view about our particular field of specialization. To give an example; the man concerned with corrosion could resolve some of his problems by specifying silicon iron but that only creates a problem for the metallurgist, an impossible nightmare for the welding engineer and for all the other people down stream from him. The essential thing for a service of this type is to act as a team and to have respect for the difficulties which each one of us can create for other members of the team by insisting that what is the ideal way to resolve our bit of the problem is also the ideal solution for the task as a whole. This is the way we try to resolve the problem associated with the interfaces between the various specialist technologists but there is perhaps a more important interface to think about and that is the one between the scientist and research technologist on the one hand and the engineer on the other. From things said during this meeting one does detect a love–hate relationship across this interface and perhaps a degree of ignorance about the method of crossing it.

To understand the problem it could help to remember that the engineer, who is the man who ultimately gives physical expression to the knowhow of technologists, needs to think convergently if he ever is to get the hardware built in a reasonable time and to the customers expectation and much as he may wish to understand every bit of technology he uses he seldom can do so not because of an intellectual weakness, but because life is too short. The scientist and technologist on the other hand is a divergent thinker and especially so if involved in R. and D. and is interested in some of the detail behind the broad facts with which the engineer is concerned.

He has to be like that if he is to be any good at his job. There is therefore a great deal of scope for misunderstanding between the man who simply wants to know what he has to do next and the man who wants to explain why things are as they are and to provide the detailed information about the pros and cons associated with each one of the available options, so that the engineer can use his judgement and make a decision.

One can argue that people should not fall in either of these categories and should be able to switch from one way of thinking to the other but why demand this versatility of people if this is not a real need and especially since it is uncomfortable for most people to make the switch. This is especially so since there are some people who seem to be especially skilled at making this very switch in the way of thinking and can and do cross the interface quite happily. I like to think of such people as being 'brokers in technology' and I think we ought to recognize them, and accept that like other brokers, they have an important role to play. Once you use such a broker you will find that it is easy to cross the engineer/technologist interface if the man is any good.

That is not quite the end of the problem however, because we who live at this interface between engineering and technology have a lot to learn from each other and we need to develop our skills and perhaps formularize our methods of bridging the gap between science and technology. I do not know if this is possible, but we could improve our skills so that science could find its way into hardware a bit more rapidly than it now does.

If I may suggest, this could be a surer way of making progress than by attempting to teach metallurgists how engineers design, estimate, procure, inspect, construct and generally manage a project or by teaching engineers the mysteries of dislocations, chemistry of corrosion and physics of g.r.p.

M. E. HARGREAVES (*University of Melbourne*)

It is rather strange for me to hear this discussion because in our University these various branches of the subject have never diverged. Since 1925, when we established a metallurgy course within the engineering faculty, all students have been taught, not only chemical extraction metallurgy, electrometallurgy and physical metallurgy, but also ore-dressing as well. I think this is a very good thing coupled with the ability to specialize in one of these branches in the final year. Now the wheel goes a full circle some times because in lecturing to my final year class in physical metallurgy last year, I was able to point out to them a paper in an ore-dressing journal which showed the influence of vacancy content in galena on its surface activity and therefore its behaviour in an ore-dressing process. So I think it is possible to teach an integrated course in metallurgy if specialization allowed and we have jogged along quite happpily like this for 50 years.

E. R. Petty (*National Institute for Higher Education, Limerick, Ireland*)

There seems to be a general feeling that physical and chemical metallurgy should reunite. I hold the view that even this is a somewhat introverted outlook in this modern age and I make the plea for a further breaking down of interdisciplinary barriers.

Metallurgists now tend to identify themselves with the whole range of materials available to the engineer and, as engineering is our raison d'être, surely we should think of ourselves as *materials engineers.*

Materials engineering could legitimately include mining, mineral processing and extraction, through chemical engineering and process engineering into production engineering. This latter area includes the organization and management of industry; one hears so many pleas for scientists and engineers to come down to earth and accept the basic fact that industry has to make money and be competitive. This means a rudimentary understanding of such topics as industrial psychology, cost and inventory control, operations research, at least to a level which allows 'experts' in different areas to communicate intelligibly with each other.

The training of a materials engineer would entail a broadly based first degree with a foundation in physical and chemical principles together with some of the more important engineering and technological applications, followed by a limited number of selected topics taken to greater depth in the final year. Greater specialization, if required, would be done in post-graduate and/or post-experience programmes of study. This 'continuing education' is a vital part of the training of any scientist/engineer now that the half-life of technological knowledge is less than ten years.

It seems to me that the flexible system outlined is more relevant to the modern world than the over production of narrow specialists, directed in the main towards university research and teaching, to which we have become accustomed.

N. L. Parr (*Ministry of Defence, Procurement Executive*)

As a move towards a common language between materials science and technology and engineering practice, more research is required on sensors for assessing the microstructural and macrostructural performance of components and structures under operational conditions. This would be of equal importance to materials development, design, cast and safety of ownership. A start has been made on acoustic emission analysis for structures, and monitors for tribological deterioration; but more requires to be done in this highly complex area of realistic characterization. This was stated by Sir James (Kenyote) in terms of a greater need for physical studies of materials under dynamic conditions.

M. G. Gemmill (*Central Electricity Generating Board*)

I was very cheered by the beginning and the end of Sir James' lecture (Keynote) because, having been subject to a University education where I was expected to study not only the chemical and physical metallurgy but also physical and inorganic chemistry I see no problem frankly in what he now claims as the educational requirement for the future metallurgist. If a person is incapable of dealing with that scope he or she should not be there in the first place. So we come back to the general argument which Sir James has put forward. After years of worrying about the over-specialization of our metallurgists, it would be very good indeed if we lose that.

Earlier two of my C.E.G.B. colleagues, Wyatt and Barford, made very honest admissions of the difficulties facing R. and D. people in transmitting their knowledge in a way that allows me, as the Boards materials engineer, to translate it into our manufacturing industry, so that we achieve more reliable components. The situation can be considered in terms of Cohen's chain of events from physical metallurgy, to the engineering end product. What we have to ensure, I would suggest, is that all the links in that chain are strong, so that we end up with an effective communication system.

The problem is how to deploy the sources of knowledge in sensible relation to the vested interest. For example, and this is quite hypothetical, it might be that organizations such as the British Steel Corporation and C.E.G.B. could find themselves in a situation where they carry more resource to contribute to the technical solution of a problem facing a component producer who lies between them, and bears the commercial responsibility. Rather than complain about the fact that he had misplaced manpower resource, I would suggest that we should be able to find ways of transferring this knowledge that give an individual both career and job incentive. This will require device, such as staff secondment or funding either by primary metal producers or ultimate users of an under-resourced component producer.

In the example mentioned, this would allow B.S.C. to promote the use of their very capably researched findings on steel metallurgy and C.E.G.B. to see for example, practical return for their equally well researched work on the effects of impurity elements in regard to welding technology.

M. Cohen (*Massachusetts Institute Technology*)

I am rather dubious that placing physical, chemical, and process metallurgists in the same department will achieve the unity and the interplay required for the two-way flow of knowledge between metal science and human needs. There are already too many instances where this juxtaposition, although desirable in itself, has not proved sufficiently attractive to students or effective in industry.

Perhaps the time has come to consider a systems approach to metallurgy or to materials more generally (at least for some students) in which the field is viewed with 'telescopes' instead of 'microscopes'. At M.I.T., we have begun to work on this different focus, taking advantage of the impressive progress that has already been achieved in other branches of engineering, where some of the core subjects now emerge as applied probability, optimization theory, and microeconomics.

J. H. van der Veen (*Hoogovens, Holland*)

Within the scope of the discussion about interface between 'process' and physical metallurgy and the merits, or otherwise of combining these disciplines into one, I would like to draw attention to the factor 'motivation'. I think, in practice, more attention could be paid to motivation of 'process' metallurgists in the direction of making steel for future products; future performance of a product. Being far away from the final product, in terms of processing stage, and far away from the customers' needs, process metallurgists otherwise perhaps might pay too much attention to steel making efficiency and other (though important!) more 'local' problems.

G. G. Saunders (*Whessoe Ltd*)

Following up the point of cost effectiveness, may I offer a suggestion towards the future role of physical metallurgy in relation to the technological objectives of this country during the next 25 years. If one studies the many published papers on physical metallurgy, one observes that these are concerned mostly with what one might consider as high technology; such as might be associated with projects like Concorde. Since the bulk of this country's prosperity relies very heavily on the 'low technology' associated with the fabrication of iron and steel, may I suggest that those responsible for directing future studies in physical metallurgy, consider the considerable benefits to be gained in knowing more about such seemingly ordinary materials as mild steel.

B. A. Bilby (*Sheffield University*)

May I first comment on the multi-disciplinary approach needed to make progress with complex problems. As more disciplines are demanded, the teacher may seek to encompass them by a course which is both more diluted and more theoretical. However, not only is the proper theory often not yet available, but a false picture of the discovery process itself may thereby be created. A thorough discussion of fewer topics in their correct historical perspective is more likely to instil that competence in the critical handling of the interaction of theory and experiment which is necessary for progress. To tackle the problems then, there must be team work, with specialists in different fields cooperating, although ever team must, of course, have a leader.

Despite these strictures on theory, I believe it has an important role to play not only in the microscopic fundamentals of materials science, but also in the interaction with the engineer (the ubiquitous dislocation is also relevant to both fields). One of the great thoughts of this conference I find in the maxim of Francis Bacon quoted by Brown (2.7) 'Truth will sooner come out of error than confusion'. This is apt for some of the activities of Rosenhain himself, as we heard from Kelly (Centenary lecture). A prime task of the theoretician is to reduce the confusion; if I may exaggerate Willard Gibbs a little, theory is supposed to make a thing so simple that you wonder why you have not already looked at it this way yourself (and that is always the feeling I have after hearing a lecture by Sir Alan). A classic example of what I mean by making things simple is to be seen in the rise of linear elastic fracture mechanics. The important simplifying concept here is the stress intensity factor K, which is the leading parameter in the characterization of the crack tip stress field. A critical K criterion for crack extension includes the necessary energy balance condition of Griffith, but in a more convenient and powerful form. The engineer can use it to infer that, other things being equal, K determines the physical processes at the tip, so that if fracture occurs at a critical K in the laboratory, so also will it do so in a structure. Thus a basis for practical testing is established. When fracture occurs only after considerable plastic deformation, the situation is more complicated. Here we have a current problem, with its attendant confusion. This is not the place to discuss it in detail, but theory can help by specifying precisely under what conditions the present attempts to characterize the fracture process will be valid, and by looking for new methods of doing so. Finally, it is clear from the discussion that an important concern for the future in the theory of fracture must be to pass from a fracture 'mechanics' to a fracture 'thermodynamics'; or, rather, to an integrated treatment which takes account not only of thermally activated processes, but also of electrochemistry.

W. H. Eaton (*University of Surrey*)

At the end of the conference I had an impression of pessimism about the future of physical metallurgy; a sensation of the end of an era. It appeared that the only path for the physical metallurgist was to take his well tried methods of structure-study and apply it to the correlation of structure and properties in non-metallic materials.

This is certainly a field that offers an enormous scope for his talents and would enable him to assist the chemists and the engineers who develop and use these materials.

However, while the mechanisms in structures leading to properties in metals may be understood, there are still large areas in the physical metallurgy field to be studied, particularly in persuading the 'understood' mechanisms to operate reliably in commercial steels and non ferrous alloys.

In the past, the physical metallurgy of steels was extended by the automobile and aircraft industries. Parts for these industries were small and could be loaded into furnaces for thermal treatments at various stages of manufacture. This initiated studies into the improvement of mechanical properties by controlled solution and precipitation of particles. It was necessary for the particles to be unstable over a range of temperature change otherwise the thermal treatment could not be applied.

The development of the gas turbine, the high temperature boiler and the atomic power station moved the study into the direction of stable steels and non ferrous alloys in which the structure did not change over longer periods of time at elevated temperatures.

In the immediate future the gas and oil pipeline technologies will demand large quantities of low carbon weldable steel leaving a high yield stress, good crack arrest properties and (in some cases) low temperature ductility.

There is still scope for the physical metallurgist to develop steels suitable for pipelines and pressure vessels, and to examine the behaviour of those steels already in service in these structures, to determine if they can be economically improved.

Other participants have already commented on the effects of residuals on the ductility of weldable steels and there is an interesting area of investigation in reducing their effects at low cost.

N. Hansen (*Danish Atomic Energy Commission*)

A large field opening up for metallurgists and engineers has not been touched much upon today: energy production. This field presents us with a number of complicated problems, just a few examples: gas–metal reactions in gas cooled reactors for process heat, materials for hydrogen production and energy storage and temperature and radiation resistant materials for the potential fusion reactor. Due to the pressing energy problem, there may be good reasons to believe that we have a promising future in this area of energy production. With reference to Cohen (this discussion) we might hope that good results in this field in a positive way will increase our interaction with society.

5. THE TASK FOR THE EDUCATOR

By Sir Alan Cottrell, F.R.S. (*Jesus College, Cambridge*)

Unreal attitudes

The first task for the educator is to get someone to educate. It is extraordinary how, in a time of shrunken classes and empty places in academic departments of engineering and metallurgy, this elementary point is overlooked in discussions about the teaching of these subjects. Arguments all too often become unrealistic because they are built on an assumption that there exist captive audiences available to the educator to mould as he wills, an assumption that is as far from reality in this permissive and feather-bedded age as the Victorian mechanics institutes are from the student unions.

Even when they are supposed to address themselves to the problem of the empty places, such discussions tend to drift away into a make-believe world. A voice goes up: 'a year in industry, between school and university, would do the students good'. That may well be so, but the setting up of an extra hurdle between the student and his degree, in the applied sciences, would inevitably deflect many school-leavers away into the pure or social sciences, or the arts, where the ambitious ones could maintain their academic momentum without interruption and the more cautious ones could continue to enjoy the familiar comforts of an academic environment. The cry 'a year in industry first' pre-supposes that the school-leavers are already committed to eventual entry into industry, an assumption which for the most part is quite untrue in the conditions of today. I believe that if such a requirement were introduced, it would in fact cause them to make up their minds not to go into industry or subjects related to it.

Once a discussion on empty places is side-tracked in this way it usually then ends up totally derailed by calls to 'make university courses more practical', or 'teach more about the widget-making industry' or 'train them in the techniques of knurdling!' I am glad to say that the professors of the applied sciences, who are usually the only audience for such remarks, generally adhere robustly in the face of this to their views that you cannot apply science until you have some science to apply; and that general potentiality rather than specific skill is the quality they want to develop in their students. But nevertheless, such remarks do not make their task any easier when trying to increase the attractiveness of their courses, in competition for students against their colleagues teaching in the other sciences or the arts.

To see the real position, let us turn to some figures. The problem is one for the applied sciences generally but, in this conference dedicated to the memory of Dr Rosenhain, I will deal only with metallurgy and materials. Over the period from 1969 to 1974 the total number of students admitted annually to degree courses in metallurgy and materials dropped, from nearly 700 initially to only about 530 in 1973, with a slight recovery last year to nearly 600. Over this same period, the number of available places rose, from about 860 initially to about 960 last year, so that on average over one-third of the places are empty at the present time. While the subject 'materials' has made headway, increasing its intake from about 110 in 1969 to slightly over 200 last year – undoubtedly a result of the development of 'materials science' as a new academic field – that of 'metallurgy' has dropped away alarmingly. In 1974 it attracted only about 200 students, compared with 440 in 1969.

Why have the numbers been declining so much? Is industry overstaffed with metallurgists and materials engineers? If that were so, one might expect to find their salaries to be lower than in comparable other professions, but I do not believe that they are, to any significant extent. Furthermore, other countries such as Japan seem to employ many more, in proportion, than we do. And in any case the Chairman of the British Steel Corporation has said that in the course of a generation it will be necessary to find 180000 metallurgically minded people to run the Corporation's new processes and plant. Even a Methuselah could barely span a generation long enough to square this demand against the supply.

The answer becomes clearer when the problem is taken in context. Some of the decline is undoubtedly a reflection of that in the exact sciences generally, which are too sharp, hard and clear for today's indulgent tastes; and much of the rest can only be part of the general reaction against industry, which still bears the image of the cloth cap and sweat rag, as well as being the main battleground of today's social and political upheavals. But there is, I think, a third factor which influences the choice of the more discerning and serious student; and this is concerned with the range and scope offered by the various subjects he might elect to study.

General education or vocational training

The range of intellectual horizons opened up by a course of study, and the scope for branching off into various careers afterwards, raises again the question of education versus vocational training. I believe strongly that undergraduate education should be very broadly based and aimed to develop general potentiality, not particular specializations. In the physical and engineering sciences I believe that a good understanding of mechanics, electrodynamics, thermodynamics and atomic structure are absolutely essential and that, in an undergraduate course, these should never be skimped to make room for more directly technological subjects. The concepts underlying the physical and mechanical sciences, such as the nature of strain, or entropy, or electrodynamics, or wave functions, are difficult to grasp and, if not mastered during one's student days, when both the opportunity and the ability to absorb new ideas are at their most favourable, it is unlikely that they will ever be really understood later, in which case one will be equipped with an incomplete and fragile set of intellectual tools for tackling the work of a technological career. By contrast, it is I believe much less difficult to add on, later, to a sound general scientific foundation, the detailed knowledge by which one becomes an expert in a specific technical field.

This is fairly well understood in academic circles but, from the standpoint of the teaching departments in the applied sciences, it raises problems because it so often leads school-leavers to choose instead to read the 'basic' subjects such as physics, chemistry and mathematics, at least for their first degrees.

There is in fact a great deal of sense in this. In many ways, I believe that the ideal course would be to start with a first degree in the basic sciences – perhaps an 'ordinary' degree taken after two years – followed by a second degree – which could be an 'honours' degree – which in the case of those aiming at the technological professions would be taken in the appropriate applied and engineering subjects.

Needless to say, such a proposal faces various objections and difficulties. I do not take seriously those, both in industry and in the applied sciences, who argue that there would not then be enough time for the students to learn their technological subjects to the required professional depth. As I have said, the aim should be to provide future potential, not present skills, particularly as the graduate will supplement his few years of degree studies with something like forty years thereafter of direct experience 'on the job'. In any case the charge of lack of time cannot be seriously maintained in this country, when we spend so much on postgraduate training in research, an outpouring of time and money which, for all but a rare few of students, could be much better used in extending the advanced formal education to something more like that in most European countries, or in the U.S.A. where an 'ordinary' Bachelor's degree is usually followed by an 'advanced' Master's one. There is a great and general problem for our system of higher scientific education here. It will not be solved readily and it may be that, for some time at least, we shall have to settle for lesser solutions that lie within the framework of the present academic structures of the country.

A second problem is, how do the students ever become acquainted with the applied and engineering sciences, sufficiently to consider at some stage transferring into them and making their careers in industrial technology? By and large, there is still very little awareness of these subjects among the schoolmasters who guide their school-leavers; and, after the failure of the many campaigns to create such awareness, I think we must accept this as an enduring fact of life. In any case, even the aware schoolmaster may advise his pupils to 'get a sound education in the basic sciences first, and then branch out from there' and from their point of view it must often be excellent advice.

If, then, we have first degrees in the basic sciences for all science students, may not the above state of affairs simply persist in the universities and such courses become in their range of interests merely school 'seventh forms'? The answer, I think, is to have *all* the scientific and technological departments participating in those basic scientific courses. I do not mean by this that they would be teaching their own technologies (although they might want to refer to them briefly to illustrate the working of basic scientific principles in practice, such as for example the principles of surface tension as used in froth flotation of metallic ores). No, I would want them to be teaching those parts of basic science, integral to the fundamental courses, of which they have particular experience. For example, the mechanical engineer might teach classical mechanics, the materials scientist teach the structure of matter, and the metallurgist teach thermodynamics. In this way the first-degree students would become acquainted, if not with the technological subjects themselves, at least with representatives from those subjects. The ice would be broken and the students would have a wider range of people from whom to seek advice on their future choices of subjects and careers.

If I may speak here briefly in support of my University, I do think the Cambridge Tripos system has great advantages from this point of view, with its division into a broadly-based Part I and a specialized Part II, and with its general entry into a common set of courses for undergraduates in Natural Sciences Part I, including the first-year course on crystalline materials of which the Department of Metallurgy and Materials Science shares a good part of the teaching. The majority of the physical scientists in Cambridge go through that course and a satisfactorily large number of them elect to continue with metallurgy and materials science in their second and third years.

Metallurgy and materials science

A special problem about metallurgy and materials science, as an integrated discipline, is that it has equally important links with three different major subjects; physics, chemistry and engineering. Of course, a lot of its fascination stems from this, and a good deal of the justification for having metallurgy as a distinct academic subject rests on its equidistance from these other subjects, but it has rarely been able to form close alliances with any of them without weakening its links with the other(s). There is thus a tendency for it to break up into materials science associated with physics, chemical metallurgy associated with chemistry and chemical engineering, and materials engineering associated with general engineering. I do not think these problems will be solved until we have single Schools of all the Sciences with much more commonality between the courses. In such Schools the students would all take a selection of 'general purpose' courses – taught by representatives from all appropriate subjects – as well as some 'single subject' courses, at all stages of their undergraduate education.

There is a second, and related, problem about metallurgy and materials science, which concerns the career of the graduate in industry. Is metallurgy a primary or a supporting technology? The answer depends on what kind of metallurgy we have in mind. Chemical metallurgy is undoubtedly a primary technology, since there is a major industry – that of basic metal production – devoted to its output, so that it is entirely reasonable for an outstanding chemical metallurgist to expect that he might eventually have charge of a major enterprise in this industry and be responsible for its general policy. But the physical metallurgist, the materials scientist and the materials engineer, is not usually in such a position. Of course there are industries that specialize in metal and alloy manufacture where he can play a leading role, but in most parts of the industrial sector his position is more likely to be that of assistant to someone else who is producing engineering goods and services. The materials man can contribute a lot to, say, nuclear reactors, or aircraft, or computers, or chemical plant, but is unlikely to be given overall responsibility for the project, which generally and rightly belongs to the engineer. He is unlikely to reach the top of the engineering tree in such industries, unless he abandons his specialist position and converts himself, both professionally and by experience and interest, into a general engineer.

I think this may be a reason why some of the modern physical metallurgists and materials scientists have, like so many physicists, preferred to seek careers in academic life; they may have felt that the road to the very top, where they could be in charge of projects and technical policy, was not really open to them in most of industry.

Relation to engineering practice

The relation of physical metallurgy to engineering practice raises other problems for the educator. How should one present the science of materials to the engineer? It is important to remember that the 'mainstream' engineer – whether civil, mechanical, electrical, chemical or aeronautical – is a *user* of materials. All his talents and interests are devoted to engineering goods and services. For him the problem of an engineering system is to achieve a satisfactory performance, to have it respond with good outputs to the stimulus of its applied inputs. So long as this is achieved, the inside of the system can be a 'black box', so far as he is concerned. Thus, for him, materials are things that fill the space of the system with certain properties, preferably expressed in terms of a few simple numbers, and which cost money.

I think that this point is sometimes overlooked in courses on the science of materials for engineering students. It is only natural for an enthusiastic scientist to wish to explain his subject, as it excites him, to his audience; for the physical metallurgist and materials scientist to explain from the phase diagram how alloys are formed, how crystals diffract X-rays, how atoms move about in solids, why metals conduct electricity and insulators do not; but are these the things the general engineer really wants to know?

They are the things important for the designer of *materials*, but the engineer designs *goods*, not materials. Most of the science of materials, when it has been concerned with applications, has been aimed at new alloys, improvements by heat-treatment, new microstructures such as fibre-reinforcement, and new fabrication processes. What we may have overlooked is that there is a potential new science, hardly developed yet but of direct relevance to the engineer, which is the science of the performance of materials in engineering service. Perhaps the best example of this is fracture mechanics, developed from A. A. Griffith's pioneering scientific investigations into the nature of brittle cracks, which has now grown into a standard practice in engineering design, based on the requirement that the material in an engineering component should have a minimum fracture toughness, the value of which is governed by the geometry of the component and the conditions of its service.

Need for a science of materials in service

If it were possible to deliver to the engineer various design procedures as explicit as this, for dealing with other problems of materials in service – such as metal fatigue, stress-corrosion cracking, or fretting failure – physical metallurgy and materials science would then be able to make new great contributions to engineering practice. But we are still a long way from achieving this. The necessary science is not yet there.

Many research metallurgists accept this but then go on to delve more deeply into the basic atomic processes of fatigue, or corrosion, etc. Admittedly, we do not yet have a complete understanding, at the atomic level, of the above complex forms of failure in engineering materials, but it may be that such an atomic understanding will not help to solve the practical problem and that the more important thing that is missing at the present time is a rigorous and sophisticated applied science of the uses and service of materials. Let us remember that the application of fracture mechanics does not depend on atomic concepts but on an understanding of the mechanics of a crack in a medium characterized only by its general elastic properties and its measured work of fracture. Similarly, from the practical point of view, the most important scientific discovery about metal fatigue was made long ago, i.e. that this is a type of plastic failure at the surface. A simple knowledge of this alone leads to many useful points in practice, e.g. the importance of hard surfaces, avoidance of soft zones, and minimization of tensile stresses in the surface.

I believe that there is great scope now for taking the existing 'pure' science of materials, as we already know it, and building from it a large new applied science of the performance of materials in service, one that will enable the engineer to get the best out of the materials available to him, even to the extent of encouraging him to adapt and transform his traditional design and constructional procedures to suit the properties of these materials, and to ensure that he will not be disappointed in practice through the unwonted failure of his constructions.

I believe that one of the most exciting challenges now, to the educator in the materials field, is to establish and teach to engineers a 'science of materials in service', comparable to the quite different 'science of materials for service' which he now teaches to metallurgists and materials scientists.

Discussion (Chairman M. E. Hargreaves (*Melbourne University*))

A. Kelly (*National Physical Laboratory*)

For a few more days I can speak as a user, rather than as a producer of trained metallurgists. I want to make a point which I have not yet heard made in this discussion; I want to approach the question of motivation, not in terms of personal ambition of the student as Sir Alan (Keynote) approached it, but in terms of a more altruistic motivation which I think is more important, to the young nowadays. One of the great motivations for people doing engineering is the pull of creativity; the joy of making things; this can be appreciated by the architect or by any of us who try to draw things. There is a real attraction in engineering for the creativeness of a man. Science I believe is motivated by inquisitiveness and so, if one accepts that definition of science, metallurgy is in a very peculiar place, shared only by one other science of which I know.

Consider as you look outside yourself, the stars, the earth you stand on, the rocks underneath, there is perhaps a fluid core in the middle of the earth, there are living things upon this planet – those are things a scientist studies, and none of those was made by man. The difference I think is that metallurgy arose to look at synthetic things. Now that carries with it a very important qualification. Astronomy may be limitless, geology less so, biology probably almost as limitless as astronomy. Metallurgy is not, because the metallurgist looks at synthetic things and those man-made things are made – why? Because the engineer wishes to have them to make other things. On the whole we support a metallurgist because his science supports the engineer.

Now there are two consequences from this which are peculiar to our subject. One, if the young are less motivated to do engineering, then clearly they are less motivated to follow the subjects which support engineering. The second is that since the subject is limited, it does not have the horizons of biology, nor yet of geology, or geophysics. Its (metallurgy's) very own success removes a need for more of us. We have seen in this conference, quite brilliantly, I think, although hedged by details that we do not understand, magnificent achievements. The steel men have told us that they very nearly understand all the factors of that once complex and invisible microstructure; that black box. They can now tell us what are most of the factors involved in producing a strong steel. That success removes the mystery, and once a subject loses the challenge of mystery, it may go the way of other worked-out disciplines.

We are all young as scientists. We are not yet used to the rise and subsequent decay of once-useful and hence fashionable fields of enquiry.

That being said, I see three routes which this corpus of knowledge embracing thermodynamics, microstructure and mechanics, which we call metallurgy, can follow. One of these has been chosen by Sir Alan and that is the last one I would list; it is to support the engineer by producing a prediction, and that prediction will be best used in the forms of constitutive equations. But there are two other routes and, depending on the funding and societal influences (as Cohen calls them) there is still a pure road. That corpus of knowledge has much to say to the geologist, to the person studying planetary structure, particularly the cold planets. It has much to say to the biologists and that is happening in part (e.g. in biomechanics). So there is still a pure road for that corpus of knowledge to follow, and then there is an intermediate one between those two and that brings me to what is the other subject like metallurgy. The answer of course is polymer science. Again a man-made material, but several years behind us in the provision of the microscopic techniques which provide means of unravelling microstructure. Polymer science may join with metallurgy and continue in much the same way as we have seen metallurgy develop.

J. G. Ball (*Imperial College, London*)

I would agree, as a result of direct experience, that it is not a lack of knowledge in schools which results in a lack of students. I think we have to live with such ignorance or such knowledge as exists and I do not think we can much influence the situation. If we look back to the time when we were on the upward grade of recruiting, in the mid 1960s, the factors influencing the flow of students included an overspill from the pure science departments and a desire of many students to keep up their maths, physics and chemistry by taking a subject which included all of these branches of science. Many recruits came because of family connections and there were also people who were intrinsically interested in the subject. In recent times I think that all of these influences have diminished, even the one of family connections and intrinsic interest in the subject. Certainly we no longer get the overspill from pure science or engineering departments because they themselves are looking for entrants. I think Sir Alan has analysed some of the reasons why people go into pure science rather than into our subjects.

What worries me is that not very long ago we considered it to be an absolute necessity, for the students on entry to have a competence in all three basic sciences and that still is expressed in most propsectus', but on the whole it is a forlorn hope that all our students will be prepared in this way. It seems now that our requirement for breadth of scientific understanding is a deterrent to entry into the subject and I do not know what we can do about it.

I feel the situation is going to get worse. The trends of sixth form studies are towards scattered A level subjects, and we might have to be dealing with people coming in with one A level. Such trends are also likely to reflect a lack of rigour in teaching which we have seen developing already. I do not feel that we are collectively approaching this problem, and I think we should be doing so.

D. McLean (*National Physical Laboratory*)

In connection with the development of a science of engineering strength, or performance as Sir Alan (Keynote) calls it, I would like to make three remarks. First, to accomplish such a development material scientists must move on from their preoccupation with uniaxial testing under very constant conditions – constant load, constant temperature and so on – and study the behaviour of materials under conditions which epitomise those of engineering service, such as under multi-axial stress, variable load, combinations of conditions etc. To some extent this involves going over the work which engineers did about the turn of the century and expanding it to suit modern conditions with the outlook of a modern materials scientist. Secondly, it is easier for materials scientists to learn engineering than the reverse, the reason being that a lot of engineering is well codified, is well digested, well condensed and well written up in text books. Quantitative materails science, on the other hand, is quite a young science which has developed during the last 25 years, and is not nearly so well digested and well written-up in text books. Thirdly, the time for such a development is ripe, partly because the problems for which engineers have to design are so complicated that what we might call the existing empirical methods are often unsatisfactory. I think there is a parallel with the relation that existed a century and a half ago between elastic science and engineering, in that the science of elasticity had then been developed by Coulomb, Poisson, Young and others, and had just begun to reach engineers, whose constructions eventually gained immensely. However, the process of being assimilated by the engineers took a very long time. Realization of the parallel should help to shorten the delay in the application to engineering design of quantitative materials science, to produce a science of engineering strength.

R. W. K. Honeycombe (*Cambridge University*)

Much has been said about the present day student and civilization by Sir Alan (Keynote) and others, but I would like to make one point about the modern student that I think points to a paradox, and that is, despite the challenges to authority, students are, on the whole, more dedicated to service of the community. There is a greater social awareness among students these days, than say, 20 years ago. Now this means that students want to do work which is clearly of need to the community, for example, they are prepared more readily to embark on applied science if they have the opportunity. They no longer automatically travel the road of pure science as a first choice. I have talked to many students in Cambridge who have reached the end of their second year, when they have to choose what to do in Part 2, their final undergraduate year. I find out from them why they often change from physics to metallurgy, and the answer is very frequently that they want to do a subject which has direct relevance to the community at large. If this is so, why are the numbers of applied scientists falling in many places? At Cambridge we are suffering much less, so I conclude that we must attribute some of the blame to the mechanics by which the options are presented to students. The students become aware of applied science not at school, but by the time they reach the middle of their University courses, which is the most appropriate time they have for making a decision whether to be an applied or a pure scientist. I think the conclusion from such observations is that we must displace this decision from the sixth form at schools. Sir Alan has suggested one way of delaying the decision namely until after the first degree. In ideal circumstances that would be the answer, as it was in the States for many years, where many of their distinguished applied scientists were originally pure scientists. However it seems to me that we cannot afford, at the moment at least, to have as the main route to technology a first degree in pure science followed by further training in applied science, furthermore, we might lose people in the transition. We have got to structure more courses in a way to encourage a good entry of scientists, and allow technological options to be decided on in mid-course involving a smooth transition from pure to applied science.

Sir James Menter, Treas.R.S.

Bilby (discussion 4) remarked that the areas of science and technology we are dealing with here can be tackled only through large teams from a variety of disciplines working together.

Speaking as an employer of substantial numbers of graduates I can confirm that but wish to point out that this sytem of working through committees and the like is forced on us by the intense specialization of the individual and, with a relative scarcity of graduates with the ability to comprehend a variety of disciplines, design and project work tends to be hindered by all the well known shortcomings of the committee system. I am therefore attracted to Sir Alan Cottrell's idea (Keynote) for creating large schools of science (and presumably also engineering) in Universities where students can study a selection of specialist disciplines in some depth. This system has greater potential than existing arrangements for producing a greater proportion of people with broader competence, having the character of the Prince Charming to which I referred in my talk.

J. F. LANCASTER (*Kellogg International Corporation*)

The hydrocarbon processing industry is a major consumer of fabricated products and, in the United Kingdom, if we make proper use of North Sea oil and gas, will become increasingly important in this respect during the next few decades. It is therefore essential that metallurgists, both in the academic and in the practical fields, understand and respond to the particular metallurgical needs of these industries.

Although the industry's problems are common to those of the engineering industry as a whole, special considerations arise in the specification and construction of the process equipment itself, where inflammable and sometimes toxic fluids are handled and processed. To keep up proper standards of safety in this area required the appropriate techniques of material selection, of quality assurance in manufacture and of operation and maintenance of equipment. It says much for the diligence of all concerned in this industry that the safety record is so good. For example, in the chemical industry the number of fatal accidents per employee is 3.5 per 10^8 h as compared with 4.0×10^8 for British industry as a whole, 40×10^8 in coalmining, and so forth.

There is, of course, considerable incentive to maintain or improve such high safety standards, quite apart from concern for the safety of individuals. The scale of single stream process units increases with time, and incidents that can lead to an unsafe situation may also lead to periods of increasingly costly downtime. Major incidents, such as the Flixborough disaster, also have an unfavourable political effect on the industry, making it more difficult to obtain suitable sites for process units and potentially attracting restrictive legislation regardless of the fact that the industry's general level of safety is good.

Metallurgists and materials engineers can make an important contribution to safety in hydrocarbon processing. Not all unsafe incidents are caused by metallurgical defects: indeed, the majority of such incidents are complex in origin and it is rare to be able to isolate a single well-defined cause. Nevertheless, the failure of a piece of metallic equipment is a feature of most accidents and it is most important to ensure that both materials and fabrication are to standards that are sufficiently high to exclude the metallurgical factor as a primary cause of failure.

This is an easy generalization to make. It is more difficult to define precisely what steps should be taken in metallurgical education, research and development to promote higher standards of reliability in this field. To do so would require much better information about failures than we presently have. Nevertheless, I think the metallurgical profession might profitably give attention to the problem of training in materials engineering.

Materials engineers in the hydrocarbon processing industry are usually metallurgists, chemical enginers or mechanical engineers who have elected to specialize in materials. On-the-job training is the normal rule, but it is not particularly easy to acquire the necessary information. This is especially the case with material selection. Chemical engineers are sometimes given a course in this subject, but it is at present tending to be crowded out of the syllabus. It is not, so far as I know, included in any of our undergraduate courses in metallurgy, nor are there any short courses available in material selection in the United Kingdom. Furthermore, there is no worth-while textbook on the subject. With the exception of the lectures included in corrosion courses organized by such bodies as U.M.I.S.T. and the City University, the materials engineer must learn his trade as best he can, obtaining information about the experience of others, mainly from U.S. journals, such as *Corrosion*, *Materials Protection* and *Hydrocarbon Processing*.

The process industries do not at present employ many materials engineers but the demand for such technologists is going to increase in the future. It might be profitable, therefore, for the metallurgical profession to consider how a better transmission and exchange of information on materials engineering and materials selection can be achieved.

The following two contributions were written after the meeting:

K. E. Easterling (*University of Lulea, Sweden*)

Of five technological universities in Sweden, four have departments of engineering materials, i.e. for the teaching of engineers with specialization in materials, and there is only one department of physical metallurgy. Thus the important interaction between engineers and materials specialists is strongly recognized. Engineers in most cases follow general engineering courses in the first two years and specialize in the area of their choice in the final two years. Every individual course is followed directly by an examination, so there is not a general examination session covering a whole year's work. There are no honours degree systems (as followed in Britain); the philosophy here is that graduation engineers have all broadly acquired the standards expected of the school concerned, and as such start out on their careers on a more or less equal basis. On the other hand, individual examinations are graded, so some recognition is made of talent.

E. R. Petty (*National Institute for Higher Education, Limerick, Ireland*)

We should remember that the word 'generalist' has come about to some extent because of the particular historical way in which we have packaged knowledge. Were we to start reorganizing the information available to mankind we might well never use the words metallurgy, polymer technology, production engineer. In fact, words like chemical engineer or systems engineer might well say it all. Or possibly, different groupings of topics might be more appropriate and we might talk of departments of tribology or thermodynamics. In view of the cries from industry and the ways in which Britain has fallen behind in the technological race, the time is probably long past that we should have stormed the barricades between rigid disciplines.

Further advantages of the broader degree scheme are related to the greater breadth of topics being taken at G.C.E. A level these days and also the necessity to give students more time to make up their minds about a chosen career path.

One major criticism of the generalist approach is based on an implied reduction of intellectual level – the so-called 'depth' of treatment. In the system proposed this may be readily avoided by cutting down on the total number of pieces of information considered – after all any degree scheme in however 'narrow' a subject can no longer hope to pack in the total sum of human knowledge in the field, so a selection is always necessary. In the generalist approach one simply broadens the range of options from which the selection is made. While on the topic of depth, it has always been my experience that students find it more difficult to assimilate and correlate material from diverse fields than to go more deeply into one specialist area. Deeper specialization can always come about through continuing education, either formally or during normal self-education at the place of work.

At the National Institute for Higher Education we have set up such a system of education. Being a new educational establishment we are not bound by tradition and, while this has many disadvantages, it does allow us the freedom to experiment with new ideas. One of these is the setting up of a degree programme in Materials and Industrial Engineering administered by an interdisciplinary committee with representatives from Business Studies, European Studies, Electronics and Mathematics as well as the engineers. We have produced a four year sandwich-type course (ten academic terms plus a total of 12 months in industry) with the first two years devoted to basic science and engineering. The two final years cover a range of elective topics many of which are sequential and interrelated.

6. SUMMARY AND CLOSING REMARKS

Mrs Kirsner

As the only representative of the Rosenhain family here at this conference, I take this opportunity of thanking so many splendid people who have made possible a conference, dedicated to the memory of my father. Most of all I want to thank Dr Anthony Kelly whom I know has been thinking about this for many years. If you heard his fine Centenary Lecture I don't know whether you would agree with me but, being the daughter, it was very moving to hear my father presented in such reality, the character and his personality as well as some of his work. May I make one point about my dad, as he is to me still, I think had he known that such a conference as this were contemplated and in fact carried through, he would have been a little bit surprised, and very delighted, because as you may know his work was paramount. Scientific honours came to him in plenty, although he did not seek them, but this kind of honour, he would have been very delighted with. And so Mr Chairman, ladies and gentleman, I wish to convey my most sincere thanks and appreciation to all those people who have contributed, to all those people who have supported Dr Kelly and all the Societies that have given their support to this the Rosenhain Centenary Conference.

A. Kelly, F.R.S. (*National Physical Laboratory*)

Mr Chairman, Mrs Kirsner, Ladies and Gentlemen. This session was planned in order to summarize rather quickly, the points made in the discussion. Sir James has covered so well in his address a summary of the points made in the papers. I think that I have to summarize extemporaneously and rather quickly, and I should like to finish by thanking all those people who have contributed.

I must say I feel rather like the young churchman who was summoned early one morning, one Sunday, and told 'You are going to have to preach in the cathedral, because the Bishop has been taken ill and so have his various vicars; not only have you got to give the address in the morning, but you also have to give an address in the evening.' So he said, 'Oh, well, oh dear dear, so I have got to speak almost immediately without notes.' And they said, 'Yes, that's right.' So he said, 'Well, the only thing I can do then is just to get up and let God speak through me, and I will see if I can do anything better this evening.'

What struck me was that in the sessions on engineering requirements, there was pointed out (most particularly in the discussion) the need for tests for stress corrosion cracking, corrosion fatigue, and particularly for fretting; it is clear that there are needs for tests in these areas, and particularly tests in corrosive environments coupled with some form of rubbing of the surface. Professor Wells summarized a long discussion about what was needed beside the yield stress to define mechanical behaviour, and there were various suggestions made. For instance, ductility at the root of the notch (which one guesses must automatically involve some size effect, not always the same for different structures). Mr James in passing said we do need to consider costs, and costs then fell almost out of consideration until the last session. The only point I would make is that as everyone becomes more aware of the necessity to consider hard economic facts, we do need more detail on those hard economic facts. We do need to know in detail the factors influencing the final cost.

Mr Cotton brought home, very vividly to me at least, that there are three types of engineering structure being considered, when he said that with offshore rigs, one did not know the service environments. These three types of structures are: one, exemplified say by a bridge, discussed by Harper, where the general design requirements are known and the new types come about by economic pressure, say for instance, box girder bridges. The general engineering requirements are known, and the innovation or substitution comes essentially I think from cost effective motives. The second type would be the one discussed by Imrie; the undercarriage, where very precisely defined and clearly enunciable requirements have to be met, e.g. these were strength, toughness, cost, each very precisely defined. Now the third one, which I don't know whether as metallurgists, we had really realized before, are the offshore installations which are being put into service in a completely new set of environments. It was brought home that here one designed and had to build the thing, but did just not know many of the conditions under which such an enormously costly structure would be operating.

Then we went back and forth over high toughness and high strength, speaking of these as material parameters. But of course there are ideas emerging from mostly Professor Morley of Nottingham, who is not here, on designing toughness into a structure as an integral part of that structure; not just using a material parameter to obtain it.

Some of the later discussion was a little confused because people, when talking about tests, were not distinguishing clearly between quality assurance tests and tests for design.

In Session 2 which dealt with some of the design implications, there was a fascinating discussion of probability. I have been unable to see Flint again, who came up with it. I could not follow his argument. But he certainly amused us at the time, when he said that the greater spread in values of a parameter, e.g. strength, gave the greater reliability in the structure. I could not understand that, and I am afraid I have not been able to speak with him since. Hirsch intervened and supported the idea of the probability concept of failure, because, as he very clearly put it, it is (*a*) quantitative, and (*b*) can then be subject to risk analysis. But of course that must be predicated upon knowing the routes to failure, and as Leckie said, we must know the mechanism and it is best to be able to present that mechanism in terms of some form of constitutive relation.

We then passed on to consider corrosion monitoring and that leads me to suggest that there had been an earlier suggestion by Meetham from Rolls Royce that may be relevant here, namely that by watching the change of a microstructure, one may deduce something about service condtiions. I wondered whether this would be useful either on the offshore oil rigs or in other applications. We have now the ability to watch even the effect of atomic motions, so that we are not limited only to high temperature changes; I wonder whether in fact by examination of failures and even of pieces put in there specially so as to be taken out later, one could not obtain a better idea of some of the campaigns conditions for a particular engineering structure.

One other point that emerged is that design, while very important of course, is not everything. One has to be sure that the design is in fact carried out, that the right material is used, and that mistakes are not made. One must design against failure to carry out the designer's specifications, and of course in one sense fracture toughness does that because besides specifying a stress, it also specifies certain types of inspection which must go with that stress in order to make it a sensible design parameter.

We then passed on to materials development, and there I was struck by the fact that the properties of steel are really remakably well understood, in terms of grain structure, dislocation structure and the different types of hardening mechanism. The locus of properties can be stated, although we may need tests for material in specific applications, as came out of Mercer's discussion – and also when Mr Clay answered a question from Mercer about how he had measured the stress in a linepipe, which involved Baurchinger effects in the steel.

In many cases the envelope of strengths is understood, and it is in other areas that more advances will be made. It seemed to me, as I said later in the meeting, that in that particular area of predicting static strength, physical metallurgy really has been very successful in (after 50 years or so) setting down what are the main properties contributing to strength. We still have not got such a detailed knowledge of fracture toughness.

I was struck by a remark from Pugh, who said some steels contain $\frac{1}{2}$ % of copper as impurity, and if I remember my metallurgy right, my extractive metallurgy, copper ores are exploitable if they contain 1 % of copper, so certain steels cannot be far off being useful sources of copper for extraction.

We then had a session on materials present, and future, and there one got the impression, or I did, from the papers and the discussion, that the very well developed titanium and aluminium alloys have in fact been developed – I think it was Nicholson who said it – using a phrase of Cohen's – by using the bank of knowledge, although one can not trace in many cases specific use of that bank. On the other hand, in the case of nickel alloys and the steels, the design seems to follow from recently discovered and clearly enunciated principles.

In addition, in that session, the effect of trace elements came out very very clearly, and the question of strength and toughness figures came up again. I wondered whether one could now predict limits for toughness in the same say as limits can be set on the strength under certain conditions.

Lastly, interface mechanics were noted as not being understood, and I think that interface mechanics (the concept was introduced in terms of grain boundary interaction) is of course central, not only to metallurgy but also to engineering because of joints, both welded joints and joints in structures. This leads into composite materials, and all of these subsume problems in the mechanics of interfaces.

We then passed to what I might call, if I may, a rather more philosophical part of the meeting, with discussions on education and I do not think I will attempt to summarize those, except to take one point from Sir James, when he emphasized the need for teams. That is what this meeting is all about; getting to know one another in order to form teams. I think there is, and I am sure a lot of other people do so too, one element leading to this necessity and that is the extreme complexity of modern technology. Because of the complexity, we are inevitably going to have to learn to live across one another's traditional boundaries.

Well, that is my imperfect summary of what I got from the discussion.

Now I think it just remains for me to thank everybody who has been involved in the conference. Personally I should say, despite Mrs Kirsner's very kind words, that this conference could not have been mounted at all without the help, particularly, of Dr Baker, and Dr Miles, and the many colleagues at N.P.L. who helped us. I would particularly like also to thank Miss Lambert from the Metals Society, Mr Tillotson from the Royal Society and Mr Hall of the Metals Society, as well as my own Director. We should thank Sir James Menter for the support he gave us from the start of the planning of the conference, and particularly, being such a busy man, for his presence for most of it and also for his paper. No man in his position could have done more to support us. I should thank all the caterers, and I must thank Mr Carrington and

Dr Miles for running the historical exhibition, and the Royal Society who will, in addition, produce the final proceedings. I should like to thank publicly Professor Hargreaves, who has presented the N.P.L. with a plaque to Rosenhain, which was in fact in the exhibition. I should thank the authors and the Chairmen. The authors have been most kind. They have done what authors seldom do at conferences nowadays. They have spent all the labour of writing excellent papers and then they have sat patiently and heard others discuss their work. I think for that alone we should give the authors one tremendous hand.

We should also thank the Chairmen who guided the discussions, and the Keynote Speakers. They did a great job. When the Keynote Speaker sits down the Chairman has to make sure that the meeting goes on for another hour and a quarter, but there has been no difficulty in that, good though the Chairmen and Keynote Speakers were, because you have been such a magnificant audience. And lastly, thanking Mrs Kirsner for coming all this way to see us, and reminding us so well that Rosenhain is not a long way away in time, I formally declare the meeting closed.

AUTHOR INDEX

(Contributions to discussion are indicated by italic page numbers)